MATHEMATICAL MODELLING (ICTMA 12):
Education, Engineering and Economics

"Mathematics possesses not only truth, but supreme beauty – a beauty cold and austere like that of sculpture, and capable of stern perfection, such as only great art can show."
Bertrand Russell (1872-1970) in *The Principles of Mathematics*

ABOUT THE EDITORS

Christopher Haines is Professor of Mathematics Education (Higher Education)  at City University, London with experience of teaching in schools, colleges and universities. Research in oceanography and theoretical physics led to the award of his PhD. His current research is in mathematical modelling, its practice and the assessment of complex tasks in mathematics. He was some-time Pro-Vice-Chancellor for Teaching and Learning and Dean of Students. He directed the UK Mathematics Learning and Assessment project, chaired the steering group of the Helping Engineers Learn Mathematics project, and was a long-time panel member for the National Teaching Fellowship Scheme

Peter Galbraith, Adjunct Reader in Education at the University of Queensland,  has an MSc in fluid dynamics and a PhD in simulation modelling of problems in educational systems. He has taught mathematics and science at secondary level, mathematics and mathematics education at undergraduate level, and in pre-service teacher education. His research interests include mathematical modelling and applications, the use of technology in mathematics instruction, and collaborative learning practices. He is a past president of the Mathematics Education Research Group of Australasia, current President of ICTMA and a member of the Executive Committee of ICMI.

Werner Blum gained his PhD in 1970 in Pure Mathematics and is Professor of  Mathematics Education, University of Kassel, Germany. His research interests include mathematical modelling and applications, mathematical literacy and quality of instruction, as well as international comparisons. He has been Co-Chair of the Kassel-Exeter Project and is currently a member of the PISA Mathematics Expert Group, Chair of the ICMI Study on Applications and Modelling in Mathematics Education, and Co-Chair of several empirical studies into the teaching of mathematics and teachers' professional expertise.

Sanowar Khan is Reader in Measurement and Instrumentation in the School of  Engineering and Mathematical Sciences at City University, London. He leads the Computer Aided Modelling and Design Group which has a reputation for innovative and applied research in collaboration with UK industry, especially in areas of computer aided design and mathematical modelling of sensors, actuators and devices for over two decades. This sustained activity was recognised by the Worshipful Company of Scientific Instrument Makers with the presentation of their 'Achievement Award'. He is a Fellow of the Institute of Measurement and Control, a member of the IEEE, and a founder member of the International Compumag Society.

MATHEMATICAL MODELLING (ICTMA 12):
Education, Engineering and Economics

Editors:
Christopher Haines
City University, London

Peter Galbraith
University of Queensland, Brisbane

Werner Blum
University of Kassel, Germany

Sanowar Khan
City University, London

Proceedings from the Twelfth International Conference on the Teaching of Mathematical Modelling and Applications.

WOODHEAD PUBLISHING

Oxford   Cambridge   Philadelphia   New Delhi

Published by Woodhead Publishing Limited,
80 High Street, Sawston, Cambridge CB22 3HJ, UK
www.woodheadpublishing.com

Woodhead Publishing, 1518 Walnut Street, Suite 1100, Philadelphia,
PA 19102-3406, USA

Woodhead Publishing India Private Limited, G-2, Vardaan House, 7/28 Ansari Road,
Daryaganj, New Delhi – 110002, India
www.woodheadpublishingindia.com

First published in 2007 by Horwood Publishing Limited
Reprinted by Woodhead Publishing Limited, 2011

British Library Cataloguing in Publication Data
A catalogue record for this book is available from the British Library

ISBN 978-1-904275-20-6

PREFACE

MODEL TRANSITIONS IN THE REAL WORLD

The Chancellor of City University and Lord Mayor of London, welcomed participants to the 12th International Conference on the Teaching of Mathematical Modelling and Applications (ICTMA12) held from 10-14 July 2005. City University, the University for Business and the Professions, and Cass Business School in particular, was an ideal location for this conference – models, modelling and applications for education, business and the professions feature strongly in its academic programme.

As elaborated at the Conference, and in this Volume, the effective practice, teaching and learning of mathematical modelling and applications play major roles in enabling successful activity within industry, business and commerce, and education. Mathematical modelling permeates society and so it is very appropriate that ICTMA contributions cover the whole spectrum of mathematicians, engineers and scientists, modellers in industry, government and finance, and teachers and, researchers in schools and universities.

In 1983, at ICTMA1 in Exeter, Henry Pollak, from Bell Laboratories NJ, USA, pointed out that society provides time for mathematics to be taught in schools, colleges and universities, not because mathematics is beautiful, which it is, or because it provides a great training for the mind that it does, but because it is so useful.

Mathematical modelling and applications, the transition, freely between real world problems and mathematical representations of such problems, being an enduring and important feature of industry, business and commerce, is thence also of major importance in mathematics education. Teaching mathematical modelling, through tasks, projects, investigations and applications embedded in courses, and of mathematics itself through applications, helps learners to understand the relationships between real world problems and mathematical models. This activity has a significant role to play at all levels of education and stages of learning, from the primary school through to college, university, and beyond.

ICTMA12 attracted participants from more than 30 countries worldwide. This volume contains 49 selected papers, by 83 authors, from 21 countries, presented in seven key sections. The papers were peer-reviewed by a distinguished international panel acknowledged below. The research, developmental practice, and reviews covered in these papers provide insights into mathematical modelling from many and varied perspectives.

We begin with "Modelling and Reality" in which Julian Hunt FRS presents a comprehensive overview of some of the big problems faced by modellers in applied mathematics. Kate Barker, a member of the Monetary Policy Committee of the Bank of England, discusses economic modelling and its impact on financial policy, business decision making, and government policy. These two papers help to place modelling in a context of society's needs on national and global scales and they provide additional motivation for the development of illustrative and practical models within education.

Our second major section deals with "Modelling Constructs in Education". Peter Galbraith takes us on a journey through mathematical modelling as it has developed and been implemented in education, motivated by its links with problems located in the extra-mathematical world. He compares and contrasts a multiplicity of practices, embedding his discussion in contemporary research. Katja Maaß focuses on activities in the secondary classroom, whilst Celia Hoyles and Richard Noss discuss the construction of models through European web interactions between younger children. These modelling constructs enable the reader to place the five ensuing sections in context and to appreciate their impact on mathematical modelling and applications and, in particular, transitions between real world problems and constructed models.

Assessment issues and identifying expertise in mathematical modelling is addressed in Section 3: "Recognising Mathematical Competencies". Section 4: "Everyday Aspects of Modelling Literacy", provides a brief snapshot of oft neglected issues around how people use mathematics and link that mathematics with reality. Section 5: "Cognitive Perspectives in Modelling", includes major contributions that seek to understand how pupils and students model. Particular models judged appropriate for education and those used in industry are in Section 6: "The Practice of Modelling", whilst reporting on how models and modelling are embedded in education is the subject of our final section, Section 7: "Behaviours in Engineering and Applications".

My co-editors Peter Galbraith, Werner Blum, Sanowar Khan, and I believe that this volume is a significant contribution to the knowledge and understanding of the research, teaching and practice of mathematical modelling and applications.

Enjoy it!

Chris Haines

Chairman ICTMA12,
Education and Lifelong Learning
City University
Northampton Square
London EC1V 0HB, England
Email: ictma12@city.ac.uk

ICTMA12 Organising Committee:
Debbie Durston, Steven Haberman, Chris Haines, Jasvir Kaur, Sanowar Khan, Tony Mann, Abdul Roudsari, ManMohan Sodhi.

ICTMA 12 - THE CONFERENCE:

ICTMA12 provided me with an opportunity to meet and to talk with others from all over the world who are teaching modelling and are interested in the same pedagogical issues in which I am interested...
Susann Mathews - Wright State University, USA

The wide range of topics covered in ICTMA12 was extremely beneficial and the networking aspects invaluable...
Nigel Atkins - Kingston University, UK

ICTMA12 was inspiring...
Adolf Riede - Ruprecht Karls University, Heidelberg, Germany

First-time participants at ICTMA12 immediately felt part of this stimulating group...
France Caron - University of Montreal, Canada

Discussing examples of Big Themes of applied mathematics and statistics might encourage mathematical scientists to realise their discipline fits them for the most responsible positions in societies ...
Julian Hunt - University College London, UK

There is a great deal of science in the construction of economic models – which are attempts to capture considerable complexity...
Kate Barker - UK

Reflecting modellers have a positive attitude towards mathematics itself as well as towards modelling examples...
Katja Maass - University of Education, Freiburg, Germany

12th International Conference
on the Teaching of Mathematical Modelling and Applications (ICTMA12)
10th - 14th July 2005

It could be argued ... that it would be impossible for everyone to specialise in mathematics to obtain a deeper understanding of how it is used in our society...
Jussarra Araujo - UFMG, Brazil

Although...personal ownership of problem situations can be fostered by contexts found desirable by learners, it does not imply that learners should not deal with situations to which they accord low priority...
Cyril Julie - University of the Western Cape, South Africa

The practice of word problem solving is relegated to classroom activities, having meaning and location, in terms of time and space, only within the school. Rarely will students encounter these activities outside of school...
Cinzia Bonotto - University of Padova, Italy

The interdisciplinary nature of mathematical modelling brings together mathematicians, engineers and people in other fields...
Andrei Kolyshkin - Riga Technical University, Latvia

A quality assessment strategy for mathematical modelling complements quality teaching approaches to mathematical modelling.
John Izard - RMIT University, Australia

Ask any adult with a job that doesn't involve mathematics when they...last [used] any piece of mathematics they were first taught in secondary school.
Hugh Burkhardt - University of Nottingham, UK

The day to day practical elements of life...provide[s] useful contextual support for the acquisition of computational skills and understandings...
Yvonne Hillier - City University, London, UK

...resulting in placing boys and girls in an active situation where they could experiment, guess, formulate, solve explain predict and contrast...
José Ortiz - University of Carabobo, Venezuela

...pupils are supposed to see the 'point of mathematics' ...with the help of reality based tasks and lessons.
Rita Borromeo Ferri - University of Hamburg, Germany

...the [ICTMA] conference was outstanding...
Jerry Legé - California State University, USA

Beneficial aspects of the [ICTMA] conference included strong content centred talks in a very friendly atmosphere...
Jochen Ziegenbalg - University of Education Karlsruhe, Germany

...our children [must] become familiar with renewable energies...this energy supply requires ... more personal effort...
Astrid Brinkmann - University of Trier, Germany

I really ... benefited from the [ICTMA] conference
Jamal Hussain - Sikkim Manipal Institute of Technology, India

In China, more than 40,000 students participate in [modelling] competitions each year...learn[ing] a lot of mathematical modelling...
Jingxing Xie - Tsinghua University, Beijing, China

...there is no consensus among researchers about the characteristics of modelling in mathematical education...
Regina Lino Franchi - Methodist University of Piracicaba, Brazil

ICTMA

The International Study Group for the Teaching of Mathematical Modelling and Applications (ICTMA) is an Affiliated Study Group of the International Commission on Mathematical Instruction (ICMI). The group emerged from the International Community of Teachers of Mathematical Modelling and Applications (hence the acronym, ICTMA), which has been concerned with research, teaching and practice of mathematical modelling since 1983.

The history of ICTMA began with concerns about the undergraduate preparation of students who were later required to solve real problems, often collaboratively, when employed as graduates. Early conferences had a particular emphasis on the design and delivery of courses to address graduate needs. The emphasis has since considerably broadened to include learning settings that involve all levels of schooling, from primary to upper secondary, undergraduate and teacher education, together with professional and workplace environments, in which achieving and nurturing the ability to apply mathematics to real problems is a valued goal.

Mathematical modelling research has been a feature of ICTMA from its inception: that research focus has developed to recognise the importance of establishing a robust knowledge base from which to address problems that continue to emerge. The overall academic focus involves the communication of research and practice, and encompasses themes such as the design and delivery of programs, analysis of modelling competencies and student performance, development and improvement of effective methods of teaching and assessment, and group and individual problem-solving expertise.

From the outset, ICTMA has maintained the integrity of its focus, which has both a mathematical and an educational component, recognising the close links between them. This makes a distinction from a purely mathematical problem focus on the one hand, and a mathematics education context in which the mathematics need have no connection with applications and modelling. Thus a distinctive aspect of ICTMA is the interface it provides for collaboration between those whose main activity lies in applying mathematics or in the fields of applications of mathematics, but who have an informed interest in educational issues, and those whose principal affiliations are within education, but who have a commitment to promoting the effective application of mathematics to problems outside itself.

ICTMA conferences have been held biennially since 1983:
ICTMA1 - University of Exeter, England (1983), ICTMA2 - University of Exeter, England (1985), ICTMA3 - University of Kassel, Germany (1987), ICTMA4 - Roskilde University, Denmark (1989), ICTMA5 - Freudenthal Institute, Netherlands (1991), ICTMA6 - University of Delaware, USA (1993), ICTMA7 - University of Ulster, Northern Ireland (1995), ICTMA8 - University of Queensland, Australia (1997), ICTMA9 - University of Lisbon, Portugal (1999), ICTMA10 - Beijing Institute of Technology, China (2001), ICTMA11 - Marquette University, USA (2003), ICTMA12 - City University, London, England (2005)

Peter Galbraith,
ICTMA President, University of Queensland, Brisbane, Australia

ICTMA BOOKS

The work of ICTMA has been published in its dedicated series of books listed here as well as in various professional journals; this volume contains the ICTMA12 proceedings.

Berry JS, Burghes DN, Huntley ID, James DJG and Moscardini AO (1984) *Teaching and Applying Mathematical Modelling* Chichester: Ellis Horwood. [ISBN: 0-85312-728-X]

Berry JS, Burghes DN, Huntley ID, James DJG and Moscardini AO (1986) *Mathematical Modelling Methodology, Models and Micros* Chichester: Ellis Horwood. [ISBN: 0-7458-0080-7]

Berry JS, Burghes DN, Huntley ID, James DJG and Moscardini AO (1987) *Mathematical Modelling Courses* Chichester: Ellis Horwood. [ISBN: 0-85312-931-2]

Blum W, Berry JS, Biehler R, Huntley ID, Kaiser-Messmer G and Profke L (1989) *Applications and Modelling in Learning and Teaching Mathematics* Chichester: Ellis Horwood. [ISBN: 0-7458-0355-5]

Niss M, Blum W and Huntley ID (1991) *Teaching of Mathematical Modelling and Applications* Chichester: Ellis Horwood. [ISBN: 0-13-892068-0]

De Lange J, Keitel C, Huntley ID and Niss M (1993) *Innovation in Maths Education by Modelling and Applications* Chichester: Ellis Horwood. [ISBN: 0-13-017351-7]

Sloyer C, Blum W and Huntley ID (1995) *Advances and Perspectives in the Teaching of Mathematical Modelling and Applications* Yorklyn, Delaware: Water Street Mathematics. [ISBN: 1-881821-05-6]

Houston SK, Blum W, Huntley ID and Neill NT (1997) *Teaching and Learning Mathematical Modelling* Chichester: Albion Publishing Ltd (now Horwood Publishing Ltd.) [ISBN: 1-898563-29-2]

Galbraith P, Blum W, Booker G and Huntley ID (1998) *Mathematical Modelling, Teaching and Assessment in a Technology-Rich World* Chichester: Horwood Publishing Ltd. [ISBN: 1-898563-X]

Matos JF, Blum W, Houston SK and Carreira SP (2001) *Modelling and Mathematics Education ICTMA 9: - Applications in Science and Technology,* Chichester: Horwood Publishing Ltd. [ISBN: 1-898563-66-7]

Ye Q, Blum W, Houston SK and Jiang Q (2003) *Mathematical Modelling in Education and Culture ICTMA 10* Chichester: Horwood Publishing, 330 pp. [ISBN: 1-904275-05-2]

Lamon S, Parker W and Houston K (2003) *Mathematical Modelling: A Way of Life: ICTMA 11* Chichester: Horwood Publishing, 267 pp. [ISBN: 1-904275-03-6]

Haines C, Galbraith P, Blum W and Khan S, (2006) *Mathematical Modelling (ICTMA 12): Education, Engineering and Economics* Chichester: Horwood Publishing. [ISBN: 1-904275-20-6]

Werner Blum
Series Editor ICTMA Proceedings, University of Kassel, Germany

ACKNOWLEDGEMENTS

The editors have been greatly assisted in their task by an international panel of referees. Their considered and independent review of the submitted papers, in which criteria of mathematical application and educational relevance have been central facets, was incisive in deciding on those that should be accepted, thus ensuring the quality of these published works. The panel included:

H. Abel (Germany), S. Abramovich (USA), A.Ahmed (UK), M. Anaya (Argentina), J. Berry (UK), D. DeBock (Belgium), G. Bowtell (UK), H. Burkhardt (UK), R. Crouch (UK), W. Van Dooren (Belgium), R. Gerrard (UK), N. Gruenwald (Germany), S. Haberman (UK), H-W. Henn (Germany), Y. Hillier (UK)), K. Houston (UK), T. Ikeda (Japan), J. Izard (Australia), T. Jahnke (Germany), C. Julie (South Africa), G. Kaiser (Germany), O. Kerr (UK), S. Klymchuk (New Zealand), D. Lawson (UK), J. Legé (USA), T. Lingefjärd (Sweden), P. Osmon (UK), S. Quinsee (UK), A. Riede (Germany), A. Roudsari (UK), M. Sodhi (UK), P. Speare (UK), M. Stephens (Australia), B. Tuladhar (Nepal), I. Verner (Israel), L.Verschaffel (Belgium), G. Wake (UK), S. Williams (USA)

We thank Minh Phan for his help in the run up to ICTMA12. We are particularly grateful to Jasvir Kaur, who ran the ICTMA12 office prior to the conference, and together with Najm Anwar for their commitment and support in helping delegates at ICTMA12. We were fortunate to be able to hold the conference at Cass Business School, by the good offices of Steven Haberman and the invaluable practical and professional services of Debbie Durston. Accompanying persons benefited from a short programme put together by Margaret Haines.

The conference ended in some style in glorious weather - with our excursion to Greenwich and the lecture from Julian Hunt. We are indebted to Tony Mann and The University of Greenwich for making this possible and hosting the final sessions of the ICTMA12.

By coming to London for ICTMA12 in the aftermath of the terrorist bombings of 7 July 2005, with a disrupted transport infrastructure, delegates and their accompanying persons showed great resolve and strong commitment to ICTMA. We thank all who came for doing so in such difficult circumstances.

ICTMA12 is grateful for the support of:
The Institute of Mathematics and Its Applications, Taylor & Francis Group, Royal Bank of Scotland, Horwood Publishing Ltd, Chartwell-Yorke Ltd, Elsevier, Virtual Image Ltd.

TABLE OF CONTENTS

xviii

Section 1
Models and Modelling in Reality

1.1

COMMUNICATING BIG THEMES IN APPLIED MATHEMATICS

Julian Hunt FRS
Lighthill Institute of Mathematical Sciences and
University College London, UK
(and JM Burgers Centre, TU Delft, The Netherlands)

Abstract–*This paper reviews, primarily for mathematical educators, how decisions taken in government and the private sector make use of the concepts and techniques of mathematical sciences, particularly those related to sudden and extreme events in natural, social and technological systems. The insights of applied mathematics, statistics, and computational modelling are shown to be relevant to understanding and managing the outstanding problems concerned with these systems. The main themes highlighted here include predictability, accuracy, extreme values, singular events, and patterns in system behaviour, especially in relation to new approaches in deterministic and statistical modelling and to the increasing use, through computer networks, of the vast volumes of data that are now available to scientists and decision makers. These developments also exploit improved optimisation methods which make better use of approximate data and, most recently, the 'multi-centre' combinations of model results obtained at different institutions, which benefit from the slightly different assumptions and numerical methods at each centre. Suggestions are made about how these concepts and techniques could be used more effectively in government and industry to guide both tactical and strategic planning. The worlds of media and politics also need to understand better how mathematically-based predictions are made and how to question them. Universities could provide general 'appreciation' courses to survey the general ideas of mathematical science and their wide applicability in society. Graduating mathematical scientists might be encouraged to take on more responsible roles which should be of benefit both to their organisations as well as to their own careers.*

1. MATHEMATICS AND SOCIETY'S MAJOR CHALLENGES

Applied mathematics and mathematical modelling are increasingly vital for understanding and dealing with society's major challenges. This is slowly being recognised, even by policy makers and industrialists, as they base more of their decisions on the concepts and calculations of mathematical sciences. Correspondingly, they rely less exclusively on expedience, experience, and intuition. This transition has been helped by the increasing use of graphical presentations, for example, with PowerPoint displays; though rhetoric still has its place. (Of course, through public opinion polling, even the most ancient of political techniques can be improved through mathematics!) The general public, as well as journalists, are increasingly using mathematical language and images, the current favourites being chaos and tipping points. These may change as pattern recognition, now part of the school curriculum in year 6, and self organisation become more relevant, for reasons discussed in this paper. The increased public recognition of the conceptual importance of mathematics has not come about by chance. The whole community of mathematical scientists throughout the world is making progress in communicating advances in research and their practical implications, but also how these developments affect the way we look and think about the world around us. In the UK, many organisations are contributing, particularly the mathematical societies (the Institute of Mathematics and its applications, the London Mathematical Society, the Royal Statistical Society) and institutes of mathematical sciences in universities, such as the Lighthill Institute of Mathematical Sciences, Smith Institute etc. However governments, while recognising the value of mathematics for industrial applications and statistical analysis (for example, Hunt & Neunzert, 1993), have still not understood their wider implications for societal problems.

Two major themes of mathematical science are discussed in this paper, where progress in research has made a particular contribution. The first is the analysis and description of extreme and singular behaviour, particularly how, where and when sudden changes and isolated events can occur in human and natural systems. The second is the behaviour and the limits to the capabilities of computation and data systems as they affect the accuracy and reliability of prediction of multi-component, complex systems. Related advances in mathematics and computer science have also led to the improvements of the control of such systems, especially in complex and changing circumstances.

The issues of greatest public concern can be identified where these themes are particularly relevant for policy and public understanding:

- Natural and technological disasters and consequences for communities, including tsunamis, earthquakes and hurricanes (2005 was a particularly bad year; Hunt, 2005, 1995; Huppert & Sparks, 2006).
- Large changes in natural systems where key components change more or less suddenly; from one state to another or through a major perturbation which eventually decays, for example, climate change events (with permanent or longterm global consequences) or global epidemics (which usually decay) (Schellnhuber et al., 2005).
- Isolated perturbations to local systems (strikes, acts of violence, etc.), for example, analysed qualitatively using catastrophe theory (Zeeman, 1987).
- Large, possibly sudden, changes to weakly-constrained or malfunctioning interacting systems (for example, Enron business collapse), or the initiation or the cessation of conflict, as occurred with the end of the Cold War, which LF Richardson predicted by his simple mathematical model in 1953. (see Hunt, 1993, 2001). However, Richardson warned that excessive reliance on mathematical models can be very misleading. The best known failure was that of the US Defence Department's models of 'pacification' in the Vietnam War.
- Improvement of performance of economic and social organisations in response to political and financial inputs and operational constraints. Mathematical analysis can test the statistical measures used by policy makers and the effects

of specific targets on complex organisations, especially where the organisations, including governments, are aiming to integrate or 'join up' their activities. (Hunt, 1997; Gallivan, 2006).Concepts and models of organisations models have to allow for the unpredictability caused by random and chaotic effects, replacing earlier technocratic concepts of society running like a well oiled engine-the image of the German engineer Reuleaux (1885)

• Prediction and design of network systems (capacity, performance, robustness etc.; transport; internet…), where new solutions to new problems and new demands have to be considered (for example, on line response, security, data compression/ communication) (Kelly, 1979).

• Understanding and redesign of physical systems (houses, cities, products) to meet new constraints; how to invent and define total changes in their structure and patterns of performance both conceptually and by massive computational modelling (Lautso et al., 2004). In electronics, new physical and mathematical concepts are developing both for the 'end' of silicon technology and of deterministic, non-quantum computing, as the size decreases and the speed increases of current solid state devices (House of Lords, 2002).

2. EXPLAINING THE INSIGHTS AND THE LIMITATIONS OF APPLIED MATHEMATICS AND MODELLING

Mathematical analysis and modelling is evermore widely used by policy makers and industrialists for guiding and even making their decisions. However, by asking some basic questions they can understand better what reliance to place on the results, and how to interpret them. They also want to know the likely future trends of these techniques.

For mathematical scientists, identifying and explaining the general concepts that emerge from their particular studies is an essential part of making progress in research. As Henri Poincaré (1914), probably the best philosopher of applied mathematics, first explained, generalising the results of a particular study enables scientists to move from one hypothesis to the next and to relate the research to wider scientific developments. In expounding his philosophy as a biologist, Haldane (1935) argued, following Kant, that the connectedness of scientific concepts means that science is not simply a subjective creation, like art, but has an objective validity. This philosophical insight is related to the sociological fact that people's familiarity with scientific data and concepts (for example, weather forecasts) enables them to make their assessments through their engagement with scientific data. One might conclude that only through such familiarity can people understand scientific concepts and be more inclined to accept the validity of controversial science.

In the UK, government funding of research implicitly recognises that this generalisation and interconnection of scientific understanding, and its popularisation, is necessary for building up an integrated structure of scientific knowledge. It is also necessary for ensuring that over the long term there is continuing support for science by society as a whole.

2.1 How are Different Types of Models Constructed, Corrected and Rejected?

Reductionist models used for predicting well-defined physical systems are built up from the basic laws of physics, chemistry. However, for predicting complex and larger scale systems, such as rockets or the weather, the models generally involve approximations and some empirical assumptions. The reductionist approach is also used for predicting the behaviour of certain types of economic, social and biological systems. However, here the models are based on ad-hoc hypotheses deduced from empirical observations, for example, about animals' and people's greed, aggressiveness or submissiveness as in predator-prey models (Volterra, 1931) or driver reactions in traffic models.

The alternative approach for modelling the behaviour of complex systems is statistical, which is based on the assumption that it can be calculated from data describing its previous behaviour. Generally past statistics do not allow for changing conditions. But in some cases when the future is not the same as the past (for example, river floods in a changing climate (Hunt, 2002, 2005) corrections can be introduced, sometimes involving a combination of statistical and reductionist models.

Both approaches essentially began in the 17th Century, when Newton introduced dynamical deterministic models and Pascal, Leibniz and Bernoulli invented probabilistic prediction (Favre, 1995; Hacking, 1974). These models, especially the former, have steadily improved, and grown more complex mathematically as they represent and predict more complex processes, more detail, more components and more dimensions in space, time and phase space. As more computer power becomes available, the speed, complexity and data input are continually increasing. However, many processes have a chaotic aspect of their behaviour associated with extreme sensitivity to small changes, which may be imposed by external conditions or may be intrinsic to the model or the actual process. These aspects may or may not be predictable in advance! However with recent developments in mathematical science and increased computer power, the calculation of all the theoretical possibilities of ideal systems becomes possible. This provides increasing insight into chaotic behaviour of processes, their representation by models and their dependence on data.

Although there is still no rigorous framework for combining different methods, such as statistical and deterministic analyses for modelling complex systems, empirical approaches are being developed for many types of practical problem. In particular statistical science, which is largely based on methods for analysing empirical data, is now being applied to analysing the computed outputs of deterministic models whose randomness results from the measured and assumed distributions of the inputs to these models.

Despite theoretical and conceptual problems arising from such hybrid methods, Maxwell and Boltzmann in the 19th century pioneered this approach with their study of molecular dynamics and random systems. In the 20th century, Wiener, Kolmogorov and others brought statistical concepts into dynamical models, for example, for turbulent flows (for example, Landau & Lifshitz, 1960).

Reductionist models with empirical elements are widely used in oceanography and meteorological models for forecasting particular events. But they cannot be run for long enough to provide reliable statistics of future likely events at any particular location. Nevertheless, in combination with empirically based statistical models, (for example, of future storm events in a changing climate), reductionist methods provide computer estimates for the probabilities of future sea level variations around the UK coast. This combined approach is regarded as more comprehensive and more reliable than simply making predictions based on past tidal records (Butler et al., 2005).

Both reductionist and statistical models are now being combined in financial and risk models. The analytical results and numerical simulations provide statistics that contribute to decisions in insurance, investment. (Turfus, 2006).

For calculating specific results in the 'real world' both of these types of models require data to initiate the calculation (for example, today's stock-price or weather) and to define the range of conditions where they are relevant. The essential point is that in the 'real world' since there is insufficient data, assumptions have to be made about the missing data.

Since the 1980's, new techniques of 'data assimilation' and data optimisation have been developed which have utilised data much more effectively to improve the accuracy of deterministically based models, and reduce errors caused by missing data. The first step was to make the optimum <u>adjustment</u> of the predictions as new data is received. Eventually the 'best line' is drawn through the data points consistent with the 'dynamics' of the system. In more advanced procedures, the previous computations are <u>repeated</u> as new data is provided. These techniques enable the vast quantity of data, now available for monitoring and scientific modelling, to be used much more effectively, notable for weather forecasts using satellite data (Hunt, 1999; Hunt & Coates, 2002).

These data assimilation concepts are beginning to be applied to the prediction and control of flows in aeronautical engineering (Bewley, 2003). Many other applications are also possible.

Understanding the errors of models and having estimates for the reliance that can be placed on their prediction, are both essential for their use in decision making.

Model predictions have markedly different patterns of growth with time. They fall into four main categories.

a) A nearly perfect type of forecast, such as these associated with regular planetary motion where the errors do not increase with time.
    LF Richardson (1922) thought wrongly that weather forecasts might reach this ideal given sufficient precision in numerical methods and in the measurement data that initiate the calculations. Nevertheless some geophysical systems are highly predictable as the recent calculation of the global movement of the tsunami waves in 2005, were remarkably accurate as they travelled on curving paths half the globe.

b) For a forecast of a typical non-linear system of limited complexity with time scale T_s (i.e., with a few degrees of freedom), errors tend to grow exponentially (i.e., proportional to $\exp\left(\dfrac{t}{T_s}\right)$). Therefore, the calculations are very sensitive to initial conditions, as first explained systematically by Lorenz (1963) though hinted at earlier by Poincaré (1914). The chaotic nature of the time variation of simpler mechanical systems is described by

Moon (2004).

c) The predictability of complex non-linear systems (with very large numbers of spatial and temporal degrees of freedom) is different when they organise themselves into coherent and persistent patterns of behaviour. Examples of discrete and continuous systems range from galaxies and atmospheric eddies in the physical universe and from societies to biological organisms in the living world. Although these patterns are randomly distributed in time and space, they have a finite life -time and spatial scales (defined mathematically by correlation functions).Time scales are shorter if each element of the system is changing rapidly or is dissipating a lot energy (whether physical or human). The persistence of such patterns means that errors in prediction grow more slowly with time than in type (b) (typically in proportion to some power of time t^{α}). For the errors of location of large scale weather systems, as well as magnitude of variables such as pressure and temperature, the value of the exponent $\alpha \sim 1$ (Met Office, 2003; Hunt, 1999).

d) The most feared and least well predicted features of the overall or average behaviour of systems occur when they have a large and sudden change, typically from one state to another (see § 2.4). These 'catastrophes' or 'bifurcations' can occur over periods much shorter than the time-scale of the usual fluctuations of the system. The errors in predicting such changes also tend to have a 'step- function' variation with time. Examples abound in both in physical systems (for example, earthquakes, volcanoes) and socio-economic systems where the events were not predicted at all, as with the dramatic ending of the Cold War in 1989.Their timing and magnitude tend to be equally erroneous. In many situations there are indications that changes are about to occur, but when warnings have been given, they have often been ignored, perhaps because they have been uncertain or so unexpected. Furthermore, they are as likely to be ignored whether they are optimistic as pessimistic, as was the view of Richardson's prediction in 1953 that could be a sudden end of the arms race and the Cold War. (Hunt, 2001).

Even in well defined laboratory systems these kinds of bifurcations can be unpredictable. Typically, they are found when the initial and boundary conditions are symmetric and do not define a preferred orientation, such as occur in buckling a solid structure or in thermal convection produced by heating initially stationary liquid in a long horizontal container. In this case, turbulent motions tend to induce a persistent mean flow whose direction (Krishnamurthi & Howard, 1981; Owinoh et al., 2005) is unpredictable. Bifurcations are inevitably sensitive to small external perturbations. Since similar kinds of mechanism can affect large regions of the ocean, the global climate could also change rather suddenly (as it has done in the geological past).

There is no doubt that in recent research on non-linear systems there has been considerable progress in their prediction using deterministic (and in some cases statistical) models, especially where the systems are the self-organising type (c). As a result of more detailed and accurate computations, these models now include more physical processes and more variables (for example, some aerosol effects in global climate models), and also more comprehensive measurements to initiate and correct the calculations, for example, through satellite monitoring of weather systems, or survey data of economic and social system. Indeed, as with the example of the tsunami wave, some calculations of type (c) are now so successful that they now fall into type (a).Another example of progress is that weather forecasts for 3 days are now as accurate as 1 day forecast 20 years ago in the North Hemisphere and about 10 years ago in the South Hemisphere (when reliable satellite data were not available) (for example, Hunt & Coates, 2002).

Empirically based models are used for predicting the economic growth rate for 12 months in advance. Although the variations between models are as much as 100%, in recent years, they have become more reliable (at least for developed countries). Perhaps this only because there have been no major economic perturbations (such as occurred in Asia in 1998 and in UK in 1992)

The history and philosophy of science has only recently begun to describe how scientific concepts and models really evolved. The basic idea taught in textbooks is that hypotheses are tested and then accepted or rejected depending on how well they compare with experimental data; provided, as Karl Popper pointed out, the hypotheses are in a form that can be disproved. This traditional interpretation of science dominates most journalists' discussions of scientific developments, leading them to the naïve conclusion that any discrepancy with data should lead to rejection of scientific theory. In fact, as Lakatos first explained, science proceeds more smoothly, with models being progressively corrected to allow for discrepancies that are bound to occur. This is an essential part of scientific progress (Chalmers, 1982). An important recent example was the correction of the first global climate models around 1990 which predicted an excessively large rise in the global temperature as a result of the increase in green house gases emitted by human activities. The model was corrected to allow for the industrial aerosols which reduced the incoming solar radiation at ground level (Hunt, 1999; Houghton, 1994).

Occasionally new data becomes available that can only be explained by a new conceptual approach and a rejection of current models, as occurred with the revolutionary changes in Newtonian and Einsteinian physics. Kuhn defined these events as 'paradigm shifts' in science. Lorenz's (1963) demonstration of the unpredictability of quasi-deterministic systems was indeed a paradigm shift in applied mathematics.

2.2 Extremes

In the practical use of models questions can arise as to whether the models are also applicable when the system experiences extreme conditions. If the answer is no, is it possible to extend models designed for 'normal' conditions to include these events? Or will it be more accurate and more computationally efficient to rely on special models for extreme conditions?

The usual calculations for the safety of engineering structures are based on the analysis of small static deflections. These are clearly inappropriate for assessing the effects of large unsteady loading and intense temperature caused by an aircraft crashing into it and exploding. Arguably, accurate calculations for such rare disasters are unnecessary. But they do matter if the destruction has damaging consequences on the surrounding environment – an essential consideration for designing nuclear power stations, when extreme conditions have to be modelled in some detail. Indeed, in this case, the extreme conditions are so important that they determine the overall design. The same dichotomy is apparent in modelling many types of extreme situation; whether they are natural disasters such as floods, and large earthquakes, or economic/ social calamities such as wars, slumps, riots etc.

It is a remarkable fact that for extreme events, which occur rarely, randomly and in an unconnected way in completely different system, their frequency of occurrence and magnitude tend to be described by general statistical laws. LF Richardson showed how the Poisson distribution (which describes frequency of horse kicks killing Prussian soldiers), also describes the occurrence of wars (see Ashford, 1985). For these and most other extreme events, determining the appropriate form and parameters of the statistical distribution depends to a great extent on the analysis of limited data sets of extreme events. Recent research has shown how sensitive the resulting distribution is to the analysis; for example, estimates for the return period of floods may change from 50 years to 10 years when data is taken only from the most extreme previous events (Cox, Isham & Northrup, 2002; Embrechts et al., 1997).

While statistical models provide guidance about the frequency and therefore the likely occurrence of an extreme event, what actually happens during the events can only be predicted by deterministic models. In wind storms, floods, and tsunami waves, the intense air and water motions vary over much shorter distances than in normal conditions. In hurricanes the winds vary over a few kilometres, whereas in usual low pressure systems they vary over hundreds of kilometres. Special models now forecast hurricane trajectories in real time (usually for 5 days ahead but up to 10 days in some situations). However, it is not necessary to solve the basic equations to make a forecast, because the events are sufficiently similar (as satellite images demonstrate so clearly) that the same special (vortex) model can be used for the inner structure of hurricanes, while the wind far from the centre is the forecasting model used for normal conditions (for example, Heming et al., 1995). Progressively, these special methods may be replaced by models used for normal conditions as computer power increases, and general calculation methods adapt to events on small length and time scales (Nikiforakis & Hubbard, 2003).

The frequencies of extreme events in complex systems tend to have characteristic statistical distributions such as the occurrences of hurricanes or tropical cyclones for each area of the world. These frequencies differ from the distribution of high winds in general. However, extreme events of one kind of phenomenon may be correlated with other related phenomena. In the case of hurricanes, physically based models show that their movement in any given season is affected by the prevailing winds. Using this information predictions for annual hurricanes reaching land have recently become significantly more accurate (with errors of less than 20% in any given zone) (Saunders, 2005).

However, as with all statistical models they are based on measurements of past events, so they have to be modified if the conditions affecting the random events are changing with time. A combination of deterministic models for global climate prediction (described above) and these statistical models are now being used to allow for the effects of climate change on probability estimates of extreme weather events. But the results depend crucially on the modelling of physical variables in the statistical models. The result of these studies show how in a warming global climate some of these factors increase the strength and frequency of storms, while others reduce them. The consequences for the insurance industry are obviously considerable!

2.3 Patterns, Shapes and Approximate Representations

Nowhere has the impact of mathematical science on modern society been so marked as its provision of techniques for defining characteristic features of signals of every kind. Fourier's series representations of signals in terms of oscillations in the early 19th century (see Farge et al., 1993) and Luke Howard's (1803) verbal and physical classification of cloud shapes were break-through concepts we still rely on. But equally revolutionary changes in the 20th century occurred with the discovery of the fractal dimensions of wiggly shapes (for example, Mandelbrot, 1982) and the quantification of the information contained in a signal and its efficient transmissions by Shannon & Weaver (1949). Scientists, engineers, economists and social scientists were able to educe or detect critical features hidden within complex signals and at the same time transmit the signals more efficiently by focussing on these features. These developments enable models of complex systems to be assessed more efficiently in terms of other models or their accuracy against measurements; rather than analysing many cases and huge volumes of data, only the key aspects of the measured or calculated signals have to be compared.

Even our aesthetic appreciation of shapes and surfaces has been changed by mathematical signal analysis, especially through the computer graphics of fractal boundaries, and the depiction of complex spiral shapes. Patterns of fractal laser light beams reflected off clouds at parties round the world show how the aesthetic aspects of mathematics can be a universal language enjoyed by everyone!

In every academic discipline major discoveries have resulted from developments in the mathematical analysis of data. The results may well be quite controversial depending on the choice of the techniques applied and in some cases how they are combined.

In geo-physics, the modern understanding of the structure of the earth's core resulted from analyses of earthquake waves, In astrophysics by analysing the pattern of microwave radiation beyond the outer-most stars it has been possible to deduce that the earliest stage of the universe was not formless - a hypothesis with the profoundest cosmological and even religious implications!

The analysis of complex data gave rise to an equally revolutionary change in ideas in social science in the 1920's (Spearman, 1904; Garnett, 1921). From sets of cross-correlations of the results from various psychological tests, matrices were constructed. It was postulated that their eigenvectors, or principal components, provided an overall measure of 'intelligence', in a similar way that the eigen vectors of oscillating or deforming mechanical systems define intrinsic properties that are independent of any particular reference frame for the measurements. The controversy surrounding of the application of these standard mathematical methods continues to this day. Eigenvectors of climate patterns are now relied upon to predict seasonal weather for several months ahead (Colman & Davey, 1999).

Since different mathematical education techniques tend to focus on different features of a signal, there have been controversies about what the most critical features are for any given signal under varying circumstances. Some techniques analyse the entire data set, while others select only certain data (for example, above a certain threshold).This is typically what animals -including humans- do in processing sight and sound. Fourier (or spectral) methods compare the signals to sinusoidal oscillations, while other (for example, fractal or capacity dimension methods) describe these in terms of the gaps (in time or space) where the magnitude of the signals are below a certain threshold. Research (for example, Bonnet & Glauser, 1993) has shown both theoretically and practically how the relative merits of different identification techniques are better understood, so that when applied to fluid mechanics, they now tend to agree about the general form of the characteristic three-dimensional time-varying 'eddy' patterns in the turbulent flows - in a sense vindicating Howard's hypothesis of 1803.

Much of modern technology depends on rapid communication of huge quantities of data using miniaturised transmissions and receiving systems. In the mobile phone systems, images, speech and visual data are digitised and effectively Fourier analysed into waves or on-off signals. Such data used to be 'compressed' rather crudely with the aim of retaining only its essential features by simply excluding high or very low frequency components of the signals. But a more discriminatory approach is to follow the information system the eye and brain and focus on the 'edges' in a signal where sharp changes occur in the amplitude (Farge et al., 1993; Silverman & Vassilicos, 1999).

Dealing with sharp gradients, such as shock waves and cracks in solids, is also a challenge. In the computational models, the methods developed earlier for concentrating numerical meshes around these critical regions are now being extended to more complex problems where the critical regions are randomly distributed and also move randomly in time (from chemically and biologically active media to weather and climate models (for example, Hubbard and Nikiforakis, 2003).

2.4 Singular Events and Sudden Changes

People have to decide what to do about sudden or even singular events in their lives, whether they are managers or employees in organisations, or driving a car through traffic jams, or more seriously when confronted by an accident or natural disaster. Well prepared individuals and organisations need to know in advance what kinds of events might occur in 'their' system, and when and how they happen, and what they should do.

A mathematically informed approach can certainly help, especially as governments and communities become more aware of all the kinds of risks they need to prepare for. They also have to consider the possibility of several hazardous events occurring close together in space and time – a precaution that is becoming increasingly necessary in modern communities whose compactness and interdependence makes them more susceptible to such events.

In some situations (discussed already in §2.3), precise modelling of the events may be possible. Also they may occur with sufficient regularity that their future probability of occurrence can be estimated quite reliably. But for events in a changing environment (for example, climate change or larger life span), systematic studies of the underlying trends and mathematical modelling may be the best way of predicting the nature and frequency of extreme events.. In some cases these fundamental studies have indeed predicted the form of changing extreme events before they were observed (as a hurricane has now occurred in the South Atlantic, in 2005, for the first time).

Mathematical analyses of events begin by defining a variable V as a function of one or more independent variables., There might be a 'step' change in traffic flow along a road (in fact over the few seconds corresponding to drivers' reaction times). In an idealised mathematical analysis this period is considered to be 'infinitesimally' small, so that the variation of V is approximated by H(t) the Heaviside step function.

This implies that the rate of change of V, i.e., the acceleration or deceleration, becomes very large. Over the infinitesimally small period dV/dt is now approximated by the 'spike' or Dirac's 'delta' function $\delta(t)$– probably the most famous development of mathematics in the 20th century (Lighthill, 1953).

New types of highly oscillatory singularity that have been conceived mathematically also correspond to observations. Random and multi-dimensional functions are other features of more complex singularities. Such idealised mathematical functions are particularly useful for describing measured variables near the points at which the functions are singular, (for example, $x = x_o$) in the above example. But very close to the singularity itself at $x = x_o$ (where V and/ or its derivative tend to infinity) the measured variables cannot exactly correspond to the idealised function. Many extreme events occur are singular in both space and time, i.e., the magnitude of the spatial and temporal derivatives, $\partial V/\partial x$, $\partial V/\partial t$ are very large at x_o and t_o. Earthquakes form through the sudden rupture of solid materials along lines of weakness. In such cases, predicting the location (x_o) of where sudden large displacements $V(x,t)$ might occur is usually more certain than the moment (t_o) when they might occur, as are the recent major earthquakes of Kobe, Sumatra and Kashmir so tragically demonstrated. In these events, the amplitude V (i.e., speed of the displacement) rapidly, but monotonically, increased to a maximum and then decreased.

But there are many types of singularity where the sudden amplification of V (or dV/dx) is associated with high frequency oscillations and waves. Deep vibrations within the earth, which were observed to amplify before the huge volcanic eruption of Mount Pinatubo in 1992, helped provide a vital warning which saved many lives (Hunt 1995). In London the notorious wobbles of the new footbridge across the Thames in 2000, which delayed its opening, was a vivid example of how oscillatory systems and people's behaviour can interact strongly as one effect feeds 'back' on the other – phenomena familiar to economists and politicians.

Many sudden events are also followed by waves and oscillations, as occurs in the Tsunami and hurricane induced waves in 2004/5 (Hunt 2005; Huppert & Sparks, 2006). The moving blockages on motorways and in production systems

(Armbruster, 2006) (both modelled by Lighthill & Whitham's (1952) kinematic wave theory) are notable examples of how physical systems and people interact, sometimes with similar patterns (not unlike Richardson's concept of conflict).

Of course most sudden and singular events are multi-dimensional; the beautiful, but potentially dangerous events of huge breaking waves or atmospheric cloud billows, as they develop sharp crests that curve into spiral forms, have to be described in two space dimensions (x,y) and time (t). Thom's catastrophe theory (for example, Saunders, 1986) is a mathematical framework for describing similar types of multi-dimensional singularities, but for most physical systems it has not provided any predictive insights. However, Zeeman (1987) suggests that certain singular events in prison break out and in hostage situations (described by a set of behavioural variables) might be predicted using these concepts. This approach has not been taken further.

In economics, oscillations and singularities are also analysed in great detail. In the 1990's some national economies experienced sudden and sharp reductions in output. For major economies, there is a continual debate as to whether the recovery of national economies from minor depressions are monotonic or oscillatory, or as the journalists say are 'double dip' events. The lessons from these extreme and sometimes sudden events, in many types of systems, is that through continued and close monitoring and studying of data, sudden events can lead be predicted. The growing fluctuations in share prices before the October 1987 'crash' and in the information 'chatter' on the internet preceding the terrorist event in New York in 2001 demonstrated such linkages. They were noted afterwards, but the data were not heeded beforehand.

When information about impending natural disasters is disseminated and used proactively it can save lives and it costs much less than dealing with the aftermath. Damage caused by these events is generally reduced when warnings have been issued in advance. This is why governments are finally accepting the need to improve the international warning systems and follow up arrangements (ISDR, 2005).

The most disruptive disturbances to complex systems lead to changes to different states of the system; subsequently they may or may not revert to the previous state. If they do, how long does this take? These are questions that can be examined using the ideas of mathematical dynamical system's theory, which for example can define particular situations when large changes are likely and what form they might have. Policy makers need to study these questions very thoroughly, since too often it is assumed that the state of a system before the disturbance is the only possible and desirable state, despite the numerous counter examples in engineering, biological, geophysical and political economy.

The mathematical description of a system's behaviour can be expressed using a phase-plane diagram in which one dependent variable (say V_x) is plotted as a curve against another (say V_y) (such as a variable V and its derivative dV/dt) over the range of values of the independent variables x, y or t or the parameter p (for example, Bondi, 1991; Thompson & Stewart, 2002). The mean or characteristic features of a complex system could be described by the statistics of these curves (for example, average values or values in certain particular or 'conditional' situations).

The simplest examples are the circular curves of V_x against V_y describing flow pattern of eddy motion in the laboratory, the cooking pan or the atmosphere. Similar curves of V against dV/dt describe time varying oscillations. In some cases these flows can flip from one direction to another, either unpredictably and spontaneously, or as the parameter p changes. By using the powerful concept of symmetry, these types of bifurcations or global instability can be characterised as a jump from one region to another in the phase plane diagram (for example, Holmes et al., 1996). Many models of biological systems such as those for population of predators and their prey show how their behaviour can fluctuate around an equilibrium state (as in the case of arctic foxes and hares). But when external conditions of a system are disturbed too severely, the system changes utterly to a new state (possibly with either only predator or only prey or neither).

In politics and economics, the language of non-linear dynamical system has gradually tended to supercede the earlier linear control theory or cybernetics (or steering) models. President Nixon famously described his economic policy method in 1968 as a 'light touch on the tiller'. In the 1960's and 1970's when it was argued by the politicians in the UK (on both the left and the right) that the economic 'system' should and could move to a new 'state'. In reality the quite sudden fluctuations and 'stop-go' policies of national economies of various countries indicated the limitations of central control. By the 1990's the concept of a lightly controlled, but largely self-equilibrating market with its characteristic oscillations (i.e., the business cycle) became the accepted model. Prof Desai of London School of Economics has remarked on the fundamental difference between forecasts of weather and of economic systems, because governments that make the forecasts about the economy, can subsequently influence it. So an accurate forecast probably needs to allow for the deliberate actions of people and organisations or the system that might be affected by the forecast. On the other hand, governments cannot control the weather (and, so far as is known, gods of weather do not listen to the forecasts!). In fact, the previous lack of success of economic planning and the inaccuracy of the associated forecasting led the government in a number of countries (including the UK) to reduce their control of key elements of the system (i.e., bank rate).The forecasts seem to have become more accurate as a result of this decoupling of forecasting and control.

2.5 Optimisation

For designing and controlling the behaviour of systems to satisfy some optimum criteria, for example, of maximum performance or minimum cost subject to certain constraints, it may not be necessary to calculate every one of its possible states as a means of finding the relevant ones to meet the objectives. Other more direct methods may be available for directly calculating the 'optimum' behaviour. These methods are especially relevant when input data are insufficient to define the behaviour of the system; some data may also be mutually inconsistent.

For perfect systems, with perfectly prescribed data, optimisation calculations can lead to results that are equivalent to those obtained by modelling deterministically the detailed behaviour of the system. However, for imperfectly defined systems and imperfect data, optimisation methods can have marked advantages for predicting system behaviour, essentially because such calculations integrate the overall behaviour and are much less perturbed by small errors. But where the system is in a state that is very sensitive to initial conditions, imperfect data can certainly lead to large errors for example, in predicting localised extreme events.

Optimisation methods were first used for analysing continuous systems determined by physical laws (for example, fluid, solid or electromagnetic). In the 18th Century, Euler and Lagrange discovered that the Newtonian differential equations models for mechanical or planetary systems governed by law of dynamics (for velocity and force variables) were mathematically equivalent to finding the minimum of an integral of the kinetic and potential energy expressed as function of the same variables as for the differential equation. These integral minimisation methods were applied in the 19th century (for example, Rayleigh - Ritz method) for obtaining approximate solutions for vibrating systems. But with modern computers, these methods are applied to vibration of very complex and interconnected systems from molecules to aircraft and satellites. But there are limitations to applying these methods for dissipative systems which exhibit chaotic behaviour, for example, with turbulence and friction. Usually, there are no general integral constraints comparable to those that that exist for simpler systems, though hypotheses have been suggested that certain dissipative systems might minimise or maximise dissipation.

Optimisation methods have had their widest application in the management of organisations, especially in enabling them to meet a range of objectives with least resources and usually as fast as possible. Large organisations have networks of sub organisations distributed spatially. Planning the tasks of organisations and individuals, can, like computer programs, be represented as networks of sequential activities. The mathematical description and analysis of these space-time networks goes back to the 18th century. By analysing his Sunday walks over all the bridges of Konigsberg in terms of a conceptual network of nodes, Euler showed in the 1740's whether it was or was not necessary to cross any bridge twice in order to cross all the bridges (Bondi, 1991). This began the generic mathematical description of shapes and curves on bodies with differing topological forms (for example, planes, spheres, doughnuts etc) in terms of edges, surfaces, singular points and how the shapes and curves do or do not change when the shapes and curves are distorted.(i.e. corners, junction of lines, centres of spirals etc.) (for example, Flegg, 1974). This analysis led engineers and scientists to developing very general and efficient methods for describing and classifying many types of real and conceptual patterns of curves and shapes. This application of simple topology can help explain the geometrical influences of the boundaries on processes such as flows through blood vessels, and around and through building complexes (Hunt et al., 1978).

In the 20th Century there was a growing interest in using network theory for planning resources and human activities while satisfying policy or economic constraints. The varied types of behaviour of such networks is modelled by sets of mathematical functions for example, calculating the total time to travel along a network depending on the path taken.

Mathematical methods were derived for calculating the optimum performance of both the linear and non-linear systems by minimising certain mathematical quantities derived from output variables for example, cost, speed etc. (for example,

Bellman & Kalaba, 1965). The performance could be optimised 'on -line' as new data arrives. These developments came after World War II to coincide with the widespread use of electronic computers. They ensured the success of the hugely complex space exploration missions in the 1960's, with their on board 'mini-computers'.

In the emerging discipline of management sciences, these techniques contributed to the extraordinary improvements in the cost-effectiveness and greater reliability of construction projects and production systems ranging from engineering to agriculture.

With these optimisation and control methods, the frequency of failure in most operational systems has been steadily declining, such as in aviation and railways (for example, Evans, 2005).

By the 1980's in Europe and USA there was increasing interest in applying the concepts of system optimisation beyond their original well defined application in technology and production processes to the less well defined problem of optimising investment and strategy for individuals and corporations, and the management of very large commercial and governmental organisations.

Politicians and managers thought that targets for improved performance could be imposed on most types of organisation (for example, in terms of profit or the speed / quality/choice of certain deliverables). It was implicitly assumed, though seldom demonstrated, that the functioning of the organisations could be understood and even modelled well enough that the implications for reaching 'targets' could be known and had been worked out. Although the discipline of such targets doubtless improved certain aspects of the effectiveness of many organisations, analysis has shown that non-targeted outputs can suffer, for example in speed, quality and reliability.

Whether this was the reason or whether there was deliberate malfeasance (not in the model) some large international corporations and government agencies have actually or nearly failed operationally and/ or financially. The consequence is that governments now have to regulate organisations more closely, relying on external and transparent mathematical modelling to do so.

Clearly the models have to evolve to reflect the changing constraints and output variables, as well as the changing nature of organisations and their financial operations and (for example, Gallivan, 2006). Some of the most risky and profitable operations have been stimulated by the very fast processing of data made possible with modern computer systems.

However optimisation of systems is often limited in practice by the capacity and speed of available computers, since their complexity of system is growing so rapidly. Designing and/or predicting the performance of complex systems of continuous field variables (electromagnetic, fluid, solid, chemical etc) which are dependent on vast fields of data, are particularly demanding. In such modelling, even with the largest computational resources, the smallest scale processes, such as cracks in metals or turbulent eddies, have to be approximated in order to speed up the calculations or even to represent them at all. But approximations that smooth out local gradients can lead to large and sometimes chaotic errors; as occurred in the first experiments using numerical methods for weather forecasts. Understanding the qualitative nature of these errors has been a great break through in applied and computational mathematics (for example, Broomhead & Iserles, 1992).

Approximations are even more essential where calculations have to be repeated many times in order to optimise the accuracy for given input data or to find the optimum design by varying the design parameters of the system. As the capacity and speed of computers increase (beyond the present maximum level of 10^{14} operations per second of the Japanese 'Earth Simulator' systems) calculations are becoming progressively more elaborate. Input data and output predictions are more detailed and therefore more extensive in space and time, (typically 10^9 units of input data per day are provided for numerical weather prediction systems) and more detailed processes are being modelled in ever more detail. Effectively there is a 'Parkinson's law' of computing capacity, i.e. the demands for more detail, larger domains and system complexity always outstrip the increase in capacity. This means that new approximations are always needed, as new processes and new constraints are introduced into the models. The use in global climate models of modern computing systems exemplifies these trends; more realistic simulations of aerosols and associated chemical processes are being incorporated and the duration of the model runs become ever greater as the demand grows to simulate prehistory and various, and sometimes alarming, scenarios for the distant future!

For optimisation of engineering designs, the aim of mathematical algorithms is primarily to provide efficient methods for calculating the effects on the performance of varying the design parameters, including possible variations in the shapes of the components. From nano-technology devices to buildings, even the topology of the system might be changed. The shapes of aeroplane wings have been optimised to minimum the aerodynamic drag given certain constraints on the shape of the wing (for example, lift force to carry the pay load, volume to carry fuel, strength etc.).

Since aircraft cruise at speeds close to the speed of sound, shock- waves form on the wings. However through optimised redesign of the wings on Boeing and Airbus aircraft, the position and amplitude of the shock waves were changed leading to substantial reductions (30-50%) in the drag (for example, Jameson, 2000). The power of similar optimisation techniques should lead to quite new designs of integrated engineering systems and their more effective operation. Targets of 50% to 70% reduction in automobiles' use of energy, and of an 80% decrease in ground level noise from aircraft are believed to be practical. Researchers and industrialists have argued that the pressure from government to make these environmental improvements needs to be stronger to make optimum use of current and emerging technological capabilities including those of mathematical modelling (Hunt, 2004).

Optimisation methods are particularly valuable for ensuring that computer models make best use of approximations and incomplete input data, even if it is not always reliable or accurate. These are critical issues for improving numerical predictions of weather and climate. .It was natural that this branch of computational mathematics should be the first to pioneer the methodology of calculating an 'ensemble' of solutions for a given model, using a range of input data corresponding to estimates of its incompleteness (for example, Hunt, 1999). The second pioneering innovation in complex computation probably resulted from the regular exchange of measurements around the world by meteorological services. They also exchange computational data, algorithms and 'advisories' to alert each other about extreme environmental conditions. Building on this tradition, these centres invented methods of combining the results of different approximate models (using approximately the same input data).The results showed, as expected, a wider distribution of calculated solutions for the future weather. 'Outlying' predictions in the ensemble are usually made by one or two models. These have provided useful guidance about the possibility of extreme conditions (Harrison et al., 1999).

A less expected benefit of using wider ensembles is that the mean values of the ensemble predictions prove to be, in general, more accurate than those of any particular model (provided they are all of comparable level of resolution).

Calculations of an enormous ensemble of climate predictions are now also being run by thousands of interested members of the public using their own personal computers (climate.net, 2005). This is a new approach to engagement of the public in scientific policy issues; as well as providing valuable data on possible climate scenarios.

These consensus approaches to modelling are being applied to the socially important problems of predicting changes to global and local climates, especially the even more controversial problems of predicting and planning how to reduce the adverse effects of human activities.

It is not yet clear whether this emerging methodology of multi-model multi-centre ensembles will be taken up in other fields where comparably large computer models and data inputs are needed for operational prediction, for example in geophysics, or engineering; or in modelling of biological systems. An alternative approach is being developed (Noble, 2002) for modelling the functioning of the healthy, and unhealthy, human heart; and how to deal surgically or pharmacologically with its defects. The various components (for example, nerve, fluid flow, walls of the vessels) are developed in different centres and then integrated into one 'super' model.

There is also enough open exchange of techniques and data in the field of geophysics, and enough urgency about the need for predictions that these collaborative methods could be tried to improve those models that have a credible level of predictability, such as those for tsunami waves, wide area flood forecasts (Hunt, 2005; Laguzzi et al., 2001), solar bursts and their interaction with the other magnetosphere (for example, Hunt & Coates, 2003).

In engineering some of the largest computer modelling systems in the world have been constructed in the recent Accelerated Strategic Computing Initiative programme of the US government. One such system included all the components in an aero engine and another for a rocket (McMillan, 1999). Although these models use the best research from other centres they were not multi-centre models, but firmly located at particular universities (Stanford, Illinois). The lessons learnt were shared by other institutions of the department of Defence who had funded the project.

So far there are no plans in Europe to develop multi-centre computational models for engineering systems, even though this might be possible using the new GRID for very rapid data exchange. No 'grand challenge' projects have been emerged comparable to these in the USA (other than the continuation of the very successful human Genome project).

The development of large computer models for complex systems with wide social implications might be highly suitable projects. There could be further development of collaborative models of the human heart and other organs, because these show great promise for improving understanding and for guiding treatments. They might also reduce the requirement for human or animal experimentation that is necessary at present (House of Lords, 2003).

The future operation of transport and energy network systems, need to be improved because their reliability and safety is essential to a modern society (as events in Europe and USA since 2001 have painfully demonstrated). Currently, different elements of their performance are predicted by many organisations using different models and different inputs of data. Similar multi-centre ensembles of predictions could be combined using the approximate models operated by government agencies and commercial organisations, and data for the calculations that is incomplete (as much because of secrecy and slow communications as lack of measurements).

The same lack of collaboration afflicts economic planning and forecasting. As far as is possible to learn from public pronouncements, national and international government agencies and commercial organisations are each running their own predictive economic models.

It is surely likely that multi-model centres economic models will be tested before long, and may well become as useful as in weather/ climate prediction.

3. CONCLUSION AND LOOKING FORWARD

Lancelot Hogben's marvellous book 'Mathematics for the million' and many similar books across the world were aimed at popularising mathematics for the intellectually curious general public. Hogben (1936) particularly focused on the evolution of mathematical culture through the ages. Whereas this paper is addressed to those educating the millions of decisions takers across the world whose roles as senior managers, politicians, civil servants and university teachers involve them in applying the techniques of computing, modelling statistical analysis etc, but also using the broad ideas of mathematical science as. These ideas lie behind the ways that they reach their decisions and explain them. Consequently if journalists, politicians, and the public were more familiar with these concepts they would be able to question organisations more effectively. (In my experiences as a manager mathematical system concepts certainly gave some insight into how far ahead pre-planning is likely to be possible before chaotic influences take over!)

Examples about the growing use of applied mathematics concepts have undoubtedly been helped by their popularisation and by progress in mathematical research (especially in dynamical systems). However, other powerful concepts which have not yet been so well popularised may well also have extensive use in practice. New pictorial images, which were so effective in stimulating interest in fractals and chaos, may be the most effective way to develop interest and understanding in the equally important themes of risk, sudden change, optimisation, new images are needed

Almost all of these concepts arose where mathematics was being applied in diverse fields of natural and social sciences and technology. The ever increasing speed and capacity of computers for calculations and for storing, sorting and transmitting of information of data systems are almost outpacing the rate at which research is producing new concepts for testing on the new computer and data systems. As we have shown, weather forecasts, the Genome project and modelling of biological system are introducing new approaches to modelling computation and interpretation. Perhaps government, research agencies and industry need to consider how these techniques and capabilities could be applied more widely for frontier research problem that are of great practical importance, such as natural disasters and complex socio-economic environmental systems.

Institutes and academic societies are exploring these new areas of mathematical modelling and also disseminating them through workshops, booklets and software, not only for instruction and sharing experience in modelling and computations (for example, Brown & Leese, 2005). They have led to publications of guide-lines for the techniques and also for the strategy of modelling, for example, the appropriate application of modelling and measurement data in the controversial field of computational fluid dynamics and environmental prediction (for example, ERCOFTAC, 1998; Olsen 1998).

There seems to be a plenty of scope for decision makers and mathematical scientists to explore new ways in which the broad ideas of mathematics might be relevant, both in terms of particular techniques and in terms of strategic concepts for decision making about the many enormous problems facing the world.

Finally those teaching in universities might like to consider whether the 'big themes' of applied mathematics and statistics should be discussed in a few lectures to the final-year students in mathematical sciences before they leave, and to graduate students. With plenty of examples, this might encourage mathematical scientists to realise that their discipline fits them for the most responsible positions in societies, just as much as, say, the study of history or economics.

Also, providing such courses might stimulate more broad-minded mathematical scientists to share their insights with those based in other academic disciplines about how modern societies function and are organised. From my experience, few disciplines are as relevant to decision-making today as those of applied mathematics and statistics.

4. ACKNOWLEDGEMENTS

I am grateful to Chris Haines of City University for inviting me to lecture on this unusual topic, and to Frank Moon, Meghnad Desai, Steven Bishop, Steven Blinkhorn and other colleagues for their comments and insights, especially the visiting speakers at the LIMS evening lectures, and the political discussions of modelling in the House of Lords. This paper was partly written at Cornell University where I was Mary Upson visiting Professor in Mechanical Engineering.

REFERENCES

Allen, M. and Stainforth, D. (2002) *Towards objective probabilistic climate forecasting*, Nature 419, 228.
Armbruster, D. (2006) Private Communication. Arizona State University.
Ashford, O. (1985) *Prophet or Professor; The life and work of L. F. Richardson*. Bristol: Hilger.

Belman, R. and Kalaba (1965) *Dynamic programming and modern control theory.* New York: Academic Press.

Bewley (2001) Flow control: new challenges for a new Renaissance. *Aerospace Science*, 37, 21-58.

Bondi, C. (ed) (1991) *New applications of Mathematics.* London: Penguin.

Bonnet, J.P., and Glauser, M.N., (1993) *Eddy structure identification in free turbulent shear flows.* Dordrecht: Kluwer Academic Publishing.

Broomhead D. S., and Iserles A. (eds) (1992) *The dynamics of numerics and the numerics of dynamics.* Oxford: Clarendon Press.

Brown, M. and Leese, R., (eds) (2005) New and emerging themes in industrial and applied mathematics. European Commission, Directorate-General for Research, Rep 21797.

Butler, A., Hefferman J.E., Tawn, J.A., Flather, R.A. and Horsburgh K. (2005) Recent trends in North Sea surge elevation. Elsevier Science pre-print.

Casulli V and Stelling G. (2001) Numerical simulation of the vertical structure of discontinuous flows. *International Journal of Numerical Methods in Fluids*, 37, 23-43.

Chalmers, A.F. (1982) *What is this thing called science?* Milton Keynes: Open University Press.

Colman A.W. and Davey, M.K. (1999) Prediction of summer temperature, rainfall and pressure in Europe from preceding winter North Atlantic Ocean temperature, *International Journal of Climatology.*

Cox, D.R., Isham, V.R. and Northrop, P.J. (2002) Floods: Some probabilistic and statistical approaches. *Proceedings of the Royal Society A*, 360, 1389 – 1408.

Davies, T. and Hunt, J. C. R. (1995) New Developments in Numerical Weather Prediction. In K.W. Morton and M.J. Baines (eds) *Proceedings of ICFD Conference on Numerical Methods for Fluid Dynamics V.*, Oxford: OUP, 1-18.

Embrechts, P., Klueppelberg, C. and Mikosch, T., (1997) *Modelling Extremal Events: for Insurance and Finance. Applications of Mathematics.* Berlin: Springer, vol. 33.

ERCOFTAC 2000 (2000) Best Practice Guidelines for industrial CFD. Available from Prof A. Hutton, Qinetiq, Farnborough UK.

Evans, A.W. (2003) Accidental fatalities in transport. *Journal of the Royal Statistical Society*, 166, 253-260.

Farge, M., Hunt, J.C.R. and Vassilicos, J.C., (eds) (1993) *Wavelets, fractals and Fourier transforms.* Oxford:Clarendon Press.

Favre, A. (ed) (1995) *Chaos and Determinism.* Baltimore: Johns Hopkins Press.

Flegg, G. (1974) *From Geometry to Topology.* Milton Keynes: Open University Press

Gallivan, S. (2006) Is the National Health Service subject to hidden systems effects? LIMS Seminar, March 2006.

Garnett, J.C.M. (1921) *Education and World citizenship.* Cambridge: CUP.

Hacking I. (1975) *The emergence of probability.* Cambridge: CUP.

Haldane, J.B.S. (1935) *Philosophy of a biologist.* Oxford: Clarendon Press.

Harrison, M.S.J., Palmer, T.N., Richardson, D.S., Buizza, R. and Petroliagis, T. (1999) Analysis and model dependencies in medium-range ensembles: two transplant case studies. *Quarterly Journal of Research Meteorological Society.*

Heming, J.T., Chan, J.C.L. and Radford, A.M. (1995) A new scheme for the initialisation of tropical cyclones in the UK. Meteorological Office global model. *Meteorological Applications* 2, 171-184.

Hogben, L. (1936) *Mathematics for the Million*. London.

Holmes, P.J., Lumley, J.L. and Berkooz, G. (1996) *Turbulence, Coherent Structures, Symmetry and Dynamical Systems*. Cambridge: Cambridge University Press.

Houghton, J. L. (1994) *Global Warming: The complete briefing*. Oxford: Lion.

House of Lords (2002) Chips for everything. Select Committee Report. London: House of Lords.

House of Lords (2003) Scientific Experiment on Animals. Select Committee Report. London: House of Lords.

Howard, L. (1803) On the modifications of clouds. Essay to the Askesian Society. London.

Hunt, J.C.R., Abell, C.J., Peterka, J.A. and Woo, H.G.C. (1978 & 1979) Kinematical studies of the flow around free or surface-mounted obstacles: applying topology to flow visualisation. *Journal of Fluid Mechanics,* 86, 179-200; corrigendum 95, 796.

Hunt, J.C.R. (1993) Life and Work of L.F. Richardson. In P. Drazin and I. Sutherland (eds) *Collected works of L.F. Richardson*. Cambidge: CUP, 1-27. (Also (1998) in *Annual Review of Fluid Mechanics*, 30.)

Hunt, J.C.R. and Neunzert, H. (1993) Mathematics and Industry. In: *Proceedings of theEuropean Mathematics Congress Paris 1992*. Birkhauser. (Also in (1994) *IMA Bulletin*, 29, 164-171.)

Hunt, J.C.R. (1995) Forecasts and warnings of natural disasters and the roles of national and international agencies. *Meteorological Applications*, 2, 53-63.

Hunt, J.C.R. (1997) Rounding and other approximations for measurement records and targets. *Mathematics Today*, 33, 73-77.

Hunt, J.C.R. (1999) Environmental forecasting and modelling turbulence. *Physica D*, 133, 270-295, (Proceedings of Los Alamos Conference on Predictability).

Hunt, J.C.R. (2001) Mathematical Model could clarify arms race. *Nature*, 411, 737.

Hunt, J.C.R. (2002) Floods in a changing climate. *Philosphical Transactions of the. Royal Society*, 360, 1531-1543.

Hunt, J.C.R. and Coates, A. (2002). Developments in space engineering and space science. *Philosphical Transactions of the Royal Society*, 361, 205-218. (See also *Nature*, 427, p13, 2004).

Hunt, J.C.R. and Coates, A. (2002) Developments in space engineering and space science. *Philosophical Transactions of the Royal Society*, 361, 205-218. (See also (2004) 'Joint efforts needed to forecast space weather' correspondence in *Nature*, 427, 13.

Hunt, J.C.R. (2004) Conclusions, Key Issues, Highlights – 13th World Clean Air and Environmental Protection Congress and Exhibition. *IUAPPA Newsletter*, November, 2-4.

Hunt, J.C.R. (2005) Inland and coastal flooding: developments in prediction and prevention. *Philosphical Transactions of the Royal Society A*, 363, 1261-1491.

Hunt, J.C.R. (2005) Tsunami waves and coastal flooding. *Mathematics Today*.

Huppert, H.E. and Sparks, S. (2006) Proceedings of the Royal Society Conference on Natural Disasters. *Philosphical Transactions of the Royal Society A*, August.

International Strategy for Disaster Reduction (2005) Kobe International Conference. Geneva: United Nations.

Jameson A. and Alonso J.J. (2000) Future Research Avenues in Computational Engineering and Design. *Proceedings of ICIAM 99*. Oxford: OUP.

Kelly, F.P. (1979) *Reversibility and Stochastic Networks*. Chichester: Wiley.

Krishnamurthi, R. and Howard, L. (1981) Large-scale flow generation in turbulent convection. *Proceedings of the National Academy of Sciences USA*, 78, 1981–1985.

Landau, L. and Lifshitz, L. (1960) *Statistical Physics*. Pergamon.

Lautso K, Spiekermann, K. Wegener M, Sheppard I, Steadman P, Martino A, Domingo R and Gayda S. (2004) PROPOLIS: Planning and Research of Policies Land Use and Transport for Increasing Urban Sustainability. Brussels: European Commission.

Lighthill, M.J. (1953) *Generalised functions* Cambridge: CUP.

Lighthill, M.J. and Whitham, G.B. (1955) On kinematic waves. II. Theory of Traffic. Flows on Long Crowded Roads. *Proceedings of the Royal Society A*, 229, 317-345.

Lorenz, E.N. (1963) Deterministic non-periodic flow. *Journal of Atmosphere Science*, 20, 130-141.

Mandelbrot, B. (1982) *The fractal geometry of nature*. San Francisco: Freeman.

McMillan, C. (1999) Introduction to the Accelerated Strategic Computing Initiative Academic Strategic Alliance Program (Lawrence Livermore National Laboratory, Livermore, Ca) AIAA Fluid Dynamics Conference, 30th, Norfolk, VA, June 28-July 1.

Moon, F. C. (2004) *Chaotic Vibrations*. New Jersey: Wiley.

Nikiforakis, N., and Hubbard, M.E. (2003) A three-dimensional, adaptive, Godunov-Type Model for Global Atmospheric Flows. *Monthly Weather Reviews*, 131.

Noble, D. (2002) Modelling the heart: from genes to cells to the whole organ. *Science* , 295, 1678-1682. Merlin Press.

Oslen, H.R. (1998) Local scale regulatory dispersion models; initiatives to improve modelling culture. 10th Conference on Applications and Air Pollution in Meteorology. Boston: American Meteorological Society.

Poincaré H. (1914) *Science & Method* (English translation) (Preface by Bertrand Russell) London: Nelson.

Reuleaux F (1885) The influence of the Technical Sciences upon General Culture. *School of Mines Quarterly* VII(1), October.

Richardson, L.F. (1922) *Weather prediction by numerical process*. Cambridge: CUP.

Saunders, M.A. and Lea, A.S. (2005) Seasonal prediction of hurricane activity reaching the coast of the United States. *Nature,* 434, 1005-1008.

Saunders, P.T. (1980) An introduction to Catastrophe Theory, Cambridge: CUP.

Schellnhuber, H.J., Crutzen, P.J., Clark, W.C. and Hunt, J.C.R. (2005) Earth system analysis for sustainability. *Environment*, 47(8), 10-27.

Shannon, C.E. and Weaver (1949) *The mathematical theory of communication* Urbana: University of Illinois Press.

Silverman, B.W. and Vassilicos, J.C. (ed.) (1999) *Wavelets: the key to intermittent informations.* London: Royal Society.

Spearman, C. (1904) General Intelligence, Objectively Determined and Measured. *American Journal of Psychology*, 15, 205-260.

Thompson, J.M.T. and Stewart, H.B. (2002) *Non-linear Dynamics and Chaos.* London: Wiley.

Thomson, W. and Tait, P.G. (1879) Treatise on Natural Philosophy, Part 1. Cambridge: CUP.

Turfus, C. (2006) Mathematical tools in finance. *Mathematics Today*, April.

Volterra, V. (1931) Lecons sur la théorie mathématique de la lutte pour la Vie. Paris: Gauthier-Villars.

Zeeman E.C. (1987) *On the Psychology of a Hijacker*: In P.G. Bennett (ed) *Analysing Conflict and its Resolutions*. Oxford OUP.

1.2

ECONOMIC MODELLING: THEORY, REALITY, UNCERTAINTY AND DECISION-MAKING

Kate Barker
Member of Monetary Policy Committee
Bank of England, London, UK

Abstract–*This paper discusses the range of issues which arise from the use of economic models, and the importance of recognising the context when selling a particular model to its clients.*

1. INTRODUCTION

Economic models are a set of relationships based on economic theory and captured by equations. Any model seeks to give a coherent and consistent framework, to help organise thought processes. However, they are of many types for example, one key distinction is between those which attempt to model the whole economy (macro models) as a central bank or government would seek to do, and those which model a sector, or only particular relationships (micro models), as a company might seek to do in considering future demand for its products.

Apart from scope, the other main distinction is the purpose of the modelling exercise. Modelling may be an exercise in analysis – trying to understand the past better, or to establish the likely effect of some policy taken in isolation when nothing else changes (for example, the impact on wages and employment throughout the economy if the minimum wage is increased). Other models are designed for use in forecasting – estimating likely future trends given past history and the present economic conjuncture.

All models have to address some basic issues if they are to be utilised successfully. These include: the situation in which they are developed and the audience which needs to be persuaded to rely on them; the tensions between theory and real-world relationships; and uncertainty, including data uncertainty. These are the questions which are discussed here. None of them are new; Keynes (1938) wrote in a letter: 'Economics is a science of thinking in terms of models, joined to the art of choosing models which are relevant to the contemporary world'.

For many economists, the audience is other economists, and the purpose is to improve understanding or advance theory. In contrast, my involvement with modelling has revolved around wanting to support or influence decision-making by non-economists. In this case, the models clearly have to be persuasive to the decision makers and consumers of economic results are unlikely to be happy to be presented with a blackbox approach. There is, frankly, little appeal in saying 'trust me, I'm an economist'.

I have been both a provider and a consumer of models during my career. Two examples of where I have acted as the persuader are in influencing senior management in the car industry, or more recently politicians and the wider public, having been asked by the UK Government to look at the performance of UK housing supply. However, at the Bank of England, I am on the other side of the fence – and have to be persuaded by the models which the staff have estimated and operate on our behalf, in order to inform our monthly judgment about the appropriate level of UK interest rates. So what makes models useful?

2. USING MODELS WITH NON-ECONOMISTS

The complexity of a model can affect both its goodness-of-fit and also its usefulness in debate. Clearly, theoretical coherence, and the ability to fit the data, will in many cases be strengthened by having a more complex model, both in terms of the variables considered and the dynamics, such as lag structures. But as models become more complicated, while this may increase explanatory power, the model may also become less transparent and more difficult to understand – if the results are unexpected it is less clear why that is so. A complex model is also harder to explain to an audience unused to economic debate. This may make for reluctance to take decisions based on its outputs.

2.1 Influencing Business Decisions

Looking at this in a business context, while working in the automotive industry, it was necessary to provide projections of the demand for passenger cars across different European markets. The most basic relationship that might be expected to hold was that the trend in demand would vary with the trend in private consumption. (Of course, this begs the question of how to forecast private consumption.)

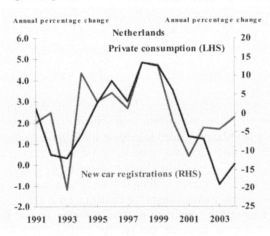

Figure 1a. Simple relationships (OECD Economic Outlook No.76)

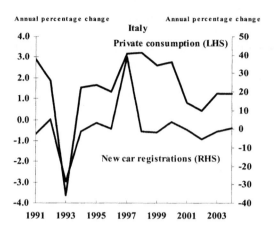

Figure 1b: Simple relationships (OECD Economic Outlook No.76)

Figures 1a and 1b show the growth of car sales plotted against consumer spending growth for the Netherlands and Italy – the kind of graph that is very persuasive in business debate. For the Netherlands, the fit is visually quite good for much of the period. For Italy, however, the relationship is much less clear. During that period, there were clearly other factors which affected the pattern of car demand. Certainly in 1992 registrations in Italy were distorted upwards, due to grey imports (cars being registered in Italy to avoid German taxes, an activity made possible by the advent of the EU's single market). Subsequently, tax changes tended to shift sales between years, so that the underlying relationship with consumer

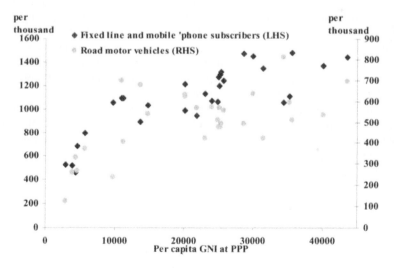

Figure 2. OECD saturation curves: the ordinate axes are per thousand population.
(Sources: OECD and World Bank Data)

spending becomes obscured, and the model needs to be more complicated. This would hardly concern an economist, but could reduce the model's credibility for those taking decisions based on it.

Figure 2 takes a different approach to the same sort of data. It plots the demand for vehicles per thousand population against per capita income, across different OECD countries. It also shows the same plot for fixed line and mobile phone subscribers. For telephone lines, there is a fairly clear saturation curve, suggesting that as income rises beyond a certain point the growth of demand declines relative to the growth of income. For car sales the picture is less clear-cut, once incomes have reached a certain level. Again, taxation differences are significant (one low outlier for cars is Denmark where taxes have been high) as is the density of population (the highest level of car ownership is in the US). So the key challenge is working out how to communicate plausible complexities around underlying simple relationships.

2.2 Influencing Politicians and Public Debate

Public policy work raises different issues of understanding and communication. Public perception often differs from reality - a familiar example of this is migration. Public polling has shown that many people believe the proportion of foreign migrants to be much higher than it actually is – across Europe the perception is that immigrants account for 18.5% of the population, and the reality is half of that level[2]. So as well as addressing the real issues raised by migration – assimilation of different cultures, provision of public services – this perception issue has to be tackled too.

In the course of conducting an independent, Government-commissioned review of why the new supply of UK housing seemed to be inadequate, I encountered a surprising, but widely-held, view. An apparently ingrained belief among many planners and local politicians is that the price of houses is driven only by demand factors, and that supply is irrelevant. Of course, in the short-term demand is undoubtedly the dominant influence on price changes, as new supply each year is generally less than 1% of the stock. But from there it is a large and unwarranted step to assert that supply is simply irrelevant, and that planners need not concern themselves with what is happening to prices.

If a market is working well, then a sustained increased in demand should act to generate new supply, otherwise prices will have to rise. Figure 3 shows the number of new houses built each year in the UK, and the annual rate of increase in real house prices. (The earliest period, in the 1960s, is difficult to compare to the more recent past, primarily because at that time there was a higher annual rate of demolition.) During the 1970s there were sharp peaks in house price inflation, related to peaks of economic activity – these were generally too short-lived to alter new supply. The longer period of house price inflation in the late 1980s, on the other hand, did result in higher supply. More strikingly, the latest period of house price inflation, although lengthy, caused no change at all in the rate of new supply up to 2003. Planners, who often only use demographic projections to determine the requirements for new housing, may be uninterested in this.

So, for this topic, a debate on policy therefore had to start with a discussion not just of the facts, but of the right way to analyse them in order to draw policy conclusions.

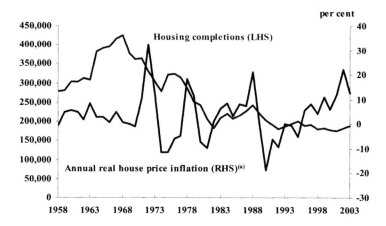

Figure 3. UK house prices and new supply

Sources: Office of the Deputy Prime Minister (National House Building Society, National Assembly for Wales, Scottish Executive, Department of Social Development)
(a) House Prices (Nationwide) adjusted for retail prices. This uses the ONS 'Retail Price index' to convert nominal prices to real prices.

3. UK MONETARY POLICY – WHAT IS THE QUESTION MODELS NEED TO ANSWER?

So far, the account has been of using relatively simple models to try to explain underlying economic relationships in contexts where these ought to be vital guides to decisions. Formulating monetary policy means discussing the use of much more complex models in an entirely different environment; although the fundamental issue is the same - how to gain a shared understanding of economic relationships which can inform good decisions.

Perhaps first, it would be useful to explain my present role. Since 1997, the Bank of England's Monetary Policy Committee (MPC) has had the job of achieving the Government's target for consumer price inflation in the UK. One good thing about this job is that the objective is therefore admirably clear, although this is compensated by the complexities of the issues the Committee has to consider.

The key points of the UK's present monetary policy framework are: firstly, that the Bank of England, on instruction from the MPC, sets the UK's short-term interest rate independently. Secondly, that the inflation target is confirmed or changed each year by the Chancellor; at present it is 2% for annual CPI inflation. Thirdly, that deviations on either side of our point target are treated symmetrically; the target is not a ceiling and we care as much about undershooting the target as overshooting it. Figure 4 is included to show that so far we have been quite successful. Until 2004, we were targeting a different measure of inflation, RPIX, and the target was 2.5% - you can see that inflation after 1997 remained pretty close to the target. Since 2004, the CPI has generally been below target – though as a matter of fact the latest month (June 2005) is exactly 2%. But in setting policy we are always looking forward, so this record gives absolutely no grounds for complacency.

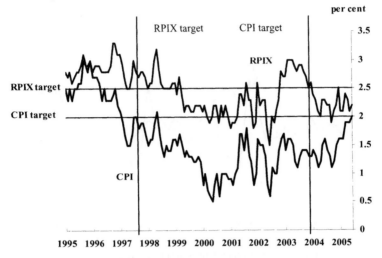

Figure 4. UK inflation performance: ordinate axis is percentage change on a year earlier. Source ONS.

There are nine members of the MPC, all of whom have some economic technical expertise (and some of whom are highly proficient). Importantly, we are individually accountable for the votes we cast each month on what the interest rate should be, and these votes are made public.

Our votes are based on our judgement about where inflation is likely to be in around two years time – as this is about the point at which a change in interest rates today would be expected to have most impact on inflation. To guide this forward-looking activity, the Bank produces and publishes an economic forecast once a quarter. This represents the best view of the Committee collectively (although of course there will be individual differences, which, if very significant, are made public). Unlike the US Federal Reserve, the forecast belongs to the MPC, and not to the Bank staff.

Being held individually accountable, and being part of the group which determines the forecast, means that all nine of us feel closely concerned in the process of producing the projections. We care greatly about the model we use, which therefore has to be one which makes the Committee feel comfortable with its representation of economic relationships and causality.

What are we seeking to do in our management of the economy? A simple way to describe it (Figure 5) is that the economy has at any one time a capacity to supply a certain amount of output (trend output). So perfection would be to set interest rates to get inflation on target, the economy producing at full potential, and an outlook in which demand was expected to grow in line with supply. Of course, even if we reach this happy point, some economic disturbance is bound to push the economy off in a new direction. A further difficulty is that the MPC might not realise until some time after the event that this balance had been achieved, due to data uncertainty.

Figure 5. The Economic Cycle

3.1 Developments in Modelling at the Bank of England

When considering large economic models of the type needed to capture relationships across the economy, the view of the 'best' model will develop as economic research progresses. In 2002, the Bank asked Professor Adrian Pagan to comment on the MPC's decision to move on from the model which had been used to guide its decisions since 1997 (Pagan, 2003). His report contains the diagram shown in Figure 6. It seeks to capture the concept that there is a 'best practice' frontier for economic modelling, and that moves along this frontier entail trading-off theoretical coherence for empirical coherence – fitting the facts.

It may be surprising that such a trade-off should exist. But any model can only represent imperfectly the complexities of a national economy with its vast array of individual actors. So theory is a necessary abstraction from the noise of reality. The approach taken by a model-builder will determine where the model sits on this diagram. Theoretically-driven models, such as the Dynamic Stochastic General Equilibrium (DGSE) approach, may not fit the short-run movements of the data well. On the other hand, Vector Auto Regression (VAR) models primarily rely on the data to determine short-run relationships, and only include long-run theory-driven conditions if these are confirmed by the data.

What exactly are we trying to model? Well, Figure 5 suggests that one thing we need to know is how changes in interest rates will affect demand over the next two years or so. Our task is essentially forward-looking. The economic linkages from interest rates to inflation are referred to as the transmission mechanism. The initial effects, shown in Figure 7, may occur fairly quickly. As the arrows imply, they are interrelated. Moves in the official base rate lead to changes in the market rates paid by borrowers and received by savers – not just short-term rates but possibly right along the yield curve to five, ten or even small changes in 25 year rates.

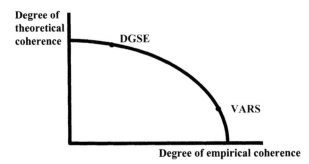

Figure 6. Pagan "modelling frontier" (1)

In addition, asset prices may change. Equity prices are likely to rise if interest rates fall. House prices may be similarly affected. And the expectations of businesses and of households about the future may be boosted by lower rates (or dampened by higher ones). Finally, the exchange rate is generally expected to appreciate at the point when domestic interest rates rise (at least if this rise is unexpected) – however of all the responses to a rate rise the change in the exchange rate, at least in the short-term, is one of the most uncertain, due to the many international influences on this rate.

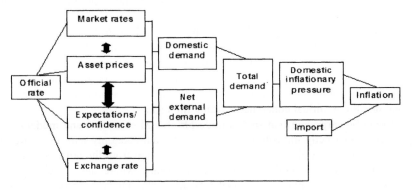

Figure 7. The Transmission Mechanism

Over time the initial changes in interest rates and asset prices go on to drive shifts in the pace of demand growth. Using the example of a rate *increase,* domestic borrowers, who may be credit constrained, tend to cut back on their spending. Firms may trim investment plans as demand prospects deteriorate and the cost of capital rises. A higher exchange rate will tend to reduce demand for the UK's products, and increase demand here for foreign ones, which are now less costly. These falls in the demand for domestically produced goods reduce the pressure on the capacity of capital and labour in the economy. Together with the expected decline in import prices, this should ease back any upward pressure on prices, and in time the rate of inflation will fall back from what it otherwise would have been. It should be apparent that for many of these linkages, there will be a delay, or lag, before the full response to changing conditions occurs.

The nature of what is being modelled is complex – Figure 7 is a significant simplification of a web of linkages and lags. At any time, it is likely that there will be particular sectors of the economy which pose puzzles. For this reason a variety of models are used in order to ensure that we do not become too bound into one way of thinking about the economy. But even so, the demands on the main model remain significant. In addition to having at its core a set of theory-based relationships which capture the key parts of the transmission mechanism, it needs to be able to forecast, and also to answer a whole range of 'what if' questions. The most obvious of these is *What if a different path of interest rates is set?* but there are many others.

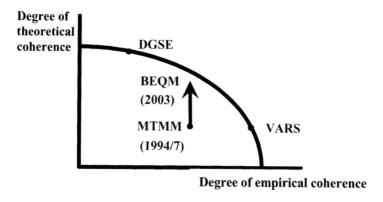

Figure 8. Pagan "modelling frontier"(2)

In common with other central banks we believe that there is always progress to be made in modelling, and continually seek to improve what we do. So, returning to Adrian Pagan's chart, when the MPC was given the task of setting interest rates in 1997, the Bank's then model (called the Medium-Term Macro Model, or MTMM) became a greater focus for attention, and over the following five years proved a valuable tool which had good empirical coherence.

However, it was considered that changes in economic understanding, and also improvements in the ability of computers to handle complex estimation, meant that the MTMM ceased to be state of the art. Rather more importantly, the MPC believes that it supports our credibility if we are able to demonstrate that we are ensuring we have the best tools. Other central banks too have changed and developed their models over recent years, and all are committed to aiming at best practice. So the main motivation for developing a new model, in terms of this diagram, was to move closer to the theory frontier; see Figure 8. It might be added that the diagram itself is a source of controversy – as some economists argue that a good theory should be able to explain the world.

3.2 The New Bank of England Quarterly Model (BEQM)

The new model therefore needed to be more transparent and explicit about the underlying theory, which in turn needed to capture a view of the world which the Committee, broadly, could collectively hold (Harrison et al., 2005). In particular, aspects which needed to be improved from the previous model included a different role for expectations – to make these more explicit as well as model-consistent. This means, for example, that agents will have expectations for the path of interest rates which are consistent with the model, and these expectations of course will influence their behaviour today.

In addition, the stronger theoretical backing means that there is an underlying steady state for the economy. An example here is that in the long-run the UK's stock of net foreign debt, relative to the UK's output, will be held constant so that the level of foreign debt cannot rise unsustainably. The steady state for the economy is where we would get to when all shocks (such as big oil price changes) had worked through – it is not therefore a situation we would ever expect to arrive at! But it will be working in the background to determine where we go.

The model also has to enable the MPC to tell stories about what is happening in the economy – to yield plausible economic accounts of how a shock to a particular variable will work through the system.

And alongside the better theoretical foundation, the requirement was for a model which fitted the data at least as well as the previous one, and would also be able to handle the short-run constraints on data which are used for the forecast (for example, rather than allowing the model to forecast the exchange rate, it is set by a forecasting convention over the next two years).

How has the combination of theory and empirical fit been achieved? The key is that the new model has a two-tier structure, in which the core consists of a coherent set of structural relationships which have been theoretically determined. This approach produces a future path for the economy in which the key variables move back towards their long-term consistent equilibria. Of course, as far as possible the structural parameters used in this part of the model are calibrated to come as close as possible to the data. But used on its own, this core is unlikely to yield plausible short run forecasts, as the simplification of reality involved in constructing the theory means it is unlikely to be empirically robust.

So in practice the forecast combines these theory driven paths with dynamics around the core which have been estimated in order to fit the data. This non-core model supplements the core in two significant ways:

- Firstly – it allows for the introduction of some elements of theory which have not been accommodated in the core model; for example credit constraints, which prevent households from optimising their consumption decision over their lifetime, are not part of the core.
- Secondly – and this is key in light of this model's purpose – the non-core model allows adjustments to be made to parameters when producing the quarterly forecast. The Committee may believe that there will be longer lags than usual in a particular relationship, and in that case this can be reflected in the model for that round. Or that structural change is occurring, which models may not pickup – a current example being changes in the behaviour of UK import prices relative to world export prices, as we increasingly import from cheaper sources of supply, primarily China. This ability to allow a role for judgement is vital to the model's practical use.

But in making these adjustments and adding these dynamics, the consistencies between flows in the economy, and between stocks and flows, are all maintained.

The imposition of these long-term conditions in the economy is one of the major gains from the introduction of BEQM.

3.3 The Structure of BEQM

Figure 9 shows a much stripped down picture of the main economic sectors modelled in BEQM, and describes the basis of the approach to modelling their behaviour – a heroic attempt to explain the working of the whole economy as a thumbnail sketch. In a nutshell, households act to maximise their utility, subject to the budget constraint which they face. Firms, faced with the level of demand for their product at different prices, and their set of technology, seek to maximise profits by their decisions on prices and on employment.

Households and firms encounter each other in two ways – firms supply goods to households and purchase labour from them – the labour market mediates the wage in the latter case. A key feature of wage-setting behaviour is that nominal wages are 'sticky' – that is, that they do not immediately reflect changes in the economy (such as a rise in productivity, or in imported inflation) because in any time period not all firms are engaged in wage settlements. This stickiness, or lag-structure, is an important part of the transmission mechanism between changes in the interest rate and changes in the rate of inflation.

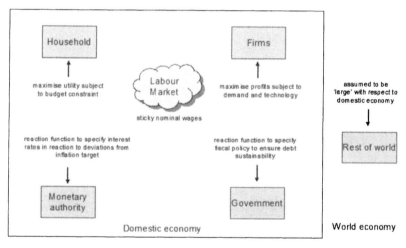

Figure 9. Sectors in BEQM

It is also worth noting from this chart that the rest of the world is assumed to be 'large' – that is, world interest rates or inflation rates are not much influenced by developments in the UK. And the model includes the behaviour of the MPC itself – economic agents base their behaviour on the assumption that monetary policy will deliver the inflation target. This may sound a little odd, but a key feature of a successful inflation-targeting regime is that households, firms and unions believe that the monetary authority will act to achieve the target – and this then affects the way in which all of these sectors respond to changes in inflation pressure. If the monetary authority is really credible, then the expectation that inflation will come back to target means that if a shock, such as a rise in oil prices, pushes the inflation rate up, it is less likely to lead to wage rises in order to compensate workers for the loss of purchasing power (because of the belief that interest rates would then be raised to bring inflation back down).

Figure 10 illustrates in summary some important issues about stocks and flows of funds. In the long-run, there are constraints on stocks. For example the government cannot increase its debt holdings, relative to overall national income, beyond a certain point. This point is when interest payments on the debt rise so high that the fiscal position becomes financially unsustainable. Further, when government debt is rising, a consistent model has to ensure that it identifies who is likely to be holding that debt, and what effect that will have on their behaviour and on the value of assets.

The picture of the flows looks complicated enough – but the model of course has to deal with the further complication that over time the *value* of the stock of assets and liabilities will also be changing. Handling these questions successfully is very helpful for understanding what is going on in the economy, because of the spotlight it shines on the sustainability of present government, household or firm behaviour. In the present UK context, the focus is very much on the sustainability of an increasing level of household debt.

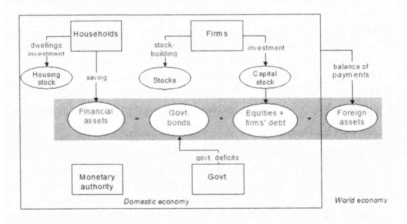

Figure 10. Savings/investment equilibrium

4. TWO ISSUES FOR MODELLERS AND POLICYMAKERS

Having given a necessarily brief overview of how the Bank's model is constructed, I now want to discuss two issues faced by modellers and policymakers, and describe the Bank's general approach and my own views.

4.1 Problems of Asset Prices

The first of these is asset prices. If in general models can only capture the world imperfectly, this issue is even more acute in relation to asset prices such as exchange rates, house prices and equity prices. The theoretical beliefs which economists hold about the determination of asset prices are frequently called into question by the actual paths these prices follow. From time to time, so-called bubbles in asset prices develop. A general definition is that a bubble is a sharp movement in the price of an asset which is not warranted by economic fundamentals – but based on extrapolations of rising prices which attract new buyers. Identifying bubbles is not easy; agreement even on the existence of a bubble is usually possible only after the event.

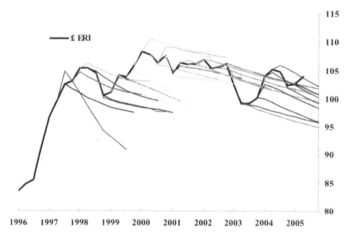

Figure 11. UIP based forecasts of exchange rates (Source: Bank of England)
Note: The different lines departing from the main £ ERI line
show respective UIP based projections for £ERI

Clearly, a model will be of little help in predicting something which is not related to fundamentals – so BEQM instead uses theory to determine the long-run value of asset prices, but in the short-run looks to empirical evidence. Taking the example of exchange rates, one widely-used theory here suggest that the future path should be governed by uncovered interest parity (UIP) – which means that arbitrage opportunities will be eliminated by the market (so if domestic interest rates are above foreign rates, the exchange rate is expected to depreciate over time). In practice, however, a path for the exchange rate in the short-term is imposed on the forecast. Why?

Figure 11 illustrates the most obvious reason for not relying on theory, which is that the UIP based projections for sterling effective rate index (£ ERI) are so often wrong, not just in terms of the size of the movement, but frequently in its direction.

The empirical evidence supports reluctance to use either UIP or economic fundamentals as a guide to future exchange rates. Meese and Rogoff (1983) found that of all the models available, the best forecast was simply to assume that exchange rates followed a random walk, or in other words simply to assume the rate remained at today's level. More than twenty years later, despite much further effort and research, this result has not been conclusively overturned.

But just using this result would leave a large gap in our projections. It leaves unanswered the question of what the exchange rate is likely to do when interest rates are changed. So theory has to be brought into play, in order to consider how our policy action is likely to work through. So while in the short-term the forecast is based on a theoretically uneasy combination of UIP and a random walk – the long-term equilibrium in the model reflects economic theory only – an example of a pragmatic solution.

4.2 Problems of Uncertainty

The second issue is uncertainty. It is not a startling revelation to observe that economic forecasting is plagued by uncertainty. So, how can the different forms of uncertainty be tackled in practice?

4.2.1 Uncertainty about the model

Uncertainties arise not only about whether we have got the form of the model right, but also about its parameters, and indeed about the data used to estimate it. This leads to inevitable uncertainties about the levels of the key underlying features of the economy which drive our decision-making.

Figure 9 sets out in simple form the basic assumptions which BEQM makes about behaviour of different economic agents. But even these basic assumptions are not uncontroversial. For example, some models of household consumption are based on the assumption of a representative household which lives for ever – whereas BEQM takes a model in which households have a probability of survival in each period. This latter type of household would tend to respond differently to higher government spending, as they do not assume that the burden of future tax payments will necessarily fall on them.

Such model uncertainty is clearly very pervasive. Therefore, while BEQM plays a central role in the MPC's forecast process, it is supplemented by the use of a range of models, partly to develop a sense of how important these uncertainties are at any one time, and also to avoid being dominated by one particular view of how the world works. But there is no clear guidance from modelling on how this kind of uncertainty should affect policy.

On the other hand, there has been more helpful work on the question of uncertainty around the parameters of the model (how much will consumption change following a particular shock). Here, it is argued that policy-makers should not respond to a sharp change in economic conditions by seeking to offset all of this immediately, according to a literal reading of their model. The reason for this is that policy actions in these circumstances can add to uncertainty about future inflation, and so tend to increase the risk of missing the inflation target by a wide margin. Therefore it is preferable, for a policy-maker who seeks to minimise the variance of inflation, to take a gradualist approach to interest rate changes, and this is one reason why a move up or down in interest rates is usually implemented in a series of small changes over several months.

In sitting down each month, to discuss our decision on interest rates, inevitably the meeting includes an overview of the latest data. But it is important not to become too focussed on analysing relatively small changes in trends, especially if these are only apparent over the more recent time period. The reason for this is that the data itself is uncertain – and this is hardly surprising given the vast number of outputs and transactions throughout the economy that these figures seek to summarise.

4.2.2 Uncertainty about data

In judging data series, we have to have something which we think of as being 'truth', although of course in this case the real truth is not really knowable. Data revisions come from two main sources – either improvements in the analytical methods, or a larger survey becoming available for the same series (it is frequently the case that large annual samples are updated through the year by the use of smaller samples). So one way of defining the truth is to take the estimates for a series published some time after the event – we tend to use around two years later.

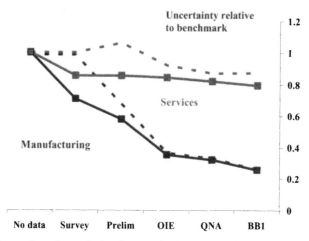

Figure 12. Uncertainty through the data cycle (Source: Ashley et al., 2005).
The vertical axis shows the level of correspondence to a benchmark based on the variance of final data out-turns. Survey, refers to survey data. Prelim, OIE, QNA and BBI are successive ONS estimates. Dotted lines are 'ONS estimates' and solid lines are 'Estimates using surveys'.

The dotted lines on Figure 12 show how successive official estimates of the output of the services industry, and of manufacturing (perhaps some of the most basic data in the economy) can be judged in terms of the level of uncertainty compared to the final 'truth'. As far as manufacturing output is concerned, uncertainty falls pretty steadily as updates to the national accounts are published. But for services, even by the time that the first so-called 'Blue Book' is published, around one year after the data period concerned, there is still considerable uncertainty.

This should not at all be taken as criticism of the UK's statisticians. These issues arise in all countries, and we are far from the worst. But, while it is not surprising that services output is the more uncertain, given the inevitable measurement problems (how, for example, do you measure the output of a bank?) – it is unhelpful, as this sector now accounts for more than two-thirds of the economy.

In order to improve our certainty about the data, the early estimates can be supplemented with the evidence from the business surveys, and from past experience it is the case that some improvement in certainty can result from this. The solid lines on this chart show how this can improve the estimates. But just as significant is using common sense, and being careful not to put too much weight on any particular recent data series.

4.2.3 Uncertainty about equilibrium

Even if all the data problems just mentioned were resolved, judgements about what level of output (or supply) can be produced without running into inflation pressures are not obvious. For example, an important part of this judgement is knowing how far unemployment can fall before it triggers rising wage inflation. The data for unemployment themselves (if based on the right collection method) are pretty reliable – so worries about this being measured incorrectly are not the problem. Rather, in the UK (and indeed in some other countries) there is clear evidence that, as a result probably of a range of labour market policies which have improved flexibility, the level of unemployment consistent with stable wage pressure has fallen in recent years. But it is very uncertain just how much lower this level is – and during my four years on the MPC the broad estimate we use has changed in the light of experience.

It is possible to place some bounds on this uncertainty, using for example Kalman filter techniques (for example, Driver et al., 2003). But the range this results in is too wide to discriminate between the usually rather fine policy decisions which are under discussion – so once again judgement needs to be used.

4.3 How the Bank communicates uncertainty

In the light of all the caveats to economic projections, it is not surprising that one of the features of our central projections is that there is a negligible probability that they will coincide with the outcome. But that broad point is not of much help to our task, or to ensuring that the outside commentators understand the basis for policy.

Uncertainty is clearly a rather pervasive feature of the modelling and forecasting process – and dealing with it therefore part of the regular routine. However, in producing our quarterly forecasts we will also discuss specific risks to the central projection which are considered to be significant at that time. Examples of these in recent quarters have included concerns about the exchange rate, about the course of house prices and about how consumers might respond to falling house prices.

These of course may not all be biased in the same direction. So while the forecast which is produced can be thought of a mode, the mean of our expectation may well be different. In addition, the experience of past forecast errors is used for a basis for estimating the uncertainty around each forecast, although these are not used mechanically. Each quarter's discussion includes consideration of whether uncertainty is judged to have increased or decreased, independently of the message from analysis of the past. This is probably the only part of our discussions where it would be right to say that the MPC may have a bias. It always seems more comfortable to say that uncertainty has risen, than to observe smugly that it has fallen.

This question of uncertainty is not just presentational. It can drive what is decided. Figure 13 illustrates the point for February 2005. The mode of the inflation forecast suggested that the target would be overshot on the upside. But the downside risks led to a majority of the MPC voting to leave interest rates unchanged.

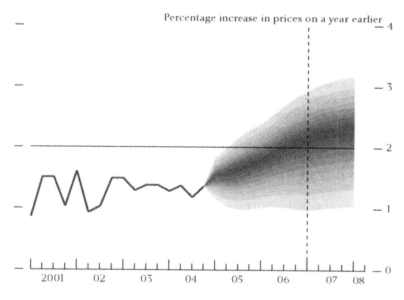

Figure 13. CPI inflation projections (Source: Bank of England)

Showing the current CPI inflation projection in February 2005 based on market interest rate expectations. The fan charts depict the probability of various outcomes for CPI inflation in the future. If economic circumstances identical to today's were to prevail on 100 occasions, the MPC's best collective judgment is that inflation over the subsequent three years would lie within the darkest central band on only 10 of those occasions. The fan charts are constructed so that outturns of inflation are also expected to lie within each pair of the lighter red areas on 10 occasions. Consequently, inflation is expected to lie somewhere within the entire fan charts on 90 out of 100 occasions. The bands widen as the time horizon is extended, indicating the increasing uncertainty about outcomes. See the box on pages 48–49 of the May 2002 Inflation Report for a fuller description of the fan chart and what it represents. The dotted lines are drawn at the respective two-year points.

The chart, illustrates the presentational device which we use to indicate the dangers of too much focus on a point forecast. The darkest central band is the modal projection. This and each pair of bands on either side indicates a 10% probability that the inflation outturn will be in this range. This chart, usually presented in red, nicknamed the 'rivers of blood' by economists in the City of London, drives home the scale of the many imponderables which we have to face.

5. CONCLUSIONS

Discussion on the range of uncertainty is intended to be a realistic, rather than negative, ending to this talk. I have sought to convey the range of issues, including questions around persuading the audience, which have to be confronted in any attempts at economic modelling, and particularly modelling of whole economies. On the way, I hope to have convinced the reader that there is a great deal of science involved in the construction of models – which are attempts to capture considerable complexity and will often need to deal with the interplay of multiple factors as well as parameters which shift over time. This means that the techniques have to be very sophisticated if the necessary coherence is to be ensured. Indeed they have become increasingly so – hence by my own admission I am no longer able to claim to be an expert.

But all of this technical sophistication runs up against limitations – it cannot hope to be a full depiction of reality; or to understand behaviour by households or firms which by economic standards would be considered irrational; or to predict structural change. This means that there is space for the exercise of judgement – to return to Keynes again – 'economics is essentially a moral science and not a natural science'.

ACKNOWLEDGEMENT

I am very grateful to Rebecca Driver and Miles Parker for research and advice in preparing this paper.

NOTES

1. The views expressed here are personal and should not be interpreted as those of the Bank of England or other members of the Monetary Policy Committee.
2. See the 2005 CEPR report on "Immigration, Jobs and Wages: Theory, Evidence and Opinion" by Christian Dustmann and Albrecht Glitz. The public consistently overestimate immigrants' share of the total population. On average immigrants account for 9.4% of the actual population in Europe (where immigrants are defined as foreign born residents), compared to a perceived share of 18.5%. This gap is common to all European countries and does not depend on the actual size of the immigrant population. Including second generation immigrants narrows but does not eliminate this gap.

REFERENCES

Ashley, J., Driver, R., Hayes, S. and Jeffrey, C. (2005) Dealing with data uncertainty. *Bank of England Quarterly Bulletin*, Spring, 45 (1).

Driver, R., Greenslade, J. and Pierse, R. (2003) The role of expectations in estimates of the NAIRU in the United States and United Kingdom. *Bank of England Working Paper*, No 180.

Harrison, R., Nikolov, K., Quinn, M., Ramsey, G., Scott, A. and Ryland, T. (2005) The Bank of England Quarterly Model. London: Bank of England.

Keynes, J.M. (1938) *letter to Roy Harrod, July 4, 1938.*

Meese, R.A. and Rogoff, K. (1983) Empirical exchange rate models of the seventies: Do they fit out of sample? *Journal of International Economics*, 14, 3-24.

Pagan, Adrian (2003) Report on modelling and forecasting at the Bank of England. *Bank of England Quarterly Bulletin*, Spring: 60-88.

Section 2
Modelling Constructs in Education

2.1

DREAMING A 'POSSIBLE DREAM': MORE WINDMILLS TO CONQUER

Peter Galbraith
University of Queensland, Brisbane, Australia

Abstract–

> Oh, East is East, and West is West, and never the twain shall meet,
> Till Earth and Sky stand presently at God's great Judgment Seat;
> But there is neither East nor West, Border nor Breed, nor Birth,
> When two strong men stand face to face, though they come from
> the ends of the earth. (Rudyard Kipling, 1889)

Approaches to real-world problem solving, in the world beyond the classroom, offer attributes that can enrich the teaching and learning of mathematics. Such attributes stand to develop the capacities of individuals to apply mathematical knowledge in situations involving social or personal decision-making or in other academic pursuits, as well as for vocationally-specific purposes. The meeting of the outside world with the mathematics classroom (metaphorically East and West), challenges both sides to enhance mutual understanding by finding common ground at the interface. We have identified a number of meeting points and structures where this need is highlighted, including: properties of real-world problems that differ from those traditionally met in mathematics curricula; identification of problem types that occur outside the classroom, and the construction of parallel school-level problems; teaching and learning conditions necessary for the solution of real problems; and associated tensions and inconsistencies within existing didactical practices.

1. THE CHALLENGE GOES ON

A dream implies unfulfilled ideals or circumstances and so we begin with a brief history lesson, for the benefit of readers who may not be familiar with some of the salient background. Over 30 years ago, Henry Pollak (Pollak, 1969) challenged the mathematics education community to engage more seriously with genuine applications of mathematics. Since that time, there has been some notable progress, but this is mitigated by the knowledge that significant challenges remain. To illustrate, we begin by revisiting some of the problematic examples used by Pollak, and point to similar examples drawn from contemporary curriculum materials.

1.1 Pretend Applications

Problem 1a

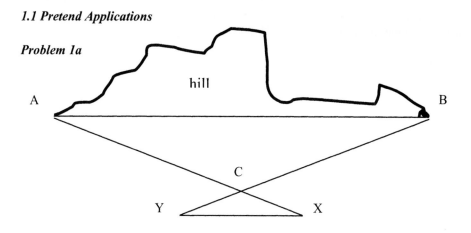

Figure 1. (after Pollak, 1969)

To find out how far to tunnel through a hill, a surveyor lays out AX =100, BY=80, CX=20, and CY=16 (distances in yards). He finds by measurement, that YX = 30. How long is the tunnel, AB?

Comment: "Word problems tend to be exercises in translation, and in the subsequent mathematical techniques, the reality of the application is often neglected" (Pollak, 1969). (Note: solution implies terrain is flat – why build a tunnel?)

Problem 1b A take-away food shop sells hamburgers, sausages, and pizzas. On one day the owner noted that the number of hamburgers sold was three times the number of pizzas, and the number of sausages sold was five times the number of pizzas. The number of hamburgers and pizzas sold was in total 176. How many of each type of food was sold? (2005 source)

Comment: Again, this word problem does not represent the 'reality' of calculations a shop owner would undertake for the purpose of monitoring the demand for particular items. While the problem has a number of useful features from a curricular perspective, real-life applications of mathematics is not one of them.

1.2 Problems of Whimsy

Problem 2a Two bees working together can gather nectar from 100 hollyhock blossoms in 30 minutes. Assuming that each bee works the standard 8 hour day, 5 days a week, how many blossoms do these bees gather nectar from in a summer season of 15 weeks?

Comment: "The function of such problems (is to) provide comic relief in the Shakespearean sense, and (they) probably do a lot of good – although not as applied mathematics." (Pollak, 1969).

Problem 2b Two meatballs roll off of a pile of spaghetti and roll toward the edge of the table. One meatball is rolling at 1.2 m/s and the other at 0.8 m/s. They fall off the table and land on a $5000 Isfahan carpet. If the table is 1.2 m high, how far apart from each other do the meatballs land? (2004 source).

Problem 2c Pythagoras was sitting in calculus in the 80-degree weather of mid-April, wishing he were at the beach. While daydreaming, his terrible case of senioritis took over and his grade quickly began to plummet. When he got his final report card, he saw his grade had decreased. The amount it decreased is equal to the volume of the solid bounded in the first quadrant by $y = 2 - x^3$, revolved about the x-axis. If he started the 4th quarter with a 69, by how much did his grade decrease and will he pass the class with a 60 or better? (2005 source).

Comment: Problems like this are possibly whimsical enough to avoid the pretence of reality – but there is concern when these seem to be the only types of examples provided.

Moving from the specifics of problems to the desirable attributes that modelling courses should engage, consider the following quotes:
"In order to prepare them for future jobs, faculty in universities believe it is important for students to develop their own mathematical ideas from real-life situations, as well as to justify how they obtained their results... This requires from students two important skills: (a) to be able to develop a mathematical model from a real-life situation, and (b) to be able to explain this model to someone else." (Chalarambos, Aliprantis & Carmona, 2003)
"Objectives of the project based course include "Modelling methodology" - providing experience of the formulation/solution/interpretation/validation phases, and of empirical and theoretical approaches; and "Communication" - providing experience of teamwork, scheduling work to deadlines, written and oral reporting, and participating in meetings." (O'Carrol, Hudson & Yeats, 1987)
Expressing almost identical sentiments, these come from sources almost two decades apart, providing another illustration of the kind of 'unfinished business' that frustrates and excites the field of applications and mathematical modelling as an educational endeavour.

2. MODELS OF MODELLING

Changing the focus from individual examples to programs, we note that theoretical positioning has substantial curricular implications, and there appear to be currently two (at least) substantially different philosophies that influence the way that mathematical modelling in education is approached. The term philosophy is a strong one, but is used here to indicate that different belief systems lead to different priorities, and the approaches summarised in the following have distinct ontologies and epistemologies, that when unpacked lead to different teaching methodologies and emphases. They represent a further crystallisation of positions described some years ago in a major survey paper that reviewed the then state of the field (Blum & Niss, 1991).

Model 1: One approach treats mathematical modelling as subservient to other curriculum purposes.

"The curricular context of schooling in our country does not readily admit the opportunity to make mathematical modeling an explicit topic in the K-12 mathematics curriculum. The primary goal of including mathematical modeling activities in students' mathematics experiences within our schools typically is to provide an alternative – and supposedly engaging – setting in which students learn mathematics without the primary goal of becoming proficient modellers. We refer to the mathematics to be learned in these classrooms as "curricular mathematics" to emphasize that this mathematics is the mathematics valued in these schools and does not include mathematical modeling as an explicit area of study. Acknowledging this curricular context, we recognize that extensive student engagement in classroom modeling activities is essential in mathematics instruction only if modeling provides our students with significant opportunities to develop deeper and stronger understanding of curricular mathematics." (Zbiek & Conner, in press).

Emergent modelling represents a focus sympathetic to this emphasis (for example, Gravmejer, 2003).

Model 2: The second position views mathematical modelling as a means of reaching out to provide students with abilities that are relevant to their mathematical learning, but that also enable them, as a major goal, to learn and apply problem-solving skills to situations that arise in life outside the mathematics classroom (commonly referred to as real-world problems).

"Starting with a certain problematic situation in the real world, simplification and structuring leads to the formulation of a problem and thence to a mathematical model of the problem...It has become common practice to use the term mathematical modelling for the entire process consisting of structuring, mathematising, working mathematically and interpreting, validating, revisiting and reporting the model." (Blum et al., 2003)

The overall purpose here is to enable students to access their 'pure' mathematical knowledge for addressing problems relevant to their world, and to focus on how this can be successfully achieved. This view of the purpose of modelling is driven by convictions that it is unsatisfactory for students to 'bank' mathematical knowledge for 10, 12, or 15 years and yet be unable to 'withdraw the funds' for purposes other than answering standard questions, and performing on formal tests designed as gatekeepers to the next level of study.

Achieving success for this vision of modelling is the "possible dream" in the title of this paper. (Julie (2002) refers to approaches 1 and 2 respectively as modelling as vehicle versus modelling as content).

In practice circumstances can provide for outcomes somewhere between these poles – for example where curricular pressures dominate, but where some are able to create space for the inclusion of one or two modelling problems in the spirit of the second approach.

Model 3: While the above priorities differ in the way they view modelling in terms of its contribution to student learning, they generally agree that modelling involves some total process that encompasses formulation, solution, interpretation, and evaluation as essential components. Within these approaches links between the real and mathematical worlds are maintained, even though they may at times be somewhat strained. It has become clear however that the term mathematical modelling is increasingly being used in a much more restricted sense, to describe procedures for fitting curves to sets of data points, and the increasing use of technology means that this issue will become increasingly pervasive. The following example written for a current Australian project illustrates implications for the integrity of models.

Living Daylights

The amount of daylight affects many aspects of life, economic, agricultural, social, and personal. For example, it helps to determine growing seasons, impacts strongly on the tourist industry, and is a factor influencing personal lifestyles and choices.

Dates	Day no	Melb Mins	Bris Mins	Dates	Day no	Melb Mins	Bris Mins
01-Jan	0	884	831	15-Jul	196	588	634
29-Jan	28	846	806	12-Aug	224	634	665
26-Feb	56	784	765	09-Sep	252	697	706
25-Mar	84	716	719	07-Oct	280	764	751
22-Apr	112	650	675	04-Nov	308	828	794
20-May	140	598	640	02-Dec	336	876	826
17-Jun	168	573	624	30-Dec	364	884	831

Table 1. Data on minutes of daylight in Melbourne and Brisbane (2004).

Table 1 contains data at 4-weekly intervals for the number of minutes of daylight (sunrise to sunset) in Melbourne and Brisbane throughout 2004. January 1 is defined as day 0 and subsequent days are labelled consecutively, noting that 2004 was a leap year.

The associated modelling project involves generating models for both cities, and making various comparisons between them to do with hours of daylight, seasonal variations, effects of daylight saving, that involve model creation, solution of equations, calculation of integrals, etc. Noting that the data and plot suggest a translated and dilated cosine function of the form $y = a + b\cos(\frac{2\pi}{T}(x+c))$, the period may be reasonably taken as 365 (days), and the minimum and maximum values estimated by noting that these occur respectively on June 21, and Dec 21.

A resulting model for Melbourne has the equation $y = 730 + 158\cos(\frac{2\pi}{365}(x+11))$, and the data and model output, as can be verified by graphing on the same diagram, show a close fit between the two.

Now the TI-83 calculator has a trigonometric curve fitting facility among its regression options that generates an immediate function of best fit that is technically a closer match than the above. Expressed in the same form as the above it has the following equation: $y = 731 + 155\cos(\frac{2\pi}{377}(x+17))$.

Interpretation now infers a 'year' of 377 days, with the longest day around December 15, outcomes that fail the test of real-world validity. The model generated by this means is a purely technical artefact whose parameters vary with the particular data set, and which can be generated in complete ignorance of the principles underlying the real situation – indeed undertaken in complete ignorance of where a table of data comes from. At another level it raises a profound theoretical issue – the relative authority of empirical data versus theoretical structure. Using mathematical modelling as a synonym for curve fitting creates a dangerous aberration of the modelling concept. In particular, the subversion of the requirement of reality by choices available on the menus of calculators or computers represents a substantial distortion of the purpose of modelling, and leads to bad modelling habits as well as inappropriate outcomes.

3. THE REAL-WORLD SWAMP

Having surveyed some current educational emphases associated with the implementation of mathematical modelling, the intention is now to delve more deeply into issues concerned with its implementation within the philosophy described in 2 above. To do this we firstly consider important non-mathematical implications associated with modelling in the world beyond the classroom.

3.1 Implied Values

"Mathematical modelling is always embedded in social practice and thus it is, in the end, not possible to promote goals and a collection of examples of mathematical modelling and applications without at the same time, (implicitly) promoting a social practice." (Jablonka, 2005).

At one level this realisation might distinguish between the ways modelling examples are treated at tertiary and secondary level. Early motivations of ICTMA included the desire to make good shortcomings noted by employers about the abilities of graduates to apply their mathematics to the solution of problems. For undergraduates in the final year of their degree, (who are looking for employment in industrial settings), it is reasonable to assume that this should be a prime objective for achievement, and providing experiences involving attempts to solve identified real problems a natural and valued response. At secondary, and indeed at primary, level there may be additional considerations – for example in the Habermasian tradition of asking "whose interests" are being served by a problem at hand. (It is not intended to imply that that this would not also apply at times in tertiary courses).

As an example consider the following problem discussed in Galbraith and Clatworthy (1990).

"The State Government is considering building a controversial major road through the northern and inner suburbs of Brisbane. Resident action groups have been very vocal in their opposition, claiming it will have a major impact on their communities and lifestyles. One of the options being considered is the building of a tunnel under one of the hills in Bardon rather than above ground, thereby causing less disruption to residents."

The problem then went on to pursue the tunnel proposal and asked for recommendations about the traffic speed that would maximise peak hour flow in the tunnel.

A quite different problem could have been set. This might have asked for estimates of the number of families displaced by the resumption of land for an above ground option, the costs involved in compensation and relocation, and whether monetary provision could really make up for impacts on personal lives. Obviously the purpose of the problem, and the type of mathematics involved, would be completely different.

A point that should emerge is that every real-world problem has values embedded in its construction, and the identification (at least) of such values is an important part of the real-world context.

3.2 Problem Characteristics

Palm (2005) has drawn attention to fidelity with respect to the actual context as a challenging dimension when attempts are made to introduce real life problems into classroom learning situations. To visit this issue substantively two real-world examples are considered below.

Example 1: Train stopping Distance (Barney, Haley & Nikandros, 2001).

The purpose of the modelling described by the authors was the design of an improved, software enhanced, braking system for a state rail network.

"Most railways do not take into account the brake delay time, and further simplify the calculation by using the *average gradient* of the track on the approach side of the Limit of Authority (Stop Sign)."

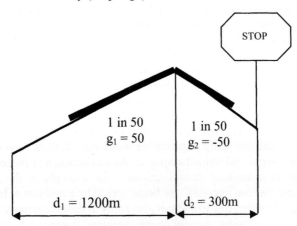

Figure 2. Average gradient concept.

Average gradient "G" = $D/(d_1/g_1 + d_2/g_2 + ... + d_n/g_n)$ where $D = d_1 + d_2 + ... + d_n$
i.e. $G = 1500/(1200/50 + 300/-50) = +83.33$ (1.2%)

The notion of 'average gradient' used in the analysis is a specific concept that impacts on the capacity to engage with problems involving braking distance and safety concerns within the railway industry.

Example 2: Teacher Supply and Demand (Galbraith, 1999)

This is an example from a dynamic modelling project that pursued issues associated with the varying supply of teachers that has been an endemic part of the Australian historical context. The most optimistic of policy makers want mechanisms that will match predicted demand with an appropriate supply – an expectation that can be proved beyond the bounds of possibility on a systematic basis. Model structure is comprised of a series of interacting feedback loops, embodying a system described in terms of a set of non-linear integral equations.

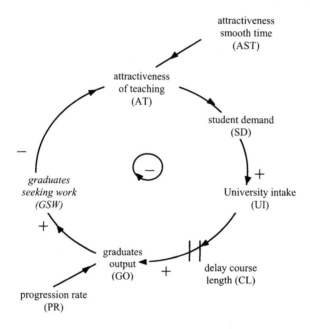

Figure 3. Supply loop (balancing).

In this section of the model 'attractiveness of teaching' is a variable introduced to reflect conditions such as, perceived future employment opportunities by graduates entering university. These perceptions change over time as reflected by the periodical waxing and waning of demand for university places in teacher education programs, and a smoothing (or averaging time) is needed to represent the rate at which this occurs. These must be represented by appropriate choices of function type and parameter values grounded in knowledge of the system and its fixed constraints. In dealing with dynamic models it is not acceptable to omit a process of significance from consideration on the grounds that 'hard data' are absent. Put another way, when dealing with systems, processes must be included because of their significance in the real world, not on the basis of the availability of data, although such should be used when available. A process deemed important must be included, for to 'omit' such a process on the grounds of insufficient data is not to omit it at all - but to include it with an assigned weight of zero! This is a far more serious structural error than getting the shape of an effect correct but its detail approximate. A central evaluation criterion in such situations is the robustness of the model output (behaviour modes) across the range of uncertainty of input data.

In both these real-world problem contexts we can identify the existence of non-standard features, whose presence must be engaged in the modelling process, and which do not represent standard pieces of mathematics likely to be encountered in conventional coursework.

4. WHEN EAST MEETS WEST – MODELLING PROBLEMS

In the previous section we have looked outside the classroom to illustrate (however briefly), some characteristics of problems that occur in the real world of applications. (To these we need to add problems that arise in the context of living that provide, for example, opportunity for informed decision-making, enhancement of quality of life, and a better understanding of the world).

Let us consider (metaphorically) *East* to represent the external world with its array of real problems, and *W*est to represent the world of the mathematical classroom. *E*ast meeting *West* then refers to the interface between these worlds.

Bonotto (2005) argues that:

"…the use of conditions that often make out-of-school learning more effective can and must be re-created, at least partially, in classroom activities. Indeed while there may be some inherent differences between the two contexts these can be reduced by creating classroom situations that promote learning processes closer to those arising from out-of-school mathematics practices."

As pointed out by Palm (2005) many (or most) real-world examples used in education cannot be exact matches with external counterparts, so we now proceed to ask in generic terms what classes of problem typify genuine real-world activity – and how these can be reasonably represented in school-based applications? Here are five problem-types (others of course exist), which I believe are found in the world beyond the classroom, that require the assistance of well-developed modelling abilities, and for which feasible representatives can be designed for classroom implementation at some level.

1. Design and optimisation problems: for example, Product design; locating positions for supermarkets, franchises, road upgrades.

School Problem: Where should speed bumps be located when designing traffic flow through a facility such as a new school or college?

2. Prediction Problems: for example, What are the future needs of a community for water storage, irrigation, reservoirs etc?

School Problem: How long will water in a farm dam last for a herd of livestock, if there is no more rain?

3. Problems that involve resolving apparent anomalies (or intuition can't always be trusted): for example, How shortages of goods in a supply chain can be traced to earlier decisions designed to avoid shortages.

School Problem: Julie and Robyn are star bowlers in their one-day cricket competition. Prizes are awarded for the best bowling averages in the semi-final and the final, and in addition there is a prize for the best average when the results for the two matches are combined. The Table below shows the results for the two matches, in which Julie wins the awards for both the semi-final and final. Who wins the combined award? When will outcomes like this occur?

Match	Semi Final			Final			Combined
Bowler	Runs	Wkts	Aver	Runs	Wkts	Aver	Average
Julie	12	5	2.4	29	5	5.8	?
Robyn	10	4	2.5	18	3	6.0	?

4. *Problems that explain and/or improve existing practices:* for example, Analysis of competitors' products, or redesign of existing products and methods.

School Problem: Which common errors will be detected by the EAN 13 (or some other) barcode system? Would other systems of calculating check digits be more efficient?

5. *Problems that identify (and perhaps help correct) injustices, inequalities, emerging social or personal needs:* for example, Analysis of tax structures and wage movements for impact on different sectors of society; the credit card trap, housing loan alternatives.

School Problem: Is the claim on behalf of Wannamutta and Weranabe (see below) a reasonable one?

SYDNEY, March 30 2000 (AFP) -- The ghosts of Australian bushranger Ned Kelly and the black trackers who helped police catch him in 1880 may be stirred by a Supreme Court action starting in Brisbane on Wednesday. Aborigines Jack Noble and Gary Owens (tribal names Wannamutta and Weranabe) had been promised 50 pounds each as their share of the then massive 8,000 pounds reward offered for Kelly's capture by the Victorian state government. They and three other Aborigines tracked the notorious Kelly gang to the Victorian town of Glenrowan where a daylong gun battle ensued between police and the armour-clad bushrangers... But because the trackers were Aborigines, the authorities refused to pay them and now their descendants are suing for recovery of the reward, which with compound interest plus damages is calculated at 84 million Australian dollars (52 million US)...

http://www.geocities.com/cpa_blacktown/20000331nedkeafpfr.htm

It is argued that the solutions to the problems above, involve modelling attributes including problem specification, making of assumptions, mathematical solution, interpretation and evaluation that are similar to those required in real-world problems.

Table 2 contains a brief indication of some of the mathematics and modelling aspects embedded in the respective problems. The middle column contains a brief summary of some of the mathematics required to address the problems. (This reflects particular approaches taken within in the solution process, and other approaches that perhaps involve different mathematics are of course possible).

*In 'cricket averages' the concept of a linear equation with integer solutions arises naturally, and while this can be handled adequately with basic algebraic skills, it does provide an avenue for the introduction of Diophantine equations which might

then be developed at an appropriately advanced level. A similar situation occurs within Barcodes with respect to modular arithmetic. In this sense these problems would fit also within the 'modelling as vehicle' philosophy, the first of the approaches discussed earlier in this paper.

The third column illustrates elements that are intricately involved with the modelling process, and hence the contextual settings of the respective problems.

Problem	Mathematics	Modelling/contextual factors
Placing speed bumps	Simple kinematics	Bump design (max speed) Simulation (real car) Restrictions (non-math)
Farm dams	Pythagoras Simple Trigonometry Integral Calculus	Evaporation rates Animal consumption Available data (distance to water line – not depth)
Cricket averages	Ratios, linear equations Integer solutions, (Diophantine equations)*	Bowling averages as rates Wickets available (10 max) Runs (bounded but unknown)
Barcodes	Divisibility, Numerical proofs (Modular arithmetic)*	Identify likely errors: single digit errors, digit reversal errors.
Ned Kelly injustice	Compound interest	Colonial history (interest rates) Compensatory and punitive damages

Table 2. Mathematical and modelling content of selected problems.

5. DESIGNING MODELLING PROBLEMS

In the search for classroom problems that represent authentic modelling contexts it is helpful to analyse and distil qualities that can be identified in problems that have worked (content analysis), and to use such information in the design of new problems, and indeed in the search for design principles. In this section some principles will be articulated and illustrated by means of the Barcode problem. Such an enterprise can be viewed as a contribution towards the development of a theoretically consistent approach to problem construction, when the purpose is to design classroom problems that provide legitimate experience for students within a program driven by a philosophy that values real-world authenticity. It is desirable that principles embody essentials, and be limited in number, and from this perspective five have been selected. In addition to these mathematical principles, one didactical principle directed particularly towards the design issue has been added (Table 3).

Principle	Enactment in **design** of EAN - 13 codes problem
Principle 1: There is some genuine link(s) with the real world of the students.	The context is a part of the everyday experience of all students.
Principle 2: There is opportunity to identify and specify mathematically tractable questions from a general problem statement.	Suitable sub-questions are implied by the general problem for example,: • What proportion of common errors will be detected by the check digit? • Is there a simpler set of weights that is as effective or better for this purpose?
Principle 3: Formulation of a solution process is feasible, involving the use of mathematics accessible to students, the making of necessary assumptions, and the assembly of necessary data.	The sub-questions require basic strategies of proof, and procedures that need only an understanding of integer arithmetic, including simple notions of divisibility. (Putting these together, as usual, increases the demand compared with the demand that would apply for each separately.)
Principle 4: Solution of the mathematics for the basic problem is possible, together with interpretation.	Having been identified and formulated, the solution of sub-questions can be addressed by students, using existing knowledge resources. (Here the successful completion of arithmetic procedures, and associated logic, and interpretation.)
Principle 5: An evaluation procedure is available that enables checking for mathematical accuracy, and for the appropriateness of the solution with respect to the contextual setting.	Checking of mathematical answers is a feasible part of the procedure. Ideally the outcomes should also be tested in their real-world setting. (Given the number of students that have part-time jobs in commercial stores this is certainly a realistic possibility.)
Didactical design principle: The problem may be structured into sequential questions that retain the integrity of the real situation. (These may be given as scaffolding hints at the discretion of a teacher, or be used to provide organised assistance by suggesting a line of investigation - often helpful at the challenging specification stage in assisting applications of Principle 2.)	Sample structuring questions (hints) for EAN codes problem: • Is the check digit unique? • Which single digit errors will be detected by the coding method? • Which transposition errors will be detected by the coding method? • Will weights of 1 and 2 do as good a job as weights 1 and 3?

Table 3: Principles for problem design.

6. WHEN EAST MEETS WEST – THE EDUCATIONAL CONTEXT

Having illustrated problems presented as authentic educational counterparts of real-world problem types, we now shift the focus to the classroom in which a modelling pedagogy is to be enacted. Again the focus is on what mathematical modelling intentions imply for the organisation of teaching and learning. By way of illustration two aspects have been chosen for elaboration:

6.1. Collaborative Activity

We recall earlier comments in which (O'Carrol et al., 1987; Chalarambos et al., 2003) independently identified (across an interval of 15 years) similar needs for students not only to develop capacities to formulate and solve real problems, but also to report and communicate effectively outcomes to third parties – involving both oral and written reporting abilities. In addition to the specifics of mathematical content, these abilities are consequences of how modelling material is taught and learned, and of the experiences provided for students – in short it is impacted strongly by pedagogy/didactics. Earlier we have argued that some content parameters need to be taken on board if the worlds of the mathematics classroom, and workplace/life contexts, are to meet effectively for mutual empowerment. We now switch our focus to the context (and cultures) on either side of this divide.

Firstly we note that in the external world the overriding motivation is to obtain a solution to a problem. To achieve this end, individuals need to be able to adapt their mathematical knowledge flexibly to the context at hand, but further need to be comfortable with team activity. The latter requires the possession of well-developed interpersonal skills, the capacity to work independently, with a partner, or in small groups, and to be adept with a range of oral and written reporting techniques. Now let us consider some of the traditions on the classroom side of the divide that bear upon this need. Regarding team work, from a sociocultural perspective, the classroom is ideally manifested as a community of practice supporting a culture of sense making, through which students learn by immersion in the practices of the discipline, so that rather than relying on the teacher or textbook as an unquestioned external authority, students in such classrooms learn to defend and critique ideas by proposing justifications, explanations and alternatives. In such a communication rich environment collaborative practices are called for, and Brandon (1999) has usefully pointed out that the 'C' in Collaborative Learning has been used ambiguously to refer to both co-operative based learning (group members share the workload); and collaboration-based learning (group members develop shared meanings about their work). While interrelated there is a clear difference in the respective emphases. Collaborative activity, in this latter sense, is characterised by equal partners working jointly towards an end (Anderson, Mayer & Kibby, 1995), as a co-ordinated activity directed towards construction and maintenance of shared meaning and understanding (Rochelle & Teasley, 1995). These characteristics of collaborative learning that emphasise the social construction of knowledge and shared conceptions of problem-based tasks, are important elements for successful learning in the field of applications and modelling. There is therefore a distancing from models of 'Co-operative learning' in which members of a group of peers are assigned individual roles (for example, recorder, checker) prior to structured group activity. In this latter model role assignment may interfere with group processes by overemphasising organisational tasks at the expense of learning processes, for role assignment stands to restrict the opportunity of individuals to engage with problems freely, and to use and share their knowledge in the widest and most relevant way to seek and test solutions. This is in fundamental conflict with the desire, transparent in mathematical modelling programs, to motivate the production of solutions through effective teamwork. Therefore we can argue that the desire for a genuine modelling context provides direct input to influence the type of collaborative learning model enacted in the classroom – that is it has provided input to influence a didactical decision.

6.2. Across the Boundary

We have seen in the railway and Teacher Supply and Demand examples, how industry-specific terms and concepts arise in problem contexts. We will also recall criticisms of the perceived ability of school leavers and graduates to display and use mathematical knowledge supposedly gained at school or university. There should be little controversy when the need is for abilities to calculate, measure, or estimate in situations where these call for the mechanical proficiency of clearly defined actions – it is most reasonable to expect these to be products of an education in mathematics. However the situation is not so simple when the issue is a perceived inability to apply a supposedly simple mathematical procedure in a situation that is industrially or otherwise context specific. For example no student would meet industry specific concepts like 'average gradient' in their conventional studies and without appropriate explanation would be perplexed by conversations which treated them as an understood part of the environment. So struggles to apply well-known mathematical knowledge should not be surprising when it is embedded in contexts that are second nature to those expecting immediate performance, but unfamiliar to those of whom such performance is expected. Careful initiation into the specifics of new environments is called for, and this is not always forthcoming when it is seen to involve the diverting of senior company time and expertise to activities not directly associated with the bottom line. A deeper tragedy is the alienation of representatives of two groups who are really on the same side, for both employers and educators are interested in the successful application of applied mathematical skills to new situations. In some extreme cases we have seen criticism of each group by the other when in practice the need is for each to learn from the other via the challenges required of students/workers on the respective sides of the boundary. While mutual sharing across the education/workplace interface has happened at times through individual initiatives, we can only imagine the benefits if it became a widespread cultural reality. An educational view of mathematical modelling that gives significant weight to applications found in the world beyond the classroom should enable this mutual purpose to be pursued with greater sympathy than perhaps is afforded by approaches maintaining a strictly curricular view. It is part of the dream.

6.3. Instructional Frameworks

Finally, in considering mathematical modelling across the *education – life context* interface we identify the need for frameworks to guide and test the development of theoretically consistent approaches to teaching and learning. Part of the primary motivation behind the development of such a theoretical system is to conceptualise the structure and impact of person-environment relationships, and to this end we have found value in theory, designed primarily as an explanatory structure within the field of human development (Valsiner, 1997). This theory adds the Zone of Free Movement (ZFM) and the Zone of Promoted Action (ZPA) to the familiar Vygotskian Zone of Proximal Development (ZPD) as frameworks for theorising and structuring teaching and learning.

• The ZPA is oriented to defining and promoting the acquisition of new skills, and hence can encompass either narrow or broad visions of what it is intended to achieve.

• The ZFM structures an individual's access to different areas in the environment, the availability of different objects within accessible areas, and the ways the individual is permitted or enabled to act on accessible objects in accessible areas.

Hence the ZFM is potentially inhibiting or enabling for student activity, as it sets boundaries for what students are able to do in pursuit of ZPA goals. Being teacher designed it provides access to sets of activities, objects, or areas in the environment (and may inhibit others) in respect of which student learning is being promoted. Both the ZFM and ZPA are thus culturally determined, and can be loosely associated respectively with means and ends. (A link between the ZFM and ZPA is provided by the ZPD, since for learning to be possible the ZPA must be consistent with an individual's capacity to learn (ZPD), while for the intended approach to learning to have a chance of success the ZPA must be encompassed by a sufficiently rich ZFM.)

We now apply this conceptual structure to the construction and analysis of teaching to achieve modelling goals, by viewing interacting zones as lenses through which to identify and interpret mathematical, methodological, and didactical attributes of teaching designs and presentations. This approach provides a means to conceptualise and communicate consistencies, inconsistencies, similarities, differences, opportunities, and extensions with respect to teaching and learning activities.

We assume that overall learning goals are consistent with the ZPDs of a student group, and that teachers/lecturers construct activities (consciously or unconsciously) within designated ZPAs to achieve these ends. However these planned activities may or may not be consistent with the actual learning opportunities and resources, both physical and intellectual, accessible to the students - the ZFMs. For example the students' may lack access to problems of an appropriate type, the range of necessary resources may be restricted, or the ways in which students are permitted or enabled to access and exploit the potential of available resources may be curtailed. Simultaneous examination of the ZPA and ZFM provides a basis for instructional design analysis or at a specific level can provide a component of lesson evaluation. Within our conception of goals for mathematical modelling programs in education it is possible to identify some critical areas of the *classroom – external world* interface, involving potential tensions between the respective cultures (Table 4). In this Table the left hand column may be thought of as capturing elements of the ZPA – aspects derived from the 'other' side of the *classroom – external world* interface that provide guidelines for achievement in programs within which mathematical modelling is intended to be *content* rather than *vehicle*. The right hand column contains beliefs about classroom practice, that when activated, substantially prescribe the ZFM available to students. It is not suggested that this column describes a typical classroom, for its entries are representative of the more conservative elements associated with mathematics teaching, and many teachers would want to distance themselves from a majority of its contents. Nevertheless the thinking represented there may be found in part in many contemporary classrooms, and the purpose here is to illustrate how a ZPA motivated by characteristics generated from a source outside the classroom, can be compromised by a ZFM generated from a conflicting set of internal assumptions. The identification of

	Modelling Culture (defines desired ZPA)	***Education Culture (controls avail ZFM)**
1	Mathematics involves both 'thinking' and 'hands – on' abilities.	Mathematics is done in the head.
2	Mathematics involves written and oral communication.	Mathematics is about calculations and bookwork.
3	Life-related mathematical activity involves both predictable and unpredictable elements.	Classroom mathematical activity occurs in a passive and controlled environment.
4	Some data are external to the classroom.	All needed data are internal to the classroom.
5	Mathematics involves both individual and team activity.	Mathematics is ultimately an individual activity.
6	Mathematics takes place where and whenever the need occurs.	Mathematics occurs only in formal scheduled sessions or through structured homework.
7	Success is measured by the solution of problems.	Success is measured by individual performance on tests.
8	Assessment involves a range of outcomes and criteria for success.	Assessment requires standardised conditions and instruments.
9	The real world is an essential component.	The real world is an optional extra.
10	Technology use is chosen to maximise problem-solving success.	Technology use is subject to local policy and availability.
11	Choice of resources is decided by what needs to be addressed.	Choice of resources is decided by curriculum detail and availability.
12	Teamwork is managed by the need to maximise problem-solving capability.	Group work is managed by organisational rather than problem-oriented decision making.

Table 4. Implications for didactics from Zone theory.

inconsistencies can then act as a source for suggesting where a program needs specific attention, or might be improved further. Put another way we can ask what a ZFM needs to look like if it is generated by the logical consequences of essential elements of a ZPA as implied by the contents of the left hand column Table 4).

Obviously this part of the discussion relates for the most part to elementary and secondary education, rather than to college and university level where there is generally more freedom to innovate and more autonomy for students. In this illustration we have provided 12 points of reference, and the corresponding pairs have been chosen to display dialectic tensions between the needs of a modelling program that prioritises real-world activity, and a classroom providing a conservative approach to mathematics teaching. A modelling friendly pedagogy can be structured by replacing each element in the right column with one able to support the didactical implications of the corresponding entry in the left column – that is to create a sufficient ZFM from a necessary ZPA. Conversely the existence of particular educational requirements, customs, or classroom procedures can be identified as potential blockages to the implementation of a modelling pedagogy. For example:

• Rules that require students to work on problems only in supervised class sessions (sometimes associated with assessment procedures) are at odds with a task-oriented culture that makes time and place subservient to the goal of problem solution.

• Rules that insist on students working individually are at odds with a workplace culture that values and encourages teamwork and group problem solving.

• A culture that regards mathematics as exclusively a mental activity is at odds with a culture that recognises and honours the interplay between mathematics and context.

• An assessment culture that is dominated by formal testing is at odds with a culture that places high value on the solution, evaluation, and defence of non-standard problems.

7. CONCLUSION

"The more things change, the more they remain the same" goes the well-known saying. We have seen that issues concerning unsatisfactory aspects of word problems, and pseudo-applications, remain active more than 30 years after their exposure. We know that learning to apply mathematics to non-standard problems is not easy, and that the situation is rendered more complex by the existence side-by-side of different enactments each carrying the descriptor mathematical modelling. While sharing some common elements there are also fundamental differences between the respective versions, and these require a careful reading of literature in the field. The version that provides the subject for this paper is significant for its drawing on structures, methods, and values across the interface between formal educational systems (schools, colleges, and universities) and the worlds of the workplace and everyday contexts of living – as complementary agencies. Both the culture of the classroom and the culture of the workplace stand to benefit from a development of such a partnership, but it is important to clarify that this vision for mathematical modelling does not see itself as preparing students narrowly for an agenda dictated by workforce demands or employer preferences. Rather the position adopted is that approaches to real-world problem solving, communication, and teamwork that occur in the world beyond the classroom, offer attributes that can enrich the teaching and learning of mathematics. Fundamental to this argument is that these abilities stand to develop the capacities of individuals to apply mathematics they have learnt, in situations involving social or personal decision-making or in other academic pursuits, as well as for vocationally specific purposes.

This paper seeks to advance arguments and provide structures for achieving crispness and theoretical clarity as a basis for action, where traditionally there has been a blurring of concepts, intentions and actions. The meeting of the outside world with the mathematics classroom (metaphorically East and West), challenges both sides to enhance mutual understanding by finding common ground at the interface. We have identified a number of meeting points and structures where this need is highlighted. These include; identifying properties of real-world problems that are different from those traditionally met in mathematics courses; identification of families of problem types that occur outside the classroom, and the construction of school level problems that legitimately represent their type; identifying teaching and learning conditions necessary for the solution of real problems, and highlighting tensions and inconsistencies with existing didactical practices. Implications for educational practice arise in the consideration of each of these challenges viewed as interfaces between the real world and mathematics classrooms.

In the sense that so much remains to be done to meet and extend challenges such as these across the boundary that divides the world outside the classroom from that within, the dream remains to be realised. But it is an important dream, a possible dream, and as Oscar Hammerstein II reminded us in South Pacific:

"You've got to have a dream, if you don't have a dream, how are you going to have a dream come true".

8. REFERENCES

Anderson, A., Mayes, J.T., and Kibby, M.R. (1995) Small group collaborative discovery learning from hypertext. In C. O'Malley (ed) *Computer supported collaborative learning*. New York: Springer-Verlag, 23-38.

Barney, D., Haley, D. and Nikandros, G. (2001) Calculating Train Braking Distance. In P. Lindsay (ed) Conferences in Research and Practice in Information Technology, Proceedings of the Sixth Australian Workshop on Industrial Experience with Safety Critical Systems and Software, Brisbane, Australia. ACS, 23-30.

Blum, W. and Niss, M. (1991) Applied mathematics, problem solving, modelling, applications, and links to other subjects – state, trends, and issues in mathematics instruction. *Educational Studies in Mathematics, 22,* 37-68.

Blum, W. et al. (2002) ICMI Study 14: Applications and modelling in mathematics education – discussion document. *Educational Studies in Mathematics,* 51, 149 – 171.

Bonotto, C. (2004) How to replace the word problems with activities of realistic mathematical modelling. In W. Blum and H-W. Henn (eds) *ICMI Study 14: Applications and Modelling in Mathematics Education: pre-Conference Volume.*

Dortmund, 41-46.

Brandon, D.P. and Hollingshead, A.B. (1999) Collaborative learning and computer-supported groups. *Communication Education, 48(2), 109-126.*

Chalarambos, D., Aliprantis, A. and Carmona, G. (2003) Introduction to an economic problem: a models and modelling perspective. In R. Lesh and H. Doerr (eds) *Beyond Constructivism: Models and Modeling Perspectives on Mathematics Problem Solving, Learning, and Teaching.* Mahwah NJ: Lawrence Erlbaum, 255-264.

Galbraith, P. (1999) *Forecasting Teacher Supply and Demand: Searching for Shangri – la ~ or chasing rainbows?* Flaxton: Post Pressed.

Galbraith, P.L. and Clatworthy, N.J. (1990) Beyond Standard Models: Meeting the Challenge of Modelling. *Educational Studies in Mathematics,* 21 (2), 137-163.

Jablonka, E. (2004) The relevance of modelling and applications: relevance to whom and for what purpose? In W. Blum and H-W. Henn (eds) *ICMI Study 14: Applications and Modelling in Mathematics Education: pre-Conference Volume.* Dortmund, 205-210.

Julie, C (2002) .Making relevance relevant in mathematics teacher education. *Proceedings of the 2nd International Conference on the Teaching of Mathematics (at the undergraduate level).* Hoboken NJ: Wiley, CD.

O'Carrol, M., Hudson P. and Yeats, A. (1987) A comprehensive first course in modelling. In J.S. Berry et al. (eds.) *Mathematical Modelling Courses.* Chichester: Ellis Horwood, 113-121.

Palm, T. (2004) Features and impact of the authenticity of applied mathematical school tasks. In W. Blum and H-W. Henn (eds) *ICMI Study 14: Applications and Modelling in Mathematics Education: pre-Conference Volume.* Dortmund, 205-210.

Pollak, H. (1969) How can we teach applications of mathematics? *Educational Studies in Mathematics,* 2, 393-404.

Roschelle, J. and Teasley, S. (1995) The construction of shared knowledge in collaborative problem solving. In C. O'Malley (ed.), *Computer supported collaborative learning.* New York: Springer-Verlag, 20-45.

Valsiner, J. (1997) *Culture and the development of children's action: a theory of human development (2nd edition).* New York: John Wiley & Sons.

Webb, N.M. and Palincsar, A.S. (1996) Group processes in the classroom. In D.C. Berliner and R. Caffee (eds.), *Handbook of Educational Psychology.* New York: Macmillan, 841-873.

Zbiek, R., and Conner, A. (2006) Beyond motivation: Exploring mathematical modeling as a context for deepening students' understanding of curricular mathematics. *Educational Studies in Mathematics,* 63(1), 89-112.

2.2

MODELLING IN CLASS: WHAT DO WE WANT THE STUDENTS TO LEARN?

Katja Maaß
University of Education, Freiburg, Germany

Abstract–*Modelling and application are regarded as an important topic for maths lessons. But so far the concept "modelling competencies" has not been described in a comprehensive manner. The aim of this paper is to supplement former descriptions of modelling competencies based on empirical data. An empirical study was carried out which aimed at showing the effects of the integration of modelling tasks into day-to-day maths classes. Central questions of this study were – among others: How far do math lessons with focus on modelling enable students to carry out modelling processes on their own? What are modelling competencies? The paper describes the theoretical and the methodological approach, the classroom settings, and the results of the study.*

1. INTRODUCTION

Modelling and applications are regarded as important when discussing didactics. The aim is to integrate modelling and applications into daily school routine. What is it that we want the students to learn? What do we mean by modelling competencies?

Recent didactical views regard the term "competence" as important. For example, the Danish KOM project is based on the question: "What is mathematical competence?" (Niss, 2003 p7). In the recent discussion about modelling we distinguish between ability and competence as well. But up to now the concept "modelling competencies" could not be described in a comprehensive manner. This can be shown for example, by some questions posed in the *Discussion Document* for the ICMI Study in Dortmund (Blum et al., 2002 p271): "Are modelling ability and modelling competency different concepts? Can specific sub-skills and sub-competencies of "modelling competence" be identified?"

This paper deals with the results of a study which tries to elucidate these questions – amongst others. The central questions of the study were:

1. How do students' mathematical beliefs change during courses of maths classes which include modelling exercises?
2. How far do such lessons enable students to carry out modelling processes on their own?
3. What are modelling competencies?
4. Which kinds of connections exist between mathematical beliefs and modelling competencies?

In this paper the focus will be on the part of the study which refers to questions 2 and 3.

2. THEORETICAL FRAME

The following concepts form an essential basis of this study. While *modelling* a real-world problem, we move between reality and mathematics. The modelling process begins with the real-world problem. By simplifying, structuring and idealizing this problem, you get a real model. The mathematizing of the real model leads to a mathematical model. By working within mathematics, a mathematical solution can be found. This solution has to be interpreted first and then validated (Blum, 1996 p18).

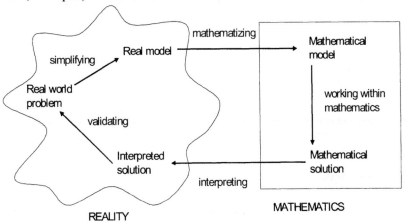

Figure 1. Modelling process (own representation, according to Blum 1996 p18).

Competence can be regarded as the ability of a person to check and to judge the factual correctness and the adequacy of statements and tasks personally and to transfer them into action (Frey, 1999 p109). Similar views can be found in the didactical discussion about modelling: "Research has shown that knowledge alone is not sufficient for successful modelling: the student must also choose to use that knowledge, and to monitor the process being made." (Tanner & Jones, 1995 p63). Based on these concepts, I define the term "modelling competency" as follows: *Competencies for modelling* include abilities of modelling problems as well as the will to use these abilities.

A further important basis of this study is different sub-competencies mentioned by Blum and Kaiser (1997) (c.f. Maaß, 2004 p32): Modelling competencies contain

- Competencies to understand the real problem and to set up a model based on reality.
- Competencies to set up a mathematical model from the real model.
- Competencies to solve mathematical questions within this mathematical model.
- Competencies to interpret mathematical results in a real situation.
- Competencies to validate the solution.

Another concept which seemed to be very important for this study is the concept of metacognition: Empirical studies refer to the significance of metacognition when solving problems and complex tasks (Sjuts, 2003 p26; Schoenfeld, 1992 p355). Metacognition is seen as a basic competency which is relevant for a variety of important competencies (Sjuts, 2003 p20). Furthermore the development to metacognition or self-regulated learning is regarded as one main task of educational institutions (Baumert et al., 2001 p271). Following Sjuts (2003 p18), this study employs the following concept of metacognition:

Metacognition is the thinking about one's own thinking and management of one's own thinking. There are three parts of metacognition:

- *Declarative metacognition* contains the diagnostic knowledge about one's own thinking, the judging thinking about tasks and the strategic knowledge about ways to solve a problem.
- *Procedural metacognition* contains planning, surveying and judging, which means the monitoring of one's own actions.
- *Motivational metacognition*: Necessary conditions for the use of metacognition are motivation and the willpower to do so.

3. CLASSROOM SETTINGS

During the data collection period of 15 months (April 2001 – July 2002), six modelling units were integrated into two parallel classes, age 13 – 14. For example, in two of these units, the students had to answer the following questions:

- How large is the surface of a "Porsche"? This question has to be answered when a new car type is designed. For economical reasons Porsche calculates the production costs of a car even before the prototype is built.
- Is it possible to heat the water required in Stuttgart-Waldhausen with solar collectors on the roofs?

3.1. Teaching-Unit "Sun-Energy"

As an example, the unit dealing with solar-collectors will be described in detail: In times when energy sources become shorter it is imperative to think about alternatives. What energy sources can we use in the future? To what degree can we use regenerative energies?

Being aware of the alarming facts the region of Stuttgart has engaged the University of Stuttgart for a research project. The aim of the research project was to find out about the possibilities and limits of a more extensive use of regenerative energy within the region of Stuttgart. The University of Stuttgart was asked to consider technical, economical and ecological aspects. They examined the use of water-power, wind-power and sun-energy to generate electricity. As technologies for heat generation sun-energy and the use of organic substances were examined (Bläsing et al., 2000 p1).

How can this investigation be used in maths lessons? This investigation is already finished but of course it can be simulated in maths lessons. Because of the extent of this investigation and the relatively low mathematical knowledge of students aged 13-14, only one aspect of this comprehensive study was examined:

Is it possible to heat the water required in Stuttgart-Waldhausen with solar collectors on the roofs?

At first, the students had to consider which information is needed to solve the problem. They mentioned:

1. Information about solar collectors
2. The consumption of warm water in Stuttgart-Waldhausen
3. The area of the roofs
4. The energy needed to heat water, and
5. The variation of the sunlight during the seasons of the year

In reality, this problem is very complex and so of course it has to be simplified. Customary solar collectors can produce $300 - 500$ kWh per m^2 per year. So the students decided to take 400 kWh/m^2 per year. The optimal alignment is southern; the slope of the roof should be 30-45° (Landesgewerbeamt, 1999 p4).

The students calculated the average water consumption per person by using 20 water bills. Then they multiplied the average water consumption with the number of persons in Stuttgart-Waldhausen. They found this information in the statistical yearbook of Stuttgart (Landeshauptstadt Stuttgart 1999).

To find out the size of the roof area the students measured the roof area of 20 houses. This was done by pacing out the distances a and b (Fig 2). Because the length of the lines h and s are difficult to estimate, the students estimated the angle α with the help of a set square and calculated s by using the congruency of triangles.

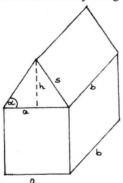

Figure 2. Sketch of a house.

Afterwards the students calculated the average roof size which was multiplied with the number of houses. This information again was taken from the statistical yearbook of Stuttgart (Landeshauptstadt Stuttgart 1999).

From an energy generating enterprise the students obtained the information that 58 kWh are required to heat 1 m^3 of water.

Finally the students decided to neglect the variation of the sunlight during the time of the year, because they had read about a new development: In the near future it will be possible to save the energy through the winter with the help of big water tanks (Landegewerbeamt, 2000 p18; Landesgewerbeamt, 1999 p18).

As a result they found out that 170,000 m^3 of water are needed, and therefore 170,000 $m^3 \cdot 58$ kWh/m^3 per year$\approx 10,000,000$ kWh/m^2 per year are required. For the roof area they obtained 143,000 m² and so 143,000 m² · 400 kWh/m² per year≈ 57,000,000 kWh per year can be generated.

This shows that in theory it is possible to heat the water required in Stuttgart-Waldhausen. However, this may not be sufficient for reality: There were many simplifications in the calculation of the water consumption and the roof area. The variation of sunlight during the seasons of the year was neglected. Furthermore, there are no means to transport hot water between the houses. But they are absolutely necessary, because on a house for one family the roof area is relatively large whereas the roof area on a house for several families might not be sufficient.

However, there are research projects which deal with the two last named aspects; for example, in Friedrichshafen. Here all houses have solar collectors and they are all connected with one big water tank which can store the energy through the winter because of its size (Landesgewerbeamt 2000, p. 18, Landesgewerbeamt 1999, p18). For this reason our calculation is meaningful for the future.

3.2. Assessment

Regarding Niss' statement "What you assess is what you get" (1993, p43) assessment was of great significance within the study. First, the students had to write a report about the modelling done in the lessons. Additionally, two weeks later, there was a written class test in which the students had to compare their own modelling with the modelling of the University of Stuttgart. In order to do so the students were given an extract of the report of the University of Stuttgart (Fig. 3).

Extract of the report of the University of Stuttgart, institute of energy economy and efficient use of energy (IER) (abridged and simplified)

Regenerative energy in the region of Stuttgart
Heat generation by solar collectors

Part A
...In the following we assume that 15 % of the buildings' roof area with sloped roofs and 25 % of the buildings' roof area with flat roofs can be used. Furthermore we assume that 95 % of the buildings have sloped roofs. So, on average 15.5 % of the whole roof area can be used for solar collectors.
In the region of Stuttgart we have 7100 ha (1 ha = 100 ar = 10000 m²) of roof area. In view of the slope of the roofs, the direction, windows and the shading solar collectors could be installed on 1100 ha. Nowadays solar collectors can produce 1800 MJ energy on average. Therefore, in theory, we can produce 20 PJ/a by solar collectors in the region of Stuttgart. ...
Part B ...

Figure 3. Extract of the test (the text given to the students was longer).

This extract shows that the University models the problem in a similar way. However it is noticeable that they do not give reasons for their assumptions.

Finally, after six weeks there was another written class test. Here the real-world context did not deal with sun energy but there was a connection to this topic because the energy question was raised again (Fig. 4).

Natural gas test

In 1993 the worldwide reserves of natural gas were estimated to be 141.8 X 10^{12} cubic metres. Since then 2.5 X 10^{12} cubic metres have been used every year on average. Calculate when the reserves of natural gas will be exhausted. Use different assumptions and models. Explain all your steps.

Figure 4. Natural gas test.

3.3. Teaching Methods

The teaching methods aimed at giving the students as much independence as possible. That is why there was a lot of group work and there were plenary discussions among the students. Furthermore the students organised their way of working on their own.

Additionally teaching methods were chosen to support the development of metacognition. For this reason the students received information about the modelling process on a meta-level, the students' different concepts of the modelling process were discussed in class and their mistakes were used in a constructive way. Different solutions were compared and discussed. The students were also requested to plan, to survey and to judge their own proceeding and positive examples of monitoring were presented to the class. If for example, some student chose a wrong way to find a solution and noticed this by himself and then found a new, successful way he was asked to present his whole proceeding to the class. If necessary the teacher helped by external monitoring.

4. METHODOLOGICAL APPROACH

Aiming at an explanation of complex relations in the context of everyday life and at a contribution to an empirically founded theory, the study is a qualitative study which methodologically starts off from qualitative social sciences (Flick, Kardorff & Steinke, 2002 p23; Strauss & Corbin, 1998 p11). Furthermore, a long-lasting incorporation of modelling tasks into everyday mathematics teaching practice became possible because in this study the researcher and the teacher were in fact the same person engaged in called Action Research (Altrichter & Posch, 1998 p15).

In order to meet the complexity of the research's object, a variety of methods in data collection has been used. Modelling competencies were evaluated with the help of tests, written class tests, concept maps and interviews (Maaß, 2004 p115).

The interviews, tests and written class tests were analysed in a detailed manner in view of the list of sub-competencies given by Blum & Kaiser (1997) (c.f. Maaß, 2004 p32) and further noticeable aspects, for example, the attitude towards modelling tasks. The concept maps were interpreted in order to gain information about metacognitive modelling competencies. A detailed report about every student and his/her specific way of modelling was written. Finally there was a subject-contrasting and subject-comparing analysis (Maaß, 2004 p138).

5. RESULTS

5.1. Modelling Competencies

One of the most fundamental results of the study was that students at lower secondary level are able to develop modelling competencies. Towards the end of the study, almost all students were qualified to model problems with known as well as with unknown problems. Differences appeared in relation to the complexity of the tasks.

To gain more information about the competencies needed to carry out modelling processes we will now look at the mistakes which occurred. This however may not be misunderstood in a way that all solutions were incorrect and poor; the following list is intended only to show in which parts of the solution process mistakes can occur:

There were *mistakes in setting up the real model*.

- The real model was inadequate, because the assumptions simplified reality too much.
- The setting up of the real model was not described.
- The students made wrong assumptions which distorted reality. A mistake of this kind is shown in Figure 5[1]. The student deals with the problem in an adequate way: He asks himself how big the head is and he answers that the head is about 1.30m to 1.50m because a child has this size. Then he asks himself what proportion between a body and a head exists. And he writes "The proportion between head and body is 1 to 31!"

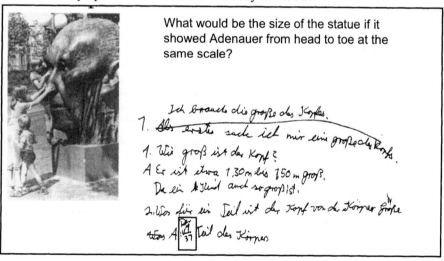

Figure 5. Assumption which distorts reality.

Then there were *mistakes in setting up a mathematical model*.

- Some students did not use adequate mathematical symbols.

- Wrong algorithms or formulas were used. An example of this is shown in Figure 6. The students were asked to describe three ways to solve the problem and finally calculate a solution by using one way. This girl describes three correct ways in detail (which are not shown here) and she decides to use a cuboid as a model. However, she has the idea to calculate the surface of the cuboid by calculating a * b and so in fact she calculates the area of a rectangle. Further more she forgets to convert the scales.

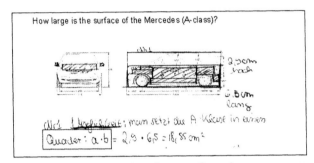

Figure 6. Use of wrong formulae.

In addition, there were **mistakes in solving mathematical questions within the mathematical model**.

- There were mistakes in the calculation.
- The work within the mathematical model was finished without any result.
- The students lacked the needed heuristic strategies.

Furthermore **mistakes in interpreting the solution** could be found.

- The interpretation of the results was missing.
- Mathematical solutions were interpreted in a wrong way. An example for this mistake is shown in Figure 7.

> Explain the statement „There are 1.2 persons in a car on average"
> **Grown-up – child, Grown-up - dog, Grown-up – shopping bag**

Figure 7. Mistake in interpreting the solution.

Of course there were also **mistakes in validating the situation**.

- The validation of the results was missing.
- The inadequacy of a model was realised but not corrected.
- The validation didn't go beneath the surface: In Figure 8 the student just wrote: The proceeding no. 2 is very exact and it is easy to calculate! He does not reflect about the simplification and he does not compare the result with other sizes.

> How large is the surface of the Mercedes (A-class)?
>
> Lösungsweg 2 ist sehr genau und ist leicht zu reduzieren

Figure 8. Part of a student's solution: Superficial validation.

Finally there were also *mistakes concerning the whole modelling process*.

- Many aspects of the real world were described without using them in the modelling.
- Some students lost track of their own proceeding. Figure 9 shows a solution of the natural gas task (Fig. 4). The girl assumes three different developments: (1) The use of natural gas remains the same during the years. (2) It rises. (3) It sinks. She gives reasons for her assumptions and then she works within her models and interprets the solution. But afterwards she corrects her calculation many times without correcting her interpretations. So she gets the same result in the first and in the second calculation, but in her interpretation she writes: The natural gas will be used up 6 years earlier than in the first case. Possibly she became confused while correcting her solution.
- Sometimes the whole modelling process was not described or only in a very short way. In Figure 10 you can see that there is no explanation at all, only a calculation.
- Some students talked about their own experiences in life without relating them to the modelling process.
- In some cases modelling was stopped without any results because the calculation was confusing or the student had chosen a proceeding which he/she couldn't follow.

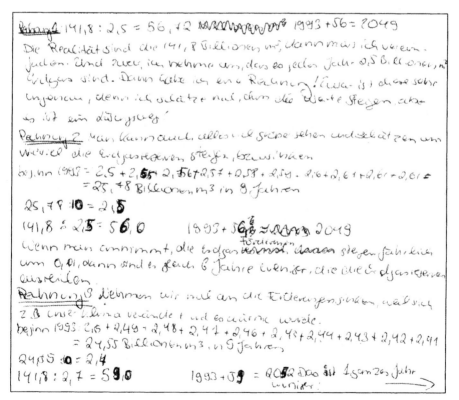

Figure 9: Students solution of the natural gas task.

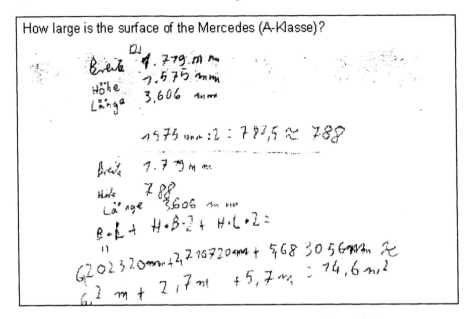

How large is the surface of the Mercedes (A-Klasse)?

Figure 10: Modelling without written argumentation.

Some kinds of mistakes could often be found together. Students who seemed to have problems in setting up a real model often had problems in validating the solution. Mistakes in setting up a mathematical model, in working within this model and in interpreting complex mathematical solutions often occurred together.

In summary these mistakes show that it requires far more competencies to solve modelling tasks than solving "standard" mathematical school tasks.

To gain more information about modelling competencies we will now look at metacognitive modelling competencies: A very important result of the study was that at the end of the study a big part of the students had an adequate metacognitive knowledge about modelling processes, some even had a very high-level knowledge. Only a few students showed only a basic knowledge about the modelling process. Although a majority of students developed adequate metacognitive modelling competencies misconceptions about the modelling process could be reconstructed which again refer to details concerning metacognitive modelling competencies:

There were *misconceptions concerning setting up the real model*:

- Some students thought that simplifying is the same as guessing.
- Some students thought they could simplify in such a way that the calculations became as simple as possible.
  "Yes, the simplifying was always easy for me, well you could simplify in a way, you find it the most simple…"

Misconceptions about setting up the mathematical model could also be reconstructed:

- Some students could not differentiate between the real model and the mathematical model.

"Well, in written class tests I was not so good, because sometimes I did not understand the difference between the real model and the mathematical model"

- The term "mathematical model" could not be explained.

Which parts of the modelling process did you find easy or difficult? "Well, simplifying reality and finding a mathematical solution was easy. But I have difficulties with the mathematical model."

Furthermore **misconceptions concerning the mathematical solutions** could be found:

- Often only a number (and for example, not a graph or a function) was regarded as a mathematical solution (Concept Map)
- Some students thought that a number always represents an exact and unambiguous result – independent of the way of calculation. (Concept Map)

In addition to that there were **misconceptions concerning the interpretation and validation**:

- Some students were of the opinion that the validation is always the same.

"Is it enough if I write for every inquest: The same as usual?"

- Some students had the impression that the validation represents a debasement of the modelling.

"Do we have to run down everything now again?"

- Some students could not differ between interpretation and validation. (Concept Map)

Finally, **general misconceptions** could be reconstructed:

- Some students thought that it is impossible to make any mistakes because every way of solution is alright.
  "Actually, you cannot make mistakes because nobody can control whether your solution is correct or not."

- Some students regarded the proceeding of experts as exact without knowing much about it whereas they regarded their own proceeding as not exact.
  "The experts from University modelled the problem in a similar way, but they had far more detailed information. Their real model is more exact than ours."

Summing up, these misconceptions refer to the complexity of metacognitive knowledge of the modelling process. Altogether it is a highly complex task to develop modelling competencies which should not be underestimated. The huge variety of possible mistakes and misconceptions shows the high performance of the majority of students who developed adequate modelling competencies.

5.2. Factors Influencing Modelling Competencies

Having identified the development of modelling competencies as a very complex challenge we have to ask what factors influence this development.

The following results are based on the analysis of the mistakes and misconceptions (see §5.1), the evaluation of the mathematical capacity with the help of a special test developed by Blum, Kaiser, Burges and Green (1994) and the evaluation of the attitude concerning modelling tasks and mathematics with the help of learners' diaries and questionnaires (Maaß 2004, p115).

Influencing factors:

1. Metacognitive modelling competencies

Relations between the meta-knowledge about the modelling process and the modelling competencies could be reconstructed for many students:

- Misconceptions about the real model could be reconstructed together with deficits in setting up the real model.
- Misconceptions about the validation occurred together with deficits in doing so.
- Parallel developments could be seen. For example, somebody who became more successful in setting up a real model also corrected misconceptions about the real model.
- There were relations between the quality of meta-knowledge and the competencies in modelling a problem. Normally, very good modellers also had a high meta-knowledge about the modelling process whereas bad modellers had a low meta-knowledge.

Furthermore many students regarded the meta-knowledge about the modelling process as helpful.

"Well, I used it for every task, I did everything by following the circulation and I did it in the way which I thought was the most sensible..."

2. A sense of direction

The results of the study showed that sometimes the modelling was terminated by the student without any result. In other cases much information about the real-world problem was given without using it and there were also students who lost track of their own proceeding. These failures show that it is obviously necessary to have an overall view and to work purposefully. A Sense of direction is necessary (see also Treilibs, 1979).

3. Competencies in arguing in relation to the modelling process

The results of the study show that some students don't argue in relation to the modelling process but in relation to their private experiences. Others don't describe their proceeding; important parts of the argumentation are missing. These failures show clearly that the students must learn to argue and to write down their argumentation. Competencies in arguing are necessary.

4. Mathematical capacity

The results of the study show a connection between the mathematical capacity and the modelling competencies of the students.

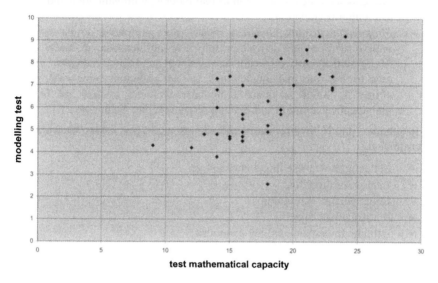

Figure 11: Connection between mathematical capacity and modelling competencies.

Figure 11 shows that for a fixed number of points in the test "mathematical capacity" the range of the number of points in the modelling test is not more than 1/3 of all possible points. This means a good mathematical capacity can have a positive impact on the modelling competencies. However the graph shows that variation is possible.

5. Attitude towards modelling examples and mathematics

The results of the study show a connection between the modelling competencies and the attitude towards the modelling tasks and mathematics. However this relation is apparently not a quantitative one. A negative attitude towards the modelling tasks seemed not to lead low modelling competencies. The relationship is rather likely to be a qualitative one. A negative attitude seems to have a negative impact on certain sub-competencies. Based on this insight reaction patterns could be reconstructed:

- *Reflecting modellers* have a positive attitude towards mathematics itself as well as towards modelling examples. They show an appropriate performance in mathematics. Deficits on modelling are hardly to be found.
- *Uninterested modellers* are neither interested in the context of real-world problems nor in mathematics itself. There are deficits in mathematical competencies. While dealing with modelling tasks, problems occur in every part of the modelling process.
- *Reality-distant* modellers have a positive attitude concerning context-free mathematics and show an appropriate performance in mathematics. They

reject modelling examples. In conclusion, an affective barrier is set up which mainly results in a lack of competency to solve problems closely connected to context-related mathematics. This means that they have problems with the construction of real models, with validation and partially also with interpretation.

- *Mathematics-distant modellers* clearly give preference to the context of real-world problems and show only low performance in maths lessons. These students are very enthusiastic about modelling examples. They are able to construct real models and validate solutions quite well. Lack of ability is found in constructing mathematical models, in finding a mathematical solution and in interpreting complex solutions.

In summary the following factors influencing modelling competencies could be reconstructed:

- Metacognitive modelling competencies
- A sense of direction
- Competencies in arguing
- Mathematical capacity
- Attitude towards modelling examples

Based on this insight we will turn back to the main question of this paper: What are modelling competencies? Let us supplement former definitions of modelling competencies. However, the definition makes no claim to be exhaustive because important aspects like linguistic competencies were not evaluated in this study.

Modelling competencies contain

- Sub-competencies to carry out the single steps of the modelling process
    - o Competencies to understand the real problem and to set up a model based on reality.
    - o Competencies to set up a mathematical model from the real model.
    - o Competencies to solve mathematical questions within this mathematical model.
    - o Competencies to interpret mathematical results in a real situation.
    - o Competencies to validate the solution.
- Metacognitive modelling competencies.
- Competencies to structure real-world problems and to work with a sense of direction for a solution.
- Competencies to argue in relation to the modelling process and to write down this argumentation.
- Competencies to see the possibilities mathematics offers for the solution of real-world problems and to regard these possibilities as positive.

6. CONSEQUENCES AND IMPLICATIONS

The results of the study show that modelling competencies contain far more competencies than sub-competencies to carry out the single steps of the modelling process. This has to be taken into consideration in maths lessons. The students have to learn to argue in the maths lessons and it has to be discussed on a metalevel how modelling processes can be carried out in an effective and purposeful manner. The teaching methods should support the development of metacognition. In order to minimize negative reactions many different contexts should be used and modelling should be integrated in maths lessons as early as possible.

Of course, these aspects have to be picked out as central topics for pre-service and in-service teacher education which is by no means a simple task. It is not sufficient to prepare teachers with modelling tasks for math classes, since they need experience in modelling problems on their own. They also have to learn about teaching methods to develop metacognition, a sense of direction and competencies in arguing and they have to become experienced with these new teaching methods. To realise these goals effective ways of teacher education are needed.

NOTE

1. This task is taken from Herget, W., Jahnke, T. and Kroll, W. (2001) *Produktive Aufgaben für den Mathematikunterricht in der Sekundarstufe I*. Berlin: Cornelson.

REFERENCES

Altrichter, H. and Posch, P. (1998) *Lehrer erforschen ihren Unterricht*. Bad Heilbrunn: Julius Klinkhardt.

Baumert, J., Klieme, E., Neubrand, M., Prenzel, M., Schiefele, U., Schneider, W., Stanat, P., Tillmann, K. and Weiß, M. (2001) *Pisa 2000, Basiskompetenzen von Schülerinnen und Schülern im internationalen Vergleich*. Opladen: Leske + Budrich.

Bläsing, J., Gerth, W., Jorde, K., Kaltschmitt, M., Raab, K. and Weinrebe, G. (2000) *Regenerative Energien in der Region Stuttgart – Kriterien und Potenziale*. Stuttgart: Verband Region Stuttgart.

Blum, W. (1996) Anwendungsbezüge im Mathematikunterricht – Trends und Perspektiven. *Schriftenreihe Didaktik der Mathematik*, 23, 15-38.

Blum, W. et al. (2002) ICMI Study 14: Application and Modelling in Mathematics Education – Discussion Document. *Journal für Mathematikdidaktik*, 23 (3/4), 262-280.

Blum, W. and Kaiser, G. (1997) *Vergleichende empirische Untersuchungen zu mathematischen Anwendungsfähigkeiten von englischen und deutschen Lernenden*. Unpublished application for a DFG-sponsorship.

Blum, W., Kaiser, G., Burges, D. and Green, N. (1994) Entwicklung und Erprobung eines Tests zur „mathematischen Leistungsfähigkeit" deutscher und englischer Lernender in der Sekundarstufe I. *Journal für Mathematikdidaktik*, 15 (1/2), 149-168.

Flick, U., von Kardorff, E. and Steinke, I. (2002) Was ist qualitative Forschung? Einleitung und Überblick. In Flick, U., von Kardorff, E. & Steinke, I. (eds) *Qualitative Forschung, Ein Handbuch*. Reinbek bei Hamburg: Rowohlt, pp13-29.

Landesgewerbeamt Baden-Württemberg, IE Informationszentrum Energie (eds) (2000) *Energie, Sonne, Stuttgart*.

Landesgewerbeamt Baden-Württemberg, IE Informationszentrum Energie (eds) (1999) *Energie, Solarthermie, Stuttgart.*

Landeshauptstadt Stuttgart, Statistisches Amt (eds) (1999) *Statistik und Informationsmanagement, Jahrbuch 1999, Stuttgart.*

Maaß, K. (2004) *Mathematisches Modellieren im Unterricht – Ergebnisse einer empirischen Studie.* Hildesheim, Berlin: Franzbecker.

Niss, M. (2003) Mathematical competencies and the learning of mathematics: The Danish KOM project. In Gagatsis, A. and Papastravridis, S. (eds): *3rd Mediterranean Conference on Mathematical Education, 3-5 January, Athens, Greece.* Athens: The Hellenic Mathematical Society, 115-124.

Niss, M. (1993) Assessment of mathematical applications and modelling in mathematics teaching. In J. de Lange, I. Huntley, C. Keitel and M. Niss (eds) *Innovation in maths education by modelling and applications.* Chichester: Ellis Horwood, 41-51.

Schoenfeld, A. (1992) Learning to think mathematically: Problem solving, metacognition and sense-making in Mathematics. In D. Grouws (ed) *Handbook for Research on mathematics teaching and learning.* New York: Macmillan, 334–370.

Sjuts, Johann (2003) Metakognition per didaktisch-sozialem Vertrag. *Journal für Mathematikdidaktik,* 24 (1), 18-40.

Treilibs, V. (1979) *Formulation processes in mathematical modelling.* MPhill Thesis, University of Nottingham.

2.3

LEARNING BY CONSTRUCTING AND SHARING MODELS

Celia Hoyles and Richard Noss
Institute of Education, University of London, UK

Abstract–*We report on a small slice of a large-scale, three-year, EU-funded research project, WebLabs, which focused on the iterative design of two systems: a programming-based environment for students to build models of their mathematical and scientific knowledge, and a set of web-based collaboration tools to share both their ideas and their programmed models. Our students were aged between 10 and 14 years, and spanned six EU countries. We focus on the iterative design of a set of activities in the context of the construction and exploration of a 'lunar lander', in which the idea was eventually to build a game about landing on the moon. Our hope was that in constructing a model of a lander, exploring its behaviour, sharing knowledge about problems and ultimately sharing the game, students would develop a deeper appreciation of the concepts of position, velocity and acceleration with the tools provided, and the relationships between them. We focus particularly on students' activities in relation to reading and interpreting position-time and velocity-time graphs.*

1. INTRODUCTION

We start from the presumption that a key objective of modern mathematics curricula should be to develop an understanding of what it means to build a mathematical model. This conviction arises from our current work in the "Techno-mathematical Literacies in the Workplace" project, a three-year research study, funded by the Teaching and Learning Research Programme of the UK ESRC, to investigate the kinds of mathematical knowledge necessary in modern workplaces. The changing nature of workplaces, and the ubiquity of computer-based systems for the automation and control of processes and the management of information, has brought about an increasing need for employees at all levels to engage with these systems, to interpret their outputs and to make sense, to some extent, of the abstract models on which they are based. This means developing competence in what we term Techno-mathematical Literacies (TmL): technically-oriented functional mathematical knowledge, grounded in the context of specific work situations (c.f. Kent, Hoyles, Noss & Guile, 2004). Thus, we have argued that employees need to be able to engage in *situated* modelling where interpretation necessarily arises from a balanced appreciation of abstract structures and workplace expertise.

In this paper, we will report on a small slice of a large-scale, three-year, EU-funded research project, *WebLabs*, which focused on the iterative design of two systems: a programming-based environment for students to build models of their mathematical and scientific knowledge, and a set of web-based collaboration tools to share both their ideas and their programmed models. Our students were aged between 10 and 14 years, and spanned six EU countries. (The interested reader will find details at weblabs.eu.com, see also Simpson, Hoyles, & Noss (2005)).

Our focus is on the iterative design of a set of activities in the context of the construction and exploration of a 'lunar lander', in which the idea is eventually to build a game about landing on the moon. Our hope was that in constructing a model of a lander, exploring its behaviour, and sharing knowledge about problems, and, ultimately sharing the game, students would develop a deeper appreciation of the concepts of position, velocity and acceleration with the tools provided, and the relationships between them. A particular focus was, as we will see, on being able to read and distinguish position-time and velocity-time graphs and to begin to understand the relationships between them.

2. OUTLINING THE APPROACH

Before we describe the design of a piece of the modelling activity, we briefly outline the modelling system that formed the technical platform for the models. Our choice of programming language was *ToonTalk,* a fully functional, concurrent programming language that has an interface modelled in the style of a video game (Kahn, 1996, 1999; see www.toontalk.com). ToonTalk has a number of unique features, most notably that the 'programmer' creates programs by directly manipulating animated objects. It is almost impossible to give a flavour of what it is like to 'write' a program in ToonTalk, but some idea can be gained from visiting weblabs.eu.com.

A substantial part, therefore, of the WebLabs research effort was spent in building and modifying the ToonTalk platform in order that students were able to express the kinds of modelling activities with which we wanted them to engage. At the same time, we developed a web-based collaboration system called *WebReports*, which allowed students to share and discuss not only their current thoughts and findings, but working models that instantiate their ideas. Students composed online reports using a *wysiwyg* editor, which includes the facility to embed files such as pictures (for example, Excel graphs), java applets, and ToonTalk objects, discuss each others' work, and share – not only their thoughts – but working models of their thoughts instantiated in ToonTalk models (see Figure 1 for an example).

In the remainder of this paper, we will illustrate the WebLabs approach by an outline of one particular activity – *Guess my Graph* – that was designed to stimulate collaboration at a distance. Our data is mainly drawn from the activity of seven students, aged between 13 and 14, in London; with some input from a distant group of 20 students in Nicosia, Cyprus of a broadly similar age (although a few were as young as 10). Students worked intensively in groups, at sessions lasting between 60 and 90 minutes.

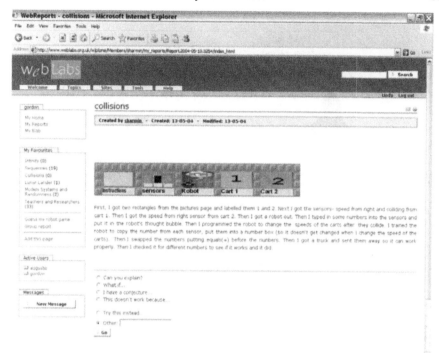

Figure 1. An example of a WebReport showing a set of boxes containing the ToonTalk model, a narrative description of the activity, and indicating how others can comment or add another model.

3. GUESS MY GRAPH

We designed a complete activity sequence for the lunar lander work. The *Guess my graph challenge* took place after various preliminary activities during which students experimented with different ways to control, observe and record motion by building ToonTalk models, and at the same time investigate the associated time-based motion graphs. Figure 2 shows a screenshot of the lander and the 'sensors' attached to it (sensors give readouts of the state of a range of parameters, and also allow the setting of that state). Position, speed and acceleration of the lander are shown in the boxes indicated, but also can be manipulated by simply typing in new values.

The basic format of *Guess my graph* was that students at one site set challenges for students at another site, and vice versa, which started threads of discussion and comparison. Specifically, students at a given site controlled their lunar lander in ToonTalk, recorded the data, and produced kinematics graphs in Excel (we engineered ToonTalk so that this interfacing with Excel was seamless). The students posted their graphs in WebReports and challenged other students to create motion of their own landers that would reproduce the challenge graphs.

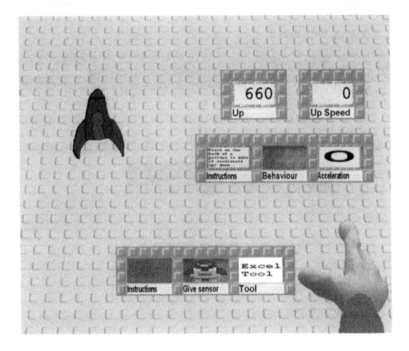

Figure 2. The lunar lander is shown at top left, with its Up (position) and Up Speed sensors at top right. Below these built-in ToonTalk sensors is the acceleration tool. Shown at bottom is the Export to Excel tool. To use this students simply take a copy of one of the sensors and give it to the "bird" which will then record the data to the windows clipboard, ready to be pasted into Excel. The programmer's hand is at bottom right.

When attempting to solve a challenge, students needed to consider a number of issues. First, they had to determine which variable was plotted and to get a feel for the type of motion that had given rise to the graph. They then had to consider how to model this motion in ToonTalk. For example, was acceleration required and what were the initial conditions of position and velocity? Once determined, students set up their ToonTalk model, recorded the data and exported it to Excel. In Excel they plotted the data on a graph and then compared it with the challenge to evaluate whether it was "correct". This in itself raised interesting questions. For instance, did the two graphs need to be identical in terms of the values or just broadly the same shape? We deliberately left this open-ended in our design as it led to interesting discussion and debate. Figure 3 shows an example of one such challenge posted by a Cypriot student.

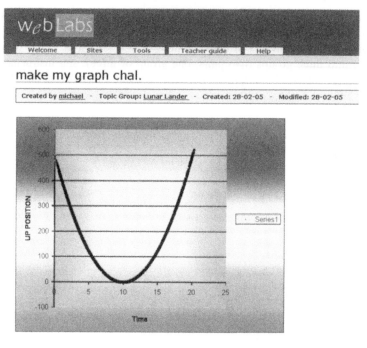

Figure 3. Guess my graph challenge posted by a Cypriot student.

At this point students may have been happy with their graph, in which case they posted it as a response on the WebReports site, ideally with an accompanying description of the relevant ToonTalk conditions used (see Figure 3 for an example graph). If they felt their graph did not match, they returned to ToonTalk to refine their model, repeating the process until such time as they were satisfied. In some instances students wanted to recreate certain values of a graph, for example reaching a peak position after, say, 10 seconds. In these cases, students adopted either a trial-and-error approach, or tried to reason algebraically about the relationship between the variables (position, velocity, acceleration, and time) if they knew how to do so. Students sometimes worked in pairs, in which case they collaborated face-to-face before sharing their products at a distance. Once a response to a challenge had been posted we hoped that the challenger would respond in turn, perhaps asking for clarification, or comparing the methods of graph production.

4. THE STUDENTS' ACTIVITIES

There were five students in London and 20 in Cyprus that participated in the *Guess my graph challenge*. There were a total of 13 challenges posted: seven from Cypriot students, three from students in London, and three from researchers in London. Nine out of these 13 challenges were responded to via the WebReports comment mechanism (significantly more than in other activities); with a total of 86 comments being posted, 48 of which were responses by students that included an Excel graph attempting to solve the challenge. We now examine some of the challenge-responses in more detail. During a full-day workshop, students in London were set the task of responding to two specific Guess my Graph challenges posted by Cypriot students. The first of these challenges is as shown in Figure 3. Students were split into three groups (two pairs, one group of three), and instructed to post their responses on the WebReports site before a group presentation and discussion was to be held.

Comment

⁊ **I have a conjecture...**

Posted by: sodapop at 08-03-05

First of all we began by seeing how much the speed affects the up position every second. We came to the conclusion that a speed value of 1 would add 1 Adward to the up postion (The same effect could happen with right speed and right position). The the was so for minus numbers which we used. From this we found out that to reach postion 0 from 500 in 10 seconds, you had to have an up speed of -50. But for it to come back up to its starting position, you had to have a positive acceleration. For it to happen in the time that you did it, you're acceleration would have to be 10.

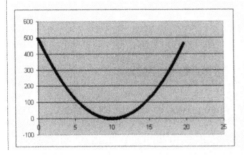

Figure 4. A response to the Cypriot challenge. Note how this group used what they termed an "Adward" for a unit of length. This had been invented by two students some weeks previously, where Adward is a concatenation of their names.

The students successfully responded to the challenge, posting both their graphs and explanations of the process they followed, as illustrated for one case, Figure 4. Both responses acknowledged the relationship between initial speed and acceleration, the sign of these variables and the relationship to time.

The third response is perhaps the most interesting, see Figure 4. Of particular note is that the graph is the same "shape" but has different actual values to the challenge, due to the fact that it starts at position 1000 rather than 500, something the students note themselves in their report. When presenting the answer to the group, Andy explains:

> Andy: So we set the initial up position as 1000 instead of 500 and we didn't actually think it made a difference and also when they, when Cyprus replied to some of ours they...
>
> Researcher: They didn't – what did the Cyprus people?
>
> Andy: They didn't use the same numbers.

What Andy is referring to here is that the Cypriot students had previously responded to the challenge that *he* had set, and their graphs had not used the same numbers as his. In fact, Andy had 11 responses to his challenge, all of which had different scales on either x-axis (time) or y-axis (position). In a sense, Andy had adopted a standard or rule based on his experience in the "guess my graph community" – a standard that was not necessarily shared by other students however:

> Mike: It's just that you doubled the amount on the y-axis, but you didn't double the amount on the x-axis, so really you didn't use the right thing. You didn't put in the right thing.
>
> Lance: Yeah, basically all your values have changed.
>
> Mike: It should have hit down – you got the right idea but it isn't actually the right graph.
>
> Peter: You've slowed it down.
>
> Andy: It's the right graph, it does the same thing.
>
> Mike: If you used what they used in your... thing... in your lunar lander...it should've -
>
> Alan: It's not mathematically similar.
>
> Lance: Exactly, if it's not mathematically similar then it's wrong.
>
> Adrian: It is mathematically similar it's just not exactly similar.
>
> Lance: No no it isn't.
>
> Mike It should have doubled the distance and doubled the time.
>
> Researcher: Just listen – shush – one at a time. You should have done what sorry?
>
> Mike: Um, if you doubled the distance –
>
> Andy: But um we, didn't actually think that making it exactly the same width or height or whatever actually made the graph wrong or right. If it does the same thing as the other graph, if it goes in the same pattern...
>
> Researcher: Yes, I have another question here?
>
> Peter: It does vaguely – well not vaguely but – it does a similar thing, but its half the speed at which it does it –
>
> Mike No it's double the speed.
>
> Peter: Double the speed, yeah.

We found this debate intriguing, and the level to which students were committed to their arguments and engaged in the discussion is hard to capture in the transcript - it is difficult to imagine such a passionate debate taking place during a typical mathematics class! The idea of *mathematical similarity* emerged, and students were divided over whether the response should be considered mathematically similar or not. Indeed, in a mathematical sense the graphs *are* similar – both being parabolas – but as Lance points out the actual values are different between the graphs (apart from when y=0) so the underlying motion was different. This is a deep mathematical idea, and one that arose spontaneously as a result of the structure of the activities. Also of interest was that the students used the physical meaning of the graph as a criterion for comparison, for example Peter stating that it is at "double the speed".

Inital Up Position set as 500, Up Speed set as -200, and Acceleration set as 20.

The lander starts on a negative decent, and after a while the acceleration counter-acts the up speed, and this then becomes a positive number, constantly rising.

We set our up position as 1000, as opposed to 500.

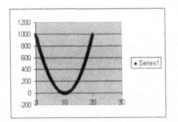

Figure 5. Third response. Note how the start and end points are different to the original challenge.

The second selected challenge from a Cypriot student is shown in Figure 6.

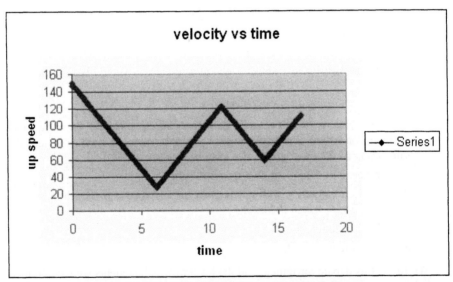

Figure 6. Second Cypriot Guess my Graph challenge.

Lance and Adrian were working together to solve this challenge, and had an argument about whether acceleration was required to recreate the graph. Note how they were discussing very explicitly the relevant variables and their relationships:

> Lance: It's not up position, it's up speed. You've got to have acceleration, yes, you do.
>
> Adrian: If it was acceleration it would be curved, not a dead straight line.
>
> Lance: No that's only if its up position, we're talking about up speed. I know I'm right. Up speed, up speed, up speed, up speed, up speed, up speed!

Adrian: You can do it a lot simpler without using acceleration.

…

Lance: You keep thinking of up position.

Adrian: I'm not.

Lance: Yes you are!

Researcher2: Just let Lance have his argument...

Lance: Because, look - if you're changing the speed you've got to have acceleration, unless you type in every single speed every millisecond. You can't type in every millisecond.

It seems clear from the transcript that Lance believes Adrian was making a "variable confusion" mistake, that is, he had interpreted the graph as position-time rather than a speed-time graph. Indeed, the reasoning Adrian applies ("If it was acceleration it would be curved, not a dead straight line") is applicable to a position-time graph, but is incorrect here. It is notable how adamant Lance is that Adrian had made a mistake. In fact this is exactly the mistake that Lance himself had made and reflected on some weeks earlier during the *Predicting graphs from motion* activity, an interesting example of learning and peer tutoring that can be stimulated when a WebReport is available that records student output for all to reflect on.

5. CONCLUDING REMARKS

Our view of modelling, illustrated above, has focused on the learner as modeller, as a participant in the process of coming to understand how variables are related, and how different representations of the evolution of the variables over time can assist in understanding the phenomenon. It is noteworthy that the students' modelling was *situated*: they could read the graphs and compare them by thinking *through* them to the motion of the lander. The Guess my Graph challenge stimulated them to do this. This mode of working in a Guess my "X" format has more generally stimulated argumentation and the sharing ideas among students in different sites.

We believe that the programming approach has much to offer, particularly given the huge evolution in the idea of what it means to program since the late nineteen-sixties when the idea was first mooted as a means to engage students with mathematical ideas (see Noss & Hoyles, 1996, for an overview of these developments). There is, however, much that needs to be done. First, we make no claim in the activities described here, that they can easily stand independently of a teacher (or, in our case, a researcher). For that to be true, we would need to embed these activities into a broader context, preferably one in which learning to program in some appropriate system was an accepted part of the curriculum. Second, it would certainly be interesting if we could adduce some evidence that learning to build models in the way described results in students who are more knowledgeable about the mathematical issues concerned. We do have evidence that they develop a language for describing phenomena, and display a confidence and willingness to conjecture, reason and generalise. We do not, as yet, have evidence that such students are easier to teach when the time comes for them finally to engage with the 'official version' of the mathematics concerned.

Finally, we cannot claim to have designed the definitive modelling system, or anything approaching it. ToonTalk itself is a major step forward in the design of programming systems, but it does not make programming "easy". Our tools and activities are similarly prototypical. We can, however, reasonably claim to have pointed a feasible direction in which to evolve at least part of a new mathematical approach, one which at the very least is both engaging and challenging.

ACKNOWLEDGEMENT

We are most grateful to Gordon Simpson, who was a collaborator on a version of this paper: Simpson, G., Hoyles, C. and Noss, R. (in press).

REFERENCES

Kahn, K. (1996) ToonTalk – an animated programming environment for children. *Journal of Visual Languages and Computing 7*, 197–217.

Kent, P., Hoyles, C., Noss, R. and Guile, D. (2004) 'Techno-mathematical Literacies in Workplace Activity'. Seminar on Learning and Technology at Work, London, March. Available online at http://www.lkl.ac.uk/kscope/ltw/seminar.htm.

Mor, Y., Hoyles, C., Kahn, K., Noss, R. and Simpson, G. (2004) Thinking in Progress. *Micromath, 20* (2), 17-23.

Noss, R. and Hoyles, C. (1996) *Windows on Mathematical Meanings: Learning Cultures and Computers.* Dordrecht: Kluwer.

Simpson, G., Hoyles, C. and Noss, (2005) Designing a programming-based approach for modelling scientific phenomena *Journal of Computer Assisted Learning,* 21, 143-158

Simpson, G., Hoyles, C. and Noss, R. (submitted manuscript). Exploring the mathematics of motion through construction and collaboration

Section 3
Recognising Modelling Competencies

3.1

EXEMPLAR MODELS: EXPERT-NOVICE STUDENT BEHAVIOURS

Rosalind Crouch[1] and Christopher Haines[2]
[1]University of Hertfordshire, Hatfield, UK,
[2]City University, London, UK.

Abstract–*We report on a study of mathematical modelling skills of final year mathematics undergraduates, studying a course in mathematical models and modelling where the students' introduction to modelling is through exemplar models. These students are found to be at an intermediate stage between expert and novice, whose modelling capabilities are affected by limitations of their domain knowledge both in organisation and in content. We describe expert-novice behaviours, in terms of novice, intermediate and expert classifications. We consider outcomes from multiple-choice coursework questions, developed in a similar style to those used in more elementary contexts. We link these outcomes, and those from other assessments, to the descriptors of behaviour. Further, we raise questions as to how these outcomes might change perceptions of applying mathematics in real situations and consider implications for education more generally.*

1. NOVICE-EXPERT MODELLING SKILLS IN UNDERGRADUATES

In our previous research (Crouch & Haines, 2003, 2004) we have studied the novice modelling skills of first-year undergraduates, finding that they had difficulty considering the demands of the real world and the model at the same time, giving them particular difficulties in moving from the real-world to the model early on in the modelling process. This behaviour is consistent with extensive reported research in mathematics and other scientific fields. Novices in general do not spend enough time analysing the problem (Schoenfeld, 1987) or making a suitable problem representation (Zeitz, 1997). They often do not pick up on which aspects of the real-world problem are relevant to the model (Patel & Ramoni, 1997) and tend to use linear, rather than cyclical, modelling strategies without validating their models (Berry & Houston, 2004). Novices also have difficulties recognizing what type of model is needed and accessing appropriate concepts and procedures to find a solution, as their knowledge-base is not yet well enough stocked and organized, whereas experts can access relevant knowledge much more efficiently, possibly in the form of templates or schemas which take time and experience to develop (Gobet, 1998). For complex tasks, and extended mathematical modelling projects and investigations are complex, it appears to take many years of motivated practice and experience to become an expert (Gobet, 1998; Simon, 1980; Shanteau et al., 2006).

In discussing a novice-expert continuum for mathematical modelling, we have constructed modelling contexts that could be readily understood by first-year students and for which multiple-choice questions, focussed on parts of the modelling cycle, could capture elements of the processes required for the successful solution of the problem posed (Haines & Crouch, 2006a). We now report on behaviours of final year undergraduate students on a course where exemplar models are discussed and analysed.

2. USING EXEMPLAR MODELS: N-STAGE ROCKETS

Following our discussion in Haines & Crouch (2006b), we note that a classical approach to teaching applications of mathematics within mathematics and in other disciplines is to use exemplar models. It is usually teacher led and often involves the presentation, discussion and analysis of applications in particular fields. Each application addresses a basic applied mathematics problem, which is to model a physical system mathematically so that it sheds light on the mechanical working of the system in the real world.

We have been concerned with modelling courses for undergraduates that have included models: kidney machine; aggregation of amoebae; road traffic flows; dimensional analysis of physical phenomena; n-stage rockets. Using this approach students are encouraged to acquire a strong understanding of specific models and in doing so learn to compare and contrast developments in complex models. Because modelling itself is not the driver, this approach needs a strong focus on aspects of modelling from the teacher but it allows more opportunity to link knowledge because of stronger engagement and motivation in students.

Recent space activity, the orbiting space laboratory, attempts to land on Mars, the Hubble telescope, GPS navigation systems, communication satellites, and the space shuttle have all raised levels of interest in rockets and satellites.

To fix ideas and to provide context for this discussion we describe simple n-stage rocket models, communicated and elaborated by Noble (1980, pp26-38). He successfully explains why rockets are built in a certain way and that for example to place a rocket/satellite in orbit we need a three-stage rocket.

We start by considering how fast a satellite must move when placed in orbit around the Earth and establish a model for the satellite speed $v = R\sqrt{\dfrac{g}{r}}$. In orbits $\approx$1000 km above the surface of the earth, taking the radius of the Earth R=6400 km, g=9.81 m/s^2 and setting r=6400+1000 km R=6400 km, this speed v is $\approx$ 7.4 km/s.

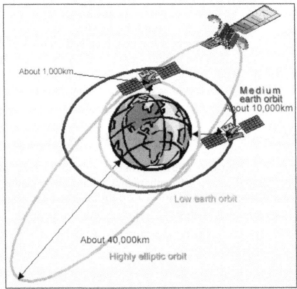

Figure 1. Typical satellite orbits. (Source: Telecommunications Bureau of the
Ministry of Internal Affairs and Communications, Japan).

Applying conservation of momentum to an idealised system we can model
the speed v of a rocket at time t, by:

$$v(t) = v_0 + u \ln\left(\frac{m_0}{m(t)}\right) \tag{1}$$

in which m_0 is the initial mass of the rocket system, $m(t)$ is its mass at time t having
burned fuel which ejects at speed u. This is a very simple result, obtained by
ignoring gravity and air resistance, suggests that the rocket's velocity depends: (i)
the velocity u of the exhaust gases and (ii) the ratio of the mass of the rocket at t=0
to the mass at time t. It is this speed v that has to at least match that of a satellite in
orbit if the rocket is to be used for that purpose.

It is easy to show that a rocket which accelerates its whole structural mass whilst
all its fuel is burned has a maximum velocity of ≈ 7 km/s leading to a conclusion
that single stage rockets cannot be used to place a satellite in orbit. In this case
equation (1) leads to

$$v = u \ln\left(\frac{m_0}{\lambda m_0 + (1 - \lambda)m_p}\right) \tag{2}$$

in which λ is the ratio of the structural mass to the structural mass plus fuel and
m_p is the mass of the payload (mass of the satellite). The limiting case where there is
no payload ($m_p=0$) gives the maximum speed ($u \cong 3$ km/s, $\lambda=0.1$).

The rocket could be more efficient if we jettison useless weight as the burning
proceeds; such a model where the rocket starts from rest indicates that

$$v(t) = (1 - \lambda)u \ln\frac{m_0}{m(t)} \tag{3}$$

This is very similar to the previous formula (equation 1), but with the crucial
difference that the final mass, when the fuel is used up, is simply the payload.

$$v = (1 - \lambda)u \ln \frac{m_0}{m_p} \qquad (4)$$

A practical way to realise continual jettisoning of mass is to build an n-stage rocket. Using equations (1, 2), we can obtain the speed at the end of the third stage of a three stage rocket which starts from rest to be:

$$v_3 = u \ln \frac{m_0}{m_p + \lambda m_1 + m_2 + m_3} \cdot \frac{m_p + m_2 + m_3}{m_p + \lambda m_2 + m_3} \cdot \frac{m_p + m_3}{m_p + \lambda m_3} \qquad (5)$$

The models captured by equations (1)-(5) and extensions and variations of them are discussed in class, through assessed coursework and more formal tests in an environment that encourages thoughtful reflection around appropriateness of models rather than focussing *per se* on the *correct answer*.

3. MULTIPLE-CHOICE QUESTIONS

The assessed coursework included multiple-choice questions (MCQs) constructed along the lines of Crouch and Haines (2003, 2004) and specifically Haines and Crouch (2006a) focussing on individual models, interpretations and understandings within a notional modelling cycle. Students were asked to indicate the option, A, B, C, D or E corresponding to their preferred answer and to give reasons for that choice. The following MCQs were used for coursework, the credit attached to each option is indicated within brackets [ ]:

MCQ1 A satellite moves in a circular orbit about 800km above the surface of the earth. The earth's radius is 6400km and the acceleration due to gravity at its surface is 9.81 m/s^2. Which of the following statements best describe the motion of such satellites?

A. This satellite is moving at about 236 kph [1]
B. For orbits 700-900km above the surface, satellites would move at about 7.5 km/s [1]
C. This satellite is moving at about 7.5 km/s [2]
D. For orbits 700-900km above the surface, satellites would move at about 236 kph [0]
E. The satellite's speed would decrease for lower orbits [0]

MCQ2 A simple rocket model gives its speed $v(t) = u \ln \frac{m_0}{m(t)}$. Which of the following statements best describe the factors affecting the speed of the rocket?

A. The velocity of the exhaust gases and gravity [0]
B. Air resistance and the ratio of the initial mass of the rocket to its mass at time t [0]
C. Air resistance and the velocity of the exhaust gases [0]
D. The ratio of the initial mass of the rocket to its mass at time t and the velocity of the exhaust gases [2]
E. The velocity of the exhaust gases and the initial mass of the rocket [1]

MCQ3 The maximum speed of a rocket satellite system is described by $v_{max} = u \ln \frac{1}{\lambda}$. If the speed of the exhaust gases $u \cong 3$ km/s and the structure is a fraction $\lambda=0.1$ of the combined mass of the structure and the fuel, then which of the following statements best describes the situation?

A.   The maximum velocity indicates that a single stage rocket could be used to place a satellite in orbit 500km above the earth
[R=6400km; g=9.81 m/s^2]   [0]

B.   The maximum velocity indicates that, on the moon, a single stage rocket can be used to place a satellite in orbit at a height of 200km
[R=1740km; g=1.62 m/s^2]   [2]

C.   The calculation for the maximum velocity allows the rocket to carry a small satellite   [1]

D.   Gravity is included in this model, but the rocket carries no payload   [0]

E.   The maximum velocity indicates that, on mars, a single stage rocket cannot be used to place a satellite in orbit at a height of 500km
[R=3400km; g=3.72 m/s^2]   [0]

MCQ4 The mass of an n-stage rocket system, can be modelled by

$$\frac{m_0}{m_p} = \left(\frac{1-\lambda}{p-\lambda}\right)^n \text{ where } p = \exp\left(\frac{-v}{nu}\right) \quad [\lambda = 0.1, u = 3 \text{ km/s}]$$

Which of the following options best describe outcomes from this model?

A.   A 3-stage rocket of more than 70 tonnes must be used to orbit a 1 tonne satellite round the earth   [1]

B.   If rocket could be built with a great many stages to orbit a 1 tonne satellite round the earth, then its mass would be about 40 tonnes   [0]

C.   A 2-stage rocket used to orbit a 1.5 tonne satellite round the earth would have a mass approximately 200 tonnes   [0]

D.   Since its mass would be only 60 tonnes, it would be better to use a 5-stage rocket to orbit a 1 tonne satellite round the earth   [2]

E.   A single stage rocket of about 300 tonnes could be used to orbit a 1 tonne satellite round the earth   [1]

MCQ5 The final velocity of a rocket system that continually sheds its mass can be modelled by $v = (1 - \lambda)u \ln \frac{m_0}{m_p}$. $[\lambda = 0.1, u = 3$ km/s]. Which of the following options best describe outcomes from this model?

A.   A payload can be accelerated up to any speed that we like   [1]

B.   A 55 tonne rocket system would be needed to orbit a 1 tonne satellite round the earth   [1]

C.   A 5 tonne rocket system launched from the moon could place a 1 tonne satellite in its orbit   [2]

D.   Both gravity and air resistance are insignificant in predicting outcomes from this model   [0]

E.   Calculating $\frac{m_0}{m_p}$ using this model gives the same result as would

be obtained using the n-stage rocket model (MCQ4) for n=10.   [0]

4. BACKGROUND TO THE DATA

We have analysed 370 responses to the 5 MCQs of §3 from 28 third year undergraduate students on the course in 2004-5. The responses were in two parts, firstly their preferred option A, B, C, D or E on each MCQ and secondly the reasons given by students for their chosen option. The chosen options were assigned credit (0, 1, 2) as indicated in §3 and the reasons, given by the student for their chosen option, attracted credit (0, 1, 2).

For example, the reasons given by students in MCQ1 might be, having chosen option: (A, D) a calculation in which g is used with units m/s², whilst the earth's radius is used with units km could attract a reason mark 1 or wrongly done, a reason mark 0. Each of these either with a mean satellite height or a range 700-900km; (B) a calculation which simply uses a mean satellite height of (say) 800km could attract a reason mark 1, one which deals with the range 700-900km might attract 2 reason marks; (C) to gain 2 reason marks the student has to calculate the speed accurately for the given satellite and note that this option describes that behaviour exactly; (E) it is difficult to envisage any reasons for this answer gaining marks, none in our data set selected this option.

For further illustration, we also refer to additional unsystematically-collected data from a later test question (TQ) on using a two stage rocket to place a satellite in orbit. The model in this case included gravity (Appendix).

We were encouraged by the fact that only 18% (30) of MCQ responses scored less than full credit (4), indicating a great deal of student expertise compared with first-year undergraduates of previous studies (c.f. Crouch & Haines, 2003, 2004). Only 2 responses did not attract credit (0), and one response was not categorized as the student gave no reason for his choice of answer. The following discussion focuses on the behaviours of students who scored less than full credit on the MCQs.

Within these MCQ and associated TQ responses, there appear to be four areas in which these intermediate students still have occasional novice-type difficulties (Haines and Crouch (2006a) and their responses were categorized accordingly:
   I.   difficulty in holding the balance between the real world and the abstract model
   II.   difficulty in choosing an appropriate model to apply
   III. difficulty deploying appropriate mathematical concepts and procedures to solve the problem
   IV.   not linking the outcomes of the model back to the real world (possibly within the problem definition).

Categories I and II were relatively rare in this dataset, but this may reflect the type of question the students were asked as these categories occurred more frequently in the students' TQ answers (which unfortunately were not accessible for systematic detailed analysis). Categories III and IV were found relatively more frequently in our data. We consider each category in more detail giving illustrative examples from the MCQ and TQ responses.

Category I:
Not holding the balance between the real world and the abstract model.

MCQ(1) and TQ(9) responses were identified in this category. The preponderance from the TQ might be due to its extended nature compared to the short MCQs. Students tended to give either inappropriate values for parameters or inappropriate units to support parameters, demonstrating some difficulty in keeping in touch with the demands of the real world and the model at the same time.

These undergraduate students found it hard to link knowledge provided and developed in course through general discussion, including newsreel and newspaper reports, to mathematical modelling contexts. This was demonstrated by students making unrealistic estimates of the time taken to burn all the fuel in the two stages of the rocket in the TQ model. Student 26 (S26), for example, had each stage of the rocket using its fuel almost instantaneously having estimated burn times 0.01s and 0.02s when burn times of a few minutes could be assured (S26, 0.01s, 0.02s). S26 failed to match his real-world experiences with his choice of suitable parametric values. Several responses of a similar nature, short burn times or rather too long, were identified: (S18, 0.1s, 0.2s), (S21, 600s, 600s), (S22, 2s, 2s), (S24, 1s, 1s), (S25, 10s, 10s). There were cases of the wrong units being used for acceleration due to gravity (S24) and for inappropriate quantities being used for burn times of the two stages in the TQ model. In one case, the latter led to the use of 1 *tonne* of fuel being used in the calculation (S2) and in another case to satellite speeds in orbit, incorrectly calculated, being used (S23).

S13 selected two answers in MCQ5, she did not justify the preferred option C. She justified option D by saying that since air resistance and gravity are ignored in the given model then these factors are insignificant in discussing outcomes from the model. She has not grasped the importance of the range of application of the model in a real situation. S22, in answering the TQ, used gravity to justify a particular design speed of the rocket, despite the fact that it had already been included in the model.

These examples indicate that intermediate students, neither novices nor experts do experience difficulty in holding the balance between the real world and the model. This conclusion applies to students whose final course grades turned out to cover the whole range: poor, average, good. It is interesting to note that S25, whilst exhibiting category I behaviour also showed signs of expert behaviour estimating suitable design speeds for the rocket in the TQ.

Category II:
Applying an inappropriate model

MCQ(1) and TQ(2) responses were identified in this category.

S19and S20 in attempting the TQ, correctly estimated the speed of a satellite in orbit 7.5 km/sec, they then used this speed in a model that does not include air resistance to estimate the size of the practical rocket system. To cater for all factors and to include air resistance, for example, in this model you would have to design a rocket to reach speeds comfortably exceeding this value. S6 attempted a similar problem, MCQ5. In calculating maximum speeds of rocket systems, she tried to make links between a model with a payload (the satellite to be placed in orbit) and a model with no payload (described in MCQ3). She compared the outcome of the given model with outcomes from the wrong model.

These students made errors in identifying the appropriate model for the context described in the MCQ or TQ. In these cases there appeared to be a broken link between the model used to calculate satellite speed (circular motion and gravitational attraction) and the various model used to design the rocket to reach a speed sufficient to place a satellite in orbit to that of the maximum speed possible in various models.

Category III:
Deploying inappropriate mathematical concepts and/or procedures

The inability to deploy appropriate concepts or procedures amongst these students accounted for 43% of all lesser credit responses for the MCQs.

Almost a third of the responses in this category (30%) showed students memory dumping, inappropriately trying to recall or estimate instead of calculating. A further 13% gave no mathematical justification for their answer or did the wrong calculation.

S1 and S2 in answering MCQ1 provided a good answer in selecting option B. Their reasoning was based specifically on calculations for an 800km orbit and then extended without justification to orbits in the range 700-800km. S3 attempted MCQ5 for which his preferred option C concerned satellites of the moon. He calculated the speed of a rocket launched from the surface of the moon (4.3 km/s) but did not calculate the speed at which satellites move in orbit, guessing or recalling a value 3 km/s. S1 and S3 offered these answers without mathematical justification as did S4, S5, S7, S8 and S9 on this question. S11 gave no justification for selected options on MCQ3 and MCQ4. S12 arrived at his preferred answers through a process of elimination of the other four distractors although no mathematical justification was given. S6 in attempting the TQ, doesn't derive the speed of a satellite in orbit, but recalls it and uses it to estimate the size of the rocket system.

The behaviours in this category might be consistent with students seeking a quick answer, attempting to arrive at a solution or a preferred option. All of these responses, except one, were observed in MCQs; a style of question that might encourage students to seek solutions with a minimum of effort.

Category IV:
Not linking the model appropriately to the real-world problem situation.

Almost half (47%) of all lesser credit responses for MCQs showed evidence of broken links between the model under discussion and the underlying real-world problem. 27% of responses in this category involved students apparently stopping short when they had finished the mathematics without proceeding to link the model back to the real world. A further 20% gave the outcomes of the model, but did not notice that they did not fully match the real-world situation.

S4 chose option B on MCQ3. He calculates correctly the speed of a satellite around the moon 1.59 km/sec and also the speed that the rocket can reach (6.59 km/sec), but he does not explicitly relate these two model outcomes to demonstrate that the rocket is good enough. S3 preferred option A on MCQ4 There are two preferred answers which are distinguished by considering the reasons given for the answer. In this case the student has said that the rocket *must* be >70 tonnes. This is so if a 3-stage rocket is used when larger rocket (about 77 tonnes) would be needed. However, if one could build a 5-stage rocket, then it could be significantly < 70 tonnes (about 60 would suffice). S3's discussion did not take account of the many possible outcomes from the multi-stage rocket model. Similar behaviour on MCQ4 is noted from the responses of S4, S5 and S8.

Option A of MCQ5 requires a deep understanding of the rocket model based on a continuous shedding of mass. It requires the student to realise that on the one hand, given a particular payload (mass of satellite to be placed in orbit) the given model determines the (fixed) maximum speed achievable. On the other hand, given a (fixed) speed to be achieved then the given model determines the maximum payload that will allow that speed to be reached. S14, a very good student, showed difficulty in interpreting the model in these mathematical terms, focussed on the limiting behaviour of the model when the mass of the payload approaches zero, and therefore could not relate the mathematics to the real terms of the option statement.

S17 in answering MCQ4 carries out mathematical procedures relevant to all distractors but provides no explanation as to why she prefers option A over option D. In doing so she is satisfied with the mathematics but does not link model outcomes to the real world.

5. CONCLUSIONS

Mayer (1998), in a review of problem-solving research, indicates that in order to foster expertise, teachers need to help students acquire skill, metaskill and will, which all work together to promote expertise. 'Skill' refers to domain-specific skills (relevant knowledge and procedures); 'metaskill' refers to skills such as picking out what is relevant in a problem (cf. Green & Wright 2003), making a meaningful problem representation (cf. Zeitz 1997), and knowing when and how to use appropriate concepts and procedures in context (cf. Weber 2001). 'Will' refers to motivation and interest in the task, which need fostering. Students need therefore to be interested in the problem situation, and get plenty of motivated practice and timely feedback (Ericsson & Lehmann 1996). Students also appear to benefit from techniques that help them make sense of the material presented, and that emphasize the importance of effort rather than innate aptitude for success (Mayer 1998).

Our intermediate-level students seem to be relatively successful modellers, although some still have intermittent novice-style difficulties, which is to be expected, as they have not yet had the opportunity to acquire the extensive experience needed to sufficiently consolidate their knowledge-base and the associated modelling skills ('skill' and 'metaskill' in Mayer's terms) to expert level. We have no data on the students' motivation and interest ('will'), and this would be a useful further line of investigation.

ACKNOWLEDGEMENT

We are grateful to the Telecommunications Bureau of the Ministry of Internal Affairs and Communications, Japan for Figure 1, obtained from www.tele.soumu.go.jp/e/system/satellit/move.htm .

REFERENCES

Crouch, R.M. and Haines, C.R. (2003) Do you know which students are good mathematical modellers? Some research developments. Technical Report No.83, Department of Physics, Astronomy and Mathematics, University of Hertfordshire. Hatfield: University of Hertfordshire 25pp.

Crouch, R.M. and Haines, C.R. (2004) Mathematical modelling: transitions between the real world and the mathematical model. *International Journal of Mathematics Education in Science and Technology,* 35, 2, 197-206.

Ericsson, K.A. and Lehmann, A.C. (1996) Expert and exceptional performance: evidence of maximal adaptation to task constraints. *Annual Review of Psychology*, 47, 273-305.

Green, A.J.K. and Wright, M.J. (2003) Reduction of task-relevant information in skill acquisition. *European Journal of Cognitive Psychology*, 15(2), 267-290.

Haines, C.R. and Crouch, R.M. (2006a) Getting to grips with real world contexts: Developing research in mathematical modelling. Accepted for proceedings from CERME 4, 17-21 February 2005, to appear.

Haines, C.R. and Crouch, R.M. (2006b) Frameworks for mathematical modelling and applications, accepted for the Springer (Kluwer) book *ICMI Study 14: Applications and Modelling in Mathematics Education,* to appear.

Gobet, F. (1998) Expert Memory: a comparison of 4 theories. *Cognition*, 66(2), 115-152.

Mayer, R.E. (1998) Cognitive, metacognitive and motivational aspects of problem-solving. *Instructional Science*, 26, 49-63.

Noble, B. (1980) Why build three-stage rockets? In J.G. Andrews and R.R. McLone (eds) *Mathematical Modelling*. London: Butterworths, 26-38.

Patel, V.I. and Ramoni, M.F. (1997) Cognitive Models of Directional Influence in Expert Medical Reasoning. In P.J. Feltovich, K.M. Ford & R.R. Hoffman (eds) *Expertise in Context*. Menlo Park, California. AAAI Press/MIT Press, 67-99.

Shanteau, J., Weiss, D.J., Thomas, P.R. and Pounds, J. (2006) How Can You Tell if Someone is an Expert? Empirical Assessment of Expertise. To appear in S.L. Schneider and J. Shanteau: *Emerging perspectives in decision research.* Cambridge University Press. Cambridge.

Simon, H.A. (1980) Problem Solving and Education. In D.T. Tuma and F. Reif (eds) *Problem Solving and Education: issues in teaching and learning.* Hillsdale,N.J. Erlsbaum. Cited in: D.L.Medin, B.H. Ross and A.B.Markham (2002) *Cognitive Psychology* (3rd ed) New York. John Wiley & Sons.

Schoenfeld, A.H. (1987) What's all this fuss about metacognition? In A.H. Schoenfeld (ed) *Cognitive Science and Mathematics Education.* Hillsdale NJ. Lawrence Erlbaum, 189-215.

Weber, K. (2001) Student Difficulty in Constructing Proofs: The Need for Strategic Knowledge. *Educational Studies in Mathematics*, 48,101-119.

Zeitz, C.M. (1997) Some Concrete Advantages of Abstraction. How Experts' Representations Facilitate Reasoning. In P.J. Feltovich, K.M. Ford and R.R. Hoffman (eds) *Expertise in Context*.Menlo Park, California: AAAI Press/MIT Press, 43-66.

APPENDIX

The test question (TQ) required the students to show that the speed of a satellite in a circular orbit at a height d above the earth's surface is given by $R\sqrt{\dfrac{g}{R+d}}$. They then were asked to use a model for the velocity of a single stage rocket of mass m(t):

$$v(t) = v_o + u \ln \frac{m_o}{m(t)} - gt,$$

in which m_o is the initial mass of the rocket system, v_o is its initial velocity and u is the velocity of the exhaust gases expelled by the rocket. With this model were expected to develop a simple model of a two-stage rocket and determine a ratio of the initial mass of the rocket system m_o to that of the payload (satellite) m_p.

The ratio is a function of several variables including v_2 the maximum speed of the rocket at the end of the second stage and $T_1 + T_2$, the times for which fuel is burned in the first and second stages respectively. Students were then expected to link the required speed of a satellite in orbit with that of the maximum speed attainable by this two stage rocket system. Taking realistic values for $T_1 + T_2$ and v_2 they were asked to estimate the size of the rocket system required to orbit a satellite of 0.5 tonnes at a height 500-800 km above the earth's surface.

3.2

A TEACHING EXPERIMENT IN MATHEMATICAL MODELLING

Toshikazu Ikeda[1], Max Stephens[2] and Akio Matsuzaki[3]
[1]Yokohama National University, Japan
[2]University of Melbourne, Australia
[3]Tsukuba University, Japan

Abstract–*When combined with group discussion and careful teacher direction, the use of multi-choice modelling tasks proved to be effective in helping to shape students' thinking about key features and stages of mathematical modelling. The problems in this pilot study were accessible and challenging to Japanese senior high school students of mathematics who had no prior teaching relating to mathematical modelling. Students initially developed their own thinking through individual problem solving and small group discussion. In whole-class discussion, the teacher helped clarify shared and different opinions from among small groups. This assisted students to re-evaluate their own ideas, or to challenge ideas from other groups. More detailed one-to-one interviewing during and after teaching will be needed to ascertain what predisposes some students to gain more from this type of approach than others.*

1. BACKGROUND AND AIM OF THIS STUDY

This study connects two themes that have been considered at earlier ICTMA meetings. Ikeda and Stephens (2001) showed that small-group and whole-class discussion of well structured questions assists students to develop their ideas about modelling. Haines, Crouch and Davis (2001) developed, for assessment and diagnosis, a set of multi-choice modelling problems focusing on distinct phases of modelling. So, what will happen if we treat these and other multi-choice modelling problems *as a basis for classroom teaching* about mathematical modelling? Is it a meaningful teaching approach for students, and does it foster students' thinking about modelling? In this pilot study, we set the following specific questions:

(1) Is it possible to elicit students' thinking about modelling through interaction between students and teacher by using multi-choice modelling problems?

(2) What transfigurations or shifts in students' knowledge of modelling can be observed as a result of teaching using multi-choice modelling problems?

(3) Can this approach be used to introduce students to some key ideas about mathematical modelling in a typically short period of time available in their busy senior high school program?

2. DESIGN OF THIS STUDY

The study comprised four distinct cycles. The following procedure outlines each empirical teaching CYCLE1 (each taking about one hour):
(1) Select and have students tackle 1 or 2 multi-choice modelling problems.
(2) Students solve the problems individually.
(3) Students discuss their answers in a small group (3 members).
(4) Students make a short report about the following points:
   Was your answer changed through discussion? Explain why?
   Justify your solution. What kinds of issues were discussed?
(5) Students then discuss their answers in a whole-class group. Teacher focuses on key issues to be discussed, and summarizes the important ideas.

Only one or two modelling problems were discussed in a CYCLE. Before CYCLE 1 and after CYCLE 4, we posed the following three "big questions" to assess any shifts in students' thinking:

[Question 1: WHAT is a mathematical model?] Is it easy or difficult to make a mathematical model? Why do you think so?

[Question 2: Which PROCESS is required?] Have you ever solved a real-world problem by using mathematics? What kind of process is required to solve a real-world problem by using mathematics?

[Question 3: What makes a GOOD model?] What ideas are important, or what points should you attend to when solving a real-world problem using mathematics?

In order to assess students' progress, we set a pre-test and post-test. The test problems comprised 12 problems developed by Haines and Crouch (2002), Treilibs, Burkhardt and Low (1979), Nagasaki (2001) and by the authors. Each test had six items, and focused on the same specific ideas but where the context is different. Students were divided into two groups as shown Table 1.

Table 1. Design of pre-test and post-test.

First group	Pre-Test A	Post-Test B
Second group	Pre-Test B	Post-Test A

3. IMPLEMENTATION IN KANAGAWA SOGO HIGH SCHOOL

The pilot study took place Kanagawa Sogo High School near Yokohama Japan. Agreement was reached with the school to participate in this study. The teacher of mathematics asked students if they were prepared to spend two Saturdays of June studying mathematical modelling. Nine students agreed to participate. One student dropped out after the first day, leaving eight students for both days. The "teacher" referred to below is one of the authors. Figure 1 shows the schedule for two days.

1st day (June 19, 2004) 13:00-16:30	**2nd day (June 26, 2004)** 13:00-16:30
1. Explanation (10min.)	1. Explanation (10min.)
2. Pre-test (30min.)	2. CYCLE 3 (65min.) Problems 4 & 5
3. Students writing about the "three big questions" (20min.)	(1) Solving by individuals
	(2) Discussion in a small group
4. CYCLE 1 (65min.) Problems 1 & 2	(3) Writing a short group report
(1) Solving problems individually	(4) Whole group discussion/teaching
(2) Discussion in a small group	3. CYCLE 4 (60min.) Problem 6
(3) Writing a short group report	(1) Solving by individuals
(4) Whole group discussion/teaching	(2) Discussion in a small group
5. CYCLE 2 (60min.) Problem 3	(3) Writing a short group report
(1) Explanation of graphic calculator	(4) Whole group discussion/teaching
(2) Solving problem by individuals	4. Post-test (30min.)
(3) Discussion in a small group	5. Writing about the "three big questions" (20min.), this time reflecting on their own short reports.
(4) Writing a short group report	
(5) Whole group discussion/teaching	
6. Summary (5min.)	6. Summary (10min.)

Figure 1. The schedule of this pilot study.

CYCLE 2: Problem 3 (Aim: Justification of an existing model) Consider the problem: On a warm summer day, some high school students decided to analyze how the cooling temperature of a cup of coffee changes over time. By examining the relation between time and cooling temperature of the coffee, they obtained the following points which they graphed, as shown in Figure 2.

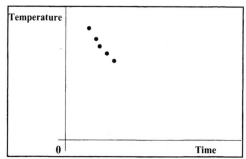

Figure 2. Graph showing time and cooling temperature.

From the graph, the students investigated whether the following functions could represent the relation between time (x) and temperature (y) of coffee where x>0.

     (1) y=ax+b (a,b: constant)

     (2) y=ax^2+bx+c (a,b,c: constant)

     (3) y=ae^{-bx} (a,b: constant)

Which **one** of the following explanations is **most** appropriate?

A. The function (1) is not appropriate. Because the temperature will become negative when the time goes by.

B. Even if we set up the range of x, all functions (1), (2) and (3) are not appropriate.
C. Even though we do not set the range of x, it is possible to use a quadratic function by transforming the formula (2).
D. The function (2) is not appropriate. Because according to this function the temperature of the coffee will eventually increase.
E. In a case of not setting up the range of x in the function (3), the temperature will converge to zero when the time goes by. However, it is possible to use the function (3) by transforming this formula.

Answer: E

CYCLE 3: Problem 4 (Aim: Generating and selecting variables)
    Consider the real-world problem (do **not** try to solve it!): The management of a large supermarket is trying to estimate how many of its checkout tills should be operating at any given time. The factors or variables that could be taken into consideration include:
    (a) The average age of customers, (b) The average bill size, (c) The efficiency of the checkout girls, (d) The maximum reasonable queuing time that can be expected of customers, (e) The number of customers in the store, (f)The average number of items bought, (g) The pay rate for checkout girls, (h)The proportion of customers using baskets rather than trolleys, (i)The working hours
Which **one** of the following sets of variables is **most** important in order to estimate how many of the checkout tills should be operating for customers?
A. (a), (c), (f), (i)
B. (c), (d), (e), (f)
C. (h), (c), (d), (e)
D. (e), (f), (g), (i)
E. (c), (e), (f), (h)   Answer: B

4. RESULTS OF PRE-TEST AND POST-TEST

    Eight students tackled both pre-test and post-test. In this pilot study, where additional items were used together with those by Haines and Crouch, students' responses to items were not scored using partial credit which Haines and Crouch subsequently applied to their items. Even so, as Table 2 shows, both groups got higher results in the post-test compared with pre-test. With a small number of students, any further statistical analysis of results is inappropriate.

Table 2. Results of pre-test and post-test.

	Pre-test	Post-test
First group (4 students)	Average of Test A: 45.8	Average of Test B: 50
Second group (4 students)	Average of Test B: 50	Average of Test A: 54.2

5. ANALYSIS OF STUDENTS' DISCUSSION IN THE WHOLE GROUP

    For the purpose of small group discussion, we set up three new groups so that each is composed of three members, however, one student dropped out after the first day. The answers of six problems derived from each group are shown in Table 3. In problem 2 and 4, the answers of three groups were the same. However, in the other four problems answers among the three groups were different.

Table 3. The answers of 6 problems derived from each group.

Problem solution	P1: D	P2: E	P3: E	P4: B	P5: A	P6: B
Group 1	D	E	D	B	C	D
Group 2	B	E	E	B	D	B
Group 3	D	E	A	B	A	B, C

Even though some students gave an ambiguous reason, through the whole group's discussion, they understood clear reasons why their answer was correct. In Problems 1 and 6, two groups had chosen a correct solution; and in Problems 3 and 5, only one group had a correct solution. In these cases, at first, the teacher guided students to distinguish shared opinions from different opinions among three small groups. The teacher's role was to help students identify the issues that need to be discussed, not telling students the correct answer. In some cases, it may be plausible to argue that more than one answer is correct. The goal is to have students reflect deeply on their own, and others', thinking, even if some disagreement still remains.

For example, in problem 1, the teacher asked why all groups had eliminated items A, C and E. After that, teacher asked each group to explain why other group's opinions were incorrect or explain why their group's opinion was correct. The following is a partial protocol of whole-class discussion of Problem 3. Here, students came to appreciate other issues and different ideas, and to evaluate other students' thinking. Important ideas that are expected to foster understanding of modeling are shown by the italic sections below.

5.1 Partial Protocol of Problem 3

Teacher: Each group has different answer in this problem; A, D and E. All groups rejected B and C. Why did you not select B and C?
Group 2: In answer B, *only if we set a small range of x, it is possible to represent this phenomena with "y = ax + b"*. Therefore, B is incorrect.
Teacher: How about answer C?
Group 1: *Function "y = ax² + bx + c" will increase when x is over a certain value. So, it is necessary to set the range of x.* C is incorrect.
Teacher: Let's eliminate other group's answers.
Group 3: We think that D is incorrect. Because when "a" of "y = ax² + bx + c" is negative number, the value of "y" does not grow up when "x" is getting larger. Therefore, the description "the temperature will go up when the time goes by" is not correct. Answer D is wrong.
Group 1: We have a question. How do you transform the function "y = ae⁻ᵇˣ " in answer E.
Group 2: The transformation means "+c". Namely *the function becomes "y= ae⁻ᵇˣ + c". If we set the adequate value of "c", the temperature of coffee will converge to a certain temperature (room temperature) that fits to a real situation.*
Group 1: Can you show this by using graphic calculator?
Teacher: There is a big screen in front. Let me show you the graph of "y = ae⁻ᵇˣ + c". (Presentation was made by the teacher)

All: Great!

Teacher: How about A and D?

Group 2: In answer D, there is no description such as "setting up the range of x". *Only if we set the range of "x", it becomes possible to represent the phenomena with "y = ax² + bx + c".* So D is incorrect. Further, *only if we set the range of "x", it is also possible to represent the phenomenon with "y= ax + b".* So answer A is also incorrect.

Teacher: We should pay attention not only the shape of functions but also the range of "x". We need to understand how the shape of function will change corresponding to changes of coefficient of the function.

As shown in this and other protocols, meaningful discussions took place between groups, and students were able to elicit important ideas that promote modelling. However, on the other hand, there were two limitations of students' discussion. First, a few students tended to consider only how to eliminate the items of multi-choice answers. As a result, students needed to be reminded that solving a real-world problem is not the same as checking and eliminating incorrect items of multi-choice answers. Second, these multi-choice modelling problems focus on a particular stage of modelling process. For example, in the following protocol of Problem 4, where the aim is generating and selecting variables, students only discussed whether each item was important or not. Students' thinking was important in generating and selecting variables (see italicized section). However, even in the stage of generating and selecting variables, we would like to expect students to clarify the meaning of given variables by envisaging what relations are generated between variables. All students failed to clarify the meaning of given variables (see bold section below) when they were solving the multi-choice problem.

5.2 Partial Protocol of Problem 4

Students first explained why some "variables" are irrelevant:

Group 1: (a) "The average age of customers" is irrelevant.

Teacher: Why do you think so?

Group 3: *The aim of this problem is to estimate how many checkouts should be operated. So, if a customer was a child or grandfather, the age is irrelevant.*

Group 1: (g) "The pay rate for checkout girls" is not important.

Group 3: *Even though checkout girl earned 1000 yen per hour or 800 yen per hour, it has no bearing on the number of checkouts.*

Group 1: (h) "The proportion of customers using baskets rather than trolleys" is also not important. *Some customers who buy small number of items choose to use a trolley. So (h) is not related to the number of items bought.*

Teacher: Are there any more ideas?

Group 3: (b) "The average bill size" is not important. *Even if the bills of two customers were same, the number of items bought could be quite different. It takes more time when the number of items bought is larger.*

Group 2: (i) "The working hours" is not related. *After determining the number of checkouts, the working hours and number of checkout girls are determined.*

Teacher: By eliminating the incorrect items; (a), (b), (g), (h) and (i), each group could select the correct answer. Namely answer B: (c), (d), (e), (f) is correct. **Is there any one who considered the relation between selected four variables?**

All: **No.**

Teacher: Let's consider the relation between four variables.

Group 2: **What is the meaning of (c) "the efficiency of the checkout girls"?**

Group 3: Let's consider that (c) means the time (seconds) to check one good. Let's ignore the time to put everything in a bag. The unit is "seconds per one good".

Group 1: **How about the number of checkouts?**

Group 2: We set the number of checkouts as "x".

Group 1: **How about (e) "number of customers in the store"? Some are selecting and taking goods into a basket and some are waiting for a checkout.**

Group 3: We should set the meaning of (e) as the number of customers who are waiting for checkouts. The number of customers who select and take goods into basket is not concerned with the problem.

The teacher then summarized these five variables on the blackboard.

Teacher: (Several minutes later) Let's explain the relation between 5 variables.

Group 3: ef/cx<d

Group 2: cef/d<x

Group 1: Same as Group 2.

Teacher: Are these two answers same or different? Which is correct?

Group 1: Different. The location of c is different.

Group 1: We considered it by substituting the concrete number in the formula.

At first, the meaning of "cf", namely multiplying "efficiency of checkout girl (seconds per one good)" by "average number of items bought" is "the time that one customer passes the checkout".

Next, the product by multiplying "cf" by "e: the number of customers who are waiting for checkouts" means "the time to wait for the checkout for the last person of all of customers". Then divide "cef" by "d": maximum reasonable queuing time".

Teacher: Do you understand the meaning of dividing "cef" by "d".

Group 3: No. ....

Group 2: *When selecting variables, we should check the meaning of variables, and estimate the relation between variables.*

Teacher: Even though you can select important variables, this has no meaning if you don't formulate a relation between selected variables.

5.3 Students' Writings on Key Questions Before and After Teaching

"Three big questions" were posed at the start and the end of the study in order to gauge to any transfiguration or shift in students' thinking through empirical teaching. Regarding question 1, all students wrote that it was still difficult for them to make a mathematical model before and after the empirical teaching. However, some students were able to explain more clearly why it is difficult to make a mathematical model after the teaching, as the following two samples of one student's writing before and after the teaching show:

[Writing **before** the teaching] I think it is difficult. I cannot image how we can express a real-world situation with numbers. Further, I don't know whether or not we can formulate a real-world problem into a mathematical problem.

[Writing **after** the teaching] I think it is difficult. Because we need to understand what is essentially required in a real-world situation. *Further, we need to take account whether or not there are any implicit assumptions.* Even though we thought that the answer is completely correct, it might be incorrect because of something that is ignored in formulating a problem.

Regarding questions 2 and 3, we found that most students made progress in their writing both before the teaching and after the teaching. However, two students seemed to confuse the process of modelling process with the process of solving multi-choice problems.

On the other hand, one student seemed to make considerable progress. This student appeared to have a vague holistic image about the modelling process before the teaching. The following are this student's writings before and after teaching:

[Writing **before** the teaching] Process to solve a real-world problem
  (1) Judging whether or not a given real-world problem will be solved by using mathematical knowledge.
  (2) If it is possible to solve it by using mathematical knowledge, clarifying what we want to know as the result and building up the formula or graph *by taking account of all of things that will be related to the problem*.
  (3) Checking the answer and considering whether or not there is any lacking point.
  (4) Solving a real-world problem by using the formula or graph.

[Writing **after** the teaching] Process to solve a real-world problem
  (1) Finding out a real-world problem.
  (2) Clarifying what we really want to know in a real-world situation
  (3) *Generating mathematical factors that will be related with the problem situation as many as possible.*
  (4) *Eliminating the factors that will be not concerned with what we really want to know.*
  (5) Generating a graph or mathematical formula from the remained essential factors.
  (6-1) **In the case that it is possible to generating a graph or mathematical formula => Go to (7)**
    (7) Solving a real-world problem by interpreting a graph or mathematical formula.
    (8-1) **Solving a real-world problem => Finished!**
    (8-2) **Not solving a real-world problem => Back to (3)**
  (6-2) **In that case it is impossible to generating a graph or mathematical formula => Back to (3)**

This student understood the important idea of generating and selecting variables (see italicized sections above), and seemed to appreciate the importance of two key ideas, namely, "Justification of an existing model" and "Eliminating errors". As a result of solving multi-choice modelling problems, he modified his previous description into a more reflexive process (see bold sections above).

6. CONCLUSIONS AND FUTURE DIRECTIONS FOR RESEARCH

When combined with group discussion and careful teacher direction, the use of multi-choice modelling tasks, such as prepared by Haines and Crouch (2001), was quite effective in helping to shape students' thinking about key features and stages of mathematical modelling in two relatively concentrated sessions. The problems in this pilot study were accessible and challenging to senior high school students of mathematics who had no prior teaching relating to mathematical modelling.

Having well designed and tested tasks on hand for teachers to use allowed students within a relatively short period of time to come to terms with some important aspects of mathematical modelling. These multi-choice modelling tasks are, of course, no substitute for actually carrying out extended piece of work involving mathematical modelling. But in many countries time available to carry out such extended tasks is often hard to find in the crowded high school curriculum. Fully elaborated modelling tasks also present challenges for many teachers. On the other hand, multi-choice tasks are familiar to teachers and students can provide students with an introduction to mathematical modeling. The teacher's role is crucial in keeping students focused on the larger picture. Group discussion and whole-class discussion can then help clarify shared opinions and different opinions from among small groups, and by drawing attention to conflicting opinions among small groups.

In this pilot case study, the teacher organized whole-class discussion so that students could discuss shared ideas at first, then asked them to consider conflicting opinions from small groups. In some cases, group discussion was able to bring all students to a correct understanding of the problem. In other cases, by criticizing others' ideas and by listening to criticism, students realized that their explanation was still ambiguous or unconvincing. Sometimes, it was necessary for the teacher to probe students' thinking further so that conflicting or opposing ideas were exposed more clearly.

Several issues need to be investigated more thoroughly in a future study. What initial ideas about mathematical modelling dispose students to learn from a teaching approach based on questions as used in this study? Those who started with some vague holistic image of modelling process, seemed more likely to clarify and develop their thinking by tackling problems that focus on a certain stage of modelling process. Others appeared to focus too closely on the task of tackling multi-choice problems, and did not appear to advance much beyond this.

In future, a larger sample of students will be needed. More detailed one-to-one interviewing will be needed to ascertain what predisposes some students to gain more from this type of approach than others. To show growth between pre-test and post-test, we will continue to use items whose comparability (in terms of specific phase of the modelling process and item difficulty) has already been established. Further exploration of partial credit scoring of items should also be considered.

REFERENCES

Haines, C., Crouch, R. and Davis, J. (2001) Understanding Students' Modelling Skills. In J.P. Matos, W. Blum, K. Houston and S.P. Carriera (eds.) *Modelling and mathematics education: ICTMA 9: Applications in science and technology.* Chichester: Horwood Publishing, 366-380.

Ikeda, T. and Stephens, M. (2001) The Effects of Students' Discussion in Mathematics Modelling. In J.P. Matos, W. Blum, K. Houston and S.P. Carriera (eds.) *Modelling and mathematics education: ICTMA 9: Applications in science and technology.* Chichester: Horwood Publishing, 381-390.

Nagasaki, A. (2001) *Relation among mathematics and society and culture.* Tokyo: Meijitosho Publishing (in Japanese).

Treilibs, V., Burkhardt, H. and Low, B (1980) *Formulation Processes in Mathematical Modelling.* Nottingham: Shell Centre for Mathematical Education.

3.3

MODELLING AND MODELLING COMPETENCIES IN SCHOOL

Gabriele Kaiser
University of Hamburg, Germany

Abstract–*The paper reports on university seminars of mathematical modelling at school. The seminars were held by the departments of mathematics and mathematics education together with various schools in Hamburg. Future teachers and students in upper secondary level carried out together modelling examples either in ordinary lessons or special afternoon groups. They tackled authentic problems proposed by applied mathematicians working in industry. In this paper we consider one of these examples in more detail and describe students' attempts at solving the problem. The paper also deals with modelling competencies and their development. These competencies are evaluated by means of a test that was applied to all participating students at the beginning and at the end of the seminar. The results of these tests are reported and they indicate which competencies were fostered through modelling examples and which were not.*

1. THEORETICAL FRAMEWORK

The relevance of promoting applications and mathematical modelling in schools is widely accepted. For example, as a goal of mathematics education the PISA study emphasises the development of students' capacities to use mathematics in their present life as well as in their future lives. It means that students should understand the relevance of mathematics in everyday life, to our environment and in sciences. In addition to that, students should acquire competencies which enable them to solve real mathematical problems from the above-mentioned areas. Within the actual discussion, it is stressed that it is not sufficient at all to deal with modelling examples in lessons but that the stimulation of modelling competencies through self-initiative is of central importance (see Maaß 2006).

Furthermore, the current discussion differentiates between modelling competencies and modelling abilities. Modelling competencies include, in contrast to modelling abilities, not only the ability but also the willingness to work out problems, with mathematical aspects taken from reality, through mathematical modelling. Efforts to develop detailed and concrete descriptions are still the topic of current controversial debate. For instance, Maaß (2006, p116ff, p139) gives in her elaborate empirical study a list of modelling competencies. In lieu of this study, and from my own unpublished research results, the following variety of competencies seems to be necessary in order to carry out modelling processes autonomously:

- Competency to solve at least partly a real-world problem through mathematical description (that is, model) developed by oneself;
- Competency to reflect about the modelling process by activating meta-knowledge about modelling processes;
- Insight into the connections between mathematics and reality;
- Insight into the perception of mathematics as process and not merely as product;
- Insight into the subjectivity of mathematical modelling, that is, the dependence of modelling processes on the aims and the available mathematical tools and mathematical competencies;
- Social competencies such as the ability to work in a group and to communicate about and via mathematics.

These descriptions can be put into concrete terms if a modelling process and its phases are taken more into consideration. For several years now, within the mathematics didactical discussion, the following ideal-typical description of modelling processes has become widely accepted. A real-world situation is the process' starting point. Then the situation is idealised, that is, simplified or structured in order to get a real-world model. Then this real-world model is mathematised, that is, translated into mathematics so that it leads to a mathematical model of the original situation. Mathematical considerations during the mathematical model produce mathematical results which must be reinterpreted into the real situation. The adequacy of the results must be checked, that is, validated. In the case of an unsatisfactory problem solution which happens quite frequently in practice, this process must be iterated (for details see for example Blum, 1996).

Based on this description and the phases described, the following part competencies can be distinguished:

- competencies to understand real-world problems and to construct a reality model:
- competencies to create a mathematical model out of a real-world model:
- competencies to solve mathematical problems within a mathematical model:
- competency to interpret mathematical results in a real-world model or a real situation
- competency to challenge solutions and, if necessary, to carry out another modelling process

Maaß emphasises in her elaborate study the necessity of meta-cognition for modelling which means to reflect about one's own thinking or to control one's own thinking (Maaß 2004, p34ff). And this is what especially distinguishes experts from beginners: beginners tend to produce assumptions for modelling without any plan and without regard to the involved complexity of the models; on the other hand, experts control their solving strategies and therefore reach their aim faster. For that reason, Maaß emphasises, for modelling-oriented mathematics teaching, the necessity of teaching meta-knowledge about modelling processes, so-called declarative meta-cognition, as well insight into the necessity of observing modelling processes, so-called procedural meta-cognition. Altogether, according to Maaß (2004, p173f), modelling competencies comprise the following competencies:

- partial competencies for conducting the single phases of a modelling process;
- meta-cognitive modelling competencies;

- competencies to structure real-world problems;
- argumentation competencies;
- competencies to judge a solution.

In order to promote a modelling-based understanding of mathematics and to develop competencies for carrying out modelling processes at school, it seems to be necessary to teach such competencies to future teachers during the course of their studies. Future teachers have to become familiar with modelling examples, because if not, later on, the barriers for the integration of such examples into lessons will be too high (see for example Kaiser & Maaß, 2006).

In the following part, I will report about a university course with future teachers through which these students could acquire competencies for implementing modelling processes in their future teaching, and through which their students could acquire competencies for carrying out modelling processes.

2. DESCRIPTION OF MODELLING COURSES IN GERMAN SCHOOLS

The project 'Mathematical Modelling in School' was established in 2000 by the Department of Mathematics (Jens Struckmeier and Claus Peter Ortlieb) in co-operation with Didactics of Mathematics at the Department of Education at the University of Hamburg. This university course project, with future teachers for upper secondary level teaching, aims to establish a conjunction between university and school as well as between mathematics and didactics of mathematics. Student groups from upper secondary level (aged 16-18 years) supervised by the future teachers are the focus of the course. Each group works independently on one modelling example, in the regular lessons or in separate after school working groups.

The main objective of the course is to change the academic curriculum of the Department of Mathematics and that of the Didactics of Mathematics, so that in future mathematical modelling and associated teaching experiences will play a central role. This project will enable future teachers to implement modelling processes in mathematics teaching in their professional work.

It is hoped that the participating students will acquire competencies to enable them to carry out modelling examples independently, that is, the ability to extract mathematical questions from the given problem fields and to develop autonomously the solutions of real-world problems. It is not the purpose of this project to provide a comprehensive overview about relevant fields of application of mathematics. Furthermore, it is hoped that it will enable students to work purposefully on their own in open problem situations, experiencing the feelings of uncertainty and insecurity which are characteristics of real applications of mathematics in everyday life and sciences. An overarching goal is that students' experiences with mathematics and their mathematical world views or mathematical beliefs are broadened. This kind of approach can be described as holistic approach, using the terminology of Blomhoj and Jensen (2003, p128f), that is, a whole-scale mathematical modelling process is carried out covering all modelling competencies described above.

Each course extends over a longer period, that is, over one or two semesters with the following structure: After a short introduction into questions of teaching modelling, in a start-up lecture an authentic real-life problem is presented by an applied mathematician. This problem will be dealt with during, more or less, three months within the framework of school lessons. First, results will be presented by students at the end of the winter semester. A continuation during the summer semester is offered, but not all schools or future teachers will participate in that continuation in which a further real-world modelling problem will be worked on. Simultaneously, a university course is taken where the students' solution attempts, problems and experiences are discussed. For example, the following modelling problems were treated: share price forecast, prediction of fishing quotas, optimal position of rescue helicopters in South Tyrol, radio-therapy planning for cancer patients, identification of fingerprints, pricing for internet booking of flights.

The central feature of this project is the usage of authentic examples, that is, most of the problems are proposed by applied mathematicians who work in industry and who have met or tackled this problem within their working environment. The problems are only little simplified and often no solution is known, neither to us as organisers of the project nor to the problem poser. For example, the problem of unique identification of fingerprint is not yet solved satisfactorily and the machines are not working on a reliable basis, which creates a strong contrast to the widespread usage of machines for fingerprint identification in many places. Quite often, only a problematic situation is described and the students have to determine or develop a question that can be solved. The development and description of the problem to be tackled is the most important and most ambitious part of a modelling process, mostly neglected in ordinary mathematics lessons. Another feature of the problems is their openness which means that various problem definitions and solutions are possible in dependence of the norms of the modellers.

The teaching-and-learning-process is characterised as autonomous, self-controlled learning, that is, that the students decide upon their ways of tackling the problem and no fast intervention, neither by the teachers involved nor the future teachers, takes place (or should take place). The future teachers and the teachers are expected to offer no more than just assistance, if mathematical means are needed, or if the students are heading into a cul-de-sac. With this kind of teaching the students experience long phases of helplessness and insecurity which is an important aspect of modelling and a necessary phase within a modelling process.

In the following paragraph, one modelling example will be described in detail in order to show the wide variety of solutions devised by the students.

3. A MODELLLING EXAMPLE: PRICE CALCULATION BY AIR BERLIN

The problem, how the low-price airline, Air Berlin, makes its pricing was presented by one participating applied mathematician. Air Berlin sells its flights predominantly by an online booking system via internet, and the prices for the various destinations are not fixed. For each flight the prices are indicated separately and they change very often. The question arises as to how Air Berlin determines its prices. The participating students were asked to develop an adequate description based on the prices announced on the internet.

The problem proved to be challenging for the participating student groups because, on the one hand the price algorithms of Air Berlin were not known by them and, on the other hand, various attempts are possible. Thus, each group developed a different attempt. However, in the following part, I will restrict my description to the attempt of the Gymnasium Harksheide course in Norderstedt because this attempt has to be regarded as the most mature one. The course is an advanced mathematics course, in year 12 (age 17-18) with 11 participating students.

At the beginning, the students collected data from the internet and observed the changes of prices. Doing that, they noticed that the changes in price obviously happened more or less arbitrarily. Some of the observed flight prices did not change over a longer period of time before; then suddenly the prices doubled. Other prices did not change at all until the end of the observation period, while for some other flights additional charges, such as the fuel surcharge, were added. The students identified the following factors that influence the development of prices: date/departure time of the flight, date of booking (referring to the flight-date), number of free seats, place of departure and destination, time of departure, meaning the day and the time of the day, number of connected bookings, flight number, and market situation.

After that, the group divided itself into two sub-groups: The first group worked on the amount of prices in relation to the number of remaining free seats for certain flights. The second group dealt with the amount of prices in correlation to the time of booking. For flights inside Germany a model about the rise in prices in relation to time left until the departure could be developed, based on a step-wise increases. For international flights no such model could be reconstructed. Likewise, a correlation of prices to the number of remaining free seats also could not be reconstructed from the data. Because of these big problems they faced with the development of a model to determine the flight prices, the students of this group decided to develop a model by their own.

For this, the group split again into two sub-groups and they followed two different attempts: The first group decided to describe the development of flight prices by means of an exponential function, due to experimental considerations with real values. Thus, as model for the development of prices this group developed the function $f(x) = e^{cx} + b$ with b as initial price and c as description of price behaviour (meaning a steep or slow rise). For the determination of the parameters, the students, assisted by the future teachers, referred to the mean value theorem of integral calculus, and for determining the price behaviour, they developed the following model: $f(x) = \int_0^t (e^{cx} + b - 1)dx = at$, in which a represents the average price of a flight. The factor -1 was apparently introduced by the students in order to adapt their description to the data.

The second sub-group agreed on a description of price trends starting 30 days before departure by means of the exponential function $f(t) = \text{basic price} \cdot a^{30-t}$, with the assumption that the prices vary between 25€ and 125€ and that no flight shall cost more than 300€. As growth factor a, based on the collected data, the group experimented with factor 1.024 and 1.03. Then, this attempt was modified further and generalised by taking the amount of remaining free seats into consideration by inserting $m(10-s)$ as additive factor with s as number of remaining free seats and m as not yet fixed lump sum. The basic price was fixed depending on flight distance for which they referred partly to the real flight distances of Air Berlin and further assumptions with $\frac{z}{16}$ = basic price , where z is the distance of the flight in km.

Both attempts were carried out simultaneously with two modifications:

- Consideration of dependence on season, modelled by means of a cosine function: $0.25 \cos\left(\dfrac{2\pi c}{100}\right) + 1.25$, where c is a factor depending on the time of flight which was determined by analysing older data
- Rounding up to the following decade

The different meanings of c was not realised by the students and did not cause major difficulties in the further development of the study.

The first sub-group produced as model for the development of prices:

$$f(c,t) = (0.25 \cos\left(\frac{2\pi c}{100}\right) + 1.25)\,(\int_0^t e^{cx} + b - 1)dx = at$$

The second sub-group produced as model for the development of prices:

$$f(c,t,s) = (0.25 \cos\left(\frac{2\pi c}{100}\right) + 1.25)(\frac{z}{16}a^{30-t} + m\,(10-s))$$

The students did not use the notations with several variables and it is unclear whether they realised the problems resulting of their model containing more than one variable or not.

Then, both models of price behaviour were compared from which the following became obvious: model 1 regards the time period from the beginning when the prices were offered first until the time of departure with an exponential rise in prices, while model 2 regards a constant price with a fir-tree-like increase over the last 30 days. Because in model 1 factor c determines the steepness of the increase of the prices, both attempts led to quite similar results which are consistent with reality.

The students presented their modelling in front of the other participating classes and they were altogether satisfied with their solution, despite some phases of discouragement and frustration (for details see Kaiser & Schwarz, 2006).

4. DEVELOPMENT OF MODELLING COMPETENCIES

As stated at the beginning, amongst other goals the modelling project is aimed at fostering modelling competencies of students. This brings up the question how far modelling competencies can be fostered by this project. Therefore, in the course of the last modelling round from autumn 2004 until spring or summer 2005 an evaluation concerning the development of modelling competencies should be carried out. As the examples discussed here are rather complex and the students had to work autonomously, a test was taken which is well known from the modelling discussion and was created especially for students from university modelling courses. The test used was originally created by Haines, Crouch & Davis (2001) and then developed further by Houston & Neill (2003) as well as Izard and others (2003). The test seemed to be appropriate, because it measures competencies along the different phases of the modelling process, although it does not consider the mentioned meta-cognitive competencies. The test concentrates on the following competencies:

1. Making simplifying assumption concerning the real-world problem;

2. Clarifying the goal of the real model;
3. Formulating a precise problem statement;
4. Assigning variables, parameters, and constants in a model on the basis of a sound understanding of the model and the situation;
5. Formulating mathematical statements that describe the problem to be addressed;
6. Selecting a model;
7. Using graphical representations;
8. Relating back to the real situation and interpreting the solution in a real-world context.

Each of the competencies described above are tested through one item which are constructed in a multiple-choice format with a partial credit system. On the whole, the test proved to be not too demanding for our test cohort apart from the item on selection a model. Each test comprised 8 items, and for each item 0-2 marks were given; thus, the maximum number of marks was 16. The test lasted 20 minutes. The test and two parallel versions were used three times, at the beginning, after the first modelling unit (three and a half months) and after the second modelling unit (after three further months).

I will concentrate on the development of competencies during the first test run because after the first modelling unit a bigger group of students as well as some of the participating future teachers did not share in the project anymore. The reasons were in most cases due to a lack of time caused by up-coming external examinations. In addition, lack of continuity as a consequence of many holy days and school exams in the second run during the summer semester was clearly reflected in the results of the tests.

About the results of the tests after the first modelling unit:

57 students from 10 different courses which some of them only with few students, participated in both tests. Altogether, 132 students from 11 courses took part in the project. However due to a lack of time, in one course the test was not performed, while in other courses the test were conducted on a voluntary base which explains the low rate of return. 30% of the tested students were from A-level courses of year 12, 61% from enrichment courses of year 11, 9% from after school activity groups of year 10, 11, 12. 67% of the tested persons were male, 33% female.

First, we only look at the results of the two first tests which were carried out at the beginning of the project and at the end of the first modelling unit. On average, 8.4 marks were reached in the pre-test and 9.6 marks in the post-test. The students achieved the half of the maximum number of marks already at the beginning which shows on the one hand that the test was not too difficult and, on the other hand that, compared to the results of the tests with undergraduate students, it is a remarkably high result. An increase of 1.2 marks proves a significant progress after the end of the first modelling unit.

If both sexes are regarded at separately, we see that the male students receive slightly higher results as the female students already at the beginning (8.6 to 7.8) and they also made better progress (1.3 to 0.9). The gender differences are consistent with results from other studies that used the same test, but the small number of students has to be considered.

Regarding the results concerning the competencies listed above, the following results, as displayed in Figure 1, were received:

First, great differences between the various competence areas become obvious, in particular the students experience specific difficulties when clarifying the goal of modelling processes (A2) as well as selecting a suitable model (A6). The problems with the last item arise due to insufficient mathematical knowledge of the students from year 11. Problems with clarifying the goals of modelling are also reported by Houston and Neill (2003) in their study.

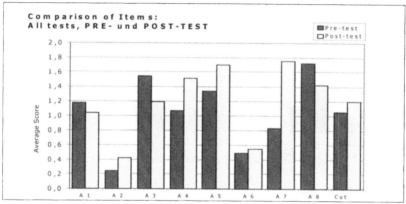

Figure 1. Comparison per item after the first modelling unit.

If one compares these results to results reported in other studies, the students' progress of development of competencies is highly gratifying and promising.

About the results of the tests after the second modelling unit:

After the first modelling unit the number of participating students decreased due to organisational reasons to 75 students in 7 courses. If only those students are taken into account who participated in all three tests, the number decreased again to 27, consisting of 37% female and 63% male students. The results of the pre-test and the first post-test hardly changed by the reduction of the sample. As shown in Figure 2 after the second modelling unit, the students' results did not improve but deteriorated about 0.6 marks which is only the half of the original progress of performance. The gender differences did not change: So, 40% of the female students in contrast to 70% of the male students made progress.

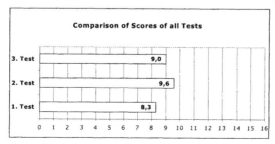

Figure 2. Global comparison for the three test times.

If one analyses the development in relation with all single competence areas, it comes out that the students made progress especially in two areas which were originally deficient, namely clarifying the goal and selecting the model. This is shown in Figure 3, where each component describes the same competency as in Figure 1.

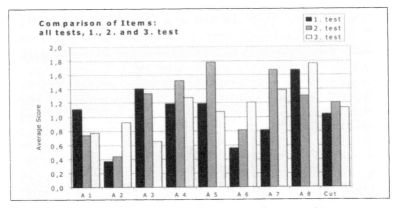

Figure 3. Comparison per item for the three test times.

The area "Formulating the problem" clearly shows loss of performance for which, of course, the equivalence of the test needs to be analysed. Altogether, a levelling of performance can be stated for all competence areas, positively and negatively as well.

5. SUMMARY AND PERSPECTIVES

To come to a close, it can be stated that the results of the tests suggest that competencies can be imparted through adequate modelling courses. However, the instability of progress indicates that longitudinal processes are necessary in order to provide a solid base for steady improvements. These results are in line with related research, for example, Haines & Crouch (2006), who emphasise in the light of the novice-expert-debate that at an intermediate level of expertise on complex tasks, people may get worse before they get better although in general they have improved. Apparently expertise tends to take time to develop and is not necessarily a stable process.

In addition, it must be noted that due to the orientation of the tests towards sub-competencies they did not test competencies which are necessary for a holistic way of conducting modelling processes. The students' presentations in front of all participants gave clear allusions about the fact that students acquired competencies especially in this area. Thus, the presentations showed clear structures for the modelling process following the phases of the modelling circle as it was shortly demonstrated at the beginning. Likewise, meta-cognitive competencies can be recognised from the way how students reflect the modelling process, especially about their assumptions and the assumptions' adequacy which can be well observed in the description of the students' modelling approaches in §3. The distinctions between atomistic and holistic modelling approaches according to Blomhoj and Jensen (2003) might lead to the assumption that students acquired competencies that are needed especially for a holistic way of doing modelling processes, such as meta-cognitive modelling competencies and argumentation competencies described by Maaß (2006). Partial competencies referring to single phases of the modelling circle might not have been fostered sufficiently or could not be fostered in the relatively short time. The balance between these different approaches as claimed by Blomhoj and Jensen (2003) has to be improved with the coming repetition of the project.

REFERENCES:

Blomhoj, M. and Jensen, T.H. (2003) Developing mathematical modelling competence: conceptual clarification and educational planning. *Teaching Mathematics and its Applications*, 22, 3, 123-139.

Blum, W. (1996) Anwendungsbezüge im Mathematikunterricht – Trends und Perspektiven. In G. Kadunz et al. (eds), *Trends und Perspektiven. Schriftenreihe Didaktik der Mathematik, Vol. 23*. Vienna: Hölder-Pichler-Tempsky, 15-38.

Haines, C.R., Crouch, R.M. and Davis, J. (2001) Understanding Students' Modelling Skills. In J.F. Matos, W. Blum, K. Houston & S.P. Carreira (eds), *Modelling and Mathematics Education: ICTMA9 Applications in Science and Technology*. Chichester: Horwood, 366-381.

Haines, C.R. and Crouch, R.M. (2006) Getting to grips with real-world contexts: Developing research in mathematical modelling. In M. Bosch (Ed.), *European Research in Mathematics Education IV. Proceedings of the Fourth Conference of the European Society for Research in Mathematics Education*, 17.-21.2.2005, Sant Feliu de Guixols, Spain. (*to appear*).

Houston, K. and Neill, N. (2003) Assessing modelling skills. . In S.J. Lamon, W.A. Parker and S.K. Houston (eds), *Mathematical Modelling: A way of life ICTMA11*. Chichester: Horwood Publishing, 155-164.

Izard, J., Haines, C.R., Crouch, R.M., Houston, S.K. and Neill, N. (2003). Assessing the impact of teaching mathematical modelling: Some implications. In S.J. Lamon, W.A. Parker & S.K. Houston (eds), *Mathematical Modelling: A way of life ICTMA11*. Chichester: Horwood Publishing, 165-177.

Kaiser, G. and Maaß, K. (2006). Modelling in Lower Secondary Mathematics Classrooms – Problems and Opportunities. In W. Blum, P. Galbraith, H.-W. Henn and M. Niss (eds), *Applications and Modelling in Mathematics Education*. New ICMI Studies Series no. 10. New York: Springer Academics. (*to appear*)

Kaiser, G. and Schwarz, B. (2006) Mathematical modelling as bridge between school and university. *Zentralblatt für Didaktik der Mathematik*, 38, 2,196-208.

Maaß, K. (2004) *Mathematisches Modellieren im Unterricht*. Hildesheim: Franzbecker.

Maaß, K. (2006) What are modelling competencies? *Zentralblatt für Didaktik der Mathematik*, 38, 2, 113-142.

3.4

EXPLORING UNIVERSITY STUDENTS' COMPETENCIES IN MODELLING

France Caron and Jacques Bélair
Université de Montréal, Canada

Abstract–*We report on the introduction of a term project in an undergraduate mathematical modelling course. Covering a wide range of real-world situations, the projects were deliberately open-ended. An exploratory study was conducted to examine which phases of the modelling process received greater attention and which competencies were displayed in each phase. Relating these observations to data on students' characteristics (educational background, modelling experience, interest in mathematics, and confidence with technology) helps identify trends suggested in previous studies, and points to possible improvements to be incorporated in future courses. The study also provides a new perspective on the relevance of examining the place given to applications, modelling and technology in university programs in mathematics.*

1. INTRODUCTION

In trying to identify the requirements for a *balanced* approach in teaching mathematical modelling (Blum et al., 2002) which would both develop students' modelling skills and enrich their knowledge and understanding of mathematics, an exploratory study was conducted in a mathematical modelling course at Université de Montréal. The course is offered to students in their final year of a mathematics, teaching of mathematics or double majors undergraduate program. With only multi-variable calculus and linear algebra as prerequisites, this makes for substantial diversity in knowledge and areas of interest among the student body.

The study revolved around the introduction of an open-ended modelling project as part of students' assignments for the course. While the main objective of such an activity was to develop the students' modelling skills, it was also used to explore connections between their modelling skills and their individual educational background. The study was conducted to improve the course in better addressing students' needs, and to identify possible implications for undergraduate mathematics programs.

2. COMPETENCIES IN MATHEMATICAL MODELLING

As a particular form of problem solving, mathematical modelling has been described as a multi-step and cyclic process (Blum et al., 2002). The number of cycles performed in the modelling process can be seen as a possible indicator of the complexity of the task and authenticity of the situation (Caron and Muller, 2005).

Trying to bridge mathematics and their applications in the teaching of modelling has led us to consider the notion of *competence*, both for defining the objectives of the course and for assessing students' performance in their modelling project. This notion of *competence* originates in industry where it captures the relationships between work performed and knowledge held by an individual. *Competencies* that distinguish individuals have been classified in three categories (De Terssac, 1996) :

➤   *communication skills*: to translate, represent, interpret what the context is, what is to be done and what has been done.

➤   *intervention skills* : to act upon a situation by using available knowledge and by transforming encountered situations into reusable knowledge;

➤   *evaluation skills* : to identify, choose and justify whatever is being or has been engaged into action.

These competencies all play a role in the modelling process and can be developed in a modelling course, especially through the inclusion of an extended project (Nyman and Berry, 2002). Since our experiment coincided with the first time a project was part of the course, we did not try to demonstrate how such an initiative helped develop these skills. We rather looked for students' differences in the competencies displayed and in the modelling paradigm adopted (*theoretical,* for understanding, or *empirical,* for predicting) (Maull and Berry, 2001), and tried to link these differences with individual characteristics and educational background. As students entered the course with different competencies, we tried to identify the areas worthy of attention, either in the project, the course or in the undergraduate mathematics core, to better address the development of students' modelling skills.

3. THE EXPERIMENT

Nine volunteers in a class of 18 students were recruited in the Fall semester of 2004. Data were collected with a 10-page questionnaire capturing the students' educational background (previous studies, courses taken in mathematics and related disciplines), and relation with mathematics, applications, modelling and technology. Work experience, experience with complex projects, and reading preferences of the participants were also documented, as was their appreciation of the interest and difficulty of the different components of the course and of the project.

In the final written reports on the modelling projects, we looked for the phases of modelling that received greater attention and assessed the competencies displayed in each of these phases.

We explored possible relationships between individual characteristics and competencies displayed in the modelling project, as well as the potential impact of these variables on the level of appreciation of the course and of the project.

4. RESULTS

The results are presented according to the phases of the modelling process.

Selecting a project

The two most popular topics were bungee jumping and alcohol absorption. In neither was a precise source of information provided, so that students had to build a model from scratch or find resources on their own. The popularity of such projects confirms the role played by the context in engaging students into a modelling activity. But interest in the context alone was not sufficient to tackle radically new problems: one student expressed a strong interest in football but dropped a project on modelling penalty shots and shoot-outs as he could not foresee the underlying mathematics. In this particular situation, evaluation skills seem to have acted as limits to the motivation to work on an interesting topic.

The projects of the nine participants are listed in Table 1 (students' names are fictitious). Wording of the problems is given in Appendix A. With all nine students rating their project as either "interesting" or "somewhat interesting", it is felt that they all found a topic that matched their own interests.

Topic	Team	Focus given to project	Members
Bungee Jumping	B	Mathematising through combination of elements found on the web.	Ben
	J	Exploring parameter effects with simulation and real data. Improving the model.	Jeanne Jules
Spreading of Minority Opinion	M	Exploring parameter effects with simulation. Interpreting the model.	Mathieu Michèle
	O	Describing and criticizing the model with respect to the situation.	Odile Olivier
Limits to Tree Height	T	Explaining the context and underlying principles; reporting the results.	Tanya Trish

Table 1. Participants, teams and projects.

Structuring the situation and defining the problem

In most real-world situations where mathematics is applied, it is very much in the dialog with non-mathematicians (who may be the ones asking for mathematical assistance) that the problem solver gains a clearer understanding of the situation, associated problem, purpose of the model and elements of structure to be retained in the simplification process. To compensate for the absence of such an expert and reflect the availability of existing models, most projects came with an initial reference, typically a paper from a scientific journal, but not of mathematics.

Understanding the situation thus required research and extended reading, tasks which are seldom part of undergraduate courses in mathematics but are typical of its application. Michèle admitted her dislike of reading and was the only one to rate as "very weak" the project's contribution to her learning of modelling: she appreciates in mathematics the "knowledge of a formula or a general method, applicable to all cases" as well as the "reuse in other disciplines of concepts and methods seen in mathematics". As the project did not fit that description, she played a passive role in the class presentation. This attitude corroborates previous observations (Caron, 2001) that the *communication* work required represents a problem for those claiming to like mathematics because it has always been less demanding in terms of reading.

For projects with a single reference, we observed a shift in the problem definition, from the initial description of the project to either the goal stated by the author of the paper or the objective of explaining the model(s) used in the paper.

For instance, Team T selected a rather complex project in which the objective was to design a model to compute a limit to tree heights in North American West Coast forests. The complexity of the context and the perceived "difficulty of finding complementary sources of information" (mentioned by Trish), however, led the team to present botanical concepts and principles of botany, physics and chemistry, with very little beyond the information already provided in the paper (Koch, 2004). Tanya, who shows a strong theoretical disposition, pointed out that the paper was hard to read, with only a partial presentation of the technical terms used. To add to the complexity of their task, neither English (in which the paper was written) nor French (in which the report was to be written) was the first language of these two students. Instead of trying to simplify the situation by staying at a level high enough to work with the main variables, they transformed their project into a diligent communication exercise.

Jules and Jeanne, both mathematical engineering majors, showed superior evaluation skills in simplifying and structuring the bungee jumping problem. After judging that they could not ignore friction, they developed two different models: one decomposed the motion into a free fall and the oscillations of a spring-mass system; the second one refined that description by bringing to zero the spring constant when the subject is at a height where the elastic cord does not exert a restoring force, and by using dry friction. They also incorporated data from a bungee cord manufacturer to validate and enrich their model. Their many courses in physics and their previous experience of complex modelling projects (spectral computation, stochastic models and genetic algorithms applied respectively to solve heat diffusion, finance and sequencing problems), may help explain such control over the modelling process.

On the same project but with a much more limited background in modelling and physics, Ben merely skimmed over elements gathered on the Internet (for example, maximum stretching factor for common elastic cords) and did not incorporate any of that information in the models he derived. He faithfully reproduced the wording of the original problem, but he did not organize his project in a way to solve it.

Teams M and O both used the contextual information provided in the paper and were the only teams to include additional sources of information. The application of concepts, principles and techniques from the physics of disorder to the study of a social phenomenon (the spreading of a minority opinion opposed to some reform) seemed to warrant such additional precautions. Yet, the familiarity with the context and the relative simplicity of the social concepts involved did not prevent these teams from moving to questioning and discussing the validity of the model.

Mathematising the problem

Mathematising a problem requires both evaluation skills to identify appropriate mathematical objects to be included in the model and communication skills to translate the structure of the situation and problem into mathematical language.

The initial reference for the project on maximum tree height gave four different one-variable regression models linking height to one of four variables: xylem pressure, leaf mass: area ratio, foliar carbon isotope composition, or photosynthetic rate per unit mass. It was expected that, using other information given in the paper and expressed qualitatively, students would try to combine these relationships into a single mathematical model in which the reduction of water potential with height would account for the reduced leaf expansion and photosynthesis, and consequently modify carbon balance. Making use of the growth rate information and looking for statistical data on natural disasters in the area could have allowed for estimate of the probability of trees reaching that limit.

Trish expressed feeling unprepared to build a new model. In addition, she may not have perceived the value of moving from an empirical to a theoretical paradigm, as her vision of modelling (shaped by her operations research background) entails "*transforming* a non-mathematical problem into something that can be *calculated*". With the four regression models already agreeing on their calculated limit for tree height, she may not have recognised the value of a single mathematical model to better understand the relationships between the different variables.

Olivier similarly describes modelling as "*transforming* a concrete situation into a system of mathematical equations and *solving* that system". His operations research background has also led him to look for precision as a crucial criterion for judging a model. Consequently, he expressed scepticism toward modelling collective human behaviour with physical laws, and he never got over the admittedly crude simplifications of the reference. Disappointed that the course had not taught him a set of techniques for creating new models, he marginally tried to improve the model from the paper, emphasizing in his presentation the flaws of the model.

For Odile, presenting an existing model was already a worthwhile exercise "to discover a new area of application of mathematical modelling, study a model in depth and understand it better". As her vision of mathematical modelling consists in "*translating* mathematically a phenomenon from observations of its behaviour", which she illustrated with the Fibonacci sequence for rabbit breeding, she appears less interested in the value of the results produced by such a translation. Making use in the writing of the report and the oral presentation of her relatively strong communication skills (observable in the questionnaire and the exam) was a way for her to work on her evaluation skills.

Mathieu's interest in modelling comes from a theoretical interest in both mathematics and its applications. An economics double major, he defines mathematical modelling as "using the available mathematical tools for the purpose of *explaining* a phenomenon in a rigorous fashion". His study of economics and finance seems to have led him to value the explanatory rather than predictive ability of a model. He did not appear bothered by the simplifications and restated clearly the purpose of the model: "this simple model cannot describe the reality faithfully, but it gives us a pretty good idea of a very complex phenomenon." Rather than trying to improve the *external coherence* of the model with the real situation it is supposed to represent (Johsua and Dupin, 1993), he made use of his evaluation skills to verify the *internal coherence* of the model by adding in the report proofs of consistency that were not present in the original paper.

With limited experience in modelling and a recent encounter with differential equations, Ben did not appear confident in mathematising the situation even if simple spring-mass systems had been covered in the course. His modelling exercise turned into an extensive search on the internet of models and solutions of related problems that he (unsuccessfully!) tried to combine and make look seamless.

With the same problem, Jules and Jeanne displayed a clear mastery of the mathematics required to model the situation. Their extensive background in physics seems to have given them a comfortable advantage in translating the situation into differential equations, as is often done in the *theoretical paradigm*.

Doing the mathematics

The mathematical work performed in this phase of modelling strongly depends on the intervention skills developed, either with pencil and paper or with technology.

Given the relative complexity of the different projects, the ability of students to use technology for exploring models and solving problems turned out to be a key factor in their deriving mathematical results; only teams using technology performed original calculations and displayed original graphics. Teams J and M discovered in the course the software suggested (Matlab), used it in the regular assignments and really became fluent with it in the course of the project. Mathieu and Jules both stated that the use of software was a particularly interesting aspect of the project.

Mathieu used Matlab to test group size effects in local debates on the value of the unstable fixed point (the cut-off point for the reform to pass) and on the time required for total polarization of opinion to occur. He particularly played with the proportion of groups with an even number of debaters as the model assumes that in the case of a tie, the group moves against the reform (from the principle of inertia).

For Jeanne and Jules, modelling is linked to technology. She describes modelling as "the simplification by approximation of concrete problems (for example, physics) or more abstract problems (for example, finance) to be able to design algorithms that allow computer treatment of these problems". He includes in his definition of modelling a computer version that corresponds to the "description of a sequence of tasks to be performed". Team J thus used Matlab in a variety of ways to work on their models. They first complemented their pen and paper resolution of their initial ODE model with graphical representations (position and velocity as functions of time, phase plane) of the piecewise function that they had found as solution to the alternating two-phase problem. For their more elaborate model, they used a Matlab ODE solver in a series of simulations. They even built a complete simulation program with the size and the weight of the jumper and the height of the bridge as inputs, to determine from that height, using a table provided by a bungee cord manufacturer, the recommended length for the cord. The intervention and communication skills they developed with the Matlab interface in this project have clearly opened for them new strategies of resolution, extended the classes of problems that they can now solve and possibly enhanced their evaluation skills.

Interestingly, all "active members" of these two teams (Jules, Jeanne and Mathieu) had described themselves as intermediate users of computer algebra systems (CAS). Conversely, the members of Teams T and O who all had reported a low level of familiarity with CAS also declared the use of specialized mathematical software as not relevant for their project. This seems to illustrate the results of Galbraith and Haines (1998) on how computing-mathematics interaction is strongly associated with confidence and motivation in computer use. Yet our study suggests refinements of these results: Olivier and Trish both reported a very strong experience in programming, and Olivier also rated himself as an expert in spreadsheets. The perceived relevance and consequent initiative of using technology for exploring and simulating a mathematical model thus appear to be linked specifically to the level at which technology has already been integrated into one's own practice of mathematics. For senior undergraduate students in mathematics, the level of familiarity with CAS appears to be a good indicator of that level of integration.

Coming back to the situation

Coming back to the situation requires a combination of communication skills (interpreting the results) and evaluation skills (validating the solution).

Team T did not have to "come back" to the situation as they almost never left it.

Ben, from Team B, tried to answer one part of the original problem, that is, how to minimize oscillations for the bungee jumper. Without an adequate model and any original work done on the differential equations, he resorted to interpreting the simple equation for the period of a non-damped spring-mass system.

With their many simulations and the graphical representation of the effect of the various parameters, Team J kept coming back to the problem, as can be observed from Figure 1, where the mapping of the consecutive 305 elements of their paper (sentences, equations, graphs) to the different stages of modelling shows frequent jumps from the Mathematical Results or Solution to the Situation or Problem.

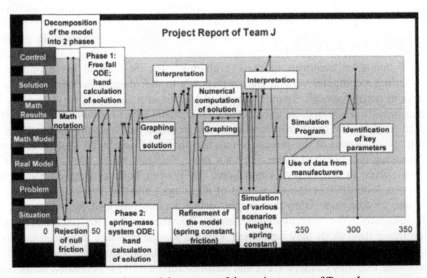

Figure 1. Sequential structure of the project report of Team J.

With comparable dedication in simulating various scenarios in the study of minority opinion spreading, Mathieu moved further away from the reality the theoretical model was supposed to explain. Although he had first stated that such simple model could not describe reality faithfully, he naively concluded his report by referring to a recent election where the model could have helped predict the results, "had we known the distribution of group sizes for local debates". And for that, he added, "we would need to take into account knowledge produced by social psychology, sociology, etc." It thus appears that the time spent playing with the theoretical model made him forget its basic radical simplifications: the possibility for Mathieu of simulating with the model, originally built to "enlighten an essential feature of an otherwise very complex and multiple phenomenon" (Galam, 2002), appears to have made him believe in its predictive ability. Although he did refer to real situations at the end, it was always to show the power of the model or the minor adjustments required to improve its realism or extend its domain of application.

With no original mathematical work performed, Team O went in more depth on the critical analysis of the model. They first noted how hard was the incorporation of the effect of the media in the model; this seemed to them a major weakness given the massive influence media may have on election results. They also looked for ways to take into account the undecided. Finally, after presenting examples from Europe (provided in the paper) that agreed with the model, they tried to apply it to the two referendums held in Quebec. They noticed that in both cases, the minority supporting the reform had increased from a poll taken one year before the day of the election, which seemed to contradict the analysis performed on the model. But they went on to "forgive" the model, since its purpose "was not to predict the exact distribution of votes between the two options but to predict only whether or not the reform would pass". An attachment to the predictive ability of a model made them miss a valid opportunity to question its explanatory ability in this particular case.

Globally, it seems that the evaluation skills demonstrated in this phase depend on the understanding of both the real-life situation and the purpose of the model.

5. CONCLUSIONS

The result of our study points to a number of considerations for the course, in particular about the project, and how it could be fitted better into the curriculum.

The diversity of topics should be maintained to address the different interests of students. More time should be spent discussing the purpose of a model: this would help students clarify the expected outcome and benefits of each stage. Some modelling heuristics should be explicitly taught. Many of the choices needed to be made in the modelling process are not concerned with mathematical abilities per se; problem formulation and identification of appropriate hypotheses on the real-world situation are not well addressed by current curricula. This also suggests that a modelling project performed over a full semester would benefit from more frequent checkpoints than was performed in our study. While enabling sounder overall assessment of the project exercise, some of these checkpoints could take the form of students' sharing of their approaches and favour emergence of valid heuristics. Also, the global modelling process should be more frequently alluded to during the course. This might help students to better position their project and assess their work.

We could not ignore the differences in treatment between social and purely scientific subjects. In this sense, there are ill recognised virtues in considering social sciences topics for modelling exercises. Since many social contexts are familiar to most students, this class of topics seems to alleviate significantly the problem of the prerequisites, which completely paralyses modelling efforts in very technical disciplines, where the formulation of the real-world situation itself requires demanding a priori knowledge. In addition, it is often much easier to get access to social data (for example, via a national statistical agency). Finally, these social contexts seem to favour a critical analysis of the model(s) used and a broadening of the research.

More generally, the superior performance of students who had prior experience of applied mathematics and/or had developed a practice of mathematics that integrates technology suggests that the development of solid modelling skills can hardly be done in isolation in a single course. Applications and technology integration should be disseminated throughout the whole course offering (algebra, analysis, etc.) so as to introduce in context the appropriate phases of the modelling process and develop the sophisticated mathematical knowledge and competencies that are required to tackle the complexity of today's applications.

REFERENCES

Blum, W. et al. (2002) ICMI Study 14: Applications and modelling in mathematics education – Discussion Document. *Educational Studies in Mathematics*, 51(1-2), 149-171.

Caron, F. (2001) Effets de la formation fondamentale sur les compétences d'étudiants universitaires dans la résolution de problèmes de mathématiques appliquées. PhD Thesis, Université de Montréal.

Caron, F. and Muller, E. (2005) Integrating Applications and Modelling in Secondary and Post secondary Mathematics. In E. Simmt and B. Davis (eds) *Proceedings of the 28th Annual Meeting of the Canadian Mathematics Education Study Group*, 63-80.

De Terssac, G. (1996) Savoirs, compétences et travail. In J.M. Barbier (ed) *Savoirs théoriques et savoirs d'action*. Paris : Presses Universitaires de France, 223-247.

Johsua S. and Dupin J-J. (1993) *Introduction à la didactique des sciences et des mathématiques*. Paris: Presses Universitaires de France.

Galbraith, P. and Haines, C. (1998) Disentangling the Nexus: Attitudes to Mathematics and Technology in a Computer Learning Environment. *Educational Studies in Mathematics*, 36(3), 275-290.

Galam, S. (2002) Minority opinion spreading in random geometry. *European Physics Journal B*, 25, 403-406.

Koch, G.W. et al. (2004) The limits to tree height. *Nature*, 428, 851-854.

Maull, W. and Berry, J. (2001) An investigation of student working styles in a mathematical modelling activity, *Teaching Mathematics and its Applications*, 20(2), 78-88.

Nyman, M. and Berry, J. (2002) Developing transferable skills in undergraduate mathematics students through mathematical modelling. *Teaching Mathematics and its Applications*, 21(1), 29-45.

APPENDIX A - PROJECTS SELECTED BY THE PARTICIPANTS

1. Bungee Jumping: The "sport" of bungee jumping consists in jumping to the void from a bridge, with feet attached to the structure with an elastic cord. Build a mathematical model of such a jump. Take into account the mechanical properties of the device, as well as the subject's characteristics. You will have to answer the following questions (among others): What minimal height from the ground does the subject reach? What height does he reach with the first rebound? How can the number of oscillations performed by the subject be minimized? [Separate the motion into two phases, the free fall and the oscillations. Adjust the description of the elastic cord as a (linear) spring.]

2. Spreading of Minority Opinion: To what extent can "crowd effects" influence the results of an election? Can we quantify the propensity to adhere to the status quo? And if so, at what level does this adhesion gain in importance? [Ref. Galam, 2002]

3. Limits to Tree Height: What is the maximum height that can be reached by giant trees (sequoias, Douglas pines) from the West Coast of North America? Establish a model in the form of a system of finite-difference or differential equations, and evaluate the associated parameters. Given the longevity of such trees, to what extent can natural disaster limit the reaching of this maximum height? [Ref: Koch, 2004]

3.5

FACILITATING MIDDLE SECONDARY MODELLING COMPETENCIES

Peter Galbraith[1], Gloria Stillman[2], Jill Brown[2], Ian Edwards[3]
[1]University of Queensland, Brisbane, Australia
[2]University of Melbourne, Melbourne, Australia
[3]Luther College, Melbourne, Australia

Abstract–*One method of engaging secondary students in mathematics classes is through taking a real-world modelling approach to teaching, beginning in the middle years (Years 8/9) and developing students' modelling abilities into the senior secondary years. This chapter documents a case study in an Australian secondary school where such an approach is being established. The purpose is to provide an approach where Year 9 students experience a sense of success pivotal to their continued engagement with the modelling process. As the aim is that students have well-developed modelling skills within a classroom context by Year 11, students' handling of transitions between phases of the modelling process is of particular interest. From intensive analysis of student responses whilst undertaking two extended tasks, a framework is developed for identifying student blockages during transitions.*

1. BACKGROUND

This chapter reports aspects of a programme in which mathematical modelling is being introduced to Year 9 students in a Victorian secondary college as the teacher concerned perceives this as a way to engage middle years students in their classroom mathematics. Primarily our approach is driven by the desire to obtain a mathematically productive outcome for a problem with genuine real-world motivation. Sometimes, this is directly feasible, but at other times, it is more a "life-like" problem that is the subject of modelling. However, the solution must take seriously the context outside the classroom within which the problem is located, in evaluating its appropriateness and value.

The world outside the classroom is a swampy place as far as problem contexts are concerned, where real data are usually messy and the mathematical methods needed to deal with them sometimes involve improvisation. To suggest that such an important part of reality should be "cleaned up" before a problem is presented is, we believe, to destroy a significant element of its authenticity. Consequently, we include the appropriate use of technology as central to our purpose, and its integration with mathematics within the modelling process as creating essential challenges about which we need to know much more. (An exception to the assumption of technology use is where technology is not available. Then the mathematics must be amenable by hand methods alone, and data provided accordingly.)

Data for this chapter have been generated within the RITEMATHS project, an Australian Research Council funded project of the University of Melbourne and the University of Ballarat with six schools and Texas Instruments as industry partners. The fourth author is leading initiatives in one of the schools. In summary, matters we are particularly interested in exploring are represented by intersections between mathematical content, technology, and modelling (Figure 1), and this chapter elaborates aspects of how we have gone about teaching and researching these interests.

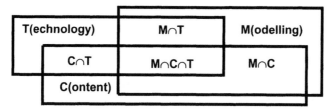

Figure 1. Interactions between modelling (M), mathematics content (C), and technology (T).

2. THE COMPONENTS

2.1 Mathematics Content

It has long been recognised that when learning to model it is unreasonable to expect students to access unaided, and apply mathematics that is at the frontiers of their experience and expertise or beyond, that is, problems need to be such that the mathematics required for solution is within the range of known and practised knowledge and techniques. It may not be clear however, just which mathematics is appropriate for the job at hand; such decisions *are* part of the requirements of the modelling process. This is a well-trodden area so we shall not elaborate further.

2.2 Technology

With respect to technology the situation is complicated, not only by the knowledge and facility required, but also by the skill and confidence with which students work with particular technologies. A *technology-rich teaching and learning environment* (TRTLE) affords new ways of engaging students in learning mathematics (Brown, 2005). The presence of electronic technologies in the classroom can fundamentally change how we think mathematically and what becomes privileged mathematical activity. Affordances are the offerings of such an environment for both facilitating and impeding learning, they are potential relationships between the teacher and/or student and the environment. Affordances need to be *perceived* and *acted* on by teachers and students alike in order to take advantage of the opportunities arising. This TRTLE includes opportunities for both teacher and students to enact a variety of affordances to support the learning of mathematics in ways well beyond those necessary simply for the production or checking of results. In the classroom reported here the teacher moves purposefully to integrate technology throughout the teaching and learning process with the goal of developing students' conceptual understanding and skills. In particular, the use of multiple representations, easily accessible with a graphing calculator, and tasks that are amenable to electronic technology, harness the opportunity for students to use technology to stimulate higher order thinking in the context of modelling real-world situations. Scaffolding of technical aspects is provided through demonstration or peer interaction when warranted, in an endeavour to ensure task cognitive demand is not increased by technology use (Stillman, Edwards & Brown, 2004).

Another factor of importance in technology use is the level of confidence and facility that students believe they possess in accessing and using technology. The existence of qualitatively different levels of confidence and expertise (see Goos, Galbraith, Renshaw, & Geiger, 2003) clearly has relevance for an introduction to mathematical modelling with technology at the middle years of secondary school. Students in the class studied came with a variety of prior experiences of scientific calculators and spreadsheets from previous classes. Year 9 was the first year that students were required to have their own laptop and TI-83 Plus graphing calculator, and the first time they used graphing calculators.

2.3 The Modelling Process

Various versions of the diagram in Figure 2 exist (for example, Edwards & Hamson, 1996) and most may be recognised as descendants from one originally used by the Open University (UK). It represents both a description of a dynamic iterative process, and a scaffolding infrastructure to help beginning modellers through stages of what can initially appear to be a demanding and unfamiliar approach to problem solving. Some such framework is desirable to support a modelling initiative, whether overtly as in Figure 2 or as a means to structure effective sequences of activities, both mathematical and pedagogical. The arrows represent fundamental transitions that depict the dynamic nature of the process, and these are associated with some of the most demanding phases of the modelling process, particularly, we would suggest the specification of a solvable mathematical problem from a messy real-world context, and its formulation in a way that will lead to an appropriate mathematical solution. In this chapter we are interested particularly in these and other transitions between phases as students engage with their first experiences of modelling activity.

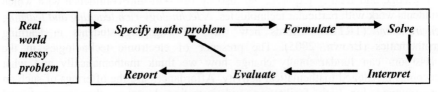

Figure 2. Modelling Process Chart.

The research being undertaken is part of a design experiment (Collins, Joseph, & Bielaczyc, 2004), which was beginning its second cycle at the time of data collection. The research focus has been fashioned by the nature of the program, and beliefs about how central elements of interest are best captured and analysed. A classroom in which mathematical modelling is being enacted is a varied and unpredictable place. There is intense activity, fallow time when impasses emerge, and spontaneous and unforeseen actions by students engaging with new material and challenges. Such occurrences trigger at times unplanned interventions by teachers, who grasp moments they could not themselves have envisaged, to capture, extend or clarify learning opportunities that have suddenly emerged. Such a culture is central to the process of teaching mathematical modelling skills, where successive implementations even by the same teacher vary substantially in their detail. This is to be celebrated even though it renders efforts to conduct controlled experiments highly questionable. Hence our focus is located elsewhere, at the level of individuals learning and applying modelling skills in a TRTLE. In particular we aim to learn more about the critical points that represent transitions between phases in the solution process. These, represented by arrows in Figure 2, refer respectively to movements from:

1. Real-world problem statement → Specification of a mathematical problem;
2. Mathematical problem statement → Formulation of an approach to solution;
3. Formulated approach → Complete solution of mathematics;
4. Mathematical solution → Interpretation in the context of the model;
5. Consequences of interpretation → Evaluation of the model quality;
6. Evaluation of model → Second or higher review and refinement of model;
7. Evaluation of model → Production of a report and recommendations.

We focus on 1 - 4 specifically, noting that 1 and 2 have long been identified as among the most difficult phases of the entire process, acting as gatekeepers to problem access. We therefore examine in detail how students approach and perform in these areas of transition, while learning modelling in an environment characterised by the interactions portrayed in Figure 1.

3. THE PROBLEMS

Below we summarise aspects of two problems (Figures 3 & 5) that have formed a focus for the introduction of modelling to a class of 28 Year 9 students (11 male, 17 female, 14-15 year olds) at an independent school in Melbourne.

3.1 Orienteering - Cunning Running (CR)

In the Annual "KING OF THE COLLEGE" Orienteering event, competitors are asked to choose a course that will allow them to *run the shortest possible distance,* while *visiting a prescribed number of check point stations.* In one stage of the race, the runners enter the top gate of a field, and leave by the bottom gate. During the race across the field, *they must go to one of the stations* on the bottom fence. Runners claim a station by reaching there first. They remove the ribbon on the station to say it has been used, and other runners need to go elsewhere. There are 18 stations along the fence line at 10 metre intervals, the station closest to Corner A is 50 metres from Corner A, and the distances of the gates from the fence with the stations are marked on the map.

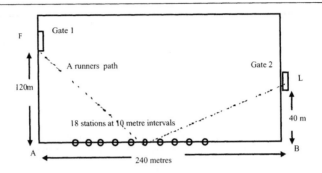

THE TASK. Investigate the changes in the total path length travelled as a runner goes from gate 1 to gate 2 after visiting one of the checkpoint stations. To which station would the runner travel, if they wished to travel the shortest path length?

For the station on the base line closest to Corner A calculate the total path length for the runner going Gate 1 – Station 1 – Gate 2. Use Lists in your calculator to find the total distance across the field as 18 runners in the event go to one of the stations, and draw a graph that shows how the total distance run changes as you travel to the different stations.

Observe the graph, then answer these questions. Where is the station that has the shortest total run distance? Could a 19th station be entered into the base line to achieve a smaller total run distance? Where would the position of the 19th station be? If you were the sixth runner to reach Gate 1, to which station would you probably need to travel? What is the algebraic equation that represents the graph pattern? Plot this graph of this equation on your graph of the points. If you could put in a 19th station where would you put the station, and why?

(Additional suggestions were provided as to how the work might be set out, and for intermediate calculations that would provide some task scaffolding.)

Figure 3. Cunning Running Task.

The solution involves the calculation of total path as the sum of two segments, followed by graphing, construction of an algebraic model, verification, interpretation, and the search for a nineteenth station optimally located. Figure 4 shows a typical graph produced by students who chose to use a spreadsheet.

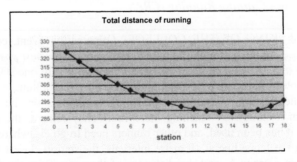

Figure 4. Typical spreadsheet graph produced in solution to *Cunning Running*.

3.2 Shot on Goal (SOG)

Many ball games have the task of putting a ball between goal posts. The shot on the goal has only a narrow angle in which to travel if it is to score a goal. In field hockey or soccer when a player is running along a particular line (a run line parallel to the side line) the angle appears to change with the distance from the goal line. At what position on a run line, does the player have the widest opening for the shot on goal? Assume you are not running in the GOAL-to-GOAL corridor. Find the position for the maximum goal opening if the run line is a given distance from the side line As the RUN LINE moves closer or further from the side line. How does the location of the position for the widest view of the goal change? (Relevant dimensions given).

Note: Because it was felt to be more inclusive in a co-educational school, and because a number of students play the game, hockey was selected to provide the specific context. There are some caveats associated with this decision. Firstly a shot on goal in hockey is only allowed from a point within the "almost" semi-circular penalty area, and hence technically only some of the run lines are feasible. This aspect can be included at a later stage by first finding the location of the best shooting position in terms of angle as is required by the question, and then checking its position relative to the penalty area. (With Football there are no such restrictions).

Figure 5. *Shot on Goal* Task.

Figure 6 shows calculations obtained using LISTs of a TI-83 Plus graphing calculator, for positions of the goal shooter at perpendicular distances from the goal line of between 14 and 20 metres along a typical run line that is 10 metres from the sideline. (Width of goalmouth is 3.67 metres). The maximum angle is highlighted at L4(17). Further accuracy can be obtained by testing nearby values, and a graph can be drawn. In practice a player often uses a zigzag run. Discussion can be used to infer that whatever the path the ultimate shot is from a position on some run-line.

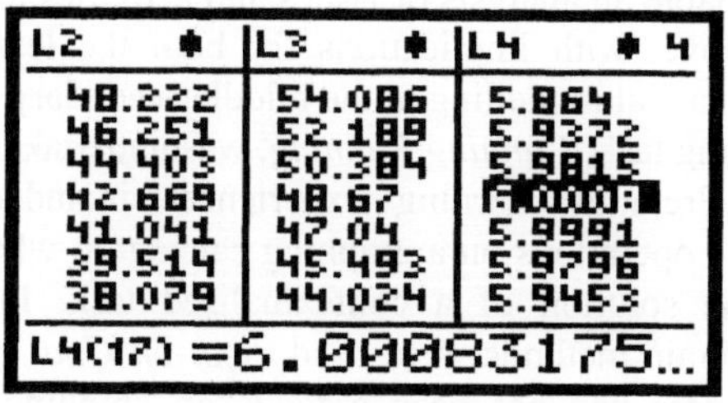

L2	L3	L4	4
48.222	54.086	5.864	
46.251	52.189	5.9372	
44.403	50.384	5.9815	
42.669	48.67	6.0000	
41.041	47.04	5.9991	
39.514	45.493	5.9796	
38.079	44.024	5.9453	

L4(17) =6.00083175...

Figure 6. Kim's LISTs for a run line 10m from the side line.
Note: Actual formulae used: $L2 = \tan^{-1}(15.67/L1)$, $L3 = \tan^{-1}(19.33/L1)$, and $L4 = L3 - L2$.

3.3 Interactions Between Modelling, Technology, and Mathematical Content

Figure 7 illustrates some of the interactions that students have to cope with between the modelling process (M), mathematical content (C), and technology use (T) as foreshadowed in Figure 1 during the solution of these tasks.

Interactions	Cunning Running	Shot on Goal
M∩C	Represent problem. To minimise sum of distances (Pythag) 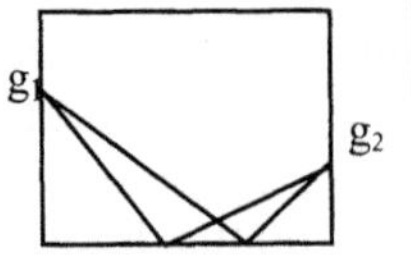	Represent problem. To maximise ∠BDC 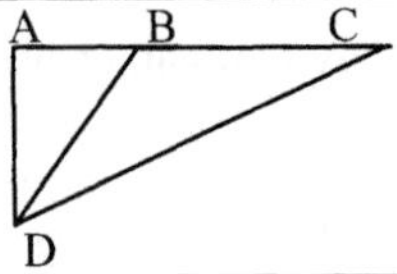
M∩T	Strategy decisions: for example, choice of GC functions, Excel.	Strategy decisions: for example, choice of GC functions, Excel.
C∩T	Computation of distances, plotting points, interpreting graphs, verification of algebraic model graphically.	Computations involving tangent, use of inverse tan to find angles for different run lines, and different positions on run lines.
M∩C∩T	Carrying through to solution: Formulate →calculate paths → plot graphs→algebraic model→ interpret output→determine implications in terms of problem requirement.	Carrying through to solution: Choice of approach → appropriate diagram → calculations of angles→ correct graphical output→ interpretation → implications.

Figure 7. Interactions between modelling process, technology, and mathematics.

4. TASK IMPLEMENTATION

The two tasks, *Cunning Running* and *Shot on Goal,* were the vehicles used to generate intensive data; collected by means of student scripts (24 and 28 respectively), videotaping of teacher and selected students, video and audiotaped records of small group collaborative activity, and selected post-task interviews (8 and 4 respectively). We seek to identify and document characteristic levels of performance; the occurrence or removal of blockages; the respective use of numerical, graphical, and algebraic approaches; quality of argumentation; and the respective interactions between modelling, mathematical content, and technology. The purpose is to enable an analysis of issues and activities impacting on transitions 1 - 4 identified above, with implications for both the learning and teaching of introductory mathematical modelling in the middle secondary years.

The first modelling task, *Cunning Running*, was undertaken in the fourth week of the school year. Previous learning experiences included the students' first introduction to LIST operations on a graphing calculator, where students worked as a class through the solution of a contextualised task. In this task numerical, graphical, and algebraic methods were used. *Shot on Goal* came two months after the first task but there had not been a focus on graphing calculator use in the intervening time. Both tasks were undertaken over about a week of class time.

5. EXAMINING TRANSITIONS

As indicated above we are interested in transitions between phases of the modelling process, and in identifying *blockages* that impede students in moving between these. In *Cunning Running* students were required to vary distances with the purpose of minimising the total distance run, while *Shot on Goal* involved looking at varying angles with the aim being to maximise the shot angle. Surface similarities of these tasks tend to obscure different levels of complexity in the respective formulations.

5.1 Real-world Problem Statement to Mathematical Problem Specification

The challenge here is in identifying the *key elements* that will form the focus for model building. This involves:

•*Firstly deciding the nature of the element (what kind of mathematical entity)*
•*Secondly identifying the specific element(s) to be focused on.*

The element was a 'distance' in *Cunning Running*, and an 'angle' in *Shot on Goal*. Both these were compound entities that needed to be constructed from other components present in the real situation (line segments or angles). Blockages occurred, particularly in the second problem, and the blockage was only unlocked for some students, when the teacher provided a supporting physical demonstration in which class members either watched or participated. This was followed by a debate by selected students using diagrams on the whiteboard about which angle was the focus. The physical enactment of the situation proved crucial for some students. "It helped to explain like what we hoped to find out. Like I didn't really get what we were trying to do. And that kind of explained what angles we were trying to find."

While *interpretation* is an acknowledged component of the modelling process it is usually associated with giving meaning to a solution in its real-world context. Interpretation in a different sense is seen here to be central in specifying a mathematical problem in the first place. No effort needs to be spared in ensuring that students receive a thorough appreciation of a problem context. This may involve direct experience, film or video, simulation, discussion and diagrammatic representation, or written description. Assuming that a mental representation will suffice is demonstrably suspect, and indeed arguably inconsistent with the purpose of solving problems grounded in real situations.

5.2 Mathematical Problem to Formulation of a Model

Key decisions with the potential to generate blockages in this transition were identified along the following dimensions:

•*Deciding how to represent an identified element*, so that known mathematical formulae could be applied (for example, in SOG how to find the angle that had been identified – two different decomposition methods were available).

•*Choosing to use technology to establish a calculation path* (for example, in SOG inverse tan calculations provided an efficient means of obtaining the required angle - provided that the mathematics within the approach was understood).

•*Choosing to use technology to automate extension of application of formulae to multiple cases* (for example, CR & SOG graphing calculator LISTs/spreadsheet). This required both technical facility and a strong grasp of the modelling requirements of the moment - recognising that generalisation via this activity was needed.

•*Recognising only one dependent and one independent variable is to be specified in an algebraic model* (for example, in CR several students used algebraic models with several variables such as $x + y = \sqrt{(120^2 + d^2)} + \sqrt{(40^2 + d2^2)}$).

•*Recognising that a particular independent variable must remain uniquely defined* throughout an application (for example,, in CR conflict occurred when '*x*' was defined as the distance from a station to corner A then from corner B in different parts of a formula).

•*Recognising that a graph can be used on function graphers but not data plotters to verify an algebraic equation* (for example, CR verification method works on graphing calculator but not on a spreadsheet).

•*Recognising when additional interim results are needed to enable progress* (for example, in attempting to place the 19th station in CR).

•*Introducing flawed problem data* into the formulation phase (for example, two students took a stepped trajectory towards the goal instead of advancing down a specified run line in 1 m intervals).

•*Selecting appropriate procedures when alternatives exist* (for example, in CR most students used an EXCEL joined scatterplot by joining numerical data generated by their graphing calculator LISTs; however a minority used formulae to generate the data - a fundamentally different approach).

5.3 FromFormulated Approach to Solution of Mathematics

As might be expected the interaction between mathematics and technology featured strongly in this transition.

•*Problems that occur as logical consequences of earlier errors in formulation* (for example, approaches giving different meanings to the same symbol in different parts of CR).

•*Problems in applying formulae correctly* (for example, in SOG using inverse tan successfully to find angles).

•*Using technology to automate extensions of application of formulae* to multiple cases (for example, CR & SOG: graphing calculator LISTs/spreadsheet).

•*Using technology to produce a graphical representation* (for example, CR & SOG: graphing calculator StatPlot, function graph or spreadsheet chart).

•*Applying algebraic simplification processes to symbolic formulae.* (for example, in CR & SOG, concatenation of graphing calculator LIST formulae to produce a function).

•*Applying the rules of notational syntax accurately.*

•*Using technology to verify an algebraic model* (for example, CR & SOG: by producing a graph or by substitution into a formula on the homescreen, or by entering a specific functional definition on a graphing calculator).

•*Reconciling unexpected interim results with real situation* (for example, in SOG one student incorrectly expected that the angle would continue to increase as the player advanced along the run line. He did not accept that his correct calculations were valid, and after debating with other students, was convinced only by another physical simulation of the activity).

5.4 From Mathematical Solution to Interpretation Within the Model Context

The following illustrate where student attempts at interpretation point to the occurrence of blockages, potential or actual:

•*Routine interpretations varied in depth from bald statements to integrated explanations* (for example,, when asked in CR "Does running via station 1, or 2, or 3 make any difference to the overall length of the run?", responses ranged from bald assertions such as, "It makes a difference," to integrated arguments such as "Yes, it does the closer you are to corner A, the further the distance you have to run.")

•*Interpretations in which the required outcome is amended to a variation introduced as the consequence of a preferred method of approach* (for example,, in finding the location of the station with shortest total run distance in CR, most students used their numerical lists to identify a particular station, rather than their graphs, which more legitimately provided the location of a minimum. The interpretation then referred to a station number rather than a location. While this did not matter here as the minimum was actually at station 14, it would be significant if the question was set so that the required station was in a position between others).

•*Differences in precision when numerical values are important* (for example, in describing the optimum position for a station in CR, most gave the station number they considered to involve minimum distance. One student added that it was "3/4 between gate 1 and gate 2", while another, seeing no need for any mathematical calculations at all simply asserted it was "towards the end").

•*Tensions in deciding how knowledge of mathematics would actually impact on the real situation.* (for example, in CR – a number of students gave credit to the sixth runner as a mathematician who would know which station to run to – others said that in the heat of a race you wouldn't really worry about distance, and just head for the 6th station).

•*Choosing criteria to use in a later aspect of a task that requires interpretation to establish a method of approach* (for example,, in CR placement of the 19th station saw the use of various criteria – rather than applying the distance criterion [correct approach], some said it should follow 18[th;] some just put it 10 m from first or last station; while one student ignored all constraints, and put it on the straight line joining the gates).

6. REFLECTIONS

Analysis of student responses during two modelling tasks has focused on the transitions between phases in the modelling process. For each transition we have been able to identify generic entries that document possible blockages halting or slowing the progress of beginning modellers. These have been illustrated by specific examples from the first such tasks implemented at the school in question. Once a mathematical approach has been formulated, students need to transform a viable approach into a solution. Our analysis has shown it is necessary to carry out mathematical and technological activities successfully to do this and the framework has been useful in identifying what prevents this happening - lack of mathematical knowledge general or specific, lack of technological skills, technical expertise, or accuracy. The complexity of interactions between modelling, mathematical content, and technology when solving modelling problems such as these in a TRTLE emphasise the importance of student perception and judicious enactment of affordances offered by such an environment. Having arrived at a mathematical solution, there are further potential blockages as students seek to make meaning of this in contextual terms - keeping in mind that the interpretation may be simple direct translation, or more demanding. The framework reveals the factors that give rise to the hierarchies of quality found in interpretation and the resultant blockages come from mistakes in interpretation, overlooking interpretation, providing superficial or irrelevant responses, and not seeing where interpretation is required for further progress to be made.

The framework has potential for identifying and documenting specifically modelling competencies with which beginning modellers need to have facility in order to successfully apply the mathematics they know in real problem settings. It also has potential as a research tool for analysing student activities during solution of modelling tasks. These considerations will guide future research activity, as students engage with further problems. The intention is to search for a higher synthesis, using additional data to clarify and illustrate conceptual structures indicative of the transitional issues provisionally identified in this chapter.

REFERENCES

Brown, J. (2005) Affordances of a technology-rich teaching and learning environment. In P. Clarkson, A. Downton, D. Gronn, M. Horne, A. McDonough, R. Pierce, et al. (eds) *Proceedings of the 28th annual conference of the Mathematics Education Research Group of Australasia*. Sydney: MERGA, 177-184.

Collins, A., Joesph, D. and Bielaczyc, K. (2004) Design research: Theoretical and methodological issues. *Journal of the Learning Sciences,* 13(1), 15-42.

Edwards, D. and Hamson, M. (1996) *Mathematical modelling skills.* Basingstoke: Macmillan.

Goos, M., Galbraith, P., Renshaw, P. and Geiger V. (2003) Technology enriched classrooms: Some implications for teaching applications and modelling. In Q-X. Ye, W. Blum, K. Houston and Q-Y. Jiang (eds) *Mathematical modelling in education and culture: ICTMA 10.* Chichester: Horwood, 111-125.

Stillman, G., Edwards, I. and Brown, J. (2004) Mediating the cognitive demand of lessons in real-world settings. In B. Tadich, S. Tobias, C. Brew, B. Beatty and P. Sullivan (eds) *Proceedings of the 41st Annual Conference of the Mathematical Association of Victoria.* Melbourne: Mathematics Association of Victoria, 487-500.

3.6

ASSESSING MATHEMATICAL MODELLING COMPETENCY

Tomas Højgaard Jensen
The Danish University of Education, Denmark

Abstract–*Focusing on the concept of competence in mathematics education has many analytical implications. Multidimensionality is a necessary, but challenging, approach and in this chapter I address the need to work with at least three dimensions to make a valid assessment of someone's possession of competence. This is demonstrated in regard to mathematical modelling competency, before the presentation of two challenges of bringing this multidimensional approach into educational practice. The two challenges are: The conflict with simple ranking as an educational goal and the conflict with the dominant focus on technical levels in mathematics education.*

1. INTRODUCTION

One of the main goals of mathematics education at all educational levels should be to assist the students' development of mathematical modelling competency. Such developments need to be assessed in a valid way, as is the case with all the main goals within educational systems framed by assessment structures, if these goals are to remain the focus of the educational endeavour. These are two not terribly controversial claims. The first is a priority in educational policy with a lot of analytical back-up in the mathematics education research literature, not least expressed in the proceedings from the ICTMA conferences (for example, Ye et al., 2003, and Lamon et al., 2003), where the sharing of this goal can be seen as the "constitutional glue" of this research community. The second claim is even less controversial as an analytical statement (the degree to which the consequence of accepting it is put into educational practice is a different issue). I do not know of any analysis arguing against it, and it is defended by many experts in the mathematics education research literature, for example, Niss (1993).

The two claims will not be defended here. They are mentioned because they jointly lead to a statement that is the premise for the arguments and conclusions put forward in this paper, namely that it is crucial to develop and implement valid assessments of the students' mathematical modelling competency as part of the development of mathematics education.

Based on this premise, the objective of the paper is to argue that a multidimensional approach is a necessary, but challenging, condition for valid assessments of someone's mathematical modelling competency. The analysis is structured as follows:

First, I propose coherent definitions of the concepts "competence" and "mathematical modelling competency" and address some of the implications of these definitions.

Then, I focus on one such implication by arguing that the proposed definition of competence suggests three dimensions to work with when assessing someone's mathematical modelling competency.

Using such a three-dimensional approach is a challenging task, and I close the paper by offering two examples of this. One example has to do with the political wish for simple ranking systems as an educational tool, and the other example has to do with the dominance of the traditional strong focus on the technical level when assessing someone's performance in mathematics education.

2. FROM "COMPETENCE" TO "MATHEMATICAL MODELLING COMPETENCY"

2.1 "Competence" as an Analytical Concept

Competence is defined as someone's insightful readiness to act in response to the challenges of a given situation (Blomhøj & Jensen, 2006).

The most important characteristic of this definition is that it makes competence headed for action. As argued in Blomhøj and Jensen (2003), "action" must be interpreted broadly, as the term "readiness to act" in the definition of competence could imply a positive decision to refrain from performing a physical act, or indirectly being guided by one's awareness of certain features in a given situation. However, no competence follows from being immensely insightful, if this insight cannot be activated in this broad interpretation of the word "action".

Secondly, all competencies have a sphere of exertion, that is, a domain within which the competency can be brought to maturity (ibid.). This does not mean that a competency is contextually tied to the use of a specific method for solving a given task. If this was the case, the attempt to define general competencies would have no meaning. Competencies are only contextual in the sense, that they are framed by the historical, social, psychological etc. circumstances of the "given situation" mentioned in the definition of competence (cf. Wedege, 1999).

2.2 A Mathematical Competency

Mathematical competence is when the challenges mentioned in the definition of competence are mathematical.

This is a both straightforward and rather uninteresting extension of the general definition of competence. To make the analysis interesting and binding for the development of mathematics education, we need to be more specific by analysing and discussing what constitutes a mathematical challenge. This leads to a focus on *a mathematical competency* defined as someone's insightful readiness to act in response to *a certain kind of mathematical challenge* of a given situation.

Hence, mathematical competence is analytically spanned by a set of such mathematical competencies, and it is a very interesting challenge to try to come up with a suggestion for, and exemplification of, the elements of such a set. Niss and

Jensen (to appear) represent an attempt to meet this challenge that has proven to promote a lot of good discussions and further analysis in the mathematics education community (cf. Blum et al., 2006).

2.3 Mathematical Modelling Competency

Mathematical modelling competency is – in accordance with the general definition of competence – here defined as someone's insightful readiness to carry through all parts of a mathematical modelling process in a given situation (Blomhøj & Jensen, 2003).

Again, to avoid arguing tautologically, there is a call for being more specific by expressing what is meant by "all parts of a mathematical modelling process". The model I use of this process, which is inspired by and not far from many of the other models presented in the literature, is depicted in Figure 1. It is analysed and presented in greater detail in Blomhøj and Jensen (2003) and Jensen (2006).

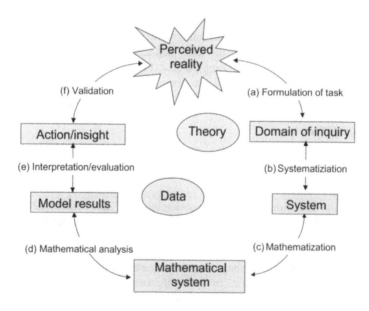

Figure 1. A visual representation of the mathematical modelling process
(Blomhøj & Jensen, 2003).

3. ASSESSING (MATHEMATICAL MODELLING) COMPETENCY – A MULTIDIMENSIONAL APPROACH

The definition, proposed here, suggests three dimensions to work with when assessing someone's possession of a competency (Niss & Jensen, to appear):

Degree of coverage, indicating which *aspects* of the competency someone can activate and the degree of *autonomy* with which this activation takes place.

Radius of action, indicating the spectrum of *contexts and situations* in which someone can activate the competency.

Technical level, indicating how *conceptually and technically advanced* the mathematics is that someone can integrate relevantly in activating the competency.

3.1 Degree of Coverage

Focusing on mathematical modelling competency, the degree of coverage addresses which part of the modelling process someone can work with and the level of the reflections involved (Blomhøj & Jensen, 2006).

One of the advantages of working with an explicitly expressed model of the mathematical modelling process in mathematics education is that it stresses the necessity of caring for this dimension: A person able to systematize an open situation in a way reflecting the wish for developing a mathematical model clearly has a higher degree of coverage than someone who can only handle pre-systematized situations. Further, a person who can enter an inner dialogue regarding the validation of a modelling process has a higher degree of coverage than someone who can only evaluate the model results and not the process leading to them.

In both these examples, someone who can carry through the various sub-processes in the mathematical modelling process, but only when prompted to do so, is more competent than someone who can not enter these processes at all, but less competent than the one who autonomously initiates the work.

3.2 Radius of Action

Focusing on mathematical modelling competency, the radius of action addresses the domain of situations in which someone can perform mathematical modelling activities (ibid.).

The need for this dimension is a consequence of the contextual nature of competencies, cf. the above discussion. Differences experienced between domains can be related to the overall mathematical approach one is "invited" to use and/or the characteristics of the extra-mathematical challenge one is trying to deal with by means of a mathematical model: Someone generally capable of modelling challenges of a geometrical nature, for example, need not be as competent when it comes to discrete mathematics or statistics, for example. Also, the fact that someone is very competent when it comes to developing and using optimization models in everyday shopping situations, for example, does not guarantee the same competence when it comes to design problems, for example.

3.3 Technical Level

Focusing on mathematical modelling competency, the technical level addresses which kind of mathematics someone can use and how flexible they are in their use of mathematics (ibid.).

This dimension represents the size and content of the "mathematical toolbox": Someone who can model a situation by means of establishing a functional relationship is more competent that another person who can only work with one variable "tied up" by an equation, but less competent than someone who can also consider using differential equations.

3.4 A Geometrical Model

The three dimensions almost suggest themselves to be visualised geometrically as in Figure 2. In this model the possession of a competence is represented by a volume and, consequently, progression is represented by an increasing volume.

This has two analytical consequences, with associated morals, that make them worth pointing out. Firstly, if the level on one of the axes is zero, that is, if the competence has not been developed at all along one of the dimensions, then the volume is also zero, that is, then the competence in its entirety has not been functionally developed either. Moral: We need to pay attention to all the dimensions when we attempt to support the development of a competence, for example, mathematical modelling competency, among a group of students.

Secondly, a significant increase in volume, that is, a significant progression in someone's competence, is easily detectable, whereas a certain volume, that is, a certain level of competence, cannot be pointed out uniquely, since it can be achieved in infinitely many ways. Moral: We can use the multidimensional approach for assessment to recognize and acknowledge progression in someone's particular mathematical competency, for example, modelling, but we cannot identify the same level of a competence across people in any simple way, and a direct ordering is impossible.

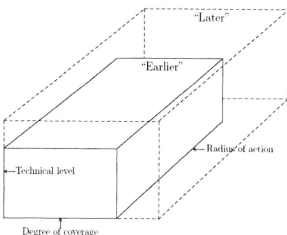

Figure 2. A visual representation of three dimensions to work with when assessing someone's competence (Niss & Jensen, 2006 ch9).

4. CHALLENGES OF A MULTIDIMENSIONAL APPROACH

As we know from mathematics, moving from one to several dimensions when analysing something is a dramatic change of perspective. It is therefore not surprising that this is also the case when it comes to addressing the challenge of assessing a mathematical competency: Difficulties are to be expected and need to be seen as challenges, of which I will shortly address two of the more serious ones.

4.1 The Wish for Simple Ranking Systems

The most fundamental challenge has already been touched upon in the discussion of the geometric model in the previous section: Traditionally, one of the main goals of including assessment as a fundamental part of the educational enterprise have been to summatively compare and rank the performance of different people by mapping this performance to a simple grading scale, and a multidimensional approach to assessment fundamentally conflicts with this "simple ranking goal", a priori.

We have stated it this way in Niss and Jensen (to appear, ch4): "It is important to emphasise that even though we have chosen terms which imply the possibility of simple quantitative measurement, there is no such assumption in the following considerations. The only thing we are implying in this regard is that each of the dimensions allows us some kind of ordering, that is, that one version of a competence can, in relation to a specific dimension, be more or less comprehensive than another version of the same competence. Since this is only a partial ordering, it may well happen that two arbitrary versions of the same competence cannot be compared in this way."

4.2 The Traditional Dominance of the Technical Level

The second challenge I want to address is a posteriori: It is my experience, both from informal discussions with many mathematics teachers and from being involved in several developmental projects, that the mathematics education community in general and the mathematics teachers in particular, have a tradition-ridden focus on the technical level when assessing someone's mathematical performance.

This tradition can make it difficult to give space, both literally and figuratively speaking, cf. the geometric model, for, and to acknowledge, the students progression in the radius of action and, not least, the degree of coverage. This is one of the results of a long term developmental research project, that focused on developing mathematical modelling competency among a group of students in the Danish upper secondary school, suggesting one possible explanation as to why developing mathematical modelling competency is not, in general, the hub of mathematics education (Jensen, 2006).

5. SUMMING UP AND LOOKING AHEAD

Let me sum up the argument given as follows: The three dimensions in the possession and progression of a mathematical competency discussed in this paper is a vocabulary for discussing quality in mathematical performance, and hence a potentially powerful contribution to maintaining the focus of mathematics education by sharpening the lenses through which we look.

The result of a multidimensional competence-based assessment process can be statements about the performance along each of the chosen dimensions, as exemplified in Niss and Jensen (to appear). Such an approach represents a more culture-based, more valid but also less reliable alternative to mark-schemes (Wiliam, 1994), known from the way large reports such as Masters Theses and PhD dissertations are often assessed.

To my knowledge, experiences with assessing the possession of explicitly expressed competencies this way are scarce (I am involved in one such ongoing project and the project documented in Jensen (2006) touched upon the idea), and to develop the idea further, both theoretically and experimentally, is definitely a challenge, which calls for more attention within the mathematics education research community.

Technically, it is of course possible to compress multidimensional statements, as the ones in Niss and Jensen (to appear), into one single grade (by assigning one grade to each dimension and then find a balanced average of these grades). This will, however, violate what I consider to be the core of the multidimensional approach discussed in this paper: This is an analytical attempt to respect the complexity of what we are striving for, when we try to assist development of mathematical competencies in general, and mathematical modelling competency in particular, and therefore in deliberate conflict with the assumption that such a development can be, and should be, assessed by means of a simple grading scale.

REFERENCES

Blomhøj, M. and Jensen, T. H. (2006) What's all the fuss about competencies? Experiences with using a competence perspective on mathematics education to develop the teaching of mathematical modelling. (*to appear in Blum et al., 2006).*

Blomhøj, M. & Jensen, T. H. (2003) Developing mathematical modelling competence: Conceptual clarification and educational planning. *Teaching Mathematics and its Applications,* 22, 123-139.

Blum, W., Galbraith, P., Henn, H. and Niss, M. (eds.) (2006) *Applications and Modelling in Mathematics Education - New IMCI Studies Series no. 10.* Springer, New York, USA. (*to appear*)

Jensen, T.H. (2006) Udvikling af matematisk modelleringskompetence som matematikundervisningens omdrejningspunkt – hvorfor ikke? Doctoral dissertation. Under preparation for publication in the series *Tekster fra IMFUFA,* Roskilde University, Denmark. To be ordered from imfufa@ruc.dk. (*to appear*)

Lamon, S., Parker, W. and Houston, K. (eds.) (2003) *Mathematical Modelling – A Way of Life: ICTMA 11.* Horwood, Chichester, UK.

Niss, M. (1993) Assessment in Mathematics Education and its Effects: An Introduction. In Niss, M. (ed.) *Investigations into Assessment in Mathematics Education.* Kluwer Academic Publishers, Dordrecht, The Netherlands, 1-30.

Niss, M. and Jensen, T.H. (eds.) (2002) Kompetencer og matematiklæring – Idéer og inspiration til udvikling af matematikundervisning i Danmark ("the KOM report"). *Uddannelsesstyrelsens temahæfteserie* 18. The Ministry of Education, Copenhagen, Denmark.

Niss, M. & Jensen, T.H. (eds.) (to appear) Competencies and Mathematical Learning – Ideas and inspiration for the development of mathematics teaching and learning in Denmark. English translation of part I-VI of Niss & Jensen (2002). Under preparation for publication in the series *Tekster fra IMFUFA*, Roskilde University, Denmark. To be ordered from imfufa@ruc.dk.

Wedege, T. (1999) To know or not to know – mathematics, that is a question of context. *Educational Studies in Mathematics*, 39, 205–227.

Wiliam, D. (1994) Assessing authentic tasks: alternatives to mark-schemes. *Nordic Studies in Mathematics Education*, 1, 48-68.

Ye, Q., Blum, W., Houston, K. and Jiang, Q. (eds) (2003) *Mathematical Modelling in Education and Culture: ICTMA 10*. Horwood, Chichester, UK.

3.7

A STOCHASTIC MODEL FOR THE MODELLING PROCESS

Michael Voskoglou
Higher Technological Educational Institute
Patras, Greece

Abstract–*We introduce a finite Markov chain having as states the main steps of the mathematical modelling process. In this way, we obtain a stochastic method for the description of this process, in situations where the teacher provides such modelling problems to the students for solution, and we also succeed in measuring the mathematical model building abilities of these students. Our stochastic method, which is compared with other reported qualitative methodologies, is actually an improved version of a model presented in an earlier paper. An application in the classroom presented in the above paper, is also reconsidered here, in order to illustrate the improvements of our model, and our conclusions are discussed in comparison with other recently reported research.*

1. MODELLING AS A TEACHING METHOD OF MATHEMATICS

As is well known, "Mathematical modelling and applications" is a process of solving a particular type of problem generated by corresponding situations of the real world. The transformation from the real word to mathematics is achieved through the use of a mathematical model, which, briefly speaking, is an idealized (simplified) representation of the basic characteristics of the real situation through the use of a suitable set of mathematical symbols, relations and functions. From 1976, when Pollak presented in ICME-3 in Karlsruhe the "circle of modelling" (see Pollak, 1979), much effort has been placed to analyze in detail the process of mathematical modelling. Summarizing the existing ideas, the main stages of the process are:

s_1 = analysis of the problem (understanding the statement and recognizing the restrictions and requirements of the real system).

s_2 = mathematising, which includes the formulation of the real situation in such a way that it will be ready for mathematical treatment and the construction of the model.

The formulation of the problem, which for many researchers must be considered as an independent stage, involves a deep abstracting process, in order to transfer from the real system to the, so-called, "assumed real system". Here emphasis is placed on certain variables that dominate the system's performance.

s_3 = solution of the model, which is achieved by proper mathematical manipulation.

s_4 = validation (control) of the model, which is usually achieved by reproducing, through the model, the behaviour of the real system under the conditions existing before the solution of the model (empirical results, special cases etc).

s_5 = interpretation of the final mathematical results and implementation of them to the real system, in order to give the "answer" to our problem.

2. THE 2 – STATE SUB-MODEL

Roughly speaking, a Markov chain is a stochastic process that moves in a sequence of steps (phases) through a set of states and has "no memory", that is, the probability of entering a certain state in a certain step, although it is not necessarily independent of previous steps, depends at most on the state occupied in the previous step. This is known as the "Markov property". When the set of its states is a finite set, then we speak about a finite Markov chain. For a good exposition of the relevant theory, see Kemeny and Snell (1976), to which we refer freely below. Markov chains have been frequently used in the past for educational purposes: on subjects concerning the counselling psychology (for example, see Lohnes, 1965 and its references); in problem solving (Voskoglou and Perdikaris, 1991); etc. (eds: *see also Chapter 6.5).*

In this paper we present a Markov chain model for the description of the process of mathematical modelling in the classroom, when the teacher gives such kind of problems for solution to the students. For this, assuming that the above process has the "Markov property", we shall introduce a finite Markov chain having as states the five main stages of the process of modelling, that we described in the previous section. This assumption is a simplification (not far away from the truth) made to the real system in order to transfer from it to the "assumed real system" (see §1). Since the conclusions obtained from our classroom experiment (see §4 below) are consistent with other recently reported research (see §5), it turns out that through the above simplification our model gives a reliable prediction of the real system's performance. Our method is actually an improved version of a model presented in Voskoglou (1995). Before the introduction of our 5-state model, we shall present a 2-state sub-model, in order to illustrate, especially for readers having barely a passing acquaintance with Markov chains, the principles involved, as well as the necessity of the introduction of the larger (and more complex) 5-state model. The two states of the Markov chain of our sub-model are:

t_1 = solution of the problem, including the stages s_1, s_2, and s_3 of the mathematical modelling process in the classroom, and

t_2=verification of the solution – answer to the given problem, including stages s_4 and s_5.

Denote by p_{ij} the transition probability from state t_i to t_j for i,j = 1,2, then the matrix A= (p_{ij}) is said to be the transition matrix of the chain.

When the process of modelling is completed in state t_2, it is assumed that the teacher gives to the students a new problem for solution and therefore the process starts again from t_1. Under this assumption it becomes evident that, in our Markov chain, it is possible to go between any two states, not necessarily in one step (such a chain is called an ergodic chain) and that $p_{22}=0$, $p_{21}=1$.

For an ergodic chain it is well known that, as the number of steps tends to infinity (long run), then the chain tends to an equilibrium situation, which is characterized by the equality Q=QA, where Q is a row – matrix called the limiting probability vector of the chain. The entries of Q, whose sum is equal to 1, give the fraction of times that the chain is expected to be in each of the states in the long run, that is, in other words they give the "gravity" of each state of the chain.

In our case we have that

$$\begin{bmatrix} a_1 & a_2 \end{bmatrix} = \begin{bmatrix} a_1 & a_2 \end{bmatrix}\begin{bmatrix} p_{11} & p_{12} \\ 1 & 0 \end{bmatrix} = [a_1 p_{11}+a_2 \quad a_1 p_{12}],$$ from which we get that

$a_1(p_{11}-1)+a_2=a_1 p_{12}-a_2=0.$

But, since the passage from state t_1 to one of the states t_1 or t_2 is the certain event, we have that $p_{11}+p_{12}= 1$, and therefore the previous relation is equivalent with the equation $a_1 p_{12}-a_2=0$ (1), while we also have that $a_1+a_2=1$ (2). The linear system of (1) and (2) has the unique solution

$$a_1 = \frac{1}{1+p_{12}}, \quad a_2 = \frac{p_{12}}{1+p_{12}},$$ while $m_{12} = \frac{a_1}{a_2} = \frac{1}{p_{12}}$ gives the mean number of

entries in state t_1 between two successive occurrences of t_2. Therefore the bigger is m_{12}, the more are the difficulties that the students face during the process of mathematical modelling in the classroom.

3. THE 5 – STATE MODEL

The results of this section are obtained by generalizing the corresponding results of §2. The solver of a problem involving mathematical modelling from the initial state, which is always s_1, proceeds via s_2 to s_3. From this state, if the mathematical relations obtained are not suitable to allow an analytic solution of the model, the solver should return to s_2, in order to make the proper simplifications – modifications to the model. Then he (she) returns to s_3, to continue the process. After the solution of the problem within the model the solver should return to the real system, in order to check the validity of the model (state s_4). If the model does not give a reliable prediction of the system's performance, (for example, if the solution obtained is not satisfying the natural restrictions resulting from the real system, or if it is not verified by known special cases etc), the solver returns from s_4 to s_2, in order to correct the model. From there he (she) will return, via s_3, to s_4 to continue the process. After ensuring that the model is valid, the solver from s_4 reaches the state s_5, where he (she) interprets the final mathematical results and applies the conclusions to the real system. that is, he/she gives the "answer" to the enquiries the problem. When the process of modelling is completed in state s_5, it is assumed that the teacher gives students a new problem for solution and therefore the process starts again from s_1.

Notice also that, a solver, who finally fails to construct a solvable mathematical model giving a reliable prediction of the real system's performance and being unable to make any other "movement" for the solution of the problem during the time

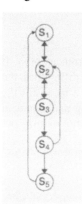

Figure 1.

given by the teacher, returns from the state s_2 to s_1 waiting for a new problem, to be given for solution (this detail has not been taken into account in our initial model in Voskoglou, 1995).

According to the above description the "flow - diagram" of the mathematical modelling process is that shown in Figure 1 and therefore the transition matrix of the chain becomes:

$$A = \begin{array}{c} \quad\begin{array}{ccccc} s_1 & s_2 & s_3 & s_4 & s_5 \end{array} \\ \left[\begin{array}{ccccc} 0 & 1 & 0 & 0 & 0 \\ p_{21} & 0 & p_{23} & 0 & 0 \\ 0 & p_{32} & 0 & p_{34} & 0 \\ 0 & p_{42} & 0 & 0 & p_{45} \\ 1 & 0 & 0 & 0 & 0 \end{array}\right] \begin{array}{c} s_1 \\ s_2 \\ s_3 \\ s_4 \\ s_5 \end{array} \end{array}, \text{ with } p_{21}+p_{23}=p_{32}+p_{34}=p_{42}+p_{45}=1 \ (1).$$

From the "flow-diagram" it becomes also evident that the above chain is an ergodic chain. Then the equation Q=QA gives that

$\alpha_1 = p_{21}\alpha_2 + \alpha_5$, $\alpha_2 = \alpha_1 + p_{32}\,\alpha_3 + p_{42}\,\alpha_4$, $\alpha_3 = p_{23}\alpha_2$, $\alpha_4 = p_{34}\alpha_3$, $\alpha_5 = p_{45}\alpha_4$. Adding the first 4 of the above equations and using relation (1) one finds the 5th equation, which therefore can be omitted.

Replacing the 5th equation with $\alpha_1 + \alpha_2 + \alpha_3 + \alpha_4 + \alpha_5 = 1$ it results a linear system of 5 equations with unknowns the α_i 's, i=1,2,3,4,5, wherefrom one easily finds the unique solution:

$$\alpha_1 = \frac{1-p_{23}+p_{23}p_{34}p_{45}}{D}, \ \alpha_2 = \frac{1}{D}, \ \alpha_3 = \frac{p_{23}}{D}, \ \alpha_4 = \frac{p_{34}p_{23}}{D}, \ \alpha_5 = \frac{p_{34}p_{23}p_{45}}{D},$$

with D= $2p_{34}p_{23}p_{45}+p_{23}p_{34}+2$.

The above equalities give a measure of the "gravity" of each state of the process of mathematical modelling in the classroom, where bigger "gravity" means more difficulties for the students in this state.

Furthermore, since the process of mathematical modelling starts again after state s_5 (as a new problem is given for solution to the class), the sum

$$m = \sum_{i=1}^{4} m_{i5} = \sum_{i=1}^{4} \frac{\alpha_i}{\alpha_5} = \frac{\sum_{i=1}^{4} \alpha_i}{\alpha_5} = \frac{1-\alpha_5}{\alpha_5}$$

gives the mean number of steps taken between two successive occurrences of s_5. It becomes therefore evident that the bigger is m, the more are the difficulties the students face during the process of modelling. Accordingly m gives a mean for the measurement of the ability of a team of solvers of problems involving mathematical modelling. In (Voskoglou, 1995, §3), we have used m_{25} instead of m, but m turns out to be a more representative measure.

4. AN APPLICATION IN THE CLASSROOM

The results obtained above can help the teacher of mathematics to estimate the abilities of his (her) students in solving problems of mathematical modelling, but also to evaluate the effectiveness of his (her) teaching. In order to illustrate the improvements of our stochastic method, but also to explain the necessity of the use of our 5-state model instead of the simpler 2-state submodel, let us reconsider the experiment in the classroom described in (Voskoglou, 1995).

I) Description and data of the experiment.

With this experiment we wanted to compare the difficulties that students from two different departments of the School of Management and Economics of the Higher Technological Educational Institute of Messolonghi face towards the solution of problems involving mathematical modelling. For this I formed two equivalent (according to their grades in mathematics of the first term) groups of 20 students, one from each department, being at the beginning of their second term of studies. Notice that they were taught the same matter in mathematics of the first term, from different however lecturers, and were examined on subjects of roughly the same difficulty. I handed to both groups 10 problems involving mathematical modelling for solution in 3 hours, explaining to the students that I was interested for all their attempts (wrong and right) to solve them. Also, in order to be helped in collecting the appropriate data from their answers, I gave to them the following instructions.

"Among the problems that you have solved:

Mark with an "A" those where your initial mathematical results didn't lead to a solution of the problem (not including cases of numerical mistakes corrected afterwards). Mark with a "B" those where you found the solution directly, without making any modifications to your initial model. Mark with a "C" those where you have made modifications to your model after realizing that the solution obtained through it didn't give a reliable prediction of the real system's performance." In order to clarify the way in which I succeeded to collect our data, let us consider a characteristic example of these problems together with the answers of the students of the first group.

Problem: We want to construct a channel to run water by folding the two edges of an orthogonal metallic leaf having sides of length 20cm and 32 cm, in such a way that they will be perpendicular to the other parts of the leaf. Assuming that the flow of the water is constant, how we can run the maximum possible quantity of the water?

Folding the edges of the longer side of the leaf (Figure 2) we have to maximize the function $E(x)=(32-2x)x$. Thus, taking the derivative $E'(x)=0$, we obtain $x=8cm$ and therefore $E(8)=128cm^2$. In this problem 14 students found the correct solution and 10 of them marked it with "B", 3 marked it with "C" and 1 with "A-C". Also 4 students folded the other side of the leaf and therefore they found $E(x)=(20-2x)x$ and

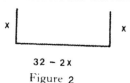

32 − 2 x

Figure 2

x=5cm. But in this case E(5)=50cm² and therefore their solution was wrong. The other 2 students didn't succeed to construct a solvable model.

We worked in the same way with the other 9 problems and then we calculated the following means per problem (for the first group):

Correct solutions 11.2, from these 5.7 obtained straightforwardly (marked with B), 2.3 obtained by making modifications to the initial model only after state s_4 (marked with C), and the remaining 3.2 solutions obtained by starting to make modifications to the model just after the first passage from state s_3 (marked with A or A-C). Notice that supplementary modifications became necessary to 1.1 from these 3.2 solutions after state s_4 (marked with A-C). This has not been taken into account in (Voskoglou, 1995). We also found 6.3 wrong solutions and 2.5 failures to construct a solvable model. The corresponding means for the second group were the following:

Correct solutions 8.1, from these 2.1 were marked with B, 0.6 with C, 5.2 with A and 0.2 with A-C. We also found 9.7 wrong solutions and 2.2 failures to construct a solvable model.

II) Interpretation of the obtained data with respect to the 2-states submodel.

Interpreting the means of the first group with respect to our 2-states submodel we observe the following (see Figure 3):

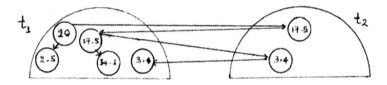

Figure 3

From the 20 students starting from t_1 the 17.5 (right and wrong solutions) proceed to t_2, while the 2.5 (failures to construct a solvable model) remain to t_1 waiting for the next problem to be given by the teacher. From t_2 the 17.5 students return to t_1 and 14.1 of them (6.3 wrong solutions, 5.7 correct solutions marked with B and 2.1 correct solutions marked with A) remain there waiting for the new problem. The other 3.4 students (correct solutions marked with C or A-C) go back to t_2, give the answer to the problem, and from there they return to t_1 to solve the next problem. Thus we have totally 20+17.5+3.4=40.9 "arrivals" to t_1 and 17.5+3.4 "departures" from t_1 to t_2. Therefore the transition probability $p_{12}=\dfrac{22.9}{40.9}=0.511$ (see section 2) and so $p_{11}=0.489$. Thus the "gravities" of the states t_1 and t_2 of the chain are given by $\alpha_1=\dfrac{1}{1.511}=0.662$ and $\alpha_2=\dfrac{0.511}{1.511}=0.338$ respectively (see section 2), while the mean number of entries in t_1 between to successive occurrences of t_2 is $m_{12}=\dfrac{1}{0.511}=1.957$ (all calculations are made with accuracy up to the third decimal digit). In the same way I found for the second group that $p_{11}=0.518$, $p_{12}=0.482$, and therefore $\alpha_1=0.675$, $\alpha_2=0.325$ and $m_{12}=2.075$.

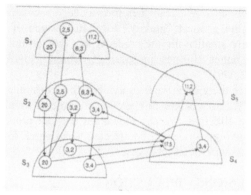

Figure 4

We lead, therefore, to the conclusion that state t_1 had the greatest "gravity" for both groups of the students. This "gravity" was slightly smaller for the students the first group (0.662<0.675), who showed in general slightly better abilities in solving problems involving mathematical modelling (1.957<2.075). Thus, we have established a method for measuring the mathematical model building abilities of a group of students. On the contrary the fact that the "gravity" of t_1 was found greater than that of t_2 for both groups does not give any significant information about the process of mathematical modelling, since state t_1 includes 3 main stages of this process and t_2 other 2 stages.

Thus the necessity of the introduction of the more complex 5-state model becomes evident.

III) Interpretation of the obtained data with respect to the 5-states model.

Interpreting the means of the first group (see Figure 4) with respect to the "flow – diagram" of Figure 1 we observe the following:

Initially all the students starting from s_1 reach, via s_2, state s_3. From there 2.5 students come back to s_2 and failing to construct a solvable model they return to s_1 in order to attempt the solution of the next problem given from the teacher. Furthermore 3.2 students return to s_2, correct their model and come back to s_3. Thus 17.5 in total students (right and wrong solutions) proceed from s_3 to s_4. Then 6.3 students (wrong solutions) come back to s_2 and from there, after failing to correct their models, they return to s_1 in order to deal with the next problem. Also 2.3+1.1=3.4 students return from s_4 to s_2 and, after correcting their models, they come back, via s_3, to s_4.

Finally, 11.2 students (right solutions) proceed from s_4 to s_5, from there they come back to s_1, to attempt the solution of the next problem. Accordingly, since we have in total 35.4 "arrivals" to s_2 and 26.6 "departures" from s_2 to s_3, we find that the transition probability $p_{23} = 26.6/35.4 = 0.751$ (see §3), therefore $p_{21}=0.249$. In the same way it turns out that $p_{34} = 20.9/26.6=0.786$, $p_{32}= 0.214$, $p_{45}=11.2/20.9 =0.536$ and $p_{42}= 0.464$. Thus we find that the "gravities" of the states s_i, i=1,2,...,5 of the modelling process in the classroom (see section 3) are $\alpha_1 = 0.176$, $\alpha_2 = 0.31$, $\alpha_3 = 0.233$, $\alpha_4 = 0.183$, and $\alpha_5 = 0.098$ respectively, while the ability of the first group of solvers is measured by the mean m=0.902/0.098=9.204 steps.

In the same way, I found for the second group of students that $p_{23}=0.688$, $p_{34}=0.71$ and $p_{45} = 0.435$, and therefore $\alpha_1=0.18$, $\alpha_2=0.343$, $\alpha_3=0.236$, $\alpha_4=0.168$, $\alpha_5 =$

0.073 and m=12.699 steps. We lead therefore to the conclusion, that the state of mathematising had the greatest "gravity" for both groups of the students. This "gravity" was slightly smaller for the students the first group (0.31<0.343), who showed in general better abilities in solving problems involving mathematical modelling (9.204<12.699).

Accordingly, since the groups were equivalent, this constitutes an indication that the teaching of the lecturer of the first group on subjects involving mathematical modelling was more effective. We say indication and not proof, because the total performance of the two groups could be affected to a degree from random factors, such as inattentions or superficial mistakes.

5. FINAL CONCLUSIONS – DISCUSSION

Mathematics does not explain the natural behaviour of an object, it simply describes it. This description however is so much effective, so that an elementary mathematical equation can describe simply and clearly a relation, that in order to be expressed with words could need entire pages. We believe that this is exactly the main advantage of our model compared with other qualitative methodologies, such as the analyses of questionnaires' collected answers by using response maps (Stillman and Galbraith, 1998), or multiple-choice tests (Crouch and Haines, 2001) and the related discussion activities.

The simpler 2-state submodel, presented before the 5-state model, provides a measure (the parameter m_{12}) for the mathematical model building abilities of a group of solvers, but it does not give any significant information about the process of modelling itself. The fact that state t_1 had greater gravity than t_2 for both groups of students of our experiment is not illustrating clearly the behaviour of the students, since t_1 includes 3 main stages of the process of mathematical modelling and t_2 includes the other 2 stages. Therefore the introduction of the more complex 5-states model becomes necessary. The data of our experiment "translated" with respect to this model show that in both groups the stage of mathematising had the greatest "gravity" among all the stages of the modelling process. This was logically expected, since the formulation of the problem involves, as justified in §1, a deep abstracting process, which is not always an easy thing to do for a non expert

Inspecting the attempts of our students towards the solution of the problems I found that in the great majority of cases the successful formulation of the problem was followed by a successful construction of the corresponding mathematical model. Another important thing to notice is that, although through the 2-states submodel the abilities of the students of the first group were found slightly better than those of the students of he second group (1.957<2.075), the same abilities were found clearly better (9.204<12.699) through the 5-states model. There is a logical explanation about this. In fact, the more are the stages that we consider in the modelling process, the bigger is the mean number of steps taken between two successive occurrences of the final stage. In other words the 5-states model provides a more representative measure (the parameter m) than the 2-states submodel does (the parameter m_{12}). Notice that our stochastic model can be also used with minor modifications to more complex "real" modelling activities, when it is necessary to make short or long run forecasts for the evolution of various phenomena in

Economics, Management, and Engineering etc (for example, see Voskoglou, 2000). We shall close by discussing our results in comparison with other, recently reported, research.

Crouch and Haines (2004, §1) report that the interface between the real-world problem and the mathematical model is really the fact that presents difficulties to the students, that is, the transition from the real word to the mathematical model (this is consistent with the results of our experiment) and vice versa the transition from the solution of the model to the real world. On the contrary the results of our experiment ("gravities" a_4 and a_5 of the states s_4 and s_5 respectively) do not indicate any particular difficulty of students across these stages of the modelling process. Our results show that the solution of the model is the stage having the second (after mathematising) greater "gravity" for both groups of students (a_3=0.233 and 0.236 respectively). This is partially crossed by Stillman and Galbraith (1998) reporting on an intensive study of problem solving of female students at the senior secondary level, where they found that more time was spent on execution (and orientation) activities with little time being spent on organization and verification activities.

Conclusively, as Haines and Crouch (2001) observe, further research remains to be done on how experts model and the relations, if any, between the processes employed by the expert modeller and by the novice (see also Crouch and Haines 2004, §2). This could provide insights on links among the several stages of the mathematical modelling process.

REFERENCES

Crouch R. and Haines C. (2004) Mathematical modelling: Transitions between the real world and the mathematical model. *International Journal of Mathematics Education in Science and Technology,* 35, 197-206.

Haines C. and Crouch R. (2001) Recognizing constructs within mathematical modeling. *Teaching Mathematics and its Applications,* 20(3), 129-138.

Kemeny, J. G. and Snell, J. L. (1976) *Finite Markov Chains.* New York: Springer Verlag.

Lohnes P. R. (1965) Markov Models for Human Development Research. *Journal of Counselling Psychology,* 12(3), 322-327.

Pollak H. O. (1979) The interaction between Mathematics and other school subjects. *New Trends in Mathematics Teaching, Volume IV.* Paris: UNESCO.

Stillman G. A. and Galbraith P. (1998) Applying mathematics with real world connections: Metacognitive characteristics of secondary students. *Educational Studies in Mathematics,* 96, 157-189.

Voskoglou, M. G. and Perdikaris, S. C. (1991) A Markov chain model in problem solving, *International Journal of Mathematics Education in Science and Technology,* 22, 909-914.

Voskoglou, M. G. (1995) Measuring mathematical model building abilities, *International Journal of Mathematics Education in Science and Technology, 26,* 29-35.

Voskoglou, M. G. (2000) An application of Markov chains to Decision Making, *Studia Kupieckie, 6,* 69-76.

3.8

ASSESSING PROGRESS IN MATHEMATICAL MODELLING

John Izard
RMIT University, Melbourne, Australia

Abstract–*Evidence of mathematical modelling skill may be obtained through performance of a number of complete tasks or by identifying the appropriate actions in a number of sub-tasks. Success is often gauged by adding the scores on each performance or sub-task. Students of different levels of modelling skill are assumed to receive different scores and students of the same level are assumed to receive the same scores. This paper reports an application of item response modelling procedures to samples from an item collection to investigate this assumption. The results inform the choice of items to meet technical and practical requirements for better assessments in mathematical modelling.*

1. CONTEXT FOR THIS RESEARCH

At ICTMA-6, held at the University of Delaware in 1993, Werner Blum asked "What needs to be done for applications and modelling in practice and in research?" (Blum, 1995 p13). Blum saw four measures as necessary for applications and modelling in the future. His first measure was *Developing appropriate modes of assessment for applications and modeling*. This paper focuses on this issue.

As Haines and Crouch (2004) have noted, teaching and learning paradigms for mathematical modelling are varied. Some approaches are holistic, commencing with simple cases and moving to an increasingly complex sequence of cases. Others focus on processes and stages that are presumed to relate to key issues in the gaining of expertise by novices. These teaching and learning paradigms assume that expertise may be developed. The task of assessment strategies is to document the qualities of that expertise as well as describe its extent.

Item response modelling (IRM) strategies were used in this research. The assumptions underlying IRM (sometimes known as IRT) are similar to those for traditional test analysis. Both assume that item marks are added to give a measure of achievement but traditional approaches usually fail to check whether this assumption is valid. As McGaw et al. (1990, p30) states, "Whenever it is believed appropriate to add results then it is appropriate to express them on a common scale before doing so". IRM aims to separate item complexity from student achievement using iterative procedures, and to reflect on the common elements in items of comparable difficulty. These IRM strategies are used widely by examination boards (Izard, 1992) and in international studies (such as TIMSS [www.timss.org] and PISA [www.pisa.oecd.org]) to interpret test results from open-ended and multiple-choice items, responses to rating scales, and other performance judgments.

IRM strategies have been applied in mathematical modelling contexts since 1989 (following ICTMA4 in Roskilde, Denmark). For example, Izard and Haines (1993) used IRM in assessing oral communications about an authentic mathematical modelling task. Houston, Haines and Kitchen (1994) reported on rating scales for evaluating major projects and shorter projects and investigations, using IRM to interpret the responses. Haines, Crouch and Davis (2001) and Haines, Crouch and Fitzharris (2003) reported on the use of partial-credit multiple-choice questions (see Figure 1) to capture evidence about student skills at key developmental stages without the students being required to carry out the complete modelling task. IRM was used by Izard, Haines, Crouch, Houston and Neill (2003) to ensure that the progress of students using such items was reported on a common scale regardless of fluctuation in test difficulty.

Consider the real-world problem (do **not** try to solve it!):
A large supermarket has a great many sales checkouts which at busy times lead to frustratingly long delays especially for customers with few items. Should express checkouts be introduced for customers who have purchased fewer than a certain number of items?

In the following unfinished problem statement which **one** of the five options should be used to complete the statement?

Given that there are five checkouts and **given** that customers arrive at the checkouts at regular intervals with a random number of items (less than 30), **find** by simulation methods the average waiting time for each customer at 5 checkouts operating normally and **compare** it with

A. the average waiting time for each customer at 1 checkout operating normally whilst the other 4 checkouts are reserved for customers with 8 items or less.
B. the average waiting time for each customer at 4 checkouts operating normally whilst the other checkout is reserved for customers with fewer items.
C. the average waiting time for each customer at 1 checkout operating normally whilst the other 4 checkouts are reserved for customers with fewer items.
D. the average waiting time for each customer at some checkouts operating normally whilst other checkouts are reserved for customers with 8 items or less.
E. the average waiting time for each customer at 4 checkouts operating normally whilst the other checkout is reserved for customers with 8 items or less.

Preferred responses are A[score = 1], B[0], C[0], D[0], E[score = 2].

Figure 1. Partial credit multiple-choice item (from Haines & Crouch, 2004).

When sampling from a collection of such mathematical modelling items it is clear that one should use at least one item from each developmental stage. But items assessing the *same* stage may differ in difficulty so more information is needed about which items serve the purpose of describing development better. Because the partial credit format shown in Figure 1 was a potential complicating factor, it was decided to use another data set where items were scored correct [score=1] or incorrect [score=0] in modelling the effects of difficulty.

2. THE REAL-WORLD PROBLEM

We know that problem tasks vary in difficulty. This variation may be a consequence of differing intellectual demands of the tasks, the quality of the information available, or the number of alternative plausible potential solutions. But the usual procedure of adding component marks when calculating scores assumes that components are equally difficult and interchangeable but the tests with different components are not equivalent (Izard et al., 2003). Further, the number of tasks correct is not necessarily a true indication of achievement. When comparing results on two *different* assessments we do not know whether results vary because the items differ, or because the students differ, or both. If the results do not vary from the initial results, we do not know whether it is due to an inappropriate range of items (ceiling effects), un-related tests differing in difficulty, or a lack of progress.) Sound evidence of achievement requires multiple demonstrations: completion of a single task is not convincing.

Implications: Students in courses assessed in traditional ways may not receive due credit for their achievements. Assessment of progress during a course is impossible when the tests differ because there is no relation between the assessment strategy for one year and the corresponding strategy for the following year: when the "rulers" are un-related and lacking common units we have no valid measure of added value (Izard, 2002a, 2002b).

3. ASSUMPTIONS MADE IN CREATING THE MODEL

Performance on real-life tasks provides evidence of achievement. Tasks used for the assessment strategy (items) are assumed to be from the same domain. (There seems little point in using irrelevant items.) All other things being equal, higher scores are assumed to indicate higher achievement. Larger differences between scores indicate larger differences in achievement. We also assume items are stable in difficulty (relative to the other items) over occasions and that learning does not occur to a substantial extent during an assessment event. Changes in achievement on comparable assessments are inferred to be evidence of learning (or lack of learning).

Implications: Assessment items on a test should be internally consistent if assessing on the same dimension. Students with high scores on the test as a whole should do better on each item than those with low scores on the test as a whole. Items that do not distinguish between able and less able students contribute little to a test. Assessment of progress implies that assessments are on the same dimension.

4. FORMULATION OF THE MATHEMATICAL PROBLEM

Assessing progress requires multiple test items (or problem tasks) as indicators of achievement over a range of difficulty consistent with the range of achievement to be measured. Sampling of items or tasks should be representative of difficulty levels as well as process and content. Higher credit should be given for difficult or more complex items than for less difficult or simple items. (This requirement can be met if more able students are successful on both the easier items and the more difficult items.)

Several hypothetical tests (with open-ended items scored right/wrong) can be generated from items of known difficulty. If a test is useful, students of the same level should receive the same scores. Conversely, students of different levels of expertise should receive different scores and students with differing expertise should not receive the same scores.

Implications: Students receive credit for their achievements through scoring on the easier items and, progressively, on the more difficult items. Students are unlikely to be correct on very difficult items when they are incorrect on easier items. Students who are correct on difficult items but who cannot achieve success on easy items are unusual. Such instances need investigation: possible explanations are carelessness, gaps in learning, students being assessed in other than their first language, or failures in test security.

5. SOLVING THE MATHEMATICAL PROBLEM

The notion of an item bank of valid and practical assessment strategies as a resource for teachers of mathematic modelling assumes that teachers can select items from the bank to construct their own valid tests. In principle this appears sound. The next assumption to be checked is the issue of sampling from the item bank to obtain valid tests – tests that can distinguish between differing levels of mathematical modeling expertise.

A rectangular distribution of item difficulties may be used to generate a series of tests, representing possibilities expected in practice. Scores on each of the tests can be calculated for hypothetical students of specified achievement levels. Apparent achievement levels from test results can then be compared with true achievement levels known from other evidence.

Earlier in this paper a useful test was described as one where students of different achievement levels receive different scores and students of the same achievement level receive the same scores. The usefulness of some tests in assessing some hypothetical students is illustrated in Figure 2 (adapted from Izard *et al.* (1983), original in Izard (2004)). For each test, each student is shown by an **X** placed on a vertical linear continuum. A numeral has been added to distinguish between students. Higher achieving students are shown at the top part of the diagram and lower achieving students are shown in the lower part of the diagram. For example, student X1 shows high achievement, and X4 shows low achievement. Items for each of the five-item tests (A, B, F and J) are shown to the right of the vertical line representing the achievement continuum. The vertical placement of each item is in terms of item difficulty. Easy items like A1, A2, A3, A4, A5, F1, F2, and J2 are near the bottom of the diagram. Difficult items like B5, F4, and J5 are near the top of the diagram. The following commentary (adapted from Izard (2004)) considers the theoretical usefulness of each hypothetical test shown in Figure 2.

	Test A	Test B	Test F	Test J
		B5	F5	
		B4	F4	
		B3	F3	J5
		B2		
		B1		
X1				
				J4
x2				
				J3
x3				
				J2
x4				
	A5			
	A4			
	A3		F2	J1
	A2		F1	
	A1			

Figure 2. Alternative possibilities for tests.

Test A would not be useful in distinguishing between the 4 students because the item difficulties are much lower than most of the student achievement levels. If an initial test, most or all of the students would get perfect scores. After a period of learning, this test would not be able to detect improvements in achievement of the 4 students because they cannot obtain any higher than a perfect score. In an evaluation of an intervention context, such lack of improvement may be interpreted as a failure of the intervention but the fault lies with the choice of Test A.

Test B would not be useful in distinguishing between the 4 students because the item difficulties are much higher than most of the student achievement levels. If an initial test, most or all of the students would get zero scores. After a period of learning, this test may be able to detect improvements in achievement of the 4 students because they can obtain a higher score than zero. But all would have to progress from their earlier level to the level above that of X4 to receive credit for their learning. In an evaluation of an intervention context, a lack of improvement may be interpreted as a failure of the intervention but the fault lies with the choice of Test B.

Test F would not be useful in distinguishing between the students. Probably all would be correct on 2 of the items and wrong on the remaining 3. Results on Test F would imply that the 4 students were the same even though they are at different attainment levels.

Test J would be useful in distinguishing between each of the students. Probably X1 would be correct on 4 items, X2 would be correct on 3 items, X3 would be correct on 2 items, and X4 would be correct on 1 item. Further, the differences between the attainment levels are reflected in the differences between the scores.

6. INTERPRETING THE SOLUTION

Did students with the same score have different achievement levels? Tests A, B and F show that all students were the same when they were not. Tests that are much too difficult or much too easy obscure the real differences between students. Combining very easy test items with very difficult test items also obscures the differences between students.

Did students with different scores have different achievement levels? Only the results of Test J reflected the actual differences between students.

Did students of the same achievement level receive the same scores? This possibility was not tested in the set of alternatives. But if we add a student X5 with the same achievement level as student X4 then both students would receive the same score on Test J and each other test would give both the same score.

Test formats and scoring procedures not reflecting true achievement should be are rejected in favor of tests that do reflect true achievement. For Tests A, B and F there were many instances where students at different achievement levels were judged to be at the same level. This denies students appropriate credit for their work and discriminates against some students. Clearly, examinations and assessment strategies like these should be replaced with tests like Test J, provided that real tests behave in a comparable way to this hypothetical set of tests.

7. VALIDATION OF THE MODEL

A test of 21 items was administered to more than 700 students and the results were analysed using item response modelling software. The 54 students who obtained a perfect score were excluded from the study because their achievement level could not be determined. (We need to know what they cannot do as well as what they can do to estimate their achievement level: with perfect scores we have no evidence of what they cannot do.) The results were sorted to obtain groups of students with the same scaled achievement level according to the overall test of 21 items. Students were sampled from these larger groups to match as closely as possible the hypothetical students shown in Figure 2 above. Generally five achievement levels were used. Items were chosen to replicate the series of tests with 5 items. Raw and scale scores were calculated for each student. Apparent achievement levels reflected by the test results were compared with the actual achievement levels. A sample set of results is presented in Figure 3. Each graph has a vertical line or scale representing the continuum of achievement as indicated by the 21 items on the overall test. The difficulty level of each item (numbered with a **bold** numeral on the right hand side of each scale) is shown by its position on the vertical scale. The position of each student sampled is shown by the score obtained, on the left of each vertical scale. For example, the top 3 students on Test A had the same achievement level according to the 21-item test and all scored 5 on Test A. They are shown at the top of the graph and are well clear of the next achievement levels as indicated by the separation on the vertical scale.

```
    Test A              Test B              Test F              Test J
  (N=15  L=5)         (N=15  L=5)         (N=15  L=5)         (N=18  L=5)
 ---------           ---------           ---------           ---------
 555 |                   |                   |               555 |
     |                   |                   |                   |
     |                   |                   |                   |
     |                   | 11                | 11                |
 555 |                   |                   |                   |
     |                   | 12                | 12                |
     |                   |                   |                   |
 545 |                   |                   |                   | 11
     |                   | 13                | 13             454 |
 555 |                   |                   |                   |
     |                   | 14                |                   |
 545 |                   | 10                |                   |
     |                   |                   |                   | 13
     |                   |                   |               433 |
     |                 . |               232 |                   |
     |                   |               222 |                   | 16
     |                   |                   |                   |
     |                   |               222 |                   |
     |               000 |               232 |               222 |
     |                   |               222 |                   |
     |                   |                   |                   | 17
     |  18            000 |                   |                   |
     |                   |                   |                   |
     |   8            000 |                   |   3           111 |
     |                   |                   |                   |
     |   1            100 |                   |                   |   9
     |                   |                   |   4               |
     |   3               |                   |                   |
     |                   |                   |               000 |
     |   4            000 |                   |                   |
 ---------           ---------           ---------           ---------
```

Figure 3. Student scores on alternative tests.

Test A was designed as a very easy test (with all items near the bottom of the scale). The majority of students achieved the same score of 5. Although they are clearly different on the 21-item test, the results of Test A imply that most are identical in achievement as shown in Table 2.

Test B was designed as a very difficult test (with all items near the top of the scale). The majority of students achieved the same score of 0. Although they are clearly different on the 21-item test, the results of Test B imply that most are identical in achievement as shown in Table 2.

Test F was designed to have very difficult items and very easy items (with some items near the top of the scale and some items near the bottom of the scale). The majority of students achieved the same score of 2. Although they are clearly different on the 21-item test, the results of Test F imply that most are identical in achievement as shown in Table 2.

Test J was designed to have a range of items spread from very difficult to very easy (with some items near the top of the scale and some items near the bottom of the scale). The majority of students achieved the scores consistent with their level of achievement on the larger test of 21 items. Students are clearly different on the overall test, and the results of Test J reflect most of those differences in achievement as shown in Table 2.

Table 2. Comparisons of Apparent and Actual Achievement Levels: 5-Item Tests Compared with Actual Test of 21 Items.

Students (n=3 for each group)	Actual Levels of Scaled Scores (logits)	Apparent Levels	Decision
Test A		on Test A	based on Test A
Group 1	3.47	13 of the 15	There is no
Group 2	2.67	students are at the	difference
Group 3	2.16	same level	between the 13
Group 4	1.76		students
Group 5	0.56		
Test B		on Test B	based on Test B
Group 1	-0.16	14 of the 15	There is no
Group 2	-1.14	students are at the	difference
Group 3	-1.43	same level	between the 14
Group 4	-1.77		students
Group 5	-2.67		
Test F		on Test F	based on Test F
Group 1	0.12	13 of the 15	There is no
Group 2	-0.12	students are at the	difference
Group 3	-0.36	same level	between the 13
Group 4	-0.61		students
Group 5	-0.87		
Test J		on Test J	based on Test J
Group 1	3.47	All scored 5	16 of the 18
Group 2	1.76	4, 5, 4	differences
Group 3	1.13	4, 3, 3	between the
Group 4	0.12	All scored 2	students
Group 5	-0.61	All scored 1	matched their
Group 6	-0.87	All scored 0	actual differences

8. USING THE MODEL TO EXPLAIN AND PREDICT

One needs to be aware of sampling fluctuations in the way representatives of each group were chosen for Tables 2 to 5. For example, for the 81 students with the highest (not perfect) score, 78 had a score of 5 on Test A, and 3 had a score of 4. Although the three students sampled had a higher probability of scoring 5, some scores of 4 were possible. Similarly, with the next highest score, 73 of the 77 students scored 5 and 4 scored 4. The next two scores split in different ways. For the higher score, 78 of the next 81 scored 5, while 13 scored 4. For the lower score, 57 of the 73 scored 5, 14 scored 4, and 2 scored 3. Similar concerns apply to the other tests. Small samples of items from a larger population need to be sampled with care to ensure that the inferences from the results are well founded. Without prior knowledge of the properties of the test items, those constructing assessment strategies cannot know whether they have tests like Tests A, B and F (with all their faults) or like Test J - capable of providing useful information at all achievement levels. Trial data must be obtained so that those assembling tests from the items can ensure the resulting tests are like Test J rather than Tests A, B and F. In assembling a bank of assessment strategies, as much care needs to be taken in ensuring that the sampling of items is appropriate and is suitable for the students being assessed, that the scoring protocols are used correctly, and that the results are interpreted appropriately, as is taken to ensure that the items are mathematically sound. For the purposes of explanation this discussion has been confined to small tests. In practice the tests should be much longer (but retain the rectangular item difficulty distribution) in order to make valid inferences about achievement and improved learning.

9. CONCLUDING COMMENTS

The traditional ways of reporting scores are *not* in terms of how well the student has satisfied each component of the curriculum. They compare students with students rather than compare each student's achievements with the curriculum intentions. We do not learn from the data collected what students know or do not know, because this information is ignored in the interpretation of the evidence. We also lack information on the progress made by students over several year levels, as a consequence of using different tests at different stages of learning without ever asking how the scores on each test relate to the overall continuum of achievement in that subject. The assessment techniques described in this paper and earlier ICTMA papers can contribute to avoiding these problems. A quality *assessment* strategy for mathematical modelling complements quality *teaching* approaches to mathematical modelling: we need the former to provide evidence about the latter.

REFERENCES

Blum, W. (1995) Applications and modelling in mathematics teaching and mathematics education – Some important aspects of practice and research. In C. Sloyer, W. Blum and I. Huntley, (eds.). *Advances and Perspectives in the Teaching of Mathematical Modelling and Applications.* Yorklyn, Delaware: Water Street Mathematics, 1-20.

Haines, C.R., Crouch, R. M. and Davis, J. (2001) Understanding students' modelling skills. In J.F. Matos, W. Blum, K. Houston and S.P. Carreira (eds) *Modelling and Mathematics Education: ICTMA9 Applications in Science and Technology.* Chichester, Horwood Publishing, 366-381.

Haines, C.R., Crouch, R. M, & Fitzharris, A. (2003) Deconstructing mathematical modelling: Approaches to problem solving. In Q-X. Ye, W. Blum, K. Houston and Q-Y. Jiang (eds) *Mathematical Modelling in Education and Culture: ICTMA10*. Chichester, Horwood Publishing, 41-53.

Haines, C.R. and Crouch, R.M. (2004) Real world contexts: Assessment frameworks using multiple choice questions in mathematical modelling and applications. Paper presented at the IAEA Conference in Philadelphia, 13-18 June 2004.

Houston, S.K, Haines, C.R. and Kitchen, A. (1994) Developing rating scales for undergraduate mathematics projects. Coleraine, Northern Ireland: University of Ulster.

Izard, J.F. (2004) Best practice in assessment for learning. Paper presented at the Third Conference of the Association of Commonwealth Examinations and Accreditation Bodies on *Redefining the roles of educational assessment*, March 8-12, 2004, Nadi, Fiji, South Pacific Board for Educational Assessment. [http://www.spbea.org.fj/aceab_conference.html]

Izard, J.F. (2002a) Constraints in giving candidates due credit for their work: Strategies for quality control in assessment. In F. Ventura and G. Grima (eds.) *Contemporary Issues in Educational Assessment*. MSIDA MSD 06, Malta: MATSEC Examinations Board, University of Malta for the Association of Commonwealth Examinations and Accreditation Bodies, 5-28.

Izard, J.F. (2002b) Describing student achievement in teacher-friendly ways: Implications for formative and summative assessment. In F. Ventura & G. Grima (eds.) *Contemporary Issues in Educational Assessment*. MSIDA MSD 06, Malta: MATSEC Examinations Board, University of Malta for the Association of Commonwealth Examinations and Accreditation Bodies, 241-252.

Izard, J.F. (1992). *Assessment of learning in the classroom*. (Educational studies and documents, 60.). Paris: UNESCO.

Izard, J.F. *et. al.*. (1983) *ACER Review and Progress Tests in Mathematics*. Hawthorn, Victoria: Australian Council for Educational Research.

Izard, J.F. and Haines, C.R. (1993) Assessing oral communications about mathematics projects and investigations. In M. Stephens, A. Waywood, D. Clarke and J.F. Izard, (eds.) *Communicating Mathematics: Perspectives from classroom practice and current research*. Hawthorn, Victoria: Australian Council for Educational Research, 237-251.

Izard, J.F., Haines, C.R., Crouch, R.M., Houston, S.K. & Neill, N. (2003) Assessing the impact of the teaching of modelling: Some implications. In S.J. Lamon, W.A. Parker, and K. Houston (eds.) *Mathematical Modelling: A Way of Life: ICTMA 11*. Chichester: Horwood Publishing, 165-177.

McGaw, B., Eyers, V., Montgomery, J., Nicholls, B. and Poole, M. (1990) Assessment in the Victorian Certificate of Education: Report of a Review Commissioned by the Victorian Minister of Education and the Victorian Curriculum and Assessment Board. Melbourne, Australia: Victorian Curriculum and Assessment Board.

3.9

AN INTRODUCTION TO CUMCM

Qiyuan Jiang[1], Jinxing Xie[1] and Qixiao Ye[2]
[1]Tsinghua University, Beijing, PR China
[2]Beijing Institute of Technology, PR China

Abstract–*The China Undergraduate Mathematical Contest in Modelling (CUMCM) is a national event held annually. Teams of up to three undergraduate students investigate, model, and submit a solution to one of two simulated real-word problems in engineering, management, etc. The aim of the contest is to expose students to the real-world challenges inherent to mathematical modelling and experimentations, and provide educational (creativity, challenge, etc.) experience unique to problem-based learning. In this paper, we briefly introduce the aims and scope, organization and achievement of CUMCM. A selected contest problem from CUMCM-2004 is discussed. We also talk about some problems and difficulties currently faced by the contest.*

1. AIMS, SCOPE AND HISTORY

The China Undergraduate Mathematical Contest in Modelling (CUMCM) is a national annual contest in China for undergraduates. The aim of the contest is to give students exposure to the modelling process and to improve students' understanding of mathematics, mathematical modelling and experimentation, thereby providing an opportunity for the students to cultivate their creativity in problem solving and improve their ability.

In this three-day (72 hour) contest, teams of up to three undergraduates students will investigate, model, and submit a solution to, one of two modelling problems, which simulate real-word problems in engineering, management, etc. The contest begins at 8:00 a.m. on the third Friday in September and ends at 8:00 a.m. on Monday of the following week. During the contest, teams are permitted to reference any data source they wish, but they must cite all sources. Failure to credit a source will result in a team being disqualified from the competition. Team members may not seek help from, or discuss the problem with, their advisor or anyone else, except other members of the same team. That is to say, inputs of any form from anyone other than the team members are strictly forbidden.

In the USA, the Consortium for Mathematics and its Applications (COMAP) first organized the Mathematical Contest in Modelling (MCM) in 1985. The Interdisciplinary Contest in Modelling (ICM) was first organized in 1999. Teams from Chinese universities have participated in the contest every year since 1989, and recently more than half of the teams of MCM and ICM are from China.

Recognizing that the mathematical contest in modelling is beneficial to students and helpful to the mathematics education reform in universities, the China Society for Industrial and Applied Mathematics (CSIAM) organized CUMCM in 1992. CUMCM is co-organized by CSIAM and the Ministry of Education of China since 1994, and from 1999, the contest has been divided into two categories: Group A for four-year university students and, Group B for two-year college students. Because of the very challenging nature of the contest, it attracts the most competitive students in China in an ever-increasing numbers. Currently, CUMCM has become the most widespread extra curricular scientific activity for undergraduates in China.

Table 1. The statistics on Chinese students participating
in MCM / ICM and CUMCM.

Year	MCM in USA		ICM in USA		CUMCM in China	
	No. of institution[*]	No. of teams[*]	No. of institution[*]	No. of teams[*]	No. of institutions	No. of teams
1989	3(143)	4(211)				
1990	4(158)	6(235)				
1991	10(161)	19(256)				
1992	13(189)	26(190)			74	314
1993	16(164)	38(259)			101	420
1994	33(192)	84(315)			196	867
1995	31(199)	84(320)			259	1234
1996	39(225)	115(393)			337	1683
1997	38(224)	107(409)			373	1874
1998	46(262)	138(472)			400	2103
1999	43(223)	155(479)	14(40)	23(60)	460	2657
2000	46(232)	169(495)	18(50)	29(70)	517	3210
2001	62(238)	198(496)	24(58)	38(83)	529	3887
2002	67(282)	216(525)	29(71)	54(106)	572	4448
2003	61(230)	204(492)	35(84)	83(146)	637	5406
2004	88(253)	297(600)	40(82)	101(143)	724	6881
2005	97(257)	389(644)	51(75)	125(164)	795	8492

Note: [*] The data are collected from http://www.comap.com. The bracketed numbers represent total participation by all institutions and teams.

In 2005, 8492 teams, from 795 institutions, participated in the contest. Almost all of the most prominent institutions, and nearly 50% of all institutions in China, were represented. It is also interesting that more than 80% of the participants are engineering, economics, management, and even humanities majors, other than the mathematics majors that one might expect. Table 1 gives the statistics on Chinese students participating in the contests MCM, ICM, and CUMCM. More details about CUMCM can be found in Li (2001), or from the website http://www.mcm.edu.cn.

2. ORGANISING AND JUDGING SYSTEM

There is a CUMCM National Organizing Committee (NOC), which is setup by the Ministry of Education of China and CSIAM. In most of the provinces or regions in China, there is a CUMCM Local Organizing Committee (LOC). Currently, 27 of the total 34 provinces (or regions) in China have established their own LOC.

Teams register, obtain contest materials, and download the problems and data at the prescribed time through the CUMCM Web site. During the three-day contest, teams of Group A (B) can choose any one from the two contest problems of Group A (B). Each team should submit a solution paper to the corresponding LOC before the contest deadline. After the contest, LOCs start judging to rank the submissions by the contestants. The top 12% of all entries will be submitted to NOC for second round evaluation, and the others will also be ranked and perhaps, awarded prizes at the regional level. After the second round evaluation by NOC, only the top 3% of the total solution papers will be awarded the national-level first prize, and the figure for the 2nd prize is about 8%. Finally, about 15 outstanding papers will be selected and published in the journal *Engineering Mathematics*, which is one of the official journals of CSIAM.

3. THE CONTEST AND ITS INFLUENCE

Most students who register for the contest have some kind of training on how to participate in CUMCM from their mathematical modelling courses, from related mini-courses, or from seminars. Some students prepare for the contest independently through studying materials related to mathematical modelling. They can also get some guidance from their advisors before the contest.

Teachers are the key to the success of CUMCM. In order to properly prepare them, with university professors as advisors, in recent years NOC have cooperated with CSIAM and several universities to organize short-term training seminars. NOC also organizes a national conference CCTMMA (China Conference on the Teaching of Mathematical Modelling and Applications) every two years as an educational forum of exchange, where teachers share information, and discuss how to mentor the students effectively, and how to prepare and teach a high quality mathematical modelling course. These activities have had great impact on the mathematics education reform and have enhanced the teaching quality of courses related to the mathematics in most universities and colleges. The contest also encourages the publication of many innovative textbooks on mathematical modelling and mathematical experiments (Jiang, 1998).

The contest is a real challenge to its participants and is much appreciated by the students. The special experience for students, provided by the contest, is very helpful in tapping their innovative potential and in strengthening their cooperative spirit. The contestants summarise their experience in one phrase "Once participated, lifelong benefit". The whole contest process consists of three stages, namely, the training and preparation before the contest, the hard work during the three-day contest, and the summing up of students' own experience and doing further work on the contest problems after the contest. Through these stages, students' creativity and overall ability are greatly improved. Indeed, most of the winners of CUMCM have done very well in their successive courses and projects before their graduation. The scientific and industry communities are getting to know more and more about CUMCM, and they are glad to accept the students who have the experience of the contest when they go to graduate schools or find jobs after their graduation. Some industry corporations, such as World-Sky Group, Netease Corporation and Higher Education Press, also sponsor the contest.

4. DIFFICULTIES AND PROSPECTS

CUMCM does encounter some difficulties as the participation continues to grow. The most important task facing NOC is about how to improve the whole quality of the contest. First of all, good contest problems are vital to the success of the contest. Contributing a good modelling problem, which is both a meaningful real-world problem, and which is also a solvable problem by most teams within three days, is a challenging task for the organizers.

Another difficulty the organizer faced is how to ensure the equity and fairness of the contest. Since the contest lasts for three days and it is essentially a completely open contest, it is not easy to ensure that teams do not violate the contest rules. As a matter of fact, some teams do violate them, for instance, by looking for help from teachers or other persons outside the team even on the internet. The organizer emphasizes the very importance of self-discipline. A firm policy is observed by NOC, namely, once we have the evidence of violating the contest rules by some teams, these teams will be disqualified from the competition.

5. AN EXAMPLE OF THE CONTEST PROBLEM

5.1. The Problem

The problem A from CUMCM-2004 is "Planning Temporary Mini Supermarkets for the Olympic Games", which is summarized in the following.

The construction work of the Olympic Games 2008 in Beijing has been in planning and implementation processes. During the Olympic Games, temporary Mini Supermarkets will be built around the stadium for supplying food, souvenir and tourist commodities to the spectators, tourists and members of staff, each Mini Supermarket (MS) consisting of variety shops. For MS around the stadium, their location, size and the amount of sales should satisfy three basic requirements:

demand for shopping, reasonable distribution and commercial profit during the Games.

The planning layout of the main stadiums for contest is shown in a figure (omitted here for saving space). For simplicity we only give, in Figure 1, the related parts and regions, such as streets (white denotes for pavement), bus and taxi stops, parking area, subway stations and restaurants etc., in which the areas marked with A1-A10, B1-B6 and C1-C4 denote the prescribed 20 shopping centres consisting of the MS.

One way to find the patterns of the consumer movement is to send out questionnaire forms to the spectators; they are the principal consumers during the preview games for investigating purchase demand and appetite of tour. Suppose that three games had been held in a stadium (Figure 2) and related data are collected and is shown in the Appendix (available at http://www.mcm.edu.cn).

Your team is asked to be a consultant for planning MS for the 20 shopping centres shown in Figure 1 according to the following steps:

●   Based on the data of questionnaire given in the Appendix find features of the spectators' routes, meals and purchase etc.

●   Suppose each spectator has two main routes each day. During the Olympic Games, one route to travel in and out the stadium and another route in going for a meal, and they always adopt the shortest route. Based on the result in 1 please calculate the distribution of consumer flux (in percent) in the 20 shopping centers in Figure 1.

●   Suppose two different sizes of MS, large and small, can be chosen. Please plan the MS for the 20 shopping centres, i.e. the numbers of different MS in each centre such that three requirements are satisfied.

●   Explain that your method is reasonable and the result is practicable.

Hints:

1. In commerce the "shopping loop" may be used to describe the covering area of shops. The main factor determining the choice of shop location is the buyer movement with their purchase demand in a shopping loop.

2. For simplicity assume that the National Stadium can admit about 100,000 spectators, the National Gymnasium about 60,000 and the National Swimming Center about 40,000, where each stand of the three buildings admits about 10,000. Assume also that each exit gate faces just one shopping centre and that all the shopping centres have the same area.

Appendix:

Send out questionnaire forms to the spectators three times with 33% reply, totally about 10,000 replies. Detailed data can be found in the attached access database, where the ages are divided into four groups: (1) less than 20, (2) 20-30, (3) 30-50 and (4) more than 50. We may design for four ways of travelling: by taxi, by bus, by subway and by car; three kinds of meals are provided: Chinese meal, Western-style

food and in marketplace (fast food); and in six price ranges (except for meals):
(1) 0-100, (2) 100-200, (3) 200-300, (4) 300-400, (5) 400-500 and (6) more than 500
(RMB).

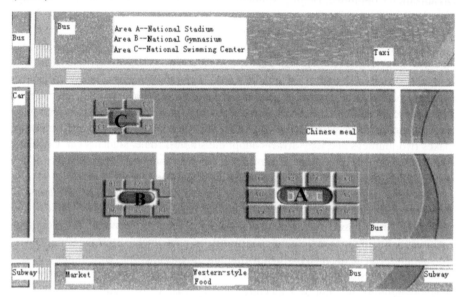

Figure 1. The planning layout of the main stadiums

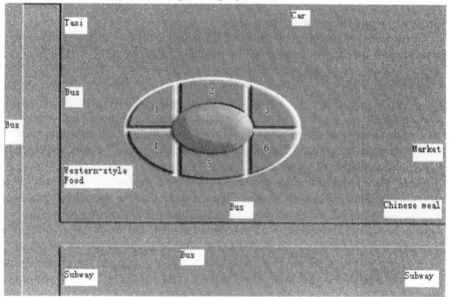

Figure 2. A Ready Stadium.

5.2. Comments on Students' Solutions

The background of this problem is the requirement of the Olympic Games 2008 Beijing. The author of the problem has received help from Beijing Olympic Games Organization Committee, Beijing Municipal Administer Committee and Beijing City Planning Committee for the provision of the design assignment of temporary Mini Supermarkets (MS) and the stadium planning Figures.

This problem needs firstly to find features of the spectators in routes, meals and purchase preference etc. from the data of questionnaire. Students may use statistical methods or data mining to achieve this task. The related rules with the consumer movement should be discovered as enough as possible. For example, the patterns that people, of either sex and of different ages, may have different personal shopping preferences should be considered. Based on the given conditions that each spectator makes two journeys and they always adopt the shortest route, students can calculate the distribution of consumer movement in the 20 shopping centres. One of the approaches is to use a network flow model for calculation.

The designs for MS of the 20 shopping centres i.e. the numbers of different size MS in each center are the primary and pivotal parts of this problem. The keys are describing three basic requirements of shopping centers (demand for shopping, reasonable distribution and commercial profit) in mathematical terms. The demand for shopping can be decided by the distribution of consumer movement, but one fact, often ignored by students, is that demand for shopping must be considered according to whole building area not only a centre. The reasonable distribution means that the discrepancy of the numbers of MS in each centre is not too large. The problem gives no information about commercial profit, so students need to collect some data about cost, profit, etc. of MS with different sizes.

The designs for two sizes of MS of the 20 shopping centres satisfying three basic requirements are actually optimization problems. The keys of establishing an optimal model are mathematical formulations of decision variables, objective functions and restricted conditions. Clearly the decision variables are the numbers of two sizes of MS of the 20 shopping centres. Many students choose rationally commercial profit as the objective function and take demand for shopping, reasonable distribution as the restricted conditions. In general, such an optimal model is a linear or nonlinear integer programming which can be solved in principle by existing methods and software.

Some students separate this problem into three independent sub-problems; *each sub-problem aims only at one building*. They consider that the spectator in National Stadium would usually not go to shop in MS of National Swimming Centre. Such a separation way not only reduces the number of the decision variables resulting in a big simplification, but also even more accords with the actual situation.

It is worthwhile to notice that some students do not analyze the rationality of computing results. For example, some papers show that there are tens, or even hundreds, of MS in a centre. This is clearly incorrect in practice.

The modelling approach for this problem can also be used in many similar situations such as planning temporary mini supermarkets at big exhibitions, state fairs, huge temple fairs in China, amusement parks etc. Another characteristic of this contest problem is that it is very open. Various mathematical models and solving methods appear in students' solution papers. These include primary calculation, statistics, shortest path algorithms, network flow models, mathematical programming, circuit simulation method and data mining.

6. CONCLUSION

This paper gives a brief introduction to CUMCM, and a selected contest problem from the contest is discussed. According to our practice, the contest provides educational experiences unique to problem-based learning and teaching.

REFERENCES

Li, D. (ed) (2001) *China Undergraduate Mathematical Contest in Modelling* (in Chinese), 2nd Edition. Beijing: Higher Education Press.

Jiang, Q. (1998). Teaching of Mathematical Modelling in China. In P. Galbraith *et al.* (eds) *Mathematical Modelling: Teaching and Assessment in a Technology-Rich World*. Chichester: Horwood Publishing, 337-344.

Section 4
Everyday Aspects of Modelling 'Literacy'

4.1

FUNCTIONAL MATHEMATICS AND TEACHING MODELLING

Hugh Burkhardt
Shell Centre for Mathematical Education
University of Nottingham, UK
and Michigan State University USA

Abstract–*Mathematical literacy (ML) is the capacity to make mathematics functional in everyday life – that is, to make effective use of it in better understanding the world and in meeting its challenges. The current focus on ML provides a unique opportunity to make modelling a reality in schools, at last. This paper will first review evidence that much of school mathematics is currently non-functional for many people at all levels of achievement. It will then look at the missing ingredients in current curriculae that are needed to develop ML, notably the explicit teaching of modelling skills. Using existing examples of proven effectiveness, it will outline what is needed to make ML an outcome of school mathematics.*

1. FUNCTIONAL MATHEMATICS – A DESCRIPTION WITH EXAMPLES

Mathematical literacy (ML) under various more-or-less equivalent names, including *functional mathematics*, *quantitative literacy*, and *numeracy,* is now becoming a major concern of mathematics curriculum improvement programmes in many countries. This movement is epitomised by the emergence of PISA (OECD 2003), which aims to assess ML, as a prime international comparison of standards of performance in mathematics. PISA is both a symptom of, and a support for, this new emphasis on mathematical literacy for all citizens as a curriculum responsibility.

> **Functional mathematics** (FM) is mathematics that *non-specialist adults, if they are taught how,* will benefit from using *in their everyday lives* better to understand the world they live in, and to make better decisions.

This reflects the PISA definition but may be less susceptible to misinterpretation as "business as usual". FM is distinct from *specialist mathematics* that prepares people for certain professions like physics, engineering, economics or accounting. It is also much more than *applicable mathematics* – that is, mathematics that is *potentially* useful in applications and modelling (as most mathematics is). Current curricula do address both these important needs; however, they lack key elements so that, *for most people, secondary school mathematics is almost entirely non-functional.*

If you doubt this, perhaps surprising, assertion, ask any adult with a job that doesn't involve mathematics when they last solved a quadratic equation, used

Pythagoras' Theorem or, indeed, *any piece of mathematics they were first taught in secondary school.*

Probing more deeply will often reveal that they use quite a lot of *elementary* mathematics in a quite sophisticated way – a key feature of ML – but they do not see it as mathematics. The powerful mathematical tools they learnt in secondary school where they spent at least a thousand hours on them, remain unused.

To see what really functional mathematics can equip you to do, consider the following tasks. (Assessment tasks are a compact and powerful way of exploring the meaning of general statements) These examples are chosen as tasks that any well-educated person should, and could, be able to tackle sensibly by the time they leave school at age 18. The comments on each task illustrate current deficiencies.

Primary teachers

> In a country with 60 million people, about how many primary school teachers will be needed? Estimate a sensible answer, using your own everyday knowledge about the world. Write an explanation of your answer, stating any assumptions you make.

This kind of back-of -the-envelope calculation is an important life skill. Here it requires choosing appropriate facts (6 years in primary out of a life of 60-80 years, one teacher for 20-30 kids), and recognizing and using proportional relationships giving $60*(6/70)/25 = 0.2$ million primary teachers (to an accuracy appropriate to that of the data). This kind of linkage with the real world, common in the English curriculum, is rare in school Mathematics (and absent in most tests).

The following task reflects key features of some high-profile miscarriages of justice, where an 'expert' witness's gross errors went unchallenged by lawyers or judge.

Sudden Infant Deaths = Murder?

> In the general population, about 1 baby in 8,000 dies in an unexplained "cot death". The cause or causes are at present unknown. Three babies in one family have died. The mother is on trial. An expert witness says:
> "One cot death is a family tragedy; two is deeply suspicious; three is murder. The odds of even two deaths in one family are 64 million to 1"
> Discuss the reasoning behind the expert witness' statement, noting any errors, and write an improved version to present to the jury.

What we expect here from a non-specialist educated citizen is not a full statistical analysis, which would also need more information, but a recognition that the reasoning presented is deeply flawed. There are two elementary mistakes in the statement, as well as others that are more subtle. It would be correct to say, for example:

- The chance of these deaths being entirely *unconnected* chance events is very small indeed – if there has been one death, the chance of two more unconnected deaths is about 64 million to one.
- What can the connection be? It may be that the mother killed the children; on the other hand, particularly since the cause(s) of cot death are still unclear, there may be other explanations. For many conditions (cancer and heart disease, for example), genetic and/or environmental factors are known to affect the probabilities substantially.

Any lawyer or judge with functional mathematics should see problems with the witness statement. Their failing was not lack of basic skills (they could, no doubt, calculate the chance of a double six on throwing two dice) but a lack of understanding of the basic and necessary assumption of independence.

Ice cream van

> You are considering driving an ice cream van during the Summer break. Your friend, who "knows everything", says that "It's easy money". You make a few enquiries and find that the van costs £600 per week to hire. Typical selling data is that one can sell an average of 30 ice creams per hour, each costing 50p to make and each selling for £1.50.

> How hard will you have to work in order to make this "easy money"? Explain your reasoning.

This task, with two more like it, was used in a research study (Treilibs et al, 1980) of the performance of 120 very able 17 year old students. Many solved the tasks, using arithmetic and, sometimes graphs. *None used algebra*, the natural language for formulating such problems. Their algebra was non-functional, despite 5+ years of great success in the standard imitative algebra curriculum

The mathematical concepts and skills needed to tackle these and similar problems are all taught and 'learnt' by age 14 – yet they are non-functional for many well-educated adults. This should not be surprising, since they have not been taught *how to use their mathematics in tackling everyday life problems* as they arise. The skills of modelling non-routine problems are not trivial, but they can be taught; currently they are not. As the following outline (and other papers in this book) will show, there are established methods, accessible to typical teachers and their students, for teaching people to use their mathematics on real problems. Any curriculum that delivers functional mathematics will have to include them.

2. CURRICULA FOR FUNCTIONAL MATHEMATICS: PRINCIPLES

What changes in current curricula seem likely to be needed to deliver mathematical literacy – mathematics that is really functional in people's lives? First, let us review the various key questions.

What does ML/FM involve?
Functional mathematics involves all the key aspects of 'doing mathematics' (see for example, Schoenfeld 1992):

knowledge of concepts and skills
strategies and tactics for modelling with this knowledge
metacognitive control of one's reasoning
disposition to *think* mathematically, based on **beliefs** about maths.

These are not, of course, independent elements but must be integrated into coherent **mathematical practices** for tackling whatever problem is at hand. (Lynn Steen (2002) has pointed out that FM involves *"The sophisticated use of (often) elementary mathematics"* whereas current curricula have it the other way round.)

Of the standard phases of modelling process:

formulation of a mathematical model of the practical situation
solution of the model by mathematical manipulation
interpretation of the model solution in terms of the practical situation
evaluation of the model.
reporting of the results and reasoning to those concerned.

Only *solution* gets serious attention in most current school mathematics curricula.

The following three similar-looking problems show the need to address *formulation*:

a) Joe buys a six-pack of coke for $3 to share among his friends. How much should he charge for each bottle?

b) It takes 30 minutes to bake 6 potatoes in the oven. How long will it take to bake one potato?

c) It takes 30 minutes to play Beethoven's 6th Symphony. How long will Beethoven's 1st Symphony take to play?

a) is typical of problems on proportional reasoning; students have been shown the model and know that *every* problem in those lessons will be of that kind. While *b)* and *c)* are superficially similar in structure, for *b)* either constant (30 mins) or proportional (~5 mins) is appropriate, depending on the type of oven (conventional or microwave). For *c)* there is, of course, no predictable connection. The essential point here is that, in current curricula, students get no teaching or experience in *formulation* – in choosing a model that fits the problem. *Interpretation, evaluation, and reporting* are similarly neglected.

Moving beyond imitation

Modelling everyday life situations is at the heart of functional mathematical literacy. Young people and the adults they become will face a range of problems where the tools of mathematics can help – common challenges in everyday life and more specialised problems in their work. These cannot all be anticipated, let alone covered, in school; we have to develop students' ability to use their mathematics in tackling non-routine problems – problems they have not analysed before or been shown how to do, like those in §1.

In contrast, mathematics teaching in most classrooms at all levels is based on imitative learning – the 'EEE' model of *explanation – example – exercises*. This works in the short term for many students learning specific skills. There is much research showing that, for many students, EEE does not lead to long-term learning, because the essential conceptual foundation of the skills is not developed. EEE does not develop students' ability to use them in other situations. For this a richer pattern of learning activities is needed. This 'teaching by transmission' is not enough. The main extra elements needed are:

active modelling with mathematics of non-routine practical situations
diverse types of task, in class and for assessment
students taking responsibility for their own reasoning
classroom discussion in depth of alternative approaches and results
teachers with the skills needed to handle these activities.

Table 1 (Burkhardt et al, 1988) shows the changes in roles and mutual expectations:

Table 1. Teacher and Student Roles.

for imitative learning	for modelling, *add*
Directive roles	**Facilitative roles**
Manager	Counsellor
Explainer	Fellow student
Task setter	Resource
with students as	*with students as*
Imitator	*Investigator*
Responder	*Manager*
	Explainer

3. FUNCTIONAL MATHEMATICS IN CLASSROOMS: TASK EXEMPLARS

What do we know about how to achieve these changes, and what more do we need to know? What resources will be needed? How far have they already been developed? How can mathematics teachers be enabled to handle the challenges, that are new to most of them?

The situation is encouraging, though there are obstacles. The last 40 years have seen the explicit teaching of modelling with mathematics move forward (see Burkhardt with Pollak 2006) from small scale explorations in the 1960s, through pilot developments in the 1970s, to developments for typical classrooms in the 1980s, to established courses, albeit in a small minority of mathematics classrooms worldwide (many of these are described in this series of ICTMA books). I shall not attempt a detailed review but seek to show through working examples what is involved.

§1 included tasks that reflected the goals of functional mathematics for adults, here I shall give some insight into what is involved in the classroom through examples of tasks that work well there. They illustrate the learning goals explicitly. The pattern of classroom activities will be exemplified in the next section.

Functional mathematics involves a much broader range of task types than current curricula, asking students to *investigate, plan, design, evaluate and recommend, critique and improve*, or simply *interpret*. (These assessment tasks are much richer in a lesson activity form – but need more space to show than I have!).

Planning tasks, like *A day out in Derbyshire*, do not have to involve the whole of the planning process. This process sequence *critique – improve – create* reflects a scale of challenge, and of time needed for meaningful work, that is useful for task and curriculum designers. In an IT-rich world the data would be provided via the web rather than on paper. We found that students enjoy finding the many faults in such a plan – especially when it comes from the examination board!

A day out in Derbyshire (p181)

| Alison and two friends have planned a cycling trip around Derbyshire on Saturday.

Here is their plan for the day >>>

Read through the plan and the information sheets (*map, timetables, brochures supplied*).

If you find a mistake, or realise something has been forgotten, write it down and say how they should change the plan. | Meet at Loughborough station at 7.23 am. Buy tickets and then catch the train to Derby. This arrives at 7.51 am.

At Derby, catch the 8.20 am train to Cromford. This arrives at 8.41 am.

Here are the instructions for getting to the Cycle Hire centre:

"Turn left as you come out of Cromford station, walk along by the river and down Mill road. Cross over the A6. Walk up Cromford hill for about 1/2 mile and you will see.

.... (*the plan continues*) |

Design is an area that is motivating for students and important in practice.

Design a Tent

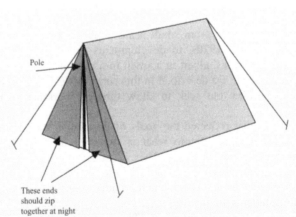

Your task is to design this tent, sketching patterns for the plastic waterproof base, for the fabric of the tent, and for the two vertical tent poles. Give all dimensions.

It must hold two adults with backpacks, with space to move around on their knees. As usual with FM/ML, this design task needs elementary mathematics, used in a flexible and reliable way. The visualisation challenge of relating three dimensions to two is not trivial.

Making a case

The spreadsheet contains 2 sets of reaction times, 100 each for Joe and Maria. Using this data, construct two arguments:

A: that Joe is quicker than Maria

and

B: that Maria is quicker than Joe

This kind of *evaluate and recommend* task (spreadsheet not shown here) builds understanding and intelligent scepticism, showing how political and marketing data is used – a crucial part of mathematical literacy.

Likewise, a lot of information is now presented in graphical form – *interpretation* tasks (see for example, Swan et al, 1986) develop and test understanding.

4. LEARNING FUNCTIONAL MATHEMATICS

What do these changes in tasks mean for the pattern of classroom activities needed to enable students to meet such performance targets. I shall illustrate this by describing a curriculum and assessment development that has enabled some fairly typical teachers in English secondary schools to develop the mathematical literacy of their students across the age range 11-16. Their achievement was assessed by a public examination, administered by a major examination board.

This development, *Numeracy through Problem Solving* (NTPS, Shell Centre 1987-89) sets out to support teachers without closing down the essentially open challenge. Developed through a process of creative design and systematic refinement in a sequence of trials, it worked well with typical teachers and students across the full ability range. It illustrates the power of well-engineered tools to help teachers (as with others) to tackle challenging tasks more successfully than they could without such help. NTPS offers a sequence of five modules that develop students' ability to use mathematics, together with other skills, in tackling problems of concern or situations of interest in everyday life. Each module provides a theme within which the students take responsibility for planning, organizing or designing.

The students work on a group-project basis, primarily guided by a student booklet, with the teacher playing a consultant role. Each module is designed to take between 10 and 20 classroom hours to complete, and has four Stages.

1. Explore the context – students explore the domain by working on and evaluating exemplars provided.
2. Brainstorm approaches – generating and sifting ideas, which are developed and implemented in detail in Stage 3.
3. Detailed design and development of the analysis and the product
4. Implementation and evaluation – each group evaluates the things that the other groups have produced.

For each module these stages take a form which fits the context. We outline this for one module, **Be a Shrewd Chooser.** Here students research and provide expert consumer advice for 'clients' in their class.

1   Students listen to a radio show on audiotape which contains a number of interviews with people who have just bought different products, and an interview with two students who have been involved in producing a consumer report on choosing orange drinks. As students reflect on and analyse the tape and the report, they begin to consider important factors that are taken into account when making a choice and different methods of making consumer decisions.

2   Students in a group now begin to work on their consumer report. They have to choose a product and decide on their research aims and methods.

3   Students develop their plan. They will be involved in conducting surveys, writing questionnaires and carrying out experiments in the classroom. They will also be considering how best to present their findings. This could involve posters and oral presentations in addition to written reports.

4   All the written reports are circulated around the other groups, and any group making an oral presentation does so. The reports are evaluated by the rest of the class, and then each group improves its own report taking into account these comments.

The four other modules – Be a Paper Engineer, Design a Board Game, Produce a Quiz Show, Plan a Trip – take a similar four-stage approach. Extracts from each module can be found at www.mathshell.com .

Two kinds of assessment are provided. Formative assessment, built into the teaching materials, is designed to check that each student in a group understands all aspects of the work, not simply those for which they have been responsible. A final examination at the end of the module assesses how well students can transfer what they have learned to more or less closely related problem situations. (The project outcomes, though evaluated by other students in the class, were not included in the formal assessment, reflecting concerns over how to assign individual credit for group work)

5. POLICY AND PROGRESS: HOW DO WE GET THERE?

If a school system wants to make functional mathematics an important goal of its *implemented* and *achieved mathematics curricula* what must it do? It is necessary to make it a well-defined part of the *intended curriculum*, as set out in policy – the description and exemplification above provide a basis for that. But this is far from sufficient; most curriculum specifications, including the National Curriculum in England, contain important elements that are completely absent in all-but-a-few classrooms. For successful implementation, everyone involved must be both *motivated* and *enabled* to make the changes. Teachers, as ever, are the key. The system challenge that this presents is generally underestimated. It requires:

pressure and *support*: Teachers and other professionals involved are busy people working under pressure; that pattern of pressure must be changed to give appropriate priority to functional mathematics. In systems with high-stakes assessment for accountability, this will only happen if the *tested curriculum* (that is, the high-stakes tests) includes FM. Equally, teachers and others have to be *enabled* to respond effectively; for most teachers this requires well-engineered teaching materials and professional development support. (Pressure is less expensive than support, which tends to be neglected by politicians – encouraged by the "we can handle whatever is required" posture of most professions.)

digestible pace of change: The professional development required of teachers and others to move beyond an imitative curriculum in the way described above is profound. It will only happen through a well-supported sequence of successful experiences in their own classroom. This requires well-engineered tools and live professional development over a period of years, encouraged by recognizable achievements at each stage.

One model that works well in practice involves (see Swan et al, 1986):

- one new task type each year in the high-stakes tests, embodied in a
- well-engineered 3-4 week curriculum unit of teaching and in-class assessment materials, supported by
- well-engineered professional development tools that work well on an in-school do-it-yourself basis on which to build the essential but expensive (and always-too-limited) live professional development.

The development of such modules, so that they work well with typical teachers and students in realistic circumstances off support, is a challenging design and development task. Thus a substantial research and development effort over a period of years will be needed to get good assessment, encouraging good learning that produces functional mathematical literacy in all students. The pay-offs in student performance in, attitude to, and take-up of mathematics as a whole will be substantial.

The challenge of developing a model of change that works is not yet recognised at policy level. Politicians and senior civil servants generally assume that once "difficult decisions" on what should be done have been made and legislated, faithful implementation will follow. In practice, outcomes that resemble the intentions are rare. The profound changes needed to make school mathematics functional will only happen if they are well-engineered – that is, skilfully designed to take into account the factors sketched here and carefully developed on a limited scale first. We now know how to teach the full range of modelling and other skills involved; it is the system implementation problems that are the difference between success and failure.

To conclude, Bill Gates has said that "the three R's for the 21st Century are *Rigour, Relevance and Relationships."*

Learning functional mathematics, built on modelling with mathematics, develops them all. Any mathematics curriculum for all students should give it priority.

ACKNOWLEDGEMENTS

The vitality of mathematics in most classrooms rests on the work of designers; the examples quoted here are largely the work of the Shell Centre team, particularly Malcolm Swan, who has led the work on functional mathematics.

REFERENCES

Burkhardt, H. (1981) *The Real World and Mathematics,* Blackie-Birkhauser; reprinted (2000) Nottingham: Shell Centre Publications http://www.mathshell.com/scp/index.htm .

Burkhardt, H., Fraser, R., Coupland, J., Phillips, R., Pimm, D. and Ridgway, J. (1988). *Learning activities and classroom roles with and without the microcomputer. Journal of Mathematical Behavior,* 6, 305–338.

Burkhardt, H. with contributions by Pollak, H.O. (2006) *Modelling in Mathematics Classrooms: reflections on past developments and the future. Zeitschrift für Didaktik der Mathematik,* 38 (2).

OECD (2003) *The PISA 2003 Assessment Framework: Mathematics, Reading, Science and Problem Solving Knowledge and Skills.* Paris: https://www.pisa.oecd.org/dataoecd/38/51/33707192.pdf .

Schoenfeld, A. H. (1992) Learning to think mathematically: Problem solving, metacognition, and sense-making in mathematics. In D. Grouws (ed.), *Handbook for Research on Mathematics Teaching and Learning.* New York: Macmillan, 334-370.

Steen, L. A. (ed) (2002) *Mathematics and Democracy: the case for quantitative literacy.* The National Council on Education and the Disciplines (NCED), USA. http://www.maa.org/ql/mathanddemocracy.html .

Shell Centre: Swan, M., Binns, B., Gillespie, J. and Burkhardt, H. (1987-89) *Numeracy through Problem Solving.* Harlow: Longman, revised (2000) Nottingham: Shell Centre Publications.

Swan, M., Pitts, J., Fraser, R., Burkhardt, H. et al. (1986). *The Language of Functions and Graphs* Manchester: Joint Matriculation Board, reprinted (2000) Nottingham: Shell Centre Publications. http://www.mathshell.com/scp/index.htm.

Treilibs, V., Burkhardt, H. and Low, B. (1980) *Formulation processes in mathematical modelling,* Nottingham: Shell Centre Publications.

4.2

MODELLING AND THE CRITICAL USE OF MATHEMATICS

Jussara de Loiola Araújo
Universidade Federal de Minas Gerais (UFMG), Brazil

Abstract–*The objective of this paper is to present a discussion about the use of mathematical formulae, or of mathematical models, by professionals from other fields, and students who work with modelling in mathematics classes, who are unfamiliar with their origin, and have not discussed and reflected on the consequences of their usage. This type of discussion is of critical concern to mathematics education. The discussion will be built around the report of an experience with a course proffered to practising and future mathematics teachers, in which one of the themes was mathematical modelling. Through the development of three kinds of activities a discussion spontaneously emerged regarding the use of mathematics in an unquestionable way. Thus, a second objective of the paper is to show how this discussion regarding the unquestionable use of mathematics (or of mathematical models) in society became increasingly important to one group of future mathematics teachers during their first contact with mathematical modelling in mathematics education.*

1. INTRODUCTION

Mathematical modelling has been an increasingly popular theme in mathematics education conferences and discussions among mathematics teachers. However, the emphasis on the theme, and the considerable discussion about it, do not signify a sizeable scientific movement. Niss (2001) points out that rather little research has been developed in this area, and that the majority of studies are related to the implementation of applications and modelling in mathematics curricula at various levels. Niss (2001) presents some themes related to applications and mathematical modelling in mathematics education that still need further research. Among them, I point out the need for research that involves a critical analysis of given models. As the author writes,

> ... mathematics education should not only equip recipients with the technical ability to successfully deal with the building and utilization of models, but also with the ability to reflect, analytically and with sound judgement, on given models (…) such matters ought to be subject of research. (p81).

Taking this into account, the main objective of this paper is to present a discussion, based on a particular experience, about the use of mathematical formulae, or mathematical models, by professionals from other fields and students working with mathematical modelling in mathematics classes who have no prior knowledge regarding their origin, and have not reflected on and discussed the consequences of this usage.

The experience in question took place in the course *Topics in Mathematics Education*. This course is optional for students in the Mathematics Teacher Education and Mathematics Graduate Programs[1] at UFMG in Brazil. Here, I will consider the session of the course proffered during the second semester of 2002 to a group of approximately 30 students (15 undergraduate and 15 graduate students). There were two teachers responsible for the course, and each one discussed mathematics education issues that were more related to her own field of research. I was one of these teachers, and I decided to study mathematical modelling in mathematics education. I had planned to approach modelling through three different activities: theoretical discussions, studies of specific teaching situations, and the development of modelling projects. Throughout the course, we observed some situations of "blind application" of mathematics that left us disturbed. That is, we found some cases of mathematics being used in other fields without an understanding of this usage.

Thus, a second objective of the paper is to show how this discussion regarding the unquestionable use of mathematics (or of mathematical models) in society became increasingly important to one of the groups (composed of future mathematics teachers) in the course, during their first contact with mathematical modelling in mathematics education.

In the next section, I will point out some aspects of the theoretical discussion about mathematical modelling (the first of the three activities in which the subject was approached). Following that, I will present the study, carried out by the group, of a modelling project developed in another course (the study of a specific teaching situation). In the fourth section, I will present some characteristics of the modelling project developed by the group itself. In these two last sections, I will point out how the question of the "blind application" of mathematics was emerging and transforming itself for this group. At the end, I will go a step beyond the experience to reflect on this issue, referring to the literature on the subject for support.

2. SOME THEORETICAL ASPECTS

In a very general way, mathematical modelling can be understood as the utilization of mathematics to approach, or to solve, a non-mathematical everyday situation or problem. In mathematics education, it takes special form, according to the desired objectives and the limitations imposed by the context, as I have discussed in a previous study (Araújo, 2002). Particularly, I consider mathematical modelling in mathematics education

> as an approach, by means of mathematics, to a real non-mathematical problem, or of a real non-mathematical situation, chosen by the students working together in a group, in such a way that issues related to critical mathematics education form the basis for the development of their work. (Araújo, 2002 p39).

In the above perspective, I cite critical mathematics education, which has been discussed by several researchers. To Skovsmose (1994), for example, critical mathematics education is concerned about the development of *mathemacy*, which is equivalent to Paulo Freire's concept of *literacy*. Here, the objective is not simply to develop mathematical calculation abilities, but also to promote students'/citizens' critical participation in society, discussing political, economic, environmental, etc., questions in which mathematics is drawn on for technical support. In this case, the critique is of mathematics itself, as well as its usage in society, and is not limited only to the teaching and learning of mathematics.

The absence of this critical vision in the development of modelling projects and/or in analysis of established models can reinforce what Borba and Skovsmose (1997) call the *ideology of certainty in mathematics education*. According to the authors, the ideology of certainty supports a vision of mathematics that imbues it with the power of the custodian of the definitive argument in any debate in society. As such, it is used to present political decisions, for example, in such a way as to suggest that the decision made represents the best path to be followed, without leaving room for argument, which characterizes its use as the *language of power*. In other words, mathematics participates decisively in the structure of political debate, which explains its political dimension in society. This being true, those who do not have access to mathematics are subject to the control and the will of those who do and who have power in society, since the impossibility of access signifies not participating in the complex political debate, also supported by this science. As a consequence, social inequalities, racism, socio-economic discrimination, etc., can be reinforced.

Thus, following my perspective of mathematical modelling, I ask students to develop projects seeking to stimulate discussions that question the ideology of certainty, and motivating them to take into account the concerns of critical mathematics education. I intend that these discussions and concerns follow throughout the development of the mathematical modelling projects, which are undertaken in four steps:

1) *choosing the theme*: the students divide into groups and define a theme to be studied;
2) *presentation of the research project:* each group presents a work plan for developing the project;
3) *development of the research project*: the groups put their plans into action, knowing ahead of time that they may undergo modifications;
4) *final presentation*: presentation of the work to the entire class and final written report.

3. FUTURE TEACHERS ANALYSE A MODELLING PROJECT

The analysis of a project developed as described above constituted part of the activity called "study of a specific teaching situation" carried out in the Topics in Mathematics Education course. This activity was proposed according to a suggestion made by Barbosa (2001), who claims that *studying the teaching cases is a stimulant to Modelling pedagogy thought, which may create dilemmas, tensions, possibilities, etc.* (p193).

The classroom teaching situation analyzed in the course focused on modelling projects in Mathematics I, a required course for geography students at UFMG which I also taught. For the analysis of this situation, the Topics students were given access to, and analyzed individually or in groups: 1) a text that had been presented to the Mathematics I students containing information about the development of modelling projects; 2) a text (Araújo, 2004) reporting on the experience; 3) the final reports of all the groups of the Mathematics I course presenting their modelling projects; and 4) the video tapes of the group presentations. The group in the Topics course, composed of Alessandra Silva, Célio Mellilo, Cláudia Lima and Deise Arruda[2], who were students in the Mathematics Teacher Education program, chose to analyze a project entitled "Environmental Monitoring and Health – water for human consumption", developed by one of the groups of Mathematics I students. The final project report and video tape of their presentation were made available to the group.

The Topics course group began their analysis by examining the description of the objectives, the justification, and some concepts, such as "environmental monitoring" and "waste water", which appeared in the Environmental Monitoring and Health project report. Then, in the sections "characterization of waste water" and "case study", the group presented a series of formulas included in the report, accompanied by definitions necessary for their understanding. These included, for example, the formulas for calculating the quota of water per capita for a given population, the average household discharge of waste water, the density index of eggs of the mosquito that transmits dengue fever, and others (Because it is not necessary to the following, these formulae will not be reproduced here). Following that, the group presented their critique which, according to their justification, focused on the *mathematical part* and on the *mathematical models used* (Silva et al, 2003a p1).

Their critique ranged from small details, such as the failure to mention the fact that a certain index was given in percentage, and the presentation of two formulas as being distinct from one another, when in fact they only differed with respect to the units of measures used, to a criticism of an extremely important characteristic: the presentation of formulas and index values without including a justification for their creation or utilization. In the words of the group:

> The formulas used to calculate the quota per capita [of water of a population] and average discharge of waste water are presented without explanation regarding the justification for their use and/or appearance (...) (Silva et al, 2003a p7).
>
> Following the report of these formulas, pre-established indexes are presented for interpreting the results; however the sources and parameters for the indexes are not mentioned (ibid, p7).

I recall that, during the presentation of their analysis to the entire class, the group stated that *the formulas seemed to have appeared out of nowhere,* and they harshly criticized the group of Mathematics I students for not being critical themselves, suggesting that the only reason mathematics appeared in the project at all was because it was a project for a mathematics course, and that the students failed to attribute importance to the course. We can look at this criticism from two perspectives. On the one hand, the group of Topics students did not seek to identify possible reasons that the Math I group may have proceeded in this manner. For example, would it not be natural for geography students to place more emphasis on geography? They might even have compared it to the attitude of their own group, students of the Mathematics Program, who focused on the mathematical part of the project. Thus, the students of the Topics course failed to take into account the fact that their own relationship with mathematics is different from that of the geography students.

On the other hand, their criticism of the use of formulas and indexes without seeking a justification for their use and creation is in agreement with the perspective of modelling adopted, which encourages the development of projects taking into account issues of critical mathematics education. In this case, the question that is most directly linked to such issues is related to the comprehension, and criticism, of the use of mathematics in society, and the acting of the same society in a more conscious manner.

The group of Mathematics I students did, in fact, present a series of mathematical formulas without explaining how they were constructed, which variables were considered, and which others were discarded to build the mathematical models, nor did they seek to reflect on the consequences of the use of these formulas to monitor environmental factors, water consumption, etc., which are relevant to the theme of their project. Thus, the modelling project did not fulfil, for this group of Maths I students, one of the most important roles in the perspective adopted, which is educating citizens in such as way that they use mathematics to act in a critical manner in society. However, one should not jump to the conclusion that this occurred simply due to lack of interest in the course on the part of the group of Mathematics I students, as will become clearer in the following section, where a discussion of the experience of this same group of Topics students developing their own mathematical modelling project is presented.

4. FUTURE TEACHERS DEVELOP A MODELLING PROJECT

The group of Topics students, whose work I am presenting and discussing, developed a mathematical modelling project as part of the experience with modelling provided in the course. The procedure for carrying out the project was divided into the same four steps presented in the section "Some Theoretical Aspects". The difference, in this case, was the theoretical discussion about the theme, and the analysis of the classroom situation, presented in the two previous sections.

The theme chosen by the group was "Obesity". They state that the objective of their modelling project was

> to evaluate whether the method which uses the measures of circumferences to calculate percentage of body fat, and the tables of ideal body weight based on age, sex, and height, lead to different conclusions regarding the determination of obesity when applied to the same person (Silva et al, 2003b p2).

This objective emerged from the theme "obesity" as a result of information collected by the group that led to some modifications in the project. The group then went on to describe the method used in the tables, which present ideal weights based on sex, height and age, and the circumference method, which involves calculating percentage of body fat based on measurements of the circumference of various parts of the body. In the case of the tables, the group explained that they are based on measurements obtained from large population samples, and questioned the adequacy of these tables for defining whether or not a person is obese.

Regarding the circumference method, the group justified its wide acceptance among professionals in the field of physical education due to its simplicity, ease of use, low cost, and objectivity. They then went on to explain how the calculation is made. They stated that a statistical analysis (multiple regression) is used to obtain a function for calculating percentage of body fat, and then provided an example of these functions, without, however, presenting how the calculation was developed.

Thus, we have a situation similar to that of the Environmental Monitoring and Health project, analyzed by the group of students enrolled in the Topics course: the emergence of formulas without presenting how they were constructed. This case is certainly different, as the group of Topics students spoke with professionals from the fields of physical education and statistics in an effort to discover the origin of the formulas, and presented some clues regarding how to construct them. What they both have in common is the failure to present this whole process. When questioned by the professors of the course, the members of the group reported their difficulties in discovering the origin of the formulas and the large number of calculations in the statistical analysis to find the formula for calculating percentage of body fat, which led them to decide not to present them. In the conclusion of their written report, the group members expressed concern about the large number of professionals in the field of physical education who use mathematical formulas and make conclusions about the obesity of people without knowing, nor even caring, about their origin.

5. A STEP BEYOND THE EXPERIENCE

A similar situation presented itself in the two examples of projects described here: the use of mathematical formulas without presenting their construction. In the project developed by the geography students, the formulas were presented, and there was no attempt to justify their origin. In the project of the mathematics students enrolled in the Topics course, the group was tenaciously dedicated to the task of justifying formulas that emerged.

In both situations, we were dealing with professionals from other fields – professionals in physical education and future professionals in geography – who used mathematics without concern for the origin of the theory they were using. But as the Topics group itself concluded based on their experience with modelling, it could be that these professionals do not do this out of mere lack of interest in mathematics. At times the level of complexity of a mathematical theory being used impedes understanding of its construction. According to Martin (1997),

> The narrow specialization involved in the [mathematical] modelling ensures that few [professionals] other than those developing or funding the application [of Mathematics] would be interested in or capable of using the model. (p166)

This fact can be noticed in the examples reported here. The future mathematics teachers struggled to obtain justifications for the formulas they encountered. Even so, they did not find these justifications among the physical education professionals; they had to seek help from professionals in statistics and yet they still had difficulties with the formulas that were constructed.

It could be argued at this point that it would be impossible for everyone to specialize in mathematics to obtain a deeper understanding of how it is used in our society. But what is to be said of the geography students who, for example, did not notice that the difference between two formulas they were presenting as "distinct" was only in the units of measure being used? Why were they not motivated to try to understand where those formulas came from, even after having been guided in this respect? The mathematics they used was, at times, apparently simple, as the group of Topics students noted in their analysis; but even so, the justifications were not presented. Perhaps the involvement of the geography students with mathematics may have been so distant that they were not willing or ready to unveil the mysteries about its use in environmental monitoring.

It is certainly not possible, nor necessary, for all citizens to achieve a high level of specialization in mathematics. As Skovsmose (1994) says, it is important to master some mathematical knowledge to support reflections about the role of this subject in society, but it is not necessary to master everything about it. However, the conception of mathematics present in our schools should not impede these citizens from knowing how to deal with this knowledge, and from having curiosity and interest in knowing how mathematics operates in our society. This conception would be in consonance with the concerns of critical mathematics education.

At the same time, according to Martin (1997), the lack of clarity caused by the excess of specialization in mathematics reinforces the impression of neutrality that surrounds this discipline. In other words, people believe in an objective mathematics, free of bias, because those who are specialists in the field present it as such, and do not struggle to make information accessible to the public in general.

Thus, we can say that the development of the modelling project by the geography students, which had as one of its objectives to question the ideology of certainty of mathematics, may have reinforced it in the end. If their relationship with mathematics was distant, the development of the modelling projects helped little to decrease this distance, as the mathematics continued to be mysterious and untouchable to them. It would have been important to provide this proximity, with the aim of helping them to become more critical citizens regarding the use of mathematics in society. Perhaps closer guidance was needed from the professor to break this barrier, which, unfortunately, was not possible.

On the other hand, I believe that the experience of the students in the Topics course is an example of how we can question this image of neutrality and objectivity that mathematics has, and which at times impedes students from seeking to understand the role of mathematics in society. In the search for justifications for the mathematical formulas used, the Topics students came face to face with choices made by the modellers, that is, subjective choices, to construct the formulas. They also encountered professionals who used mathematics blindly, which caused them to be concerned about the relationship between these professionals and mathematics. This concern, in turn, can lead these future mathematics teachers to reflect on their teaching practice, as they will have, in their classrooms, students who will become professionals in the most diverse fields and who, above all, deserve to have education in mathematics that enables them to act critically in a society that is constantly influenced by the mathematics used in it.

ACKNOWLEDGEMENTS

This study was partially funded by CNPq – Brazilian National Council for the Scientific and Technological Development.

Although they are not responsible for the ideas presented in this paper, I would like to thank Alex Jordane, Juliana Xavier, Paola Sposito, Pollyanna Milanesi and Teresinha Kawasaki, students from UFMG, for their comments on the original draft.

NOTES

1. The Mathematics Teacher Education Program is an undergraduate course, while the Graduate-level course is a *lato sensu* one.
2. These are real names, since the group authorized me to use their works and to publish their names. Then, I would like to thank them the authorization and the availability of the works, in which this paper is based.

REFERENCES

Araújo J.L. (2002) *Cálculo, Tecnologias e Modelagem Matemática:* as Discussões dos Alunos. Unpublished Doctoral Thesis. Instituto de Geociências e Ciências Exatas, Universidade Estadual Paulista.

Araújo, J.L. (2004) Matemática para Geografia: reflexões sobre uma experiência. In H.N. Cury (ed) *Disciplinas matemáticas em cursos superiores*: reflexões, relatos, propostas. Porto Alegre: EDIPUCRS, 85-109.

Barbosa, J.C. (2001) Mathematical Modelling in Pre-Service Teacher Education. In J.F. Matos, W. Blum, S.K. Houston and S.P. Carreira (eds) *Modelling and Mathematics Education: ICTMA 9: Applications in Science and Technology.* Chichester: Horwood Publishing, 72-88.

Borba, M.C. and Skovsmose, O. (1997) The Ideology of Certainty in Mathematics Education. *For the Learning of Mathematics*, 17(3), 17-23.

Martin, B. (1997) Mathematics and Social Interests. In A.B. Powell and M. Frankenstein (eds) *Ethnomathematics: challenging eurocentrism in Mathematics Education.* Albany: State University of New York Press, 155-171.

Niss, M. (2001) Issues and Problems of Research on the Teaching and Learning of Applications and Modelling. In J.F. Matos, W. Blum, S.K. Houston and S.P. Carreira (eds) *Modelling and Mathematics Education: ICTMA 9: Applications in Science and Technology.* Chichester: Horwood Publishing, 72-88.

Silva, A.P., Mellilo, C.R., Lima, C.R.S. and Arruda, D.N. (2003a) *Análise e Crítica do Projeto de Modelagem Matemática Proposto na Disciplina Matemática I para a Geografia sobre "Vigilância Ambiental em Saúde – água para consumo humano.* Unpublished Academic Work. Instituto de Ciências Exatas, Universidade Federal de Minas Gerais.

Silva, A.P., Mellilo, C.R., Lima, C.R.S. and Arruda, D.N. (2003b) *Projeto de Modelagem Matemática:* obesidade. Unpublished Academic Work. Instituto de Ciências Exatas, Universidade Federal de Minas Gerais.

Skovsmose, O. (1994) *Towards a Philosophy of Critical Mathematics Education.* Dordrecht: Kluwer Academic Publishers, 246.

4.3

LEARNERS' CONTEXT PREFERENCES AND MATHEMATICAL LITERACY

Cyril Julie
University of the Western Cape, South Africa

Abstract–*Mathematical Literacy deals primarily with the insertion of a mathematical gaze on extra-mathematical issues and situations. These issues and situations are to a large extent determined by curriculum, learning resource and test designers. The issues and contexts learners prefer for mathematical investigation is a largely under-researched area. In order to ascertain contexts that learners would find interesting to deal with in Mathematical Literacy a study, the Relevance Of School Mathematics Education (ROSME), was embarked upon. Some of the findings of this study and how they relate to current issues in South Africa are presented in this paper. It is concluded that within their Mathematics Literacy experiences learners should also be confronted with issues and situations to which they accord priority. However, this should be done with circumspection given that schools are also places where interests are to be developed. This, in turn, calls for a curriculum where the interests of learners and those fixed by curriculum, learning resource and test designers are balanced.*

1. INTRODUCTION

It is widely accepted that schools should graduate learners who are mathematically literate. Mathematical literacy is variously defined but across these variations there is a great deal of overlap. A commonly accepted definition is that of the Programme for International Student Assessment (PISA) (OECD, 2001 p22) stated as "Mathematical literacy is… the capacity to identify, understand and engage in mathematics, and to make well-founded judgements about the role that mathematics plays in an individual's current and future private life, occupational life, social life with peers and relatives, and life as a constructive, concerned and reflective citizen". It is not difficult to conceive that Mathematical Literacy deals primarily with extra-mathematical contexts. This places Mathematical Literacy within the realm of "applications and modelling of mathematics" viewed in a comprehensive sense as both "'modelling' [which] focuses on the direction reality → mathematics" and "'application' [which] focuses on the opposite direction mathematics → reality" (International Programme Committee for ICMI Study 14, 2002) since "reality" as conceived here is essentially the extra-mathematical.

The essential outcome for mathematical literacy is captured as "to make well-founded judgments and to use and engage with mathematics in ways that meet the needs of that individual's life as a constructive, concerned and reflective citizen." (OECD, 2000 p21). The outcome pivots around "needs" and is futuristic with "needs" determined by curriculum designers, learning resource developers and test constructors such as manifested, for example, by the Programme for International Student Assessment (PISA) test for Mathematical Literacy (OECD, 2000). As such the needs are assumed and considered as that for the "future" not-as-yet citizen. Although there is some likelihood that the issues and situations being dealt with and tested in Mathematical Literacy will emerge in some analogous way in learners' futures, it is also necessary, from a motivational point of view, to get some sense of what issues and situations learners would prefer to deal with in Mathematical Literacy. This paper deals with a study which focused on the contexts learners prefer to embed activities in Mathematical Literacy.

2. RELATED RESEARCH

The contexts that learners prefer to handle in Mathematical Literacy appear to be a topic that is relatively under-researched. In this regard web trawling and an ERIC database search using the key sentences and phrases such as "Contexts young people prefer for mathematics", "Contexts pupils are interested in to learn about in mathematics" and "Students and contexts in mathematics" rendered zero hits. A wider search with the key phrase "Contexts and mathematics" using the authoritative Mathematics Education database, MathDi, rendered 683 hits. However, the reported articles do not deal specifically with the contexts learners would prefer to deal with in Mathematical Literacy. Studies deal primarily with the effect of using contexts for mathematical concept formation, the effect of the use of contexts on learners' mathematical achievement and the ability of learners to identify mathematics in everyday activities. For example, the work of Dapueto and Parenti (1999) deals with contexts and the formation of mathematical concepts. The study by De Bock, Verschaffel, Janssens, Van Dooren, and Claes (2003) is representative of those ascertaining whether the use of contexts positively impacts on learner achievement in mathematics. The third kind of study is represented by a study conducted by Edwards and Ruthven (2003). They interviewed learners using pictures of five everyday English activities—dressmaking, Lego, chess, knitting and pool—with the purpose to ascertain whether the learners can identify mathematics in everyday activities.

3. INSTRUMENTATION

The study was inspired by the Science and Scientists project (Sjøberg , 2002) and its extension, the Relevance Of Science Education (ROSE) project (Schreiner and Sjøberg, 2004). As is the case for the afore-mentioned projects a survey instrument was developed around identified topics or clusters. The clusters were identified by mathematics educators from South Africa, Zimbabwe, Uganda, Eritrea and Norway and a group of mathematics teachers from South Africa. Thirteen clusters including two intra-mathematical ones evolved through the identification process. The identification of the eleven extra-mathematical clusters was in a major way informed by modules and learning materials developed by the Consortium for Mathematics and its Applications (COMAP) to ensure compliance to the possible mathematical treatment of the cluster items which were developed as possible indicators of the identified clusters. For example, the mathematics related to the item "Mathematics involved in making pension and retirement schemes" is dealt with in a UMAP Module by Ng (1987). The extra-mathematical clusters, with the number of items per cluster in brackets, decided upon through competitive argumentation were: Health (4), Physical Science (3), Technology (5), Transport and delivery (3), Life Science (3), Crime (4), Sport (2), Youth Culture (4), Politics (3), Agriculture (4) and General (7). A guiding criterion in deciding the elements belonging to a cluster was conceptual consistency. Examples of the items of some of the clusters are: health— mathematics involved in determining the state health of a person, crime— mathematics involved in setting up a crime barometer for my area and youth culture— mathematics linked to music from the United States, Britain and other such countries.

The sample comprised of grade 8, 9 and 10 learners from areas characterized as Low Socio-Economic Status (LSES) ones. The respondents were requested to indicate their preference to items on a four-point scale indicated as "Not at all interested", "A bit interested", "Quite interested" and "Very interested".

The demographic information of the sample is presented in Table 1.

School Grade	8	9	10	Not reported							
Number of learners	453	448	352	5							
Gender	Girl	Boy	Not reported								
Number	663	578	17								
Age	12	13	14	15	16	17	18	19	20	21	23
Number	2	129	284	381	264	113	50	7	3	1	1

Table 1. Demographic information for the sample.

4. DISCUSSION OF FINDINGS

In this section the findings are restricted to the extremities—the highest and lowest preferred clusters and similarly for individual items. For the individual items, only those common to the items ranked as extreme overall, by females and males are dealt with. Figure 1 depicts the rankings of the learners' preference across the different clusters.

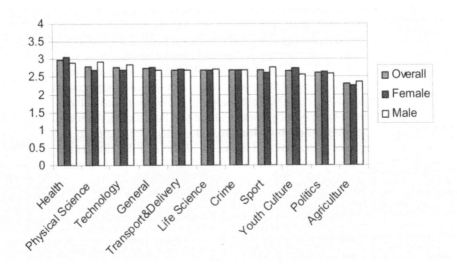

Figure 1. Cluster Preference Overall and by Gender.

This cohort of learners expressed a strong preference for issues dealing with health. The health cluster consisted of four items and these were posed as: "Mathematics to prescribe the amount of medicine a sick person must take"; "Mathematics involved in determining the state of health of a person"; "Mathematics used to predict the growth and decline of epidemics such as AIDS; tuberculosis" and "How mathematics is used to predict the spread of diseases caused by weapons of mass destruction such as chemical, biological and nuclear weapons." Of the 44 items dealing with contexts on the instrument these items have rankings of 5, 9, 10 and 19 respectively. The strong interest in health can be accorded to the knowledgeability learners have about health and its consequences. As indicated the learners surveyed were all from areas of low socio-economic status. These areas have high mortality rates due to preventable diseases such as tuberculosis, violence and possibly HIV/AIDS. Furthermore, campaigns related to these issues are regularly conducted through the media and programmes targeted at schools. This awareness of undesirable and fatal consequences of these issues, I would argue, currently excites learners to want to know more about them. The contextual interest can be viewed as a form of self-interestedness. However, there are elements of altruism included. The girls ranked the four items in the health category as 4th, 8th, 9th and 19th respectively whilst the boys ranked them as 6th, 16th, 19th and 13th respectively. Grandmothers and girls are still the major caregivers in LSES situations in South Africa. This somewhat accounts for the females ranking the items that can be linked to caring much higher than the boys and brings to the fore the element of altruism.

The agriculture cluster was ranked the lowest. It is the only cluster that is below the desirable mean of 2.5 for a 4-point scale. This is the case for both genders as is evident from Figure 1. Also, all four items comprising the cluster had means less than this desirable mean. These items and their means (between brackets) were:

How to estimate and project crop production (2.41); Mathematics needed to work out the amount of fertilizer needed to grow a certain crop (2.35); Mathematics involved in working out the best arrangement for planting seeds (2.28) and Mathematics involved for deciding the number of cattle to graze in a field of a certain size (2.15). The last-mentioned item ranked as the least-preferred item. Some of the learners were from squatter areas to which families from rural areas migrate in search of jobs and better livelihoods. The learners in these areas do have some sense of agricultural production. Other learners were born and grew up in low-cost urban housing areas and their only experience of agricultural matters is what they experience through the media. Much of media attention is given to the hardships farmers suffer, discrimination against farm workers, the unsatisfactory living conditions on some farms and the general lack of opportunity for farm workers and their families. It is speculated that these issues mitigate against a general interest in agricultural matters. However, there is wider disinterest in agricultural matters as reported by the Centre for Development and Enterprise (CDE, 2005). They report that "Far fewer black South Africans want to farm than is commonly supposed; most blacks regard jobs and housing in urban areas as more important priorities. A national survey commissioned by CDE shows that only 9 per cent of black people who are currently not farmers have clear farming aspirations...Other surveys suggest that only about 15 per cent of farm workers have aspirations to farm on their own or to farm full-time." (CDE, 2005 p16). Within South Africa high priority is given to land reform where most arable land fell into the hands of the Whites during colonialism and apartheid. The policy for land reform and redistribution is aimed at providing opportunity for agricultural production to be more equitably distributed across the country's population. If the low interest is indicative of a trend of young people's interest in agricultural matters then much motivational work at school level will have to be done to ensure a flow of new entrants into the agricultural sector to allow the South African government's policy to have the desired equity effects.

The elements common amongst the 5 highest ranked items overall, by females and males are given in Table 2. The most preferred items indicate a keen interest in the mathematics dealing with the modem things which young people perceive as indispensable. Mobile phones are a commodity which young people across the world desire to (and do) possess. Urban children from LSES environments are in this sense no different from their counterparts on the higher rungs of the socio-economic ladder

Item	Rank (Means between brackets)		
	Overall	Female	Male
Mathematics involved the sending of messages by SMS, cellphones and e-mails	1 (3.27)	1 (3.33)	4 (3.20)
Mathematics involved in secret codes such as pin numbers used for withdrawing money at an ATM	2 (3.24)	5 (3.20)	2 (3.29)
Working out financial plans for profit-making	3 (3.23)	2 (3.24)	3 (3.22)

Table 2. Common Highest Ranked Items.

The interest in profit-making can be ascribed to the government's and accompanying media propagation of self-employed entrepreneurship. The reported 35% to 40% unemployment rate in the country and the bureaucratic visibility of success of the self employed create in learners an expectation that their futures will be better secured through the avenue of self-employment. Thus the high interest in financial planning for profit-making.

From Table 3 the low ranking accorded to the issue related to national political matters can be ascribed to the dominance one political party enjoys currently and thus there appears to be little concern to engage in matters of this nature. Furthermore, there is speculation that the present-day youth are generally less interested in political matters than their counterparts were 20 to 15 years ago when the political struggle for democracy was at its height and young people from LSES environments were heavily involved in these struggles.

Item	Rank (Means between brackets)		
	Overall	Female	Male
Mathematics political parties use for election purposes.	5th lowest (2.37)	3rd lowest (2.39)	3rd lowest (2.36)
Mathematics linked to decorations such as the house decorations made by Ndebele women.	2nd lowest (2.29)	5th lowest (2.41)	Lowest (2.15)
Mathematics of a lottery and gambling.	Lowest (2.21)	Lowest (2.06)	4th lowest (2.37)

Table 3. Common Lowest Ranked Items (excluding agricultural items).

Mathematics linked to decorations such as the house decorations made by Ndebele women deals with the mathematical analysis of cultural artifacts which links to indigenous knowledge systems. This kind of work is encouraged in the curriculum for mathematics in all schooling phases in South Africa. From the perspective of learners the ethnomathematical approaches underlying the mathematical analysis of cultural artifacts and indigenous mathematical knowledge systems do not appear to carry high currency.

The national lottery and casinos are recent developments in South Africa. These activities are promoted through forms of aggressive advertising and the "rags to riches" stories, in addition to the popularization of contribution of the lottery and casinos to social upliftment, are regularly broadcast and discussed in communities. Despite this learners indicated a low preference for dealing with the mathematics related to lotteries. Various hypotheses can be generated for this seeming disinterest. Some of these are: winning has to do with luck and the perception of mathematics having to do with absolute certainty; the age for participating in gambling is 18 years and older and the learners involved were generally younger than this qualifying age and thus a possible non-declaration of interest due to fear of being caught out for illegal participation.

5. CONCLUSION

It is speculated in the previous section that the learners' preferences were influenced by the bureaucratic visibility of the issues that comprised the survey instrument. These assertions though remain at the level of hypotheses. From a retroductive research perspective the mechanisms generating these responses need to be ascertained which implies a careful investigation of the "rules, plans, conventions, images and so on that people use to guide their behaviour." (Harré and Secord, 1972 p151).

Given the strong interest in issues of direct personal appeal bolstered by a high media visibility confirms the generally accepted assertion that some preferences will differ across contexts (Jablonka, 2004). Thus context preferences must also be considered as transitive and time-dependent. Thus at a particular moment and in a particular milieu the mathematical literacy activities that are presented to learners must be tempered with what is personally relevant to them. This will improve the possibility of learners taking *control of knowledge* at the *tool level,* the *choice level* and the *goal level* (Mellin-Olsen, 1993).

Although it is argued above that personal ownership of problem situations can be fostered by contexts found desirable by learners it does not imply that learners should not deal with situations to which they accord low priority. Schooling can never be driven solely by the interests of learners. Schooling is also about the generation of interest and the uncovering of issues which learners do not as yet perceive as of interest. This should be catered for in a school mathematics literacy curriculum and as Skovsmose (1998, p419) asserts "It is essential to consider students' interest. But the interest cannot solely be examined in terms of the background of the students. Equally important is the *foreground* of the students." This foregrounding is exactly those which learners do not as yet perceive as interesting.

Lastly, the varying nature of context interests of learners puts demands on the notion of a "universal" implemented curriculum for Mathematical Literacy across countries, localities in countries and even at individual level. The major demand is that of tailoring the learning experiences to fit local interests as preferred by learners. But, as alluded to above, opportunity should also be provided for the generation of interest. This calls for a curriculum sense where the interests of learners and those determined by curriculum, learning resource and test designers are balanced.

REFERENCES

Dapueto, C. and Parenti, L. (1999) Contributions and Obstacles of Contexts in the Development of Mathematical Knowledge. *Educational Studies in Mathematics*, 39(1-3), 1-21.

De Bock, D., Verschaffel, L., Janssens, D., Van Dooren, W. and Claes, K (2003) Do Realistic Contexts and Graphical Representations Always Have a Beneficial Impact on Students' Performance? Negative Evidence from a Study on Modeling Non-Linear Geometry Problems. *Learning and Instruction,* 13(4), 441-463.

Edwards, E. and Ruthven, K (2003) Young people's perceptions of the mathematics involved in everyday activities. *Educational Research Journal*, 45(3 Winter), 249–260.

Jablonka, Eva (2004) The Relevance of Modelling and Applications: Relevant to Whom and for What Purpose. In H-W. Hen and W. Blum, Werner (eds) *ICMI Study 14: Applications and Modelling in Mathematics Education*. Pre-Conference Volume. Dortmund University of Dortmund, 127-132.

Harre, R. and Secord, P. F. (1972) *The Explanation of Social Behaviour*. Oxford: Blackwell.

International Programme Committee for ICMI Study 14 (2002) ICMI Study 14: Applications and Modelling in Mathematics Education—Discussion Document. *Educational Studies in Mathematics*, 51, 149 -171.

Ng, Ho Kuen (1987) Funding Pension Benefits: The Individual Spread Gain Method. *The UMAP Journal*, 8(1), 63 -75.

Mellin-Olsen, Stieg (1993) *Kunnskapsformidling Virksomhetsteoretiske perspektiver.* [Mediation of knowledge. Activity theoretical perspectives]. Second edition. Rådal: Caspar Forlag

Organisation for Economic Co-operation and Development (OECD) (ed) (2001) *Knowledge and Skills for Life. First Results from the OECD Programme for International Student Assessment (PISA) 2000*. Paris: OECD.

Organisation for Economic Co-operation and Development (OECD) (2000) *Measuring Student Knowledge and Skills*. Paris: OECD.

Schreiner, Camilla and Sjøberg, Svein (2004). *Sowing the Seeds of ROSE: Background, questionnaire development and data collection for ROSE (The Relevance of Science Education). a comparative view of students' views of science and science education.* Oslo: ILS og forfatteren.

Sjøberg, Svein (2002). *Science for Children? Report from the Science and Scientists Project* Oslo: ILS og forfatteren.

Skovsmose, Ole (1998) Critical Mathematics Education. In C. Alsina, J. M. Alvarez, M. Niss, A. Perez, L. Rico and A. Sfard (eds), *Proceedings of the 8th International Congress on Mathematics Education*. Seville: SAEM Thales, 413-425.

The Centre for Development and Enterprise (2005). *Land reform in South Africa: A 21st century perspective. Research report no 14*. Johannesburg: Centre for Development and Enterprise.

4.4

'REAL WORLD' INTERACTIONS FOR ADULT BASIC NUMERACY TUTORS

Yvonne Hillier
City University, London, UK

Abstract—*We are concerned with how adult basic numeracy tutors enable people to deal with the 'real world'. This chapter draws upon an ESRC funded research project (no R000239387) which explores the formative decades of adult basic skills from 1970 – 2000. The chapter explores how practitioners from the past three decades have developed numeracy practices, aimed at enabling adults to improve their basic maths and examines the tensions within basic skills that have privileged the work with literacy and language. It also identifies how the government definition of the problem of numeracy and basic maths favours a functional approach but how the professional development of tutors attempts to incorporate a much wider understanding of how maths is taught. The resulting tension continues in the current debates about how numeracy tutors should be trained and developed.*

1. INTRODUCTION: THE CHANGING FACES PROJECT

In England we have funds today to pay for the professional development of basic skills practitioners and a core curriculum for learners (BSA, 2001). We have funds to help adults acquire qualifications in literacy, numeracy or English language (ALNE), and we have a government agency, the Skills For Life Strategy Unit, to help oversee these activities and identify how best to ensure that people can improve their basic skills in the most effective ways possible. Research and development in the field is supported by the government funded National Research and Development Centre in Adult Literacy, Numeracy and ESOL (NRDC). Yet we have not always been so fortunate in being to the fore of government thinking around adults and their level of literacy, numeracy or language. The Changing Faces Project was funded through the Economic and Social Science Research Council (ESRC) to examine the history of ALNE from 1970, to help understand how the field has become central to government policy in its quest to create a skilled workforce which can contribute to the economic success of the country.

England is a densely populated country and in each of its 104 local education authorities (LEAs) there are adult and community learning centres, further education colleges, and a host of voluntary and work-based learning organisations, which are involved in the delivery of adult basic skills programmes. We used a case study approach in choosing four areas in England that represent some of the diversity of the provision: an urban area, Manchester; a county that had urban and rural areas and a community education structure, Leicestershire; a rural area, Norfolk and a London region, North East London.

We interviewed regionally and nationally 'key people' in the field. Our semi-structured interviews asked respondents to tell us how they had become involved in basic skills, what experiences they had of teaching, organising and management. We asked them about their views of the volunteers, learners and other practitioners and policy makers. We invited them to tell us about high points and low points in their careers and to identify key moments that they felt had influenced their practice and the field in general.

We also wanted to find a way to gain an understanding of basic skills from the perspective of learners. We interviewed members of a longitudinal study, the National Child Development Survey (NCDS), comprising people who were born in one particular week in March 1958. Of the original 17000 members, there are approximately 12000 still in the cohort who have been surveyed at different times throughout their lives. In 1981, when they were young adults, they were asked as part of the survey if they felt they had difficulty with literacy or numeracy and from the number who declared that they did, it was estimated how many people in the country may have this problem (Adult Literacy and Basic Skills Unit, ALBSU, 1987). A ten percent sample of the whole cohort was asked to take part in literacy and numeracy tests in a later survey of the cohort. There were four groups of respondents: those who did not self-identify in 1981 but were found to have low basic skills according to the test results, those who had not identified any basic skills problems and did not demonstrate any difficulties in the test, those who did identify basic skills problems and testing confirmed this, and finally those who did not show any difficulty with basic skills from their test results but felt that they did have difficulty. We were able to interview 78 people who were drawn from these four groups who lived as near as possible to our 4 case study regions. We asked people to tell us about their lives over the past three decades; we were interested in what learning they had done, whether they remembered any of the basic skills campaigns and what they felt about education provision for their families compared with their own experiences. These interviews provided us with an insight into the way in which any basic skills initiatives over the past thirty years have influenced their learning.

Finally, we asked practitioners to donate materials and documentation for an archive, which we have gathered for future researchers to use. This archive is being collated and stored at the University of Lancaster. We analysed our interviews by using Atlas-ti software to code our transcripts, develop themes and create families of codes. These, too, helped us to identify some of the key moments and key interactions between the three groups of stakeholders in our research; policy makers, practitioners and adult learners

We have identified four policy phases which demonstrate shifting power between the different agencies involved in the field.

- Mid 1970s: Literacy Campaign led by a coalition of voluntary agencies with a powerful media partner, the British Broadcasting Corporation (BBC).
- 1980s: Provision developed substantially, supported by Local Education Authority (LEA) Adult Education Services and voluntary organisations, with leadership, training and development funding from a national agency (Adult

Literacy and Basic Skills Agency, ALBSU, later the Basic Skills Agency, BSA)

- 1989 – 1998: Depletion of LEA funding and control, statutory status of ALNE through a more formalised further education (FE) system, dependent on funding through a national funding body.
- 1998- present: Development of Skills for Life policy: New government strategy unit created, £1.5 billion of government money is committed.

During these phases, we identified that there were major influences on the way in which basic skills was perceived, its practice and how the field was supported or ignored by government at national and regional levels. For example, during the 1980s, there was massive unemployment for young people and adults and basic skills provision was often used to help unemployed people find work. By the 1990s, information technology was affecting the world of work and basic skills was expected to help people use technology, particularly in areas of work that previously had not required such high levels of literacy or numeracy that the technology demanded.

We also identified a number of tensions that practitioners experienced. Some of these are deep seated, such as the reason why people need basic skills tuition and whether a functional or social practice approach (Barton, Hamilton & Ivanic, 2000) is the most appropriate way to help adults meet their learning goals. A functional approach attempts to enable people to manage the basic skills that are required in everyday situations, for example reading a timetable, writing a letter, reading a health and safety notice. The social practice approach views basic skills as being part of other practices which are mediated by the skills involved, so that the focus is less on knowing how to write a letter than finding out how such letter writing occurs and what it achieves. Other tensions relate to more national and international influences, including the move towards greater accountability to central government and the use of accreditation as a measure of learning outcomes. For every person we interviewed who approved of accreditation as it empowered their learners, there would be someone else who was scathing of the kind of qualifications being offered and the effect it had on people's life chances. Our research covered literacy, language and numeracy. This chapter sets out the emerging themes relating to the practice of numeracy, in particular, the way in which practitioners have developed the field in the early years and how this field is now being shaped by government policy.

2. DEFINITIONS

The definition of adult numeracy has been defined differently from mathematics during the past thirty years (Coben, 2001). Within adult numeracy, basic maths is often seen to be 'functional' in other words, it is the maths necessary to function in society for example, being able to use money, read timetables and understand weights and measures for everyday activities such as shopping. There is a counter argument to this, in that numeracy is just a subset of maths skills and that practitioners, therefore, need to understand and use maths in order to be able to teach it to those who need it for functional purposes. Partly because adults who had difficulty with basic maths need new and different ways to help them learn and partly because the ethnic backgrounds of many adult learners represent a diverse range of cultures, a wider view of numeracy has emerged which takes account of the different understandings of mathematics more generally, for example, the different ways in which people count and therefore undertake simple computation. Adult numeracy takes account of the context in which people perform basic maths, whether for functional purposes or to help them develop maths skills in ways that are more meaningful to them. For the purposes of this paper, adult numeracy encompasses basic maths skills and the application of maths skills to function in everyday situations and is therefore contextualised within a social practices view of basic skills (Barton et al, 2000).

3. A DEVELOPING FIELD OF ADULT NUMERACY

Adults want to improve their numeracy for various reasons, including needing to pass tests before progressing to further study. In the 1970s and 80s, certain government funded vocational training schemes such as the Training Opportunities Scheme (TOPS) required potential trainees to take diagnostic numeracy tests and those who failed were refused places on training schemes. Other adults wished to improve their numeracy because they wanted to help their children at school, and didn't understand 'modern maths'. There is usually a 'presenting problem' that facilitates adults coming forward to improve their numeracy. They often cite being unable to understand fractions, in particular, as an indication that they were 'no good at maths'. It was tempting, therefore, to develop a way to teach adults that simply returned to the basic computational skills that they needed to learn, decontextualised as their early experiences of learning maths had been. Yet practitioners were aware of the social imperatives that had influenced people's decisions to begin learning maths again. As one organiser noted

> I worked on a very poor housing estate in Edinburgh and the number of families that used to come to us wanting advice because they would have borrowed money from a loan shark and hadn't understood the interest they would have to pay back. I used to find it quite devastating that somebody could take advantage in that respect and get themselves into an even bigger hole. Somebody was getting very rich on the back of them (Organiser, London).

It was clear from the start that adults needed different approaches to help them improve their maths. One of the defining characteristics of teaching maths to adults is the contextualisation in everyday examples

> Well I discovered very quickly when I first started trying to teach fractions to a group of women, that teaching fractions in isolation was a waste of time, 'a quarter plus 3/6 is equal to and this is how you work it out' was a complete waste of time because it meant nothing to them. As soon as you started to convert it to decimals and put a pound sign in front of it, or as soon as you started to talk about a quarter of ham or half a pound of potatoes then you think yes they are getting it, they understand what I am talking about now (Organiser, London).

The day to day practical elements of life can provide useful contextual support for the acquisition of computation skills and understandings, as can the vocational learning that students undertake at further education colleges.

I would design the numeracy component to accompany courses electrical installation for example. I worked with trade women who were natural teachers. I was teaching them equations because it's (sic) electrical currents and stuff, and I really enjoyed that (Tutor, London).

Numeracy, because of its attention to context is a very practical subject, yet had a dearth of appropriate materials and resources to support learners in the early days of basic skills work. Tutors used everyday situations to help reinforce the concepts they were attempting to instil

I remember there was one time, like trying to get people to understand what 1-0 means in terms of 10, if you can get people over place value and actually a really good visual resource for that one is coinage, thank God we went decimal (Organiser, London)

They can't do bloody division, what are we going to do? And we would think of all the different ways and we would make worksheets and we would try them out to see if they worked. We would go out to the pond and measure, using trigonometry using the tree and the shadows...It was wonderful it really was...we did some really interesting things (Tutor, London).

A practitioner used bottles which had 'rogue' measures, cc instead of cl, as a way of helping students begin to understand metric measurement. As England has used imperial measurement until the 1980s, adults needed to understand metric even though they had been taught imperial measurement at school. Many had not been able to make the adjustment to the new system of measurement as a result of their lower levels of numeracy.

4. ONGOING TENSIONS

4.1 Cultural, Not Context Free Mathematics

The general view held by the general public is that maths is neutral, something that adult numeracy specialists in the field coherently argued against (Coben et al, 2000). Given the range of backgrounds that students came from, it was necessary to help people understand that there are different ways of doing maths particularly when mathematical problems are drawn from everyday situations. Often someone with English as a second or third language would find difficulty with doing problems, simply because of language issues rather than the computational skills required to solve the problem.

4.2 Accreditation

During the 1980s, vocational education and training became highly influenced by the competence based movement with an emphasis on outcomes. A new infrastructure of vocational qualifications was created through the National Vocational Qualification Framework (NVQF), and numeracy was included in this new structure. Awarding bodies which had previously offered qualifications in maths and numeracy continued to be used by basic skills tutors, but a new qualification created specifically for basic skills, Numberpower, was offered by City and Guilds from 1990. This proved to be a highly controversial qualification, bringing to the fore debates about the ability of an outcomes based system to facilitate and assess learning, as well as controversy over the value of such qualifications compared with their more 'academic' counterparts.

4.3 Professional Development

Practitioners have always been encouraged to improve their professional knowledge. Professional development originally stemmed from the initial training of volunteers and tutors. Although there were specialist qualifications in teaching literacy and ESOL, there was no specific qualification for teaching numeracy. In 1990, ALBSU created a new professional qualification, the Initial Teaching Certificate (ITC), to accredit the training that had previously been undertaken. Numeracy was the 'poor cousin' for this award as the original learning outcomes and assessment specifications were devised for literacy tutors.

It was not until the new Skills for Life Strategy that a set of standards against which basic skills teaching in literacy, numeracy and ESOL were created. Qualifications for practitioners based upon these standards are now mandatory for new tutors and trainers entering the field. This has led, though, to a new complication, as many tutors who already teach numeracy are not maths specialists and may be put off trying to gain a numeracy qualification because the standards that they must now meet require knowledge of maths at Level Four, the equivalent of a first year undergraduate programme. The government commitment to increasing the supply of suitably qualified maths teachers in schools, resulting from the Smith report on Post-14 Mathematics, is likely to draw qualified teachers away from numeracy for adults and into schools, particularly as there are 'golden hellos' and bursaries available for those taking their post graduate certificate in education (PGCE). The level of teaching basic numeracy, too, may be seen to be too low for highly knowledgeable and qualified mathematicians.

> it's quite interesting to see that quite a number of very high level specialised maths tutors didn't see enough of a challenge in teaching, you know adding up, giving change, almost a very basic numeracy that people need because it's just too far removed from the challenges that they enjoy, the mathematical challenges (Organiser, Norfolk)

5. FUNCTIONAL VERSUS SOCIAL PRACTICE VIEWS OF BASIC SKILLS: CONTESTED TERRAIN

We can see that the teaching of numeracy to adults has developed its own curriculum and practices through taking creative steps to help adults learn maths who have already failed as children (Benn, 1997). The need to understand the cultural significance of numeracy, as opposed to a functional mathematics for everyday life has been proselytised by numeracy practitioners for at least two decades (see Coben, 2001, 2003). However, the current attempt by the UK government to focus on functional maths is far reaching, seen in the setting up of the Advisory Committee on Mathematics Education (ACME) in 2002, and in the proposed changes to the 14 - 19 qualification system by Tomlinson in 2004. Functional maths is defined as the

> 'mathematics that people need to participate effectively in everyday life, including the workplace' (ACME, 2005 ¶5.5, Tomlinson, 2004)

Numeracy, however, is more problematic to define, and as Tout points out, it is not even a word that is recognised by spell checkers in word processing software such as MS Word (Tout, 2003). He questions whether numeracy is 'greater or smaller than maths', or just 'skills and application'. The word was first used in the Crowther Report (1959) as a term to parallel literacy and tends to be defined in relation to the social practices approach of literacy, where context is central to the use of functional skills. Thus, Tout draws upon Johnston who suggests that numeracy incorporates a critical aspect of using maths

> To be numerate is more than being able to manipulate numbers, or even being able to 'succeed' in school or university mathematics. Numeracy is a critical awareness which builds bridges between mathematics and the real world...In this sense. There is no particular 'level' of mathematics associated with it: it is as important for an engineer to be numerate as it is for a primary school child, a parent, a car drive or a gardener (Johnston, 1994, in Tout, 2003 p384)

Practitioners today are expected to focus on the more narrow *functional* definition of mathematics as they enable adults to improve their numeracy and gain qualifications, which counts as meeting the government target of improving the basic skills of 2.25 million adults by 2010. Numeracy in the Skills for Life Strategy covers the ability to understand, use, calculate, manipulate, interpret results and communicate mathematical information. Yet Coben (2004) questions how these definitions address what aspects of mathematics and mathematical information at which levels are sufficient for these purposes. She adds that numeracy

> remains a deeply contested concept and there is much terminological confusion and debate about the relationships between numeracy and literacy, and between numeracy, mathematics, mathematical literacy, quantitative literacy and a plethora of other related terms (Coben, 2004 p17)

The way in which literacy is seen as *the* basic skill means that any attempt to address the mathematical component of the field has been neglected, misunderstood or seen to be less important. We have data from numeracy practitioners who have bemoaned their lack of status compared with literacy and language practitioners, and it is well documented that people feel more able to confess to being 'useless at maths' than having trouble with reading and writing. Such a view is not new, as can be seen twenty years ago in ALBSU's *Developments in Adult Literacy and Numeracy: An Interim Report* 1984

> It has been misleadingly easy, unfortunately, to compare and contrast literacy provision and numeracy provision and the latter has not only suffered from being seen as the mathematical equivalent of literacy but also through somewhat simplistic comparisons of statistics relating to referral and take up. It is not always easy to avoid such comparisons. Differences of approach to the development of adult literacy and adult numeracy appear more significant than first thought and provision is beginning to recognise these differences. The use of simplistic arithmetic screening tests has also impeded the development of a wide ranging and relevant curriculum for numeracy and whilst ALBSU

recognises the influence of such tests, we would hope that their use will decrease in the future. A pragmatic response to narrow objectives, whilst not the answer, may provide a way forward until such time as the wider concepts of basic maths become recognised (ALBSU, 1984 p3).

We live in a world where numbers 'count' and are privileged over words when it comes to evidenced-based policy making. People eschew maths yet are then influenced by statistics which are misleading and sometimes downright false, because they do not have the mathematical knowledge to challenge what they are told, particularly in reports by the mass media. As Baynham (2003) argues

In the wider social and policy environment, since numbers were so influential could so often talk down other forms of evidence and ways of knowing, it seems highly ironical to me that Adult Numeracy should be so apparently marginal within the Adult Basic Education context (Baynham, 2003 p5).

6. CONCLUSION

We know that more people have difficulties with basic numeracy than literacy (OECD, 2001; DfES, 2004,) and the government has created strategies to improve the level of numeracy and functional maths through the creation of ACME, the first time it has addressed this issue since the days of the Cockcroft Report in 1982 (Department for Education and Science, DES, 1982). Yet in the world of adult numeracy we continue to find government policies competing with each other, where practitioners working with adults are expected to hold specialist qualifications in teaching mathematics, yet are paid less to teach as their compulsory school based counterparts. There is a danger, too, that the enormous dedication and good will of those who work in the field of adult numeracy will be eroded by the narrow target driven culture in which the Skills for Life Strategy resides. On the other hand, who can dispute that people who want to improve their basic maths should be taught by well qualified and knowledgeable tutors, who understand the context in which everyday maths is used, but also aware that a curriculum that is too narrow will stifle the enormous potential for improvement in teaching numeracy today.

REFERENCES

Adult Literacy and Basic Skills Unit (1984) *Developments in Adult Literacy and Numeracy: An Interim Report*. London: ALBSU.

Adult Literacy and Basic Skills Unit (1987) *Literacy, Numeracy and Adults: Evidence from the National Child Development Study*. London: ALBSU.

Barton, D., Hamilton, M., and Ivanic, R. (2000) *Situated Literacies: Reading and Writing in Context*. London: Routledge.

Basic Skills Agency (2001) *Adult Numeracy Core Curriculum* London: Cambridge Training and Development on behalf of the Basic Skills Agency.

Baynham, M. (2003) The Power of Numbers: Research investigations in a number saturated world ESRC Seminar Series on Adult Basic Education http://www.education.ed.ac.uk/hce/ABE-seminars

Benn, R. (1970) *Adults count too: mathematics for empowerment*. Leicester: National Institute for Adult and Continuing Education (NIACE).

British Association of Settlements (1974) *A Right to Read: Action for a literate Britain*. London: Author.

Coben, D., O'Donoghue, J., and FitzSimons, G.E. (eds) (2000) *Perspectives on Adults Learning Mathematics: Research and Practice*. Dordrecht: Kluwer.

Coben, D. (2001) Fact, fiction and moral panic: the changing adult numeracy curriculum in England *Adult and Life-long Education in Mathematics: Papers from working Group for Action 6, 9th International Congress on Mathematical Education ICME 9* G.E.FitzSimons, J.O'Donoghue and D. Coben. Melbourne Language Australia in association with Adults Learning Mathematics – A Research Forum (ALM), 125-153.

Coben, D. (2003) *Adult Numeracy: a review of research and related literature* National Research and Development Centre for Adult Literacy, Numeracy and ESOL (NRDC) www.nrdc.org.uk .

Coben, D. (2004) Research and Development – the implications for practice *Numeracy Briefing,* April 2004, Issue 1, 17 – 19.

Department of Education and Science (1982) *Mathematics Counts: report of the committee of Inquiry into the Teaching of Mathematics in Schools.* (The Cockcroft, Report) London DES/Welsh Office.

Department for Education and Skills (2001) *Skills for Life: The National Strategy for improving adult literacy and numeracy skills.* London: The Department for Education and Employment.

Department for Education and Skills (2004) *Making Mathematics Count: The Report of Professor Adrian Smith's Inquiry into Post-14 Mathematics Education* .London: The Stationery office.

Department for Education and Skills (2005) *14 – 19 Education and Skills White Paper.* London: DfES.

Moser, C. (1999) *A Freshstart: Improving literacy and numeracy. The report of the Working Group chaired by Sir Claus Moser.* London: Department for Education and Employment (DfEE).

Organisation for Economic Co-operation and Development (2001) *Knowledge and Skills for Life: the first results from the Programme for International Student Assessment.* Paris: OECD.

Tomlinson, M. (2004) *14 – 19 Curriculum and Qualifications Reform Final Report of the Working Group on 14 – 19 Reform.* Annesley: Department for Education and Skills.

Tout, D. Time for a 'great numeracy debate' RaPAL Journal 52 Autumn 2003, in M. Herrington and A. Kendall (eds) (2005) *Insights from research and practice* Leicester: NIACE.

4.5

MATH MODELLING: WHAT SKILLS DO EMPLOYERS WANT IN INDUSTRY?

ManMohan Sodhi[1], Byung-Gak Son[1]
City University, London, UK

Abstract–We analysed 401 mathematical modelling-related job advertisements from Monster.com and OR/MS Today to find out what employers want from graduates from mathematical modelling related courses. We used content analysis to analyse job advertisements by tallying up relevant phrases and keywords and phrases in the advertisements regarding (1) skill requirements, (2) degree requirements, and (3) background disciplines requirements. Our analysis shows that many skills required for mathematical modelling jobs are "soft" skills pertaining to problem solving and to communication, which are not covered by ordinary mathematical modelling curricula. Educators can help students to obtain such soft skills by adopting innovative teaching methods such as team exercises, competitive exercises, group discussions, presentations, and case analysis.

1. INTRODUCTION

We seek to find out the skills that industry employers want from graduates of mathematical modelling programmes. We take the view that employers state what they want in their recruitment advertisements, so we analyse job advertisements pertaining to mathematical modelling, in particular, those based in operational research (OR).

We started with 670 OR job advertisements and then filtered out 401 job advertisements that specifically asked for mathematical modelling. Although all advertisements are from the US, a quick check of advertisements from UK-based and Indian companies indicated that these companies are looking for the same skills and backgrounds. So our results may be applicable for other countries as well.

We have tried to make only broad qualitative inferences from our analysis. After all, we are analyzing text not numbers. Still, we believe that the results are a useful starting point in providing modelling students with the right skills for modelling jobs.

2. LITERATURE REVIEW

There have been other, similarly motivated, studies in the past. The focus of such studies was to identifying gaps between the industry needs and the actual skills of university/college graduates. In general, these studies identified that industry demands more of the soft skills such as 'inter-personal skills' and 'communication skills'.

Specifically for the technology sector, Bryans and William (1996) analysed advertisements for entry level jobs and concluded that the proportion of advertisements requiring technology and inter-personal skills declined between 1992 and 1994 but the proportion of advertisements requiring basic communication skills was on a rise. Leckey and McGuigan (1997) found that employers in the US and Europe show a preference for teamwork, communication and self-skills above knowledge, degree classification, intelligence and reputation of the institute. They identified the perception difference between students and faculty regarding the extent to which key skills were developed during their courses despite both valuing the importance of such key skills equally.

Todd, McKeen, and Gallupe (1995) analysed the content of information systems job advertisements published in US and Canadian newspapers from 1970 to 1990. Contrary to the above findings, they concluded that the demand for technical knowledge had remained high and the requirement for business and systems knowledge had remained relatively low without changing over the two decades.

As regards mathematics education including mathematical modelling, Challis *et al.* (2002) and Challis and Houston (2001) have carried out research for the curriculum design for mathematics course at the BSc level and also have suggested ways of embedding key skills. As regards OR, there are few recent studies that investigate OR jobs. But a report on a panel discussion in 1976 about skills needed for OR graduates is consistent with our findings. Schrady (1976) organised this panel with industry representatives wanting OR students to be more capable of solving real-life problems and have more soft skills such as communication.

3. METHODOLOGY

3.1 Data Collection

Our sample comprises OR job advertisements placed by US-based industry employers. Government jobs, in particular, from the defence forces and intelligence services, are also included. We collected advertisements from Monster.com and *OR/MS Today* for US placements only posted between the beginning of April'05 to the end of June'05. At Monster.com, we searched for advertisements using the keywords, "operations research" or "operational research" (i.e., the entire phrase with the "or" clause). This search returned more than 650 advertisements containing either phrase but mostly with the US spelling (operations research). We stored these in a database and deleted duplicate or otherwise irrelevant entries and ended up with 408 advertisements. Details on error checking and duplications appear in our earlier article (Sodhi and Son 2005). To these 408 advertisements in our database we added 262 industry advertisements from *OR/MS Today* posted between January 1999 and July 2005 resulting in 670 job advertisements.

Finally, we filtered out 401 job advertisements that pertained to "math modelling" by using such keywords as mathematical modelling and LP (linear programming) (Table 1). Thus, the OR mathematics modelling job advertisements analysed in this paper represent an overlap of OR jobs and mathematical jobs, which comprises 401 modelling jobs in our overall sample of 670 advertisements.

Algorithms	Modelling	Simulation
CPLEX	Modelling experience	Strong quantitative
LP	Modelling techniques	Linear programming
Management science	Modelling	Quadratic programming
Mathematical modelling	Model development	Mathematical programming
Mathematical models	Network modelling	
Matlab	Optimization	

Table 1. Keywords we used to identify job advertisements for "math modelling". Of the original 670 OR job advertisements; we identified 401 advertisements this way as relevant for math modelling.

Over 31 industries have been recruiting OR graduates for math modelling jobs. Computer service and software, consulting industries and consulting are the top three industries hire the graduates with various levels of operations research degrees for math modelling jobs (Figure 1).

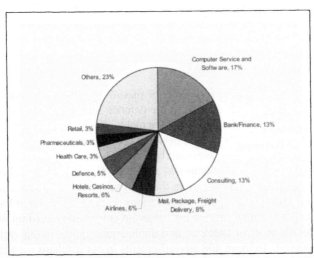

Figure 1. Each job ad is for a particular industry sector as coded by us. The pie chart shows how many advertisements belonged to each sector. The "others" category (23% of the advertisements) comprises many sectors each of which is present in less than three percent of the advertisements.

3.2 Analytical Approach

Our data comprises of one text variable that stores of each ad and one categorical variable for the industry classification that we manually coded ourselves. We analysed this text variable by tallying up relevant phrases and keywords and phrases in the advertisements. We defined subsets of these keywords and phrases as "categories". For instance, if the phrase "bachelors degree" (or variants like BS degree) appears in an ad, we tally up that ad as a relevant case in the Bachelors category. The same ad may specify "C++" so we tally it up in the Programming category as well. We have 19 such categories and our approach is to draw out inferences from the number of advertisements belong to each category and to pairs of categories.

Such analysis is called "content analysis". According to Holsti (1969), this is "any technique for making inferences by objectively and systematically identifying specified characteristics of the message". Typically, text data have a huge volume -- in our case, the total word count in our data was over a quarter million -- and analysing such volumes requires a computer-based tool. For our analysis, we used WordStat from Provalis Research.

3.3 Keyword Selection and Categorisation

The backbone of a computer based text analysis tool is the "dictionary". This dictionary is a specification of keywords and phrases under various named categories that allows the software to either exclude certain words from the analysis or, more to the point, to create counts under each category when a word or phrase under that category is found in a record. (A part of the dictionary we used appears in the Appendix.)

Degree	Discipline	Skill
Bachelors	Operation research	Managerial
Masters	Computer Science	Analytical
PhD	Business	Communication
	Economics	Database
	Engineering	Programming
	Statistics	Project management
		Basic IT
		Presentation
		Team

Table 2. We used three master categories and the various categories under each to categorise all the key words and phrases. The specification of each category and keywords associated with it, i.e., dictionary, allows the software to count how many advertisements have keywords in each category.

We first conducted an exploratory analysis of the advertisements by using WordStat to list out the frequency of all words and phrases (up to five words long) in the advertisements. We studied the words and phrases that occurred in at least 2 percent of the advertisements. From these we discarded words and phrases that we deemed to be not relevant for OR (and math modelling). Then, we put the remaining (and therefore relevant) 150+ words and phrases under 19 categories in three master categories: (1) type of degree, (2) discipline background, and (3) skills required. There are three degree categories, Bachelors, Masters, and PhD corresponding to the employers' requirements for a position. Likewise there are seven categories for the disciplinary background needed, and nine skills-based ones that the advertised jobs would require (Table 2).

The categories within a master category are neither mutually exclusive nor exhaustive. An ad can fall in more than one category within a master category. For instance, an ad seeking for someone with either a masters or a PhD degree would fall in both the Masters and the PhD categories even though the lists of keywords corresponding to the two categories are mutually exclusive. Also, an ad for an experienced person may not mention any degree at all and would not fall in any of the three degree categories. We list all the 150+ key words or phrases at the Appendix.

4. DEGREE REQUIREMENTS

The majority of the companies that placed math modelling job advertisements require a bachelor's or a master's degree. Note that an ad asking for a bachelor's or master's degree would be tallied up under both. For 28% of the jobs, a bachelor's degree was sought or was acceptable and for 40% a master's degree. The number of advertisements in which a PhD was sought is about one in seven (15%), small but not insignificant. Again, we should emphasise that an ad may belong to more than one category or to no category at all.

5. BACKGROUND DISCIPLINE

In general, companies are looking for a variety of backgrounds including OR for math modelling jobs. In our sample, almost all the advertisements (98%) require an OR background. This should not be surprising given how we selected advertisements using "operations research" as a keyword on Monster.com. Also, not all advertisements require an OR background possibly because *OR/MS Today* advertisements may not use the phrase given the reader's focus. But many non-OR backgrounds were sought. In particular, 23% of the modelling advertisements mention computer science and 5% mention business. Other backgrounds, statistics, economics and science were sought after by less than 2% of the total advertisements.

6. SKILL REQUIREMENT FOR MATH MODELLING JOBS

Companies require a variety of skills and analyzing the advertisements has been extremely useful to the authors. Of all the categories, educators, and students need to pay special attention to these. Indeed, the bulk of our analysis effort has gone into selecting over 100 skills-related keywords and phrases from which we constructed 10 categories required by employers for mathematical modelling jobs (Table 3 and the Appendix).

Analytical skills	Managerial skills	Project Management skills
Ability to conduct and interpret Analytical Analytical results Analytical skills Analytical support Analytical techniques Analyzing information Business analysis Business analyst Problem solver Problem solving Problem solving skills	Action plans from multi disciplinary perspectives Acts as liaison Attention to detail Change management Leadership Management skills Manager/supervisor of staff Organizational skills Staff leadership Strategic Strategies	Assesses project impact Develops project plans Facilitate sharing of project outcomes and best practices Multiple projects Program management Project documentation Project management Project managers Project support specialists Strategic direction of projects

Table 3. The keywords and phrases associated with three types of skills. An ad that has any of these words or phrases (among others) is categorized in the corresponding category. While the lists of keywords and phrases for the different categories do not overlap, an ad may have phrases from multiple lists and may therefore belong to more than one category. For details, see the Appendix.

Of these categories, some are what might be considered as "hard" skills like Programming and Database skills that might already be a part of many OR programs. Having such skills would benefit OR graduates given that more than half (57%) of the advertisements require programming skills (with C++ a clear favourite) and almost a third (32%) require database skills. About a quarter (21%) of math modelling job advertisements require basic IT skills – spreadsheet, word processing, basic data manipulation skills – that most OR graduates at least in the US are expected to have (Figure 2).

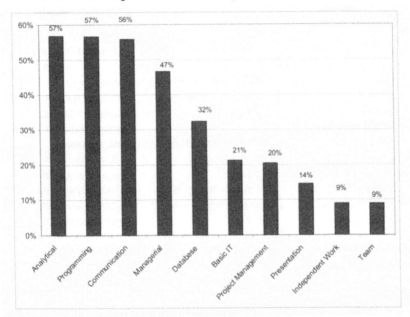

Figure 2. The percentage number of Math modelling job related advertisements that require a particular set of skills. Each "skill" comprises a number of different keywords and phrases.

At the top of list of skills are three categories corresponding to analytical skills (57%), programming skills (57%), and communication skills (56%) category that a majority of the advertisements requires as indicated by the percentage in the parentheses. The requirement of communication skills is typically qualified by the adjective "excellent" in an ad. This could suggest that the employers find these skills weak in graduates for Math modelling. Analytical skills are both "soft" in the sense that analysis involves problem-solving skills and possibly technical skills as well.

Next on the list are managerial skills (47%) and database skills (32%). About one in seven advertisements specifically requires strong presentation skills (14%) possibly going beyond the usual communication skills. Nearly 9% of the advertisements emphasise teamwork while another 9% require the ability to work independently.

7. CONCLUSION

We have shared results of text analysis of 401 modelling jobs from the sample of 671. Our analysis shows that many skills required for math modelling jobs are "soft" skills pertaining to problem-solving and communication that may not be easily gained in most modelling programmes that we are aware of. Actual work experience may be the only provider of such skills! Still, OR and mathematics programmes could do something to rectify this just as MBA programs help their students gain these skills through team exercises, competitive exercises, group discussions, presentations, and case analysis. These changes could be brought about by not necessarily changing WHAT to teach but changing HOW to teach it.

Students applying for jobs should use our results to figure out what to mention and emphasise in their resume. Students starting their programs could use this analysis to figure out how they should supplement their programme-related learning with additional skills gained on their own. For educators, our analysis may provide a discussion point in terms of not also redesigning their programs but also revisiting the age-old question of what OR is. They could also use the above to redesign homework (or "coursework" in the UK) by requiring students to work in groups and to use spreadsheets and/or database software. Also, presentations (not the same as showing MS PowerPoint or other slides) could be a norm for a class rather than something done once a year in a conference.

NOTE

1. Cass Business School, City University, London

REFERENCES

Challis N. and Houston S.K. (2001) Embedding Key Skills in Mathematics Degrees. *Proceedings of the 2000 Undergraduate Mathematics Teaching Conference.* Sheffield: SHU Press, 30-31.

Challis N. and Houston S.K. (2000) Embedding Key Skills in the Mathematics Curriculum. *New Capability*, 4, 3, December, 7-10 and 50.

Challis, N., Gretton, H., Neill N. and Houston S. K. (2002) Developing Transferable Skills - Preparation for Employment. In P. Khan and J. Kyle (eds) *Effective Learning and Teaching in Mathematics and its Applications.* London: Kogan Page, 79-91.

Holsti, O.R. (1969) *Content Analysis for the Social Science and Humanities.* Reading, MA: Addison-Wesley

Leckey J. F. and McGuigan, M. A. (1997) Right Tracks—Wrong Rails: The Development of Generic Skills in Higher Education. *Research in Higher Education*, 38, 3, 365–378.

Schrady, D. (1976) Are We Gambling on OR/MS Education? *Interface*, 6, 3, 104–115.

Sodhi, M. and Son, B. (2005) What Industry Employers Want from OR/MS Graduates – Preliminary Results from an Analysis of Job Advertisements. *OR/MS Today*, August, 2005.

Todd, P. A., McKeen, J. D. and Gallupe, R. B. (1995) The Evolution of IS Job Skills: A Content Analysis of IS Job Advertisements from 1970 to 1990, *MIS Quarterly*, March, 1-27.

APPENDIX – *A Part of the Dictionary of Keywords in Each Skills Category*

ANALYTICAL SKILLS

ABILITY TO CONDUCT AND INTERPRET
ANALYTICAL
ANALYTICAL RESULTS
ANALYTICAL SKILLS
ANALYTICAL SUPPORT

ANALYTICAL TECHNIQUES
ANALYZING INFORMATION
BUSINESS ANALYSIS
BUSINESS ANALYST
PROBLEM SOLVER
PROBLEM SOLVING

PROBLEM SOLVING SKILLS
RESEARCH ANALYST
RESEARCH PLANS FOR DATA
   GATHERING AND ANALYSIS
STRONG ANALYTICAL
STRONG ANALYTICAL SKILLS
STRONG PROBLEM SOLVING

BASIC IT SKILLS
COMPUTER SKILLS
MICROSOFT OFFICE
MS ACCESS, MS EXCEL
MS OFFICE, MS WORD
PC SKILLS, WORD PROCESSING

COMMUNICATION SKILLS
ABILITY TO COMMUNICATE
COMMUNICATIONS SKILLS
COMMUNICATION SKILLS
EFFECTIVE COMMUNICATION
EXCELLENT COMMUNICATION
EXCELLENT COMMUNICATION SKILLS
EXCELLENT ORAL
EXCELLENT VERBAL
EXCELLENT VERBAL AND WRITTEN
EXCELLENT VERBAL AND WRITTEN
   COMMUNICATION SKILLS
EXCELLENT WRITTEN
EXCELLENT WRITTEN AND ORAL
ORAL AND WRITTEN
ORAL AND WRITTEN COMMUNICATION
ORAL AND WRITTEN COMMUNICATION
   SKILLS
ORAL COMMUNICATION SKILLS
STRONG COMMUNICATION
STRONG WRITTEN
VERBAL AND WRITTEN
VERBAL AND WRITTEN
   COMMUNICATION
VERBAL AND WRITTEN
   COMMUNICATIONS
VERBAL AND WRITTEN
   COMMUNICATION SKILLS
VERBAL COMMUNICATION
VERBAL COMMUNICATION SKILLS
WRITTEN AND ORAL
WRITTEN AND ORAL COMMUNICATION
   SKILLS
WRITTEN AND VERBAL
WRITTEN AND VERBAL
   COMMUNICATION SKILLS
WRITTEN COMMUNICATION
WRITTEN COMMUNICATIONS
WRITTEN COMMUNICATION SKILLS

DATABASE SKILLS
SQL , DATABASE
ORACLE , RELATIONAL DATABASES
SQL SERVER

PRESENTATION SKILLS
FORMAL PRESENTATIONS
FORMAL PRESENTATIONS TO VARIOUS
   SENIOR LEVEL AUDIENCES
MAKES FORMAL PRESENTATIONS
PRESENTATION
PRESENTATION SKILLS

PROGRAMMING SKILLS
C+ , *C++* , *HTML* , *PERL* ,*UNIX*
XML , FORTLAN , J2EE
JAVA , JAVASCRIPT , JAVA SCRIPT
PROGRAMMING
PROGRAMMING LANGUAGE
PROGRAMMING LANGUAGES
PROGRAMMING SKILLS
SOFTWARE APPLICATIONS
SOFTWARE DEVELOPMENT
SOFTWARE ENGINEERS
VB , VISUAL BASIC
WEB DESIGNERS/ HTML XML
WINDOWS NT

PROJECT MANAGEMENT SKILLS
ASSESSES PROJECT IMPACT
DEVELOPS PROJECT PLANS
FACILITATE SHARING OF PROJECT
   OUTCOMES AND BEST PRACTICES
MULTIPLE PROJECTS
PROGRAM MANAGEMENT
PROJECT DOCUMENTATION
PROJECT DOCUMENTATION FOR
   SENIOR
PROJECT MANAGEMENT
PROJECT MANAGEMENT EXPERIENCE
PROJECT MANAGERS
PROJECT SUPPORT SPECIALISTS
STRATEGIC DIRECTION OF PROJECTS

TEAM SKILLS
INTERPERSONAL
INTERPERSONAL SKILLS
TEAM MEETINGS
TEAM MEMBERS

INDEPENDENT WORK SKILLS
ABILITY TO WORK INDEPENDENTLY
MINIMAL SUPERVISION
WORK INDEPENDENTLY
TEAM PLAYER

Section 5
Cognitive Perspectives on Modelling

5.1

HOW DO STUDENTS AND TEACHERS DEAL WITH MODELLING PROBLEMS?

Werner Blum and Dominik Leiß
University of Kassel, Germany

Abstract–*In this paper, we shall report on some of the work that has been, and is being, done in the DISUM project. In §1, we shall describe the starting point of DISUM, the SINUS project aimed at developing high-quality teaching. In §2, we shall briefly describe the DISUM project itself, and in §3 we shall present and analyse a modelling task from DISUM, the "Sugarloaf" problem. How students dealt with this task will be the topic of §4, the core part of this paper. How experienced SINUS teachers dealt with this task in the classroom will be reported in §5. Finally, in §6, we shall briefly describe future plans for the DISUM project.*

1. THE CONTEXT: QUALITY DEVELOPMENT

The teaching of mathematics in school is aimed toward supplying students with knowledge, skills, competencies and attitudes so that they are able to use mathematics in a well-founded manner when solving mathematical or real-world problems. We know from educational research that the desired effects of mathematics teaching can only (*at most*) be achieved if mathematics teaching obeys certain non-trivial criteria for "high-quality teaching". The following set of *quality criteria* constitutes *our* definition of *"Good Mathematics Teaching"* and is the basis of all our research and development activities (for details see Blum & Leiß, 2006):

I. Demanding orchestration of the teaching of mathematical subject matter
> That means in particular: Providing *manifold opportunities* for learners to acquire *competencies*, especially modelling ability, and creating manifold *connections,* within and outside mathematics.

II. Cognitive activation of learners
> That is: Stimulating permanently *cognitive activities* of students, including *metacognitive* activities (that is, a conscious use of strategies and reflections upon one's own activities), and fostering students' *self-regulation* and independence.

III. Effective and learner-oriented classroom management
> That means, for instance, *separating learning und assessing* consciously and using pupils' mistakes in a constructive way as learning opportunities.

In all aspects, the *teacher* has a crucial role to play. We can speak, in the words of Weinert (1997), of *"learner-centred and teacher-directed"* teaching.

We all know that everyday teaching is usually far from "good teaching" in this sense – also and particularly in Germany. The unsatisfactory TIMSS results induced the German government (both federal and states) in 1998 to establish a reform project that aimed at improving the quality of mathematics (and science) teaching, called *SINUS*. It started in 1998 with 180 schools; at present, more than 1500 schools are involved. The guiding principles of SINUS are:

- The *"new culture of teaching"*: Following consequently those quality criteria in all teaching activities.
- The *"new culture of tasks"*: Treating ("what?") a broad spectrum of competency-oriented tasks, notably modelling tasks, ("how?") in ways obeying the quality criteria.
- The *"new culture of cooperation"*: The new ways of teaching mathematics are embraced and supported by the whole mathematics staff of a school, not only by individual teachers.

However, classroom observations showed numerous *shortcomings* even in the ambitious SINUS programme. Some of these shortcomings are definitely *not* due to a lack of practical *realisation* of existing knowledge by the SINUS teachers, but rather to a *lack of knowledge* both

- of the actual procedures and difficulties of *students* when solving cognitively demanding tasks, and
- of appropriate ways for *teachers* to act when diagnosing students' solution processes and when intervening in case of students' difficulties;

that is, a *lack of* corresponding *research*. It is really surprising how little we know about the micro-structure of students' and teachers' dealing with such cognitively demanding tasks.

2. THE FRAMEWORK: THE DISUM PROJECT

This lack of knowledge was the starting point for the research project *DISUM* ("Didactical intervention modes for mathematics teaching oriented towards self-regulation and directed by tasks"). DISUM is an interdisciplinary project between mathematics education and pedagogy at the University of Kassel (see Leiß, Blum & Messner, 2004). It investigates how students and teachers deal with modelling problems, mainly in grade 9. A lot of activities have already been carried out in DISUM, especially:

1. The *construction* of appropriate modelling tasks.
2. Detailed cognitive and subject matter *analyses* of these tasks (constructing the "task space", based on the modelling cycle).
3. A detailed study and theory-guided description of actual *problem solving processes* of students in laboratory situations (involving pairs of students, sometimes with and sometimes without a teacher).
4. A detailed study and theory-guided description of actual *diagnoses* and *interventions* from teachers in these laboratory situations.
5. A detailed study of regular *lessons* with such modelling tasks, taught by experienced teachers from the SINUS project, and a theory-guided description of these lessons, emphasising our quality criteria.

6.   The construction of various instruments to *measure* students' achievements and attitudes.
7.   The construction of manageable and promising *tools* for
    a) the training of students in strategies for solving modelling problems, and
    b) the training of teachers in "well-aimed coaching" of modelling problems
How the DISUM project will be continued will be reported in §6.

3. AN EXAMPLE FROM DISUM: THE "SUGARLOAF" TASK

Sugarloaf

From a newspaper article:

The Sugarloaf cableway takes approximately 3 minutes for its ride from the valley station to the peek of the Sugarloaf Mountain in Rio de Janeiro. It runs with a speed of 30 km/hr and covers a height difference of approximately 180 m. The chief engineer, Giuseppe Pelligrini, would very much prefer to walk – as he did earlier on, when he was a mountaineer, and first ran from the valley station across the vast plain to the mountain and then climbed it in 12 minutes.

How big is the distance, approximately, that Giuseppe had to run from the valley station to the foot of the mountain? Show all your work.

A global cognitive analysis yields the following ideal-typical solution, oriented towards the modelling cycle (Figure 1):

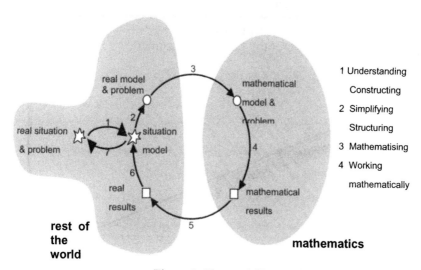

Figure 1. The modelling cycle.

First, the text has to be read and the problem situation has to be understood by the problem solver; that is, a so-called *situation model* has to be constructed, supported by the photo (Figure 2):

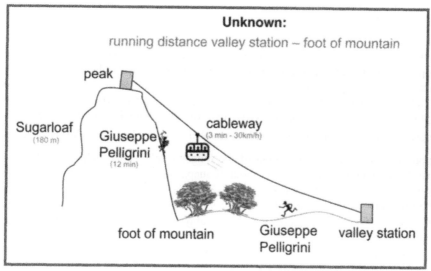

Figure 2. A situation model of the "Sugarloaf" task.

Then the situation has to be simplified, structured and made more precise, leading to a *real model* of the situation (Figure 3):

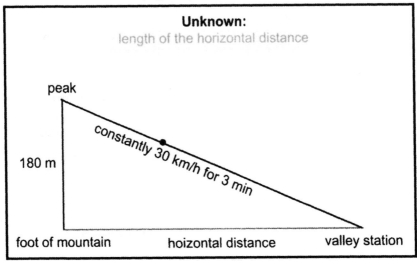

Figure 3. A real model of the "Sugarloaf" task.

Mathematisation transforms this real model into a *mathematical model,* basically a right-angled triangle where one side length is unknown. Then mathematical *tools* are activated, in the basic version of the solution process mostly the Pythagorean

Theorem, yielding the *mathematical result:* $b = \sqrt{1.5^2 - 0.18^2} \approx 1.49$ [km]

This result has to be interpreted in the real world as the *real result:* The distance from the valley station to the foot of the mountain is approximately 1.5 km. The important next step is a validation of this result: Is it reasonable? Is the accuracy appropriate (taking into account Keynes' well-known aphorism: "It is better to be roughly right than precisely wrong!")? First: For the result 1.5 km, no application of Pythagoras would have been necessary; in a triangle where the length of one side is approximately $\frac{1}{10}$ only of the length of another side, the length of the third side is approximately equal to the length of the longer side! Second: The simplification that the mountain has no width is certainly inappropriate, so something like 1.4 km will be a more reasonable answer. Third: The rope of the cableway is certainly not straight. Fourth: The speed of the cableway is definitely smaller near the beginning and near the end (which results in a considerably smaller distance). So one might go round the loop in Figure 1 several times. In any case, the whole process ends with an exposition of a final answer to the original problem (see Figure 4):

  This version of the modelling cycle has been influenced by various sources, among others by the cognitive theories of Reusser (1998) or Verschafel, Greer & deCorte (2000). We would like to emphasise that this version is more oriented towards the *problem solving individual* than the versions usually found in the ICTMA context, for instance in Blum (1995). We are convinced that it is a better model of what problems solvers actually do. In particular, step 1, reading and understanding, is strongly individually shaped, that means the resulting situation model is an *idiosyncratic construction* of the problem solver (according to constructivism and situated cognition; see, for example, de Corte et al, 1996). This version of the modelling cycle has proved extremely helpful for our purposes. It provides a better understanding of what students do when solving modelling problems and it gives teachers a better basis for their diagnoses and interventions.

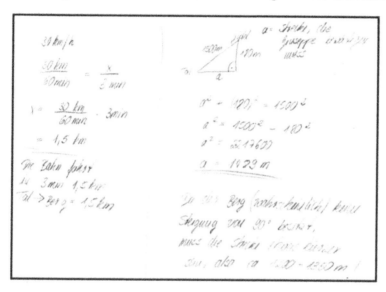

Figure 4. A pupil's solution of the "Sugarloaf" task.

Of course, actual individual problem solving processes are usually not as linear as it is suggested by this model. For instance, the second step, simplifying and structuring, is already influenced by the mathematical tools available to the problem solver, and often the process goes several times back and forth between the real world and mathematics. It is very interesting to identify students' *"modelling routes"* and to compare these (see Chapter 5.5, Borromeo Ferri).

4. "SUGARLOAF" IN THE LAB: STRENGTHS AND DIFFICULTIES OF STUDENTS

As mentioned in part 2, we have observed, videographed and interviewed 9[th]graders solving modelling tasks, including the "Sugarloaf" problem. We have selected pairs of students of each of four "competency levels", from weak Hauptschule students to strong Gymnasium students.

The most obvious strength exhibited by students was – not surprisingly in Germany – the *procedural* part of the process, that is the use of the Pythagorean Theorem and the calculation of the unknown length, often to a ridiculous degree of accuracy: ("b = 1489.16 m"). This presupposes, however, that the students got to this point. Some students got lost before; they failed already in step 1 or step 2, also because the text contains an unnecessary datum (the 12 minutes). There is a lot of evidence from research, as we know, that reading a text and understanding both situation and problem is a considerable cognitive barrier for students. In the "Sugarloaf" example, this was the *most difficult part* of the task. Excerpt 1 shows an example of this difficulty:

Excerpt 1 (Hauptschule students)

O.: How many kil... How long that takes with the three minutes, that is when it drives 30 km/h, that thing.

P.: Yes.

O.: And is over here in three minutes – how much does it cover then? How many kilometres or how many metres? How should I know that?

P.: Wait, wait. That is ...

O.: 15 km, with 15 km/h it takes half an hour.

P.: Rule of three!

O.: Rule of three!

The pupils calculate with the "rule of three" (proportions) that the distance is 1,5 km.

O.: 1.5 km.

P.: How long does it take ...

O.: 1500 metres. That is the distance that it takes approximately.

P.: That is it takes him 1.5 km.

O.: Yeah but what – perhaps the 12 minutes mean something as well.

P.: Here they ask how *far* he has run and not how *long!*

O.: 1.5 km, 1500 m.

P.: Yes.

The students can easily calculate by the "rule of three" $s = 1500$ m from the given v and t and then they are convinced they are completely done. For a moment Osman realises that they have not used the given "12 minutes" in their calculation, but Pascal's they ask how *far* he has run, not how *long*" stops the solution process. This is a well-known phenomenon, a typical strategy especially for German students: "You don't have to understand the situation, just use the data of the task in some way". In other words: The students do not construct an appropriate situation model.

Excerpt 2 (see p229) shows another section in the modelling process.

Katherina would like to apply the Pythagorean Theorem, but Christoph realises that the situation model contains no right-angled triangle. Katherina obviously feels puzzled. After some discussion, Christoph suggestion "Shouldn't we do it approximately like that, with a right angle?" helps Katherina to overcome her cognitive barrier. She realises that "the mountain is very steep", gives a reason for the assumption of a right angle and thus accepts Christoph's simplification (that is the common real model). Immediately thereafter, the students draw a triangle and begin to work successfully within the mathematical model.

These two examples demonstrate how cognitively demanding the first two steps are. So we should not simply and globally say (as it is often done) that the *mathematisation* part of the modelling process is difficult but look upon the process a bit more carefully.

Excerpt 2 (Gymnasium students)

K.: Hight difference. That has nothing to do with width.

C.: No, but if you want to do this Pythagoras now, you need a right angle here and you don't have it.

K.: Why not?

C.: Is that a right angle here, if you …

K.: Height difference, they speak of the height difference. That's here.

C.: No, height difference is no right angle.

K.: Oh, the mountain is 180 m high.

C.: But you want to calculate the distance *here.*

K.: Yes, but now one can perhaps – yeah! You only have to know how wide the mountain is.

C.: It says "approximately". Shouldn't we do it like it says, approximately, with a right angle? *The pupils laugh.*

C.: I think that's sufficient.

K.: Yeah, it says the mountain is very steep, and if the mountain is very steep then it's not so much what is more here, then we can do it like that.

C.: That's what I'm saying!

K.: Super!

Another uniform shortcoming was the lack of *validation* and of substantial *reflection*. We know from research into teaching and learning how important it is to look back and to reflect on one's own problem solving process, that is – in the words of Reusser (1998) – "to extract the relevant conceptual-schematic and processual-strategic characteristics of a problem solution in an abstracting way" and, thus, to contribute to the further development of meta-cognitive knowledge. More generally, no conscious use of *problem solving strategies* by students was recognisable; in particular, no student seemed to have the modelling cycle as a guiding tool at his or her disposal. As a consequence of the absence of validation, no student tried to improve his or her solution, they were all satisfied when they had reached *any* solution whatsoever.

5. "SUGARLOAF" IN LESSONS: STRENGTHS AND DIFFICULTIES OF TEACHERS

As we said in part 2, we have also observed and videographed *lessons* with our modelling tasks, in all types of schools, from Hauptschule to Gymnasium. In particular, in the context of a "Best Practice Study", several experienced teachers from the SINUS project included some of our tasks in their regular lessons, among others the "Sugarloaf" problem. Here are a few results of our observations.

First of all, most lessons clearly stood out positively from typical German lessons. Looking more thoroughly at these lessons with our *"quality glasses"* shows that (see Blum & Leiß, 2006)

- the students had opportunities to model, to argue, to communicate;
- mental activities were stimulated;
- students, for the most part, could work independently;
- the atmosphere was tolerant towards mistakes and free of assessment;
- there was a discussion in the end, as an element of reflection.

None of these aspects is self-evident, on the contrary. For instance, the TIMSS Video Study in six high performing countries has revealed that even in these countries there was normally no reflection at all in the end of solution processes. All teachers, including those in the Hauptschule, succeeded in helping the students to find a solution. This is again an example of the important and often neglected distinction between working *independently*, with support from the teacher, on the one hand, and working totally *on one's own*, on the other hand. Thus, the crucial role of the teacher also in more student-centred learning environments is highlighted once again.

Another important feature was to realise during the lessons how imperative it is that teachers have a detailed and intimate knowledge of the modelling cycle, both as a solid basis for diagnoses and as a rich source for supporting interventions. In one of these lessons, a teacher succeeds in leading his Hauptschule class to an appropriate real model, as an important intermediate stage between the situation model and the mathematical model (T.: "It only matters what is mathematically relevant, no trees or so", S.: "Hence a triangle").

A problem reflected even in those Best Practice teachers' actions was the absence of further reflections on the solution processes. In particular, no teacher considered the question of how certain assumptions influence the solutions and how accurate the results actually could then be. Often, also the teachers were satisfied when the students had *any* solution. Equally, the counterpart of the above-mentioned absence of problem solving strategies on the students' side was the absence of any stimulation of such strategies by the teachers. Such stimulation seems not to be a part of teachers' everyday repertoires. Consequences for teacher education are obvious.

6. PROSPECTS FOR DISUM: THE NEXT PHASE

We are convinced that both
- a conscious use of problem solving strategies by students, and
- well-aimed and independence-supporting coaching by teachers
will have positive effects on students' achievement. However, this has never been examined systematically and quantitatively in the context of demanding modelling tasks. We will compare the effects of teachers' training in well-aimed coaching with the effects of "normal" instruction or of no teaching at all (students working totally on their own). The notion of *"well-aimed coaching"* has a definite meaning; it consists of five components, one being the DISUM strategy set for students (see Blum & Leiß, 2006): goals/volition/organisation/strategy/evaluation.

We shall measure the effects on achievement by a classical pre-test/post-test design. Our tests contain modelling tasks as well as tasks from the current curricular topic (in our study: the Pythagorean Theorem and its context) and anchor items from PISA. In addition, we shall measure the effects on students' motivation, emotions and independence feeling by classical scales taken from PISA and from the German project PALMA (Pekrun et al. 2004).

Our final aim with DISUM is, of course, to improve both everyday teaching and teachers' expertise by implementing our instruments and findings in school classrooms and in teacher education programmes.

REFERENCES

Blum, W. (1995) Applications and Modelling in Mathematics Teaching and Mathematics Education - Some Important Aspects of Practice and of Research. In C. Sloyer, W. Blum and I. Huntley (eds): *Advances and Perspectives in the Teaching of Mathematical Modelling and Applications*. Yorklyn: Water Street Mathematics, 1-20.

Blum, W. and Leiß, D. (2006) "Filling up" – The Problem of Independence Preserving Teacher Interventions in Lessons with Demanding Modelling Tasks. In M. Bosch (ed) *Proceedings of the European Society for Research in Mathematics Education CERME 4 (2005)*. IQS FUNDIEMI Business Institute www.fundemi.url.edu, 1623-1633. ISBN 84 611 3282 3.

deCorte, E., Greer, B. and Verschaffel, L. (1996) Mathematics Teaching and Learning. In Berliner, D. and Calfee, R. (eds): *Handbook of Educational Psychology*. New York: Macmillan, 491-549.

Leiß, D., Blum, W. and Messner, R. (2004) Sattelfest beim Sattelfest? Analyse ko-konstruktiver Lösungsprozesse bei einer realitätsorientierten Mathematikaufgabe. In *Beiträge zum Mathematikunterricht 2004*. Hildesheim: Franzbecker, 333-336.

Pekrun, R., Götz, T., vom Hofe, R., Blum, W., Jullien, S., Zirngibl, A., Kleine, M, Wartha, S. and Jordan, A. (2004) Emotionen und Leistung im Fach Mathematik: Ziele und erste Befunde aus dem „Projekt zur Analyse der Leistungsentwicklung in Mathematik" (PALMA). In J. Doll and M. Prenzel (eds): *Bildungsqualität von Schule: Lehrerprofessionalisierung, Unterrichtsentwicklung und Schülerförderung als Strategien der Qualitätsverbesserung*. Münster: Waxmann, 345-363.

Reusser, K. (1998) Denkstrukturen und Wissenserwerb in der Ontogenese. In Klix, F. and Spada, H. (eds): *Enzyklopädie der Psychologie, Themenbereich C: Theorie und Forschung, Serie II: Kognition, Band G: Wissenspsychologie*. Göttingen: Hogrefe, 115-166.

Verschaffel, L., Greer, B. and de Corte, E. (2000) *Making Sense of Word Problems*. Lisse: Swets & Zeitlinger.

Weinert, F.E. (1997) Neue Unterrichtskonzepte zwischen gesellschaftlichen Notwendigkeiten, pädagogischen Visionen und psychologischen Möglichkeiten. In Bayerisches Staatsministerium für Unterricht, Kultus, Wissenschaft und Kunst (ed.): *Wissen und Werte für die Welt von morgen*. München, 101-125.

5.2

TEACHER-STUDENT INTERACTIONS IN MATHEMATICAL MODELLING

Jonei Cerqueira Barbosa
State University of Feira de Santana, Brazil

Abstract–*This investigation took part at a Mathematics Modelling course to pre-service teachers. The students were requested to formulate and solve applied problems, observed and monitored by a teacher through meetings. Interactions between students and teacher were video-ed as qualitative data. The results underline the notion of interaction spaces to denote the moment of encounter between the teacher and the students or the students by themselves to discuss the modelling task. The data illustrates different styles of interaction between the teacher and the students.*

1. INTRODUCTION

A very visible feature in Mathematical Modelling is the invitation to the students to work in groups (Barbosa, 2003; Goldfinch, 1992). This does not, however, mean that the teacher is not present to follow the students' work. As stated by Ikeda and Stephens (2001), the teacher needs to stimulate the students' questioning and reflection during the mathematical modelling process. A study conducted by Araújo and Salvador (2001) illustrates a case in which a student had intense contact, through frequent meetings with the teacher, which resulted in a modelling project that received a good evaluation.

Thus, my intention is to highlight here the moments in which students and teachers come together to discuss modelling activities. Here, I present part of a broader study about interactions between teachers and students during mathematical modelling activities. I will use "interactions" inspired by the sense used in discursive psychology (Lerman, 2001), assuming that the interaction is constituted of discursive practices that help people to act and organize their social worlds.

2. CONTEXT AND METHODOLOGY

To analyze teacher-student interactions in classroom tasks, I took episodes from my own classroom. The scenario was a mathematical modelling course that met two hours per week for a total of 36 hours over the semester, proffered between August and December, 2004, in a third grade class in a Mathematics Pre-service Teacher Education Program.

One of the activities for the course was a modelling project, developed over the course of the semester, where the students, organized in groups, were required to select a theme, formulate problems, and solve them using mathematics. To follow the work of the students, I requested two partial written reports, which were read and commented on in meetings with the groups (2 times per semester) in the classroom. At the end of the semester, the students presented oral seminars about their projects, followed by discussions with the teacher and the rest of the class (each students' group used 1 hour). They were then required to hand in their final written report. This is more extensive modelling activity, classified as case 3 in Barbosa (2003), wherein the students have the responsibility to elaborate and solve the problems.

To carry the research out, I chose to observe the moments of discussion between the teacher and the students regarding the partial versions of the written report and the oral presentation. Using the qualitative perspective (Bogdan and Biklen, 1998), the encounters of this nature with all the groups of students throughout the semester were filmed by a research assistant. With the filmed material in hand, I viewed them many times, seeking to identify passages, referred to here as episodes that were important for the proposal of the research. The selected passages were transcribed and attributed meanings in light of the context.

In this paper, I will take two episodes from such moments of encounters with the group of students composed of Anna, Paula, Maria, Marcelo, Allan and Katherine, who selected "cigarettes" as the theme for their project. Like their classmates, they had no previous contact with modelling, and the activity was a novelty to them.

3. PRESENTATION OF THE DATA

Following is a transcription of the episodes of the interactions between the teacher and the student groups described in this paper, referred to here as "the problem" and "the solution.

Episode 1 - *The problem*
In their first report, the group of students highlighted here presented a 2-page text about the harmful effects of cigarettes for health and the environment. The information appeared to have been collected from the Internet. They presented the following question as the objective of their project: How much carbon do smokers in the city of Salvador release into the atmosphere per year? They failed to bring the solution to the situation in their written report, but teacher had requested it when he presented the task.

In the face-to-face encounter to discuss the report, the conversation lingered on the objective of the project. The interaction was initiated by the teacher:

Teacher: *What did you select?* [looking at the report] . . . *Cigarettes!* [The students observed attentively] *So . . . you did a problem. . What is it you want to know?*

Marcelo: *We want to know how much carbon Salvadorans who smoke release into the air.*

Teacher: *But how are you going to solve this? Take the number of smokers in Salvador, is that it? You'll have to take the*

	average that a smoker releases into the atmosphere? Apply some basic operations. . and you will have a result.
Anna:	*Yeh, we were thinking like this: define a time limit. So many years, but we haven't done that yet, because we are going to discuss it.*
Allan:	*Until a given year.*
Anna:	*For example, 5 years. During 5 years, how much carbon do Salvadorans dump into the atmosphere of Salvador?*
Teacher:	*Yeh, but this isn't a problem for you guys . . . you already know how to solve it. There is another aspect: what substance do cigarettes leave in the blood? Is it nicotine that stays in the blood?*
Danielle:	*It stays in the skin, the hair . . .*
Anna:	*Cigarettes have many components that are absorbed by the organism.*
Teacher:	*Maybe you could investigate what happens with the concentration of nicotine in the blood of a smoker over time.*
Luiza:	*Then we would have to find a function of the number of cigarettes he smokes per day.*
Teacher:	*Definitely, the more the individual smokes, the greater will be the presence of nicotine in the blood . . . it depends on the number of cigarettes he smokes per day and the length of time he smokes.*
Allan:	*It depends on the group he's in.*
Teacher:	*Then we would need data from specialists.*
Danielle:	*For smokers, there are associations like Healthy Club, more or less smokers anonymous. . I could find this data.*
Allan:	*Here we are overlooking other variables, aren't we?*
Teacher:	*Yes, we are leaving out other variables and focussing on the quantity of nicotine in the blood over time.*

The suggestion given to the students was that they insert this new situation into their project: relate the amount of nicotine in the human body to the length of time the individual has been smoking and the average consumption of cigarettes. The conversation went on a bit longer, talking about how to carry out the study, and a date was established for the presentation of the second written report.

The development of the teacher-student interaction did not follow a linear path, but rather is quite dynamic. In this case, the conversation unfolds in reference to the modelling project, initially materialized in the written report produced by the students. The initial comments of the students reveal understandings about what the task is: they explain the question, for which the strategy to solve it is as yet unknown to them. The teacher, meanwhile, situates himself in the interaction with another understanding, pointing to the need for the situation to be a problem for them. It also appears to have been a moment of elucidation of the task for the students.

However, the passage described suggested that the interaction may not be reactive; that is, the enunciation may not refer to the one immediately preceding it. Its development should be subject, as pointed out above, to the understanding of the subject, which can provoke unexpected detours. In the above episode, the teacher clearly provoked a digression when he made the following comment: *Yeh, but this isn't a problem for you guys . . . you already know how to solve it. There is another aspect: what substance do cigarettes leave in the blood? Is it nicotine that stays in the blood?* With this, it ended up guiding the work of the students too much. One of the proposals, which was for the students to formulate the problems, dissolved. In practice, a transformation took place in the activity proposed to the students, since one of their responsibilities that of elaborating the problem, was no longer in their hands.

Episode 2 - *The solution*

In their second written report, the students presented more detailed information regarding the effects of the presence of nicotine on the body, probably collected from the Internet. They cited data about an experiment done with a sample of people in which dermal nicotine patches were applied of 15, 30 and 60 mg, respectively, and the level of nicotine in nanograms per millimetre of blood was measured over the next 24 hours.

In the face-to-face encounter with the students, possible approaches to the data were discussed, and they committed themselves to pursuing them. In their extra time, the group sought the teacher a number of times to discuss the construction of a mathematical model that would relate the level of nicotine in the blood with time.

On the day of the oral presentation, the team begins with a presentation of the effects of nicotine on the body. Next, they mention the experiment with the dermal nicotine patches in a sample of people, and say they will show the construction of the mathematical model specifically for the case of the 30 mg patch.

Presenting the data on the level of nicotine in the blood (in ng/ml), denoted N, as a function of time passed from the moment the nicotine patch was applied, denoted t (as in Figure 1), one of the students, Katherine, who was monitoring the projection of the computer image for the group, generated the graph of the data in Excel (see Figure 1).

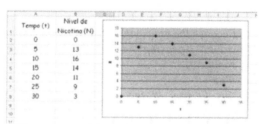

Figure 1. – Table and graph relating N and t.

Right after that, Marcelo, who was presenting at that moment, explained that the graph was inadequate:

> Marcelo: *If we observe this graph, we'll see that there will be a point at which the nicotine will become negative [pointing to the Ox axis]. So, this graph doesn't make sense.*

The student was referring here to the impossibility of making extrapolations in the case that they fit a function of the type $y = ax^2 + bx + c$ *(1)* - whose parameters are real numbers and *a* not equal to zero.

The strategy that they adopted for the case was to obtain the natural logarithm of the values of the nicotine level (denoted ln N) and time (denoted ln t), generating a graph that also resembled a parabola (see Figure 2). This time, however, they argued that ln N could be negative, which would not make real sense for N.

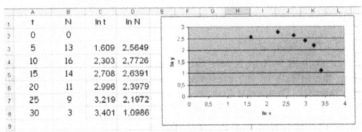

	t	N	ln t	ln N
1				
2	0	0		
3	5	13	1.609	2.5649
4	10	16	2.303	2.7726
5	15	14	2.708	2.6391
6	20	11	2.996	2.3979
7	25	9	3.219	2.1972
8	30	3	3.401	1.0986
9				

Figure 2. Table and graph relating ln N and ln t.

Then, they fitted the graph which equation is
$$\ln N = -0.9941(\ln t)^2 + 4.7013(\ln t) - 3.6444 \qquad (2)$$
from which follows:
$$N = \exp(-0.9941(\ln t)^2 + 4.7013(\ln t) - 3.6444) \qquad (3)$$
Equation (3) was validated and considered by the students to be the mathematical model that related level of nicotine in the blood to time.

When the students had finished their presentation, a discussion followed in which the class and the teacher could pose questions and/or make comments. Below is a transcription of the interaction between the teacher and the students:

Teacher: *Look, people. They developed a very interesting strategy. The data suggested a parabola, and they analyzed that, according to the situation, it didn't correspond to a parabola. So, why exactly doesn't a parabola fit?*

Katherine: *Because the concavity is below and it cannot assume negative values [speaking hesitatingly, as though elaborating her comment].*

Professor: *Why?*

Katherine remains looking at the projected computer screen for a few moments.

Katherine: *No, there is no such thing as negative time, isn't that it?*

Teacher: *OK, time is going to be positive.*

Katherine: *However, it was ln, ln can have negative values [referring to the graph in Figure 2].*

Teacher: *OK, but it's that graph, but here [referring to the graph in Figure 1] . . . why isn't it possible to adjust the parabola?*

Katherine: *Ask again!*

Maria: *Because, in this case, as time goes by, the nicotine decreases, then, there in that case, since the parabola is with the*

concavity below the nicotine acquires a negative value. So, in
this case, it can't be a parabola.
The teacher agrees with the explanation offered by Maria.
Katherine: *That's what I wanted to say!*
Next, the teacher returned to the group's strategy of considering the function
of ln N = f(ln t):
Teacher: *And not, why does the parabola work here? [pointing to the*
                graph in Figure 2].
Maria: *Well, when we put ln, a logarithmic function, it doesn't touch*
                zero, right?
Teacher: *No?*
Maria: *Ah, no, no. I'm getting confused. . . it's that the ln of nicotine*
                can be negative in this case.
Teacher: *Why?*
Marcelo: *It can. We calculated the logarithm. It can have a number*
                that, raised to a negative number, gives a positive number as
                the answer.

In this episode, a certain sequence can be noticed in the words spoken by the people during the interaction. The interventions that followed were configured as reactions to the preceding enunciation. With this, the opportunity to reflect on the construction of the relations between the variables was not lost.

Subsequently, it can be noticed, mainly in the comments made by Katherine that the teacher-student interaction served the purpose of clarifying the procedures carried out by the group. Although she participated in the project, one could notice that she did not show confidence in explaining the procedures used to construct the mathematical model. Perhaps she had accepted the steps taken by her classmates. With the teacher's questions, she had the opportunity open it up for discussion again.

In this episode, the interaction between the interlocutors has its own sequence, which ended up producing an explanation for the procedures used to construct the mathematical model for the level of nicotine in the blood as a function of time. Certainly, the participants cannot be expected to produce the same meanings that are being produced collectively in the teacher-student interaction. For example, in the above episode, at a certain point, Luiza's participation becomes more visible in the interaction, such that we have few elements to know what is happening with Katherine. However, in the perspective adopted here, we assume that the discursive interactions are the raw material for individual understanding.

4. Discussion of the data

The two episodes analyzed above portray passages of interaction between the teacher and the students during the development of a mathematical modelling activity. In the first, the conversation is about the definition of the problem to be pursued by the group of students; the second revolves around the process of the construction of the mathematical model. We can consider that people discuss a lot of things in interactions. Inspired in Skovsmose (1990), Barbosa (2004) points out that the discussions in the mathematical modelling may be, at a minimum, of three types:

- mathematical: referring to those that are about mathematical ideas;
- technical: those that refer to the manner of translating characteristics of the phenomenon to be modelled into a mathematical representation;
- reflexive: referring to the comprehension of the role of mathematics in social practices.

On the episode called "the solution", the students were debating whether the logarithmic function graphic touches zero or not, which is a good example of mathematical discussions. When they point out that ln t and ln N are fitting by the parabola, students referred to how translate the situation in mathematical terms, in other words, it illustrates a technical discussions. There are no good examples of reflexive discussions on these episodes, but students could raise questions like these: How cigarette marketing influences people? What happens with the healthy when nicotine level takes high? And so on.

What links the teacher and the students in these discussions is the modelling work, materialized in the written versions or oral presentations. With these conditions, we can consider these elements as a system that moves according to the enunciation of each of these actors, even taking into consideration their written production. Borba and Villareal (2005) present the metaphor *humans-with-media* to support the idea that media constitute our thinking. In the case of the episodes analyzed here, the teacher-student interactions were mediated by oral, written, and computer media.

Thus, we can use another metaphor, analogous to that proposed by Borba and Villareal (2005), to consider the moment of encounter between the teacher and the students to discuss the modelling work: teacher - modelling work - students. This encounter constitutes what I call here the *interaction spaces*. Analogously this notion can be used to refer to students' discussions by themselves. As suggested in the introduction, what happens in these appears to influence decisively the students' mathematical modelling activity. This hypothesis has broad resonance with the sociocultural perspective of learning, which emphasizes social factors as constituents of learning (Lerman, 2001).

In the spaces of interactions, the relations are not horizontal, as the teacher has the task of organizing and setting the tone for the form that these will. As we saw in the first episode, the teacher provoked a digression, leading the conversation in another direction. In the second, although no such detour occurred, it was the teacher's questions that instituted working norms on that space of interaction.

It appears reasonable, therefore, to consider that the voice of the teacher, because of its legitimacy, should be the focus of attention in the monitoring of mathematical modelling activities. With the suggestion that the students pursue the construction of a model that relates the level of nicotine in the blood as a function of time, the nature of that activity was altered, as well, as the formulation of the problem practically became the task of the teacher. Leiß and Wiegand (2005) analyse a similar situation when the teacher strongly guide the students' work. This style approximates what Alrø and Skovsmose (2002) call the sandwich pattern of communication, in which the authority of the teacher to correct "errors" and direct the students predominates.

If we are interested in the students having an "authentic" modelling experience, it is not fitting that the teacher be directive, but rather that he/she can direct queries to the students based on their own words. The second episode, where the students talk about the procedures for the construction of the mathematical model, illustrates this other possible way for the teacher to intervene in the spaces of interaction. For this, it was necessary to hear what was being said by the student and use it as a type of lever to formulate the next comment. Here, there is a sequence to the enunciations, a type of action-reaction. It was very similar to situation investigated by Leiβ and Wiegand (2005), which the teacher throws the ball back to the students when they asked questions.

This style approximates what Alrø and Skovsmose (2002) refer to as cooperative investigation, which has active listening as a condition. These authors use a passage from Rogers and Farson (1969), and Alrø and Skovsmose (2002) to illustrate its fundamental characteristic:

> It is called 'active' because the listener has a very definite responsibility. He does not passively absorb the words which are spoken to him. He actively tries to grasp the facts and the feelings in what he hears, and he tries, by his listening, to help the speaker work out his own problem. (p62)

In this way, in the first episode, once the teacher had suggested a new problem to the students, he could have attempted to help them formulate one based the comments made by the students. In this case, as was suggested in the second episode, the *interaction space* distinguishes itself as a *negotiation space*, as the teacher and the students, mediated by the modelling work, would have to negotiate meanings. Negotiation implies dialogue, which assumes that none of the parties wishes to impose his/her perspective, but rather put it up for discussion. This implies that the style of the interaction be of an open and reactive type, in which each enunciation is based on the one preceding it.

5. FINAL REMARKS

In this paper, I have analyzed two episodes of discursive interactions between the teacher and the students during a mathematical modelling activity. I have called *interaction spaces* all the moments in which the teacher and the students or the students by themselves get together to discuss a modelling activity. However, this idea can be equally transposed to other learning environments, such as problem-solving and mathematical investigations.

The importance of knowing what happens in these interaction spaces lies with its constitutive nature of students' understanding about the activity. As Christiansen (1997) points out, students can suffer from a conflict regarding whether the task is an exercise or a problem, which can be taken up for discussion in the meeting with the teacher. It is in the spaces of interaction that the students seek help to understand the demand made by the teacher in the activity proposed.

For this reason, it is important that the teacher consider his/her style of participation in these spaces. We can consider two styles. The first, denominated *directive* refers to that in which the teacher responds readily to questions, immediately corrects the "errors", and offers direction for the students' work. The second, denominated *open*, refers to a style in which the teacher seeks to formulate questions for the students based on what they say, approximating a reactive pattern.

The transition from a directive to an open style may not be easy for teachers. It is related not only to a choice, but is also, as stated by Alrø and Skovsmose (2002), in addition to being rooted in our attitudes, contrary to the logic of the school. We can consider it as a horizon to be pursued daily in the task of organizing and conducting mathematical modelling activities, as well as others, and it deserves the attention of teachers.

REFERENCES

Alrø, H. and Skovsmose, O. (2002) *Dialogue and learning in mathematics education: intention, reflection, critique.* Dordrecht: Kluwer.

Araújo, J. L. and Salvador, J. A. (2001) Mathematical Modelling in calculus courses. In J. F. Matos et al. (eds) *Modelling and Mathematics Education: ICTMA9 – Applications in science and technology.* Chichester: Horwood Publishing, 195-204.

Barbosa, J. C. (2003) What is Mathematical Modelling? In S. J. Lamon et al. (eds) *Mathematical Modelling: a way of life.* Chichester: Ellis Horwood, 227-234.

Barbosa, J. C. (2004) Modelagem Matemática em cursos para não-matemáticos. In H. N. Cury (ed) *Disciplinas matemáticas em cursos superiores: reflexões, relatos e propostas.* Porto Alegre: EDIPUCRS, 63-83.

Bogdan, R. and Biklen, S. K. (1998) *Qualitative research for education: an introduction for theory and methods.* Boston: Allyn and Bacon.

Borba, M. C. & Villareal, M. E. (2005) *Humans-with-media and the reorganization of mathematical thinking.* Dordrecht: Springer.

Christiansen, I. M. (1997) When Negotiation of Meaning is also Negotiation of Tasks: Analysis of the Communication in an Applied Mathematics High School Course. *Educational Studies in Mathematics,* 34 (1), 1-25.

Goldfinch, J. M. (1992) Assessing Mathematical Modelling: a review of some of the different methods. *Teaching mathematics and its applications,* 11 (4), 143-149.

Ikeda, T. and Stephens, M. (2001) The effects of students' discussion in mathematical modelling. In J. T. Matos et al. (eds) *Modelling and Mathematics Education: applications in science and technology.* Chichester: Ellis Horwood, 381-390.

Leiβ, D. and Wiegand, K. (2005) A classification of teacher interventions in mathematics teaching. *Zentralblatt für Didaktik der Mathematik,* 37 (3), 240-245.

Lerman, S. (2001) A cultural/discursive psychology for mathematics teaching and learning. In B. Atweh et al. (eds) *Sociocultural research on mathematics education: an international perspective.* London: Lawrence, 3-17.

Skovsmose, O. (1990) Reflective knowledge: its relation to the mathematical modelling process. *International Journal of Mathematics Education in Science and Technology,* 21 (5), 765-779.

5.3

MATHEMATICAL MODELLING: A TEACHERS' TRAINING STUDY

José Ortiz[1], Luis Rico[2] and Enrique Castro[2]
[1]University of Carabobo, Venezuela
[2]University of Granada, Spain

Abstract–*The application of modelling has had promising results in the field of mathematics education. It offers an organised and dynamic way of closing the gap between mathematics and the student's physical and social world. To investigate the teaching competence of student teachers, a training programme was designed and applied to incorporate mathematical modelling and the graphic calculator in teaching activities with an algebraic content. The aim of this programme was to broaden the cognitive support for participants and to help them act rationally in decision-making when designing teaching activities. The productions of ten pre-service mathematics teachers are analysed, specifically their proposals for problem situations in secondary school linear algebra. The results showed evidence that technical and didactic use of the graphic calculator and of the benefits that it offers were important for both teacher and student. Pre-service teachers reflected on the joint teacher-student work group discussion and the reporting and communication of mathematical ideas to the class.*

1. INTRODUCTION

This paper focuses on the didactic knowledge derived from implementation of a training programme involving use of the graphic calculator and the modelling in the initial training of secondary school mathematics teachers in teaching linear algebra. The analysis is based on proposed problem situations in designing didactic activities for students in linear algebra. A theoretical structure for the curricular analysis was introduced, supported by the following four dimensions: conceptual, cognitive, formative and social (Rico, 1997). This theory states that the didactic knowledge of mathematics topics should be based on representation systems (Janvier, 1987; Duval, 1995), mathematical modelling (Niss, Blum & Huntley, 1991; Houston, Blum, Huntley & Neill, 1997; Ortiz, 2002), mathematics errors and difficulties (Borassi, 1987), phenomenology (Freudenthal, 1983), history of mathematics and materials and resources.

The research chose to incorporate the use of the T1-92 graphic calculator (GC) as a didactic resource in the teaching and learning process of mathematics. The mathematics content involved was linear algebra, which offers a wealth of important applications for situations modelling the real world, as is established by Harel (1998), Brunner, Coskey & Sheehan (1998) and Dorier (2000), among others.

It is not frequent incorporate new technologies and mathematical modelling in the design of initial training programmes for mathematics teachers, despite the benefits that these curricular organizers offer to the implementation and evaluation of the programmes by its operative and integrative character. However, this study used a programme that involves the modelling, the graphic calculator and linear algebra in the development of didactic activities. Taking into account the foregoing, the following questions were formulated:

For which proposed problem situations do teachers in training apply mathematical modelling and what aspects do they consider important?

What didactic possibilities do the proposed situations offer for the establishing and relating the graphic calculator and mathematical modelling?

2. METHODOLOGY

The study was performed with ten subjects who pursued a 30-hour training programme (10 sessions) through a course-workshop. The workshop was based on the modelling and the graphic calculator as resource for the context of linear algebra in secondary school. The subjects of study were teachers in training who participated as volunteers. Participants were required to be potential mathematics teacher. In this work, they are identified with the acronym STi (i = 1,..., 10).

This research is performed using case study methodology (Miles & Huberman, 1994). The study analyzed the participants' productions concerning the use of graphic calculator and modelling in the teaching of linear algebra, the instrumental handling of graphic calculator and its connection to modelling. The use of these organizers for planning teaching and learning mathematics assignment increases the didactic knowledge of teachers in training. The participants are assigned the task of designing a didactical activity with algebraic content for potential secondary students. The participants were required to consider that their didactical activities should pertain to a subject of the mathematical secondary curriculum. The activity developed focused on the following proposal:

Design of a didactical activity: Suppose a Secondary Teacher must develop a didactical activity that illustrates the use of linear equations. To satisfy this goal we ask you to describe or propose a problem situation from the real world that fulfils this assignment. Assuming that the teacher knows the modelling process and will apply the graphic calculator with his/ her students:

a) *Formulate at least two questions whose answer requires the use of the modelling and graphic calculator;*

b) *List the sequence of the activities to be followed by the teacher, to obtain this objective;*

c) *Suggest at least two issues to be assessed in students and how such assessment should be made.*

It is important to emphasize that these tasks were assigned in the first and last sessions in order to identify changes in the didactical knowledge of future teachers in applying the programme.

The productions of each participant identify the skills development proposed, the open problems presented using modelling, and the discussion and reflection on errors and difficulties in the problem situations. The activity developed critical appraisal of each part and oral and written communication skills for working in a group (Galbraith, Haines & Izard, 1998). The support of the graphic calculator (GC), the use of several representation systems and connections among them were considered both mathematical concepts and situation proposed in the design of the didactical activities. Calculation of possibilities, experimentation, visualization, and contrast of possible results to be achieved with handling of the graphic calculator were also used (Ortiz & Rico, 2001; Kutzler, 2000).

3. RESULTS

The subjects developed their didactic proposals in their notebooks and then presented them to the group, using view screen and overhead projector.

In the initial session, the participants presented daily situations related to fruit sales (FS), article prices (AP), parking a car (PC) and other situations from physical nature or the social world, such as milk cow production (MCP), test questions (TQ), mass-spring system (MS), motorbike circuit (MC), nightclub for rent (NR) and trajectory of a boat in the ocean (TB). In general, the situations proposed by the participants pertained to algebra problems at secondary school level. The linear algebra concepts related to each model are linear equations (FS, PC, MS), linear systems of equations (MCP, TQ) and matrixes (TQ), such as linear inequality in one and two variables (NR, TB). The participants also presented two models that include elemental functions (MC, PC). The problem situation pertaining to the test question (TQ) is repeated in the linear systems and matrixes because the students' teacher presented two alternative models after developing the matrix model on the notebook. Overall, the variety of the problem situations shows skills for identifying problem situations in the physical and social worlds linked to mathematics directed primarily to students in a secondary school environment. The participants' notebooks showed that the mathematical models considered could help increase comprehension the problem situation (ST7), get data (ST7), explain the situation (ST8) and identify practical utility (ST9). It is important to note that some teachers in training (ST2, ST3, ST4, ST6) did not answer the questions related to the model's usefulness and its interest. This could be due to disciplinary worries that lead them to focus attention on the model structure rather than on results and utility.

In general, analysing the productions shows that in the first session of programme implementation, the pre-service teachers:

1. Are open to handling the graphic calculator (GC), although they maintain a moderate position about its use by the students;
2. Know the appropriate skill for proposing situations from the students' environment;
3. Maintain a scheme for the teacher's control and conducting of the class;
4. Show little initiative for proposing assessment activities.

In the last session, the pre-service teachers' productions showed the following results.

First question:
Propose a problem situation from the real world to demonstrate the use of the linear equation systems.

Here, the future teachers established situations from daily life, family, managerial, commercial contexts and a war scenario. Some of the every day situations proposed were related to traffic signs (ST1), social relations at the school (ST5) and travel for shopping (ST9). The family situations referred to similar support (ST4) and ages of the family members (ST7). From the managerial world, participants used brick manufacturing situations (ST2) and canned goods production (ST6). They also suggested a current commercial situation like the cellular phone (ST3). Another proposal suggested a situation related to spy satellites' trajectories in the war scenario (ST10). The following situations illustrate the results:

Situation linked to family life (ST4):

"A father wants to motivate his son to study and do mathematics exercises (with or without the graphic calculator). So, he proposes the following deal to his son: For each correctly solved exercise, I will give you 1 Euro, but you will have to give me 50 cents Euros for each incorrect exercise. The child agrees and he wins 11 Euros on the first assignment of 20 exercises.
Solve:
a)   To obtain the number of right and wrong answers.
b)   To study all possibilities for which the son wins money.
c)   To solve a) in different ways with the graphic calculator."

In this situation the pre-service teacher suggested as a requirement that students should have had previous training of at least an hour with the graphic calculator. The participant, with the view screen support, showed several ways of solving the problems (see Figures 1- 6). In the Figure 1, the teacher in training establishes the equation *ecu1* for considering x right and wrong answers and the total of 20 exercises proposed ($x + y = 20$). He also defines the profit function *win (x, y)* = $x - 0.5y$ taking into account that the son receives 1 Euro per correct exercise and pays 0.5 Euro per wrong exercise. The participant solves the system

$$\left. \begin{array}{l} x + y = 20 \\ x - 0,5y = 11 \end{array} \right\} \text{ with the solve command using the Scaffolding method}$$

(Kutzler, 1998). Note the difference between ordinary notation in mathematics and the graphic calculator notation, which could generate difficulties if the student is not familiarized with the connection between the two notations. In the present situation, the participant could have corrected for possible difficulty with the graphic calculator through the requirement of previous training.

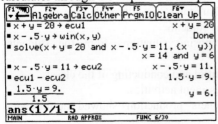 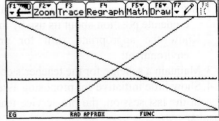

Figure 1. Algebraic solution.   Figure 2. Graphical solution.

Figure 3 shows the window visualization and configuration of the graphic calculator. In Figure 4, the *y1* and *y2* function is defined on the graphic calculator functions editor emphasising the use of different representation systems and their interconnections. This indicator shows students' alternative analyses and the effort to understanding mathematics ideas and problem situations. The table in Figure 5 thus contributes to the discussion about the ways of solving the proposed problem, illustrated from seven right answers for what the profit was.

 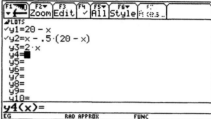

Figure 3. Window settings.   Figure 4. Function editor.

The participant also tackled the problem situation another way, using the graphic calculator as a spreadsheet to introduce the algebraic functions and variables. This can be seen in Figure 6. Data processing is another possibility that can be presented to students to increase their understanding of the problem situation and the related mathematics concepts. Using the graphic calculator as a spreadsheet also opens another way of solving algebraic problems.

Figure 5. Numerical exploration.   Figure 6. GC as a spreadsheet.

As can be observed, this participant not only illustrated one technical power of the graphic calculator he also suggested interrelations between the algebraic concepts involved in the question. He tried to show different ways to present mathematical reasoning, which opens didactical possibilities for introducing mathematical modelling to secondary students successfully. This done, students can make more connections to help them to build the model at the moment of abstraction, also fulfilling the objective for which the mathematical modelling process was applied. The way the different problem situations were tackled shows us that the participants used mathematical modelling by means of the GC in the algebraic concept context.

Second question:
Give at least two questions in which the answers require the use of mathematical modelling and the graphic calculator.

The questions formulated by the participants (see Table 1) suggest the orientation toward the activities. The differences would probably correspond to the complexity of the tasks, based on the hypothetical differences of the student's level of understanding in secondary school. We find that the questions formulated were open and can infer that the participants were oriented to the use of the mathematical modelling, following the information given in the course. We also found some closed questions.

Table 1. Questions formulated by the participants in the 10th session.

Kind of question	Example
Open	What price is more attractive? (ST3)
	To study possibilities for when the child earns (ST4)
	If a satellite that does not pass the point (2,3), how could you modify its trajectory? (ST10).
Closed (in algebra)	Make a graph [the equations system] and interpret the solution (ST7)
	Which satellites go to *(2,3)*? (ST10)
Closed	How many traffic signs can be painted? (ST1)
	How old is my brother? (ST5)

The open questions were formulated from the context of the situations and to go through the models built; they may be used to draw conclusions related to these situations. The closed questions referred to the area of algebra and also to the questions proposed by the context. The questions presented in the 10th session (Table 1) could contribute to the development of the mathematical modelling process that will generate discussions where some written and oral communication skills will be developed as well as the independence of thinking of the students.

Third question:

Order the sequence of the activities to be followed by the teacher

The sequence proposed by the participants shows changes from the sequences presented in the initial session. In general the sequences proposed considered the following:

1. Organizing the students in small groups
2. Posing a problem situation
3. Formulating (and selecting) the problem
4. Identifying variables
5. Establishing relationships between the variables (the GC can be used)
6. Building the mathematical model
7. Representing the model using different systems of representation (with the support of the GC)
8. Solving the mathematical problem
9. Interpreting the solution or solutions
10. Formulating new questions
11. Planning new situations as examples.

Organizing the students in small groups requires the exploration of the situations in a shared way. This generates discussions and propositions about how to tackle the problem. Teachers agreed that the students had to be in groups to discuss the modelling. "The professors were to help a little, but not much." (ST9).

Posing the problem situation involves the students with family, cultural and community problems. The presence of these situations could motivate the students to begin with activities to be developed by means of the application of mathematical modelling and its integration with the GC. Posing the problem has been part of mathematical modelling, so the questions must come from the given situations. The step from "the real world" to "the real model" needs to elicit new questions, prior to its mathematical formulation.

Identifying the variables, establishing the relationships between them and constructing the model forces students to centre attention on the specific properties of the given situation in their relation to the algebraic context. Along the way they develop the process of the abstraction needed to make the mathematical model. In this step from the real model to the mathematical model the GC could be used for example, in moving to the experimentation or representation of data.

Teachers also considered taking advantage of the potential of GC to represent the model using different systems of representation, and they proposed using the capacities of the GC to solve mathematical problems. Its use was widely extended among the participants.

Students' interpretation of the solutions represents the moment when mathematics connects again with the "real world", when the application of mathematics arises in the physical and social world. Despite the importance of this connection the teachers had no considered it in the first session. This awareness could be identified as a result of the course.

Fourth question:
Suggest at least two aspects to assess and how to carry out the assessment

The participants started from the idea that the evaluation was performed to "... determine whether the student learned" (ST4) or "... understood the problem" (ST7). This could mean that teachers considered the evaluation to be a search for information for the teacher. Evaluation would build toward a general mark for their students and help the teachers to make decisions related to strategies of teaching and learning. We did not consider the evaluation explicitly as a set to contribute to strengthening the student's skills and taking advantage of the possibilities that the school context offers. However, we appreciate that the participants recognized that the evaluation complemented other dimensions of the school curriculum that support and stimulate the mathematics learning process.

The future teachers considered evaluating different issues that converged on evaluating the students at every moment of the mathematical modelling process and the results of the handling of the GC as a support itself. They agreed that most of the evaluation was the formulation of the model and the interpretation of the results. From the algebraic context the participant mentioned the identification of the variables, the establishing of correct relationships between the variables, the planning of the linear equation systems and the achieving of correct resolution in different ways.

As to how they would perform the assessment, participants showed that they would handle it in written and oral questions and in discussion with their respective arguments. They say: "... the pupils should interpret the opposite and the solution...." The written questions could include correction of the notebooks, blackboard activities, test and posters. It means that the future teachers are involved in the evaluation process proposed by Galbraith, Haines and Izard (1998).

4. CONCLUSIONS

The teachers proposed some real-world situations adjusted to the levels of the secondary school and the students' immediate surroundings. The situations proposed by the participants were thus connected with algebraic concepts and process content in the secondary school programmes. The graphic calculator showed a technical and didactical domain and the options that it offers, as well as their importance to teachers and students. These domains were illustrated in the design of the activities proposed by the pre-service teachers. The resulting view toward of the teaching of mathematics is oriented to placing boys and girls in an active situation where they could experiment, guess, formulate, solve, explain, predict and contrast with both the other partners and teachers. First, examples and procedures were considered; then teachers performed these procedures; and finally, activities were oriented to development by the students. These examples and procedures did not seek to classify or thwart the creativity of the pupils. The teachers used different systems of representation and the interconnections between them, which reveal the search for alternatives to facilitate the student's comprehension. They explored ways to explain the algebra to students as a mechanism to foster understanding of the situation. The process of application to the GC, in all phases of the design of the didactical activity with algebraic context, underscores the consistent emphasis on the use of open questions.

In conclusion, the arguments given by teachers at the end of the course-workshop show changes, more in the reflection about students and teacher than in the mathematical content and the assessment. We observe some significant changes in the sequence of the activities given in the first and last sessions. In the first session, we found that the sequence was empathised only in the planning of the problem situation: make the model, solve, plan similar examples and prove results with the GC. In contrast, the last session considered students' teamwork in posing situations and selecting problems, constructing a multiple representation of the model (with the support of the GC), interpreting the solutions and the formulating new questions. This change reveals the advances, or the effect on the didactical knowledge of student teachers in the course.

One element to highlight was the incorporation of the team activities among students to encourage learning. Evaluation was seen merely as a search for information for the student's mark. However, an advance was observed in the idea of having students use the GC in their assessments. Teachers proposed different assessment strategies, some written and some oral.

In general, the participants expressed concerns that students' learning through the algebraic activities developed with the addition of mathematical modelling and GC support in different systems of representation. In every situation they encouraged algebraic knowledge without damaging nature. Finally, tackling the situations given by the teachers revealed the structure of linear algebra constituted by the definition and treatment of the variables, handling of relationships through linear equations and establishment of a linear equation system. It was good that participants used different systems of representation, making use of the potential of the GC. Thus, the situations presented illustrated the richness of algebra as a mathematical context for the description, explanation and prescription of the linked phenomenon themselves.

REFERENCES

Borassi, R. (1987) Exploring Mathematics Through the Analysis of Errors. *For the learning of mathematics*, 7, 2-9.

Brunner, A., Coskey, K. and Sheehan, S. (1998) Algebra and Technology. In L.J. Morrow and M.J. Kenney (eds) *The Teaching and Learning of Algorithms in School Mathematics* (1998 Yearbook). Reston, VA: NCTM, 230-238.

Duval, R. (1995) *Semiosis et pensée humaine*. Paris: Peter Lang.

Freudenthal, H. (1983) *Didactical Phenomenology of Mathematical Structures*. Dordrecht: Reidel Publishing.

Galbraith, P., Haines, C. and Izard, J. (1998) How do Students' Attitudes to mathematics Influence the Modelling Activity? In P. Galbraith, W. Blum, G. Booker and I.D. Huntley (eds) *Mathematical Modelling. Teaching and Assessment in a Technology-Rich World*. Chichester: Horwood Publishing, 265-278.

Dorier, J. (Ed) (2000) *On the Teaching of Linear Algebra*. Dordrecht: Netherlands: Kluwer Academic Publishers.

Harel, G. (1998) Two Dual Assertions: The First on Learning and the Second on Teaching (or Vice Versa). *American Mathematical Monthly*, 6, 497-507.

Houston, S. K., Blum, W., Huntley, I. and Neil, N.T. (1997) *Teaching and Learning Mathematical Modelling*. Chichester: Albion Publishing.

Janvier, C. (ed) (1987) *Problems of Representation in the Teaching and Learning of Mathematics*. Hillsdale, New Jersey: Lawrence Erlbaum Associates.

Kutzler, B. (1998) *Solving Systems of Linear Equations with the TI-92*. (Second edition). Hagenberg, Austria: bk teachware Series "Support in Learning".

Kutzler, B. (2000) The algebraic calculator as a pedagogical tool for teaching mathematics. *The International Journal of Computer Algebra in Mathematics Education*. 7 (1), 5-24.

Miles, M.B. and Huberman, A.M. (1994) *An Expanded Sourcebook Qualitative Data Analysis* (2nd edition). Thousand Oaks, California: Sage.

Niss M., Blum, W. and Huntley, I. (eds) (1991) *Teaching and Mathematical Modelling and Applications*. Chichester: Ellis Horwood Limited.

Ortiz, J. (2002) *Modelización y Calculadora Gráfica en la Enseñanza del Álgebra. Estudio Evaluativo de un Programa de Formación* (Doctoral Thesis). Granada, Spain: University of Granada.

Ortiz, J. and Rico, L. (2001) Graphic Calculators and Mathematical Modelling in a Program for Preservice Mathematics Teachers. In W. Yang, S. Chu, Z. Karian and G. Fitz-Gerald (eds), *Proceedings of the Sixth Asian Technology Conference in Mathematics*. Melbourne, Australia.

Rico, L. (1997) Consideraciones sobre el Currículo de Matemáticas para Educación Secundaria. In L. Rico (ed) *La educación matemática en la enseñanza secundaria*. Barcelona, Spain: ICE/Horsori, 15-38.

5.4

MATHEMATICS IN THE PHYSICAL SCIENCES: MULTIPLE PERSPECTIVES

Geoff Wake and Graham Hardy
University of Manchester, UK

Abstract-*There is much concern in the UK about the "mathematics problem": the lack of ability of students to apply mathematics when entering Higher Education courses in science and technology. Some studies suggest students' lack of facility with basic techniques as a root cause. However, as studies of application of mathematics in different settings such as workplaces suggest, the problem is likely to be deeper than this. We have found the theoretical framework of Cultural Historical Activity Theory useful in drawing attention to different factors that mediate, for better or worse, (mathematical) activity in different settings. We use this here to assist us make sense of data we have collected in a study in which we ask of students, as they apply their mathematics to solve problems in the physical sciences, "what is mathematical activity and how can this be used to assist scientific activity and understanding?" We examine this question from the different perspectives of the students, their teachers and ourselves as educators / researchers.*

1. INTRODUCTION

In the UK there is concern that too few students continue with the study of mathematics beyond the age of sixteen, and that many of those who do are ill-prepared to use and apply mathematics in different contexts in Higher Education (cf. Smith, 2004; Savage & Hawkes, 1999). Proponents of *situated cognition* (for example, Lave, 1988), suggest that the lack of ability to apply knowledge, skills and understanding is not surprising and that such 'transfer' is difficult, if not impossible, as knowledge needs to be (re-) constructed in the socio-cultural practice in which it is being used.

Many studies have sought to examine the problem of 'transfer' or 'transformation' of mathematical knowledge by focusing on its use in workplaces. Our own work in this field has led us to suggest that in workplaces, mathematics often becomes automated or "black-boxed" (Williams & Wake, 2006) to such an extent that it appears hidden to the workers themselves and that this can prove problematic to workers as they move from one workplace practice to another. Our analysis of workplace case studies suggests that workers might be better prepared to make sense of workplace situations by being encouraged to develop a range of mathematical modelling and problem solving strategies (Wake & Williams, 2003). It would perhaps seem likely that the application of mathematics within different areas of the school curriculum itself may prove less problematic, but the concerns of those who accept students into our universities to follow technical and scientific courses suggest that this may not be the case.

This paper reports on initial findings of research that sets out to extend, into the classroom, our previous investigations in which we explored the use of mathematics in workplaces. We present a case study in which we explored, as researchers and maths and science educators, the use of mathematical models within the physical science curriculum and extend our analysis by exploring this from the different perspectives of maths and science teachers and the students themselves.

2. MATHEMATICAL MODELS IN THE PHYSICAL SCIENCES

Mathematical models play an important role in the physical science curriculum, for example, being used to describe relationships between data that have been collected to make sense of physical phenomena. For example, when investigating the elasticity of a spring or string one can measure the force applied and the resulting extension. This, as Robert Hooke discovered some 350 years ago, leads one to the understanding that until the elastic limit of a spring or string is reached, its extension is proportional to the force applied. In other words, a fixed increase in applied force will result in the same extension of something like a spring or rubber band whatever the initial applied force. We can describe such a relationship, using agreed mathematical conventions, as $T = kx$, where T is the applied force, x the resulting extension and k a constant parameter which depends on the 'stiffness' of the material. A Cartesian graph, such as that in Figure 1 a, can be used to visualise such a situation. Such models of direct proportion are useful in other instances that might be considered as human constructs rather than as discovered relationships, rules or laws between measured quantities. For example, the concept of density is useful in assisting us make sense of situations in which quantity of matter is important. Density is derived from two fundamental quantities that we can more easily measure: mass and volume. Thus, density is defined as mass per unit volume: again this can be described algebraically by $m = \sigma v$, where m is mass, σ is density and v is volume. This, therefore, is another situation giving rise to a model of direct proportion, although in this case the model arises because of the development of a measure that scientists find useful rather than the discovery of a law or model linking two measurable quantities. Of course there are other common models that are found to be useful in helping one describe how one physical attribute varies with another: for example, some quantities are found to grow or decay in an exponential way (Figure 1b), and other phenomena like gravitational and electrostatic forces vary as the inverse square of the separation of the masses or charges (Figure 1c).

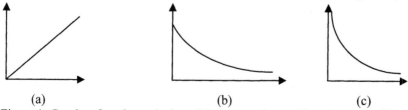

(a)   (b)   (c)

Figure 1. Graphs of mathematical models commonly used in science: (a) direct proportion; (b) exponential decay; (c) inverse square.

3. EXPERIMENTATION IN THE SCIENCE CURRICULUM

In a first phase of the research in a Sixth Form College we observed, in the ethnographic tradition, two groups, each of approximately twenty students, practising for a formal assessment of their physics practical skills, towards the end of the first year of a two year pre-university course. We video-recorded the sessions and additionally made notes of our observations. The phases of our research that followed were sparked by our interest in an experiment that we had observed that required some mathematical understanding and which appeared particularly problematic to a relatively large number of students in both groups.

The students followed written instructions from a previous assessment paper which directed them to release a pencil for a number of different heights, H, of the pencil's mid-point (centre of mass) above the surface of water contained in a long glass cylinder. For each of these values they were to observe and record the pencil tip's maximum depth of penetration, D, into the cylinder of water. Having been directed to record values of D for starting values of H of 50, 70 and 90 millimetres, the students were then asked to, "Use your data to perform calculations to test whether D is proportional to H".

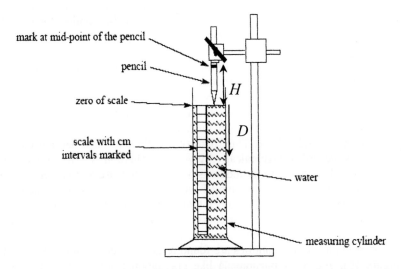

Figure 2. Experimental set up[1].

It was clear from our observations that determining proportionality proved problematic for many, with the majority not knowing how they might start to attempt this. This led us to probe the students' understanding further by asking a small sub-group to undertake an activity (loosely based in its early stages on the practical experiment) which required students to explore a range of different visualisations of models of direct proportion. Three volunteers, all able mathematicians in terms of prior qualifications and current studies, worked together sorting 16 cards into two sets: those that described a situation that was directly proportional and those that did not. These cards were set face down in a pile and the students were asked to turn each over one-by-one (in the numerical order 1–16 as in Figure 3) and decide into which group it should go justifying to each other their choice. The activity and follow-up questions were video and audio recorded.

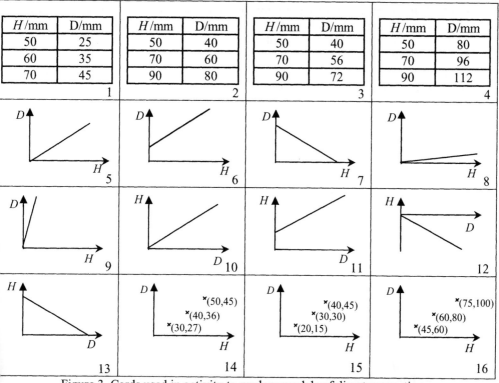

Figure 3. Cards used in activity to explore models of direct proportion.

In the space available here it is only possible to give a brief overview of the outcomes of this activity with some description highlighting the main issues. The first four cards presented the type of data that the students had collected themselves and which had proved so problematic. After turning over the first card there was a brief discussion and the group decided that in the case of numerical data of this type the key to deciding whether the data is proportional or not is to divide either H by D or D by H to see if the results for each set of data pairs are approximately equal. Having established this procedure, which the class teacher had emphasised during the practical lesson, the students quickly and correctly sorted the first four cards. Card 5 halted the students in their tracks for a while, and with little justification it was eventually decided that it should join the set of cards demonstrating direct proportionality. Again, without much justification, card 6 was decreed to signify direct proportion. Card 7 caused much discussion, still being linear, but having negative gradient. One student was convinced that this depicted "inverse proportion", presumably interpreting the change in gradient from the two previous cards as being related to the inverse function in some way. For this reason it was decided that this card did not demonstrate proportionality. Cards 8 and 9 caused some discussion, but as they were seen as being of the same type as card 5 they eventually joined that in the correct pile. The axes being reversed in cards 10 – 13 caused few problems as it was decided to ignore this detail and sort the cards in the same way as their previous matches (that is, 10 with 5, 11 with 6, and 13 with 7). Card 12, however, did not have a prior match and this did prove difficult for the group who eventually decided to place it in the "not proportional" pile. Cards 14 – 16 were correctly sorted by applying the division test. It had been suspected that for these cards students might have looked at the differences between successive values of each variable, for example noticing in card 15 that equal steps in H gave rise to different but equal steps in D which is often emphasised in mathematics classroom discussion of proportionality (although this gives linearity but not necessarily direct proportionality). This was pursued further in the ensuing discussion when the students were asked to plot a quick graph of card 15 and compare this with that depicted on card 6 which was in the "proportional" pile. It was suggested that the other graphical representations might be explored in a similar way. Although there was a great deal of uncertainty about the nature of some of the representations on cards 5 – 13 this had not been a strategy that had come to mind to the students.

4. INTERPRETING THE MATHEMATICAL ACTIVITY OF THE STUDENTS

How should we interpret the actions of the students when undertaking the activity we describe above? At this point we turn to the different stakeholders: the discussion we present here is a summary and synthesis of follow-up interviews and conversations we had with the teachers after watching together the video of the activity and a further interview with the small group of students.

Some views of the "scientists"
It is fair to report that the students' science teacher was surprised, if not "shocked" at the seeming lack of depth of understanding of proportionality displayed by the group of students tackling the card-sorting activity. It was clear to the teacher that this is potentially going to prove problematic as mathematics has an important role to play in "linking together" physical science concepts and allowing students to solve particular problems. In some ways this appears to accentuate a procedural approach in which the mathematics is subservient to the science. However, the teacher also felt that mathematics, whilst being servant to the master of science, often has a clear iconic status which militates against clarity of understanding of scientific principles. For example, students when asked, "What is density?" will turn to a mathematical formula, "mass divided by volume" rather than suggesting scientific conceptual understanding such as "mass per unit volume". Although the difference between these two possible answers may at first glance seem small the difference suggests a markedly different attention of the student's focus, with the former suggesting attention to the procedural algebraic formula and the latter attention to the underlying physical construct. It seems that mathematical signs (in the sense of Pierce), such as algebraic expressions and graphs, have strong status for students and act to shift their attention away from the science. Rather than

assisting students in their making sense of the physical phenomenon under scrutiny the mathematics (often expressed algebraically or graphically) itself becomes the focus of the students' attention.

This was perhaps exemplified by discussion with the student group about the different models of direct proportion suggested by cards 5, 8 and 9 which could be taken to represent experiments undertaken with liquids of different viscosity. The students had great difficulty in suggesting such physical interpretations with initial guesses suggesting that the experiments might have been carried out in glass cylinders with different diameters or depths.

Some views of the "mathematicians"

The ability to work with, use and understand mathematical models is important in science. This might involve students in a process that lies somewhere between full-blown "mathematical modelling" and "application" of mathematics (see Blum et al, 2006). Here we take mathematical modelling to incorporate the full, and often cyclical, process involving mathematisation of a "real-world" situation, the analysis of this using mathematical techniques, and the interpretation and validation of results. On the other hand we use "application" in the sense of the less comprehensive use of a common mathematical model (in this particular case, that of direct proportion) to make sense of "real-world" (here, physical science) phenomena.

There are common mathematical models that have many uses in different areas of the physical sciences: as suggested earlier these include models of direct proportion, exponential growth and decay and so on. The model of direct proportion has particular importance in science in that it can be used to find and validate more complex functional relationships between measurable quantities. For example, if one suspects that the time period, T, of a pendulum is proportional to the square-root of its length, l, then one would test whether or not T is directly proportional to $\sqrt{l}$.

The topic of direct proportion is also important within the mathematics curriculum itself, allowing one, as it does, to make connections, between different branches of mathematics. For example, proportionality can be explored numerically, graphically, algebraically and geometrically. This particular concept thus permeates the mathematics curriculum; it is evident in many topic areas and pupils/students can be expected to meet it at many stages of their mathematical development. It is, therefore, perhaps surprising that these mathematically able students seemed to lack any depth in their understanding of such an important concept. It appears that although many situations in physical science can be described as being directly proportional, in each case students need to reconstruct their understanding: so, for example, the student does not see similarities between Hooke's, Ohm's and Newton's second Laws (see below). Each time that one of these laws appears the student needs to reconceptualise their important features of direct proportionality. This is much as Noss and Hoyles have recognised in attempting to make sense of how workers use mathematics: they propose the construct of *situated abstraction* (Pozzi et al, 1998), which allows one to understand how workers may develop a generalised mathematical understanding, but within the situational context of their work, using a discourse other than that of standard/formal mathematics but which may be mapped to this. It would seem that our students in their science lessons are doing much the same thing: in each new physical science situation in which an understanding of the concept of direct proportion, in an abstract and general sense, would empower them to quickly make sense of the phenomenon, they have to reconstruct their understanding of what it means in terms of this specific situation.

Some views of the students

The students in a follow up discussion about the role that mathematics played in their study of Physics reflected to some extent the view of their teacher that mathematics could assist them make sense of the science, although this emphasised rules and procedures rather than seeing mathematics as giving them access to a set of knowledge, skills and understanding that can be empowering. They therefore recognised that, in terms of making sense of experimental data, they would be expected to tabulate this, plot as a graph and then draw conclusions from that graph, possibly having developed a mathematical function to describe any relationship that appeared to exist. Although, having difficulty in defining, and perhaps articulating boundaries between the disciplines, it was clear that this process of "analysis" was conceptualised as being (or doing) mathematics whereas interpretation in the real-world of physical science is not in the mathematical domain but clearly belongs to science.

During a follow-up interview further discussion about the generality of models of direct proportion ensued and the students were asked to suggest what similarities they saw in a list of physical laws: Hooke's Law, the wave equation, Ohm's Law and the definition of density. Initial responses were to suggest that they could all give rise to a formula into which one can put data values, through to the idea that these formulae all have constants and variables. It may be that the students do not have the necessary language to articulate their understanding of laws of direct proportion, but further probing suggests that they do not see similarity in their underlying structure. They are more comfortable in organising physical laws into groups that reflect the underlying physical science content area, such as equations for circular motion, or motion under constant acceleration, rather than their underlying mathematical structure such as models of direct proportion or linearity. Perhaps this is not surprising when they are operating in the Physics classroom where the organising framework is based on scientific content area.

5. DRAWING CONCLUSIONS: A CULTURAL HISTORICAL ACTIVITY THEORY (CHAT) PERSPECTIVE

It seems that the potential of mathematics to provide a powerful tool of analysis that can empower students to make sense of the physical sciences is not being developed in our young scientists and mathematicians. Why is this?

Students have opportunities in both mathematics and science classrooms to develop their understanding of how mathematical models, such as those of direct proportion, can be used to not only describe physical relationships between measurable quantities but also assist them make sense of such relationships. However, they do not seem to have a good understanding of the main and generalisable features of such models. They do not, for example, appear to understand the common and analogous features of Hooke's and Ohm's Laws, to highlight just two of the genre. It would appear from our interviews with teachers and students, and from our own experiences of working in curriculum research and design, that somehow the opportunities that an understanding of such a very useful and common mathematical model affords, are not developed in typical classrooms.

In our workplace research we found the theoretical framework of Cultural Historical Activity Theory (cf. Engestrom & Cole, 1997) useful in assisting us make sense of perhaps why mathematics is the way it is in a variety of practices. Before considering the schema representing this given in Figure 4a it is worth considering the top triangle (Figure 4b) which is due to the fundamental thinking of Vygotsky, who suggested that the action of a subject is mediated by 'instruments' which may include artefacts and tools, or in the case of communicative action, as is often the case in classrooms, by cultural tools, concepts and language genres. Vygotsky draws the distinction between physical tools that operate on or control natural objects and semiotic tools that operate on ideas and concepts. So in the case of applying mathematics, the concept of direct proportion can only make sense in the context of a specific activity. As an abstract concept, a mathematical sign, such as the model of direct proportion (perhaps visualised algebraically or graphically), has a general, abstract meaning associated with a network of other mathematical signs (in this case, for example, linear models), and many potential uses, but as yet no practical meaning. It is only when a mathematical sign is applied in a specific context that it develops meaning to the user. We have seen in the particular case studied here that not only does the algebraic or graphical representation of direct proportionality seem to need to acquire meaning in each particular case that it is introduced, but its introduction as a mathematical sign in the Physics classroom also seems to shift the attention of the student towards the mathematics and away from the physics.

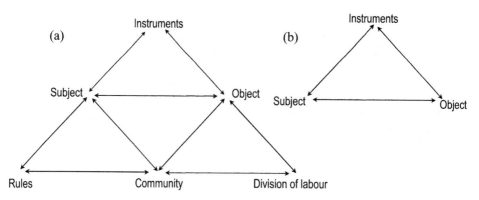

Figure 4. (a) Schema of Activity System, (b) Schema of Vygotsky's model of mediated action (of the individual).

Leont'ev suggests the other nodes (in Figure 4a) to describe the subject's action in relation to a community: this includes the ways in which the *division of labour* and associated *norms/expectations/rules* mediate the subject's activity in relation to this community. Leont'ev suggests that we should further distinguish between the *action* and its *operation*, on the one hand, and the *action* and the communal *activity*, on the other hand so that the hierarchy "activity – action – operation" is associated with the parallel hierarchies of "community – subject – instruments" and "motivation – goal – methods".

In the case of the activity of students attempting to make sense of physical phenomena in a practical scientific context, therefore, we should consider their action of testing for proportionality between two sets of values and how this becomes operationalised as dividing one set of values by their corresponding values in the other set and checking to see if one obtains a constant or near constant values as a result. We suggest that the motivation of the community of students and their teacher is their preparation for assessment (and in current classroom practice this is unfortunately often the case), and the goal of the individual is to successfully complete the individual assessment question.

This analysis leads us to suggest that the students do not have the motivation to come to understand and learn to apply the powerful mathematical models to make sense of situations in the physical sciences. The assessment in science instead leads to pressure for them to operationalise or 'black box' mathematical procedures. The motivation of meeting with success in their science assessment, and goal of being able to determine whether measured quantities are proportional or not led them to find a quick-fix test for this. It would seem that effecting a change in assessment procedures might be the only way in which to ensure that our students are able to engage with their mathematics in such a way that it allows them to develop their understanding in a powerful way. Somehow we need to ensure that the assessment is such that students are encouraged to recognise the important mathematical structure of important scientific models. Equally, there appears to be a need to break down barriers between academic disciplines in our schools. As one student succinctly, if not disappointingly suggested, "Maths takes place in maths classrooms and Physics in Physics classrooms". This "black-boxing" allows little, or perhaps no, room for students to explore the important underlying structures of important mathematical models and how these may be applied in a range of different situations.

Giving greater prominence in the curriculum to mathematical modelling as an activity certainly seems to have the potential to meet these needs. As this preliminary research suggests, there is perhaps much we can learn to inform its development and implementation in classrooms, by investigating students' mathematical activity in other subjects.

NOTE

1.   From assessment paper AQA, Physics Advanced Subsidiary Examination, Specification B, Unit 3 Practical, question 2 (January 2004)

REFERENCES

Blum W., Galbraith P.L., Niss M. (2006) *Modelling and Applications in Mathematics Education*, New York, USA, Springer Science+Business Media, inc (*in the press*)

Engestrom, Y. & Cole, M. (1997) Situated cognition in search of an agenda, in D. Kirschner & J.A. Whitson (Eds.) *Situated cognition: social, semiotic and psychological perspectives*, (pp. 301-309). NJ: Lawrence Erlbaum.

Lave, J. (1988). *Cognition in practice: Mind, mathematics and culture in everyday life.* Cambridge: CUP.

Pozzi, S., Noss, R. and Hoyles, C. (1998) Tools in Practice, Mathematics in Use. *Educational Studies in Mathematics,* 36, 105-122.

Smith, A (2004). *Making Mathematics Count.* London: HM Stationery Office

Savage, M. D. and Hawkes, T. (eds.) (2000) *Measuring the Mathematics Problem (Engineering Council).* London: London Mathematical Society and the Royal Society.

Wake, G.D. and Williams, J.S. (2003) Using Workplace Practice to Inform Curriculum Design, in S.J. Lamon, W.A. Parker and S.K. Houston (eds) *Mathematical modelling a way of life,* Chichester, U.K.: Horwood Publishing

Williams, J.S. and Wake, G.D. (2006), Black boxes in Workplace Mathematics, *Educational Studies in Mathematics (in the press).*

5.5

MODELLING PROBLEMS FROM A COGNITIVE PERSPECTIVE

Rita Borromeo Ferri
University of Hamburg, Germany

Abstract–*Looking at modelling from a cognitive perspective has largely been neglected in the current discussion regarding modelling. Using the mathematical didactical and cognitive-psychological approach of mathematical thinking styles, this study analyses the modelling performed by teachers and students in context-bounded mathematics lessons. This study is complex, and so are the results. The focus of this paper is on the depiction of reconstructed and so-called individual modelling routes of sixteen-year-old learners working in groups on modelling problems during mathematics lessons. These routes provide an insight into the learners' cognitive procedures during modelling.*

1. THEORETICAL FRAMEWORK: FOCUS AND RESEARCH QUESTIONS

This research project, in which context-bounded mathematics lessons are being analysed from a cognitive perspective, takes up issues from my doctoral thesis on mathematical thinking styles (Borromeo Ferri, 2004a), in which learners' different individual mathematical thinking styles were reconstructed. The study also refers to an already completed follow-on case-study (Borromeo Ferri, 2004b, 2004c), which discusses the influence of mathematical thinking styles on transition processes from the real world to mathematics.

1.1 Mathematical Thinking Styles – Theoretical Framework 1

At first, I will quickly outline the theoretical framework developed in my doctoral thesis in order to illustrate the mathematical didactical and cognitive-psychological approach of mathematical thinking styles which formed the basis for the research project as well as the data analysis.

In my thesis, I developed the following definition of mathematical thinking style: Mathematical thinking style is the term I use to denote the way in which an individual prefers to present, to understand and to think through mathematical facts and connections using certain internal imaginations and/or externalised representations. Hence, mathematical style is based on two components: 1) internal imaginations and externalised representations, 2) the holistic and the dissecting way of proceeding. (cf. Borromeo Ferri 2004a, p50).

In my thesis, I used a laboratory design to reconstruct and analyse different mathematical thinking styles of 12 students attending 9th or 10th grade, that is, I was able to describe the "existence" and distinctness of three mathematical thinking styles:

Visual thinking style (pictorial-holistic thinking style)
Analytical thinking style (symbolic-dissecting thinking style)
Integrated thinking style

1.2 Mathematical Thinking Styles Linked to Modelling Activity – Theoretical Framework 2

The preliminary study for the study presented in this paper consequently dealt with the question of to what extent a connection can be established between the mathematical thinking styles of individuals and the transfer processes they perform in order to translate from real model into mathematical model. Using the findings of my thesis as a basis, I developed a questionnaire on mathematical thinking styles which makes it possible to reconstruct the mathematical thinking styles of individuals in a class. Besides questions on the students' view of mathematics, it also includes a reality-based task for which the students had to write down an account of their problem-solving, or rather, thought process. This questionnaire was distributed in four classes. After the evaluation of the questionnaire, individual students were selected according to their preferred mathematical thinking style and were asked to solve two more reality-based problems. The way the students worked on the tasks as well as the interview conducted with them straight afterwards demonstrated their preference for a mathematical thinking style which had been analysed before with the help of the questionnaire. Regarding the question of whether different mathematical thinking styles also lead to different transfer processes, the following interesting connection could be established: Pupils with a preference for a visual thinking style worked longer the real model or switched from the real to the mathematical model more often than pupils with a preference for an analytical thinking style, who were able to break away quickly from the real model and who very rarely referred back to the factual context.

As these insights were generated using Word Problems, that also gave rise to the question of how, amongst others, complex modelling tasks and the entire modelling cycle of learners can be examined.

1.3 Cognitive Processes as an Aspect within the Modelling Discussion – a Short Overview

Using cognitive psychology for analysis, with mathematical thinking styles as theoretical "glasses", is a new approach in the field of modelling research.

However, in the current discussion of modelling with regard to cognitive processes one has to mention the extensive work on this subject by Richard Lesh and his team (cf. amongst others Lesh & Doerr, 2003). In his theoretical approach and explanation Lesh primarily refers to works by Piaget, Vygotsky, Dienes and other psychologists and pedagogues. Lesh's work has another emphasis than the emphasis of the study discussed in this paper, which are the following: Firstly, on a micro-process-level, the focus lies on the individual as part of a group or class. Secondly, the question of how and why actual individual modelling occurs, and what kind of constructions of meaning takes place while working on reality-based tasks are also aspects which are goals of this study.

Regarding didactic literature on modelling, the paper by Treilibs (1979) is worth mentioning (cf. Treilibs, Burkhardt & Low 1980) as his analyses also focused on the individual during modelling. However, he mainly focussed on determining how learners build a model. Consequently he did not examine the complete modelling process, but instead concentrated on the so-called "formulation phase" during which the model is formed. Treilibs did not continue his research in this area, therefore this strong focus on the individual was not taken up again as a relevant aspect in his following work.

However, Matos' and Carreira's (1995, 1997) research puts a special emphasis on 10th grade learners' cognitive processes and representations while solving realistic problems. On a micro-level, they analysed the creation of conceptual models (interpretations) of a given situation and the transfer of this real situation into mathematics. In their results, they point out the numerous and diverse interpretations which learners use while modelling:

"... most of the students' representations processes were conformed with two general reference systems. One refers to what students perceived from real situations, according to their own experience and knowledge of specific aspects involved in the problem. (...) The other reference system included ideas, concepts and mathematical procedures, most of them tuned with the computational instrument they used." (Matos & Carreira, 1995, p78)

Accordingly, they did not put the main emphasis on the analysis of the complete modelling process. Yet, coming from a cognitive standpoint equal to the one Lesh detailed in his works, they arrived at the following conclusion:

"... we came closer to the idea that the whole modelling activity has a cognitive architecture that could consist of a multiplication of micro-modelling cycles." (Matos & Carreira 1995, p78)

The above mentioned works notwithstanding, aspects of cognitive psychology were widely marginalised in the discussion of modelling. This also becomes evident in the analysis of the Discussion Document to be found in the ICMI Study 14: Applications and Modelling in Mathematics Education (Blum et al, 2002). While the question of Beliefs (cf. Grigutsch, 1996; see Maaß, 2004) has increasingly gained importance over the last years, a more intensive discussion of cognitive influences on the individual while modelling in maths lessons has yet to take place, and in this context, the role of the teacher will also have to be taken into consideration.

This study provides a coherent analysis of four different aspects from a cognitive perspective:

1) Analysing learners and teachers in contextual mathematics lessons

2) Analysing micro-processes at an individual level

3) Analysing groups of pupils during the process

4) Considering the role of the teacher at the same time

This comprehensive analysis would therefore also yield new insights for the current discussion of modelling. Especially the linking of mathematical thinking styles to modelling, or rather the investigation of the possible influence of mathematical thinking styles on the entire modelling cycle, are new aspects which are introduced into the discussion by this study.

1.4 Research Questions

The following questions were central to my study:

1. What influence do learners' and teachers' mathematical thinking styles have on modelling processes in contextual mathematics lessons?
2. Can the differences between situation model, real model and mathematical model (as described in didactic literature on modelling) be reconstructed from the learners' ways of proceeding and what role do they play with regard to understanding the relationship between mathematics and "the rest of the world"?
3. How do pupil-pupil and teacher-pupil interactions develop during lessons, depending on whether their mathematical thinking styles match or not?

2. METHODOLOGY AND DESIGN OF THE STUDY

The project was carried out within the context of qualitative research. Quantitative seemed inappropriate, given the study's focus on learners' internal cognitive processes.

The investigation was conducted in three 10th grade classes from different *Gymnasien* (German Grammar Schools) in Hamburg. The sample was comprised of 65 pupils and 3 teachers. In my dissertation as well as the preliminary study on mathematical thinking styles and Word Problems mentioned in §1, I also studied learners belonging to the 10th grade-age group. In addition, this age group allows the inclusion of more complex contextual tasks which explicitly include modelling processes.

The study's design is highly complex, as the research questions required different levels of data collection and data analysis. As far as the evaluation is concerned, the design has turned out to be a useful tool due to its multi-layeredness.

Lesson 1: <u>Carrying out the questionnaire on mathematical thinking styles</u>
At the beginning of the data collection, each individual of a class had to do the questionnaire on mathematical thinking styles. The questionnaire was evaluated independently by me and my research student by reconstructing the individual learner's mathematical thinking style.

Lesson 2: <u>In a contextual mathematics lesson, students work on one, possibly two not too complex modelling problems</u>
The division of the learners into groups (5 per group) was based on the evaluation of the questionnaire and according to their mathematical thinking styles. One group was videotaped during the modelling process. The only guideline I gave the teachers regarding their lesson and how the tasks should be dealt with was to do group work.

Lessons 3 and 4: <u>Two further, but more complex modelling tasks are worked on during mathematics lessons</u>

The second to fourth lessons were videotaped. The camera was directed at a group desk and recorded a view of the class, teacher, and blackboard during plenary discussions. Additionally, the teacher was equipped with a minidisc-recorder strapped to their body in order to record all their interactions with learners. Thus, I tried to record the teacher's help or suggestions during modelling as this could possibly influence the pupil's modelling process.

In addition, an interview was conducted with the teacher which also included biographical questions. Questions were also asked about their study of mathematics at university but also about their current view of mathematics or about reasons why their view of mathematics might have changed over the course of their teaching life. The modelling tasks selected for the learners are of central importance, as they delineate the field for the analysis. The tasks were analysed with regard to subject matter aspects and from a cognitive viewpoint, that is, on the one hand according to whether a holistic or a dissecting approach was favoured. The tasks were taken from the DISUM-project (Blum, Messner & Pekrun). See as an example the "Lighthouse-task", which I used in my study:

> In the bay of the city of Bremen, a lighthouse measuring 30.7m and called "Red Sand" was built directly on the coast in 1884. With its beacon, it was meant to warn ships that they were approaching the coast. How far was a ship still away from the coast when the lighthouse could be seen for the first time? (Round up to full kilometres)[1]

The evaluation of the data that has been looked through and analysed so far includes the reconstruction of the individuals' modelling processes in the videotaped groups as well as plenary talks and interviews of teachers. In accordance with Grounded Theory (Strauss & Corbin, 1996), codes were formed and used in order to break up and reassemble data. The data of two of the three classes has already been transcribed and to a large extent analysed.

3. RESULTS AND HYPOTHESIS OF THE STUDY

On the basis of these results, the following hypotheses could be generated:

1   The teachers' mathematical thinking style can be reconstructed and manifests itself during individual pupil-teacher conversations as well as during discussions of solutions and while imparting knowledge of mathematical facts.

1a   Teachers, who differ in their mathematical thinking styles, have the preferences of focusing on different parts of the modelling-cycle while discussing the solutions of the problems.

2   Different mathematical thinking styles of the learners result in different observable **modelling routes**.

2a   The learners' different mathematical thinking styles manifest themselves during the modelling process in such a way that the **starting-point** of the modelling route seems to occur during different phases.

I do not give any concrete examples for hypothesis 1 and 1a here, because I will discuss hypotheses 2 and 2a in great detail as these constitute the focal subject of my paper. However, I would quickly like to illustrate hypothesis 1 which is concerned with teachers:

Due to the fact that I am using mathematical thinking styles as theoretical "glasses" for my analyses, two aspects are relevant for the reconstruction of mathematical thinking styles: statements from the teachers' interviews and the teachers' actions and interactions during the actual lessons. Thus, statements made in the interviews and actual utterances can be compared.

Based on the interviews, the teacher Mrs R was reconstructed to be a visual thinker while teacher Mr P was reconstructed as an analytical thinker. Mrs R made clear that visual imagination and visual representations are integral aspects for her understanding of mathematics. In the interview, she said that while formal aspects are important, she does not put a major emphasis on this area while teaching. Understanding mathematical facts and questioning and analysing while doing mathematics are her priorities.

Making his pupils understand mathematics is just as important for Mr P, yet he puts a greater emphasis on formalisation. He also emphasises his wish to make students apply mathematics to real contexts, especially with regard to his second subject, Physics.

The analysis of their lessons shows that their presumed mathematical thinking style expressed themselves during the discussion of reality-based tasks in the plenary as well as during one-on-one talks with learners. What is more, in relation to modelling, the following interesting connection (of which only the crucial point is mentioned here) could be made:

- Mrs R as a visual thinker interpreted and, above all, validated the modelling processes with the learners. This was evident in her very vivid, reality-based descriptions she used for the learners.

- Mr P as the more analytical thinker focussed less on interpretation and validation. For him, the subsequent formalisation of tasks in the form of abstract equations is important. Accordingly, the real situation becomes less important.

Definition of the term "modelling route"

First of all, I give a working definition of the term "modelling routes", based on my analyses. This is necessary in order to illustrate the meaning of the term and also because there is no established term in current didactic literature on modelling.

"Modelling route" is the term I use to denote the individual modelling process on an internal and external level. The individual starts this process during a certain phase, according to their preferences, and then goes through different phases several times or only once, focussing on a certain phase or ignoring others.

To be precise from a cognitive viewpoint, one has to speak of visible modelling routes, as one can only refer to verbal utterances or external representations for the reconstruction of the starting-point and the modelling route.

Hypotheses 2 and 2a belong together in a certain sense, as both refer to the modelling process of the individual. However, the phase during which the visible starting-point of the modelling activity occurs and the modelling route which can be reconstructed for the pupil should remain analytically separate.

3.1 Reconstruction of Individual Modelling Routes of Students with Different Mathematical Thinking Styles

In the following text, I describe the modelling routes of two learners with regard to the "lighthouse task", which I presented earlier. Based on the analysis of the questionnaires, it could be reconstructed that the two pupils in question preferred different mathematical thinking styles. Max is an analytical and Sebastian a visual thinker. The reconstructed individual modelling routes of the two students are illustrated with the help of the modelling cycle according to Reusser (1996) and Blum and Leiss (2006). Reusser assumes that a so-called situation model exists in which an individual illustrates the situation depicted in the task through what can be called a mental picture. Blum and Leiss (2006) have adapted the situation model for their work on the DISUM-project. For the purposes of my study, I use this cycle to illustrate modelling routes and also because my research questions are aimed at the actual empirical differentiation of these phases.

Figure 1 depicts the reconstructed modelling routes of Max (analytical thinker) and Sebastian (visual thinker).

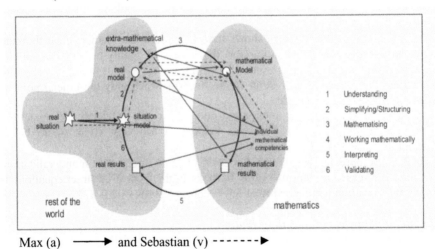

Max (a) ——▶ and Sebastian (v) - - - - - -▶

Figure 1. Modelling routes for Max and for Sebastian

The following quotes made by students during the videotaped modelling process can only be an exemplary illustration of a change of modelling phase. The modelling processes are too long and complex to give an account of all the utterances in detail.

Max's modelling route:
Max read the "lighthouse task" and expressed the following thoughts shortly afterwards:
   • M: *Okay, what shall we do, I'd say we do Pythagoras!*

Max changed immediately from the situation described in the task into the mathematical model, as he could see in his mind's eye that he could apply Pythagoras.

(real situation => mathematical model (and individual mathematical competencies)

He did not make any progress with the mathematical model, because he did not seem to have clarified the given situation sufficiently. He then changed to the real model, in order to better imagine the situation described. Doing this, he started thinking aloud and intensively about the earth's curvature, which shows that he was literally "picturing" the situation.

• M: *Actually, it's the earth's curvature that makes the lighthouse disappear; if it was a smooth plane, it would be visible all the time! (mathematical model => real model)*

After Max had got a more precise mental picture, he changed quickly to the mathematical model. He still remembered Pythagoras theorem and made a drawing.

• M: *We have to mirror this on this cathetus, can you see the length, it's the one up here. (real model => mathematical model)*

Max dwelled on the mathematical model for quite some time. He increasingly started wondering about what the earth's curvature is and asked himself and the others for this extra-mathematical knowledge.

Unlike the other group members, he held the opinion that the earth's curvature would also have to be taken into account for the calculations.

• M: *Yeah, see, we've got to include the earth's curvature in our calculations. (mathematical model => extra-mathematical knowledge)*

Leaving the question of the earth's curvature aside, Max returned to the mathematical model and remained in that phase for a long time. During that phase, he used his intra-mathematical skills (Pythagoras' theorem) as well as extra-mathematical knowledge (the earth's diameter) to reach a conclusion.

• M: *It's twenty kilometres. I've got the lighthouse to the power of two minus the radius. (extra-mathematical knowledge => (individual mathematical competencies) mathematical results)*

Max interpreted the result only to some extent and did not validate it with regard to the real situation; he assumed it to be "mathematically" correct.

• M: *I've got twenty kilometres, as the crow flies. (mathematical results => real results)*

Sebastian's modelling route:

Sebastian started immediately with a sketch and at first described the real situation given very vividly. That way, he got the situation described in the task clear in his mind and created a situation model.

• S: *Here's the ship, somewhat like this and this is the earth's curvature. (real situation=>situation model)*

Starting with his mental picture, he kept simplifying the situation further and created a real model.

• S: *We're gonna do a triangle here. (situation model => real model)*

In his further statements, an increasing mathematization became apparent, and he changed to the mathematical model.

- S: *We need an angle on this side in order to calculate the distance. (...)' Cos I need this (points at Mark's drawing), then I could hundred and eighty minus ninety minus...(real model => mathematical model)*

Sebastian didn't stick to the mathematical model for long, as he had to keep "picturing" the situation. When the group started discussing the question of whether the earth's curvature should be included in the calculations, he remained rather neutral.

- S: *The only thing which otherwise prevents us from getting a clear view is mostly our eyes, if the plane was level, and probably particles in the air.*
  (mathematical model => real model)

From the real model Sebastian returned to the mathematical model and continued to work more mathematically. As it did not occur to him to work with Pythagoras, but with Sinus instead, he only focussed on applying this individual mathematical competence.

- S: *And if we knew one angle now, then we could, we could use Sinus.*

Sebastian often switched between the real and the mathematical model because he had to transport himself into the real situation and needed to picture the situation visually in order to keep working on the task. In contrast to Max, who solved the problem, Sebastian did not reach a conclusion and was therefore stuck in the mathematical model.

The analysis of the modelling processes of the two students shows that individual modelling routes on a micro-level exist. Based on the reconstruction of numerous modelling routes, for which other tasks besides the lighthouse one were included, the following results can be recorded:

- At first it is worth mentioning once more that the data is being analysed with the theoretical approach of mathematical thinking styles. Basically, the analysis shows that an individual's mathematical thinking style influences the way in which the modelling process is carried out.
- The modelling routes of learners with different mathematical thinking styles differ from each other: Analytical thinkers usually change to the mathematically model immediately and return to the real model, or sometimes the situation model, almost of necessity in order to understand the task better. They work mainly in a formalistic manner and are better at "perceiving" the mathematical aspects of a given real situation. Visual thinkers, on the other hand, often imagine the situation in pictures and often use pictographic drawings. Switching between the mathematical and real model is frequent, as they use this technique to better grasp and visualize the situation described.
- Like I already explained in my definition of "modelling route" (which also resulted from the findings of my analysis), the different phases of a modelling cycle (cf. Reusser, 1997) can be distinguished empirically. The differences occur in different ways in each individual modelling route.

4. DISCUSSION

As already mentioned in the abstract, the study has a very complex design because the research questions address many different levels of contextual mathematics lessons. Although the analysis of the teachers' performance has not been described in great detail in this paper, it is still sufficient to deduce implications for the teaching and learning of modelling. Individual modelling routes of learners differ because of the influence of mathematical thinking styles. It can likewise be shown in the analyses that advice from the teacher and the discussion of reality-based tasks in the plenary serve to emphasise or even avoid certain phases of the modelling process. The fact that most teachers are for the most part unaware of their preference for a certain mathematical thinking style is worthy of discussion. Furthermore, one has to bear in mind that pupils are supposed to see the "point" of mathematics, as is often demanded, with the help of reality-based tasks or lessons. The latter shall make the pupils aware of the connection between mathematics and reality. Can this work if the teacher formalises a great deal and does not validate much? Or what, on the other hand, happens to mathematics, if the focus is put to strongly on reality? If teachers and students prefer different mathematical thinking styles, does this interfere with the construction of meaning during the modelling process, as their modelling routes would be too different from each other?

The hypotheses generated here give rise to new questions which will have to be addressed in the near future, especially with regard to the learning and teaching of modelling.

NOTE

1. The solution is 20 kilometres; you can solve this with Pythagoras' theorem or using the cosine rule.

REFERENCES

Blum, W. and Leiß, D. (2006) "Filling up" – The Problem of Independence Preserving Teacher Interventions in Lessons with Demanding Modelling Tasks. In M. Bosch (ed) *Proceedings of the European Society for Research in Mathematics Education CERME 4 (2005)*. IQS FUNDIEMI Business Institute www.fundemi.url.edu, 1623-1633. ISBN 84 611 3282 3.

Borromeo Ferri, R. (2004a) *Mathematische Denkstile. Ergebnisse einer empirischen Studie.* Hildesheim: Franzbecker.

Borromeo Ferri, R (2004b) Mathematical Thinking Styles and Word Problems. In: Henn, Hans-Wolfgang and Blum, Werner (eds.), *Pre-Conference Proceedings of the ICMI Study 14, Applications and Modelling in Mathematics Education.* Dortmund: University of Dortmund, 47-52.

Borromeo Ferri, R. (2004c) Vom Realmodell zum mathematischen Modell – Übersetzungsprozesse aus der Perspektive mathematischer Denkstile. – In: *Beiträge zum Mathematikunterricht.* Hildesheim: Franzbecker, 109-112.

Grigutsch, S. (1996) Mathematische Weltbilder von Schülern, Struktur, Entwicklung, Einflussfaktoren. – Dissertation, Gerhard-Mercator-Universität, Gesamthochschule Duisburg.

Lesh, R. and Doerr, H. (eds) (2003) *Beyond Constructivism – Models and Modelling Perspectives on Mathematics Problem Solving, Learning and Teaching.* Mahwah: Lawrence Erlbaum.

Matos, J. and Carreira, S. (1995) Cognitive Processes and Representations Involved in Applied Problem Solving. In C. Sloyer, W. Blum and I.D. Huntley (eds) *Advances and Perspectives in the Teaching of Mathematical Modelling and Applications*. Yorklyn, Delaware: Water Street Mathematics, 71-80.

Matos, J. and Carreira, S. (1997) The Quest for meaning in students' Mathematical Modelling. In S.K. Houston, W. Blum, I.D. Huntley and N. Neill (eds) *Teaching and Learning Mathematical Modelling* Chichester: Albion Publishing Ltd, 63-75.

Maaß K. (2004) *Mathematisches Modellieren im Unterricht. Ergebnisse einer empirischen Studie*. Hildesheim: Franzbecker.

Reusser, K. (1997) Erwerb mathematischer Kompetenzen. In Weinert and Helmke (eds) *Entwicklung im Grundschulalter*. Weinheim: Beltz, 141-155.

Strauss, A.L. and Corbin, J. (1996) *Grounded Theory, Grundlagen Qualitativer Sozialforschung*. Weinheim: Beltz.

Treilibs, V. (1979) Formulation processes in mathematical modelling. MPhill Thesis submitted to the University of Nottingham..

Treilibs, V, Burkhardt H. and Low, B. (1980) *Formulation processes in mathematical modelling*. Nottingham: Shell Centre for Mathematical Education, University of Nottingham.

5.6

EXPLORATIVE STUDY ON REALISTIC MATHEMATICAL MODELLING

Cinzia Bonotto
University of Padova, Italy

Abstract–*In this contribution we will present preliminary results of a study which is part of an ongoing research project aimed at showing how the use of suitable cultural artefacts can play a fundamental role in bringing students' out-of-school reasoning experiences into play, by creating a new tension between school mathematics and everyday-life knowledge with its incorporated mathematics. The focus is on fostering a mindful approach towards realistic mathematical modelling which is both real-world based and makes sense quantitatively.*

1. INTRODUCTION

In many current documents relating to reform in mathematics education, a strong plea is made for making problem solving in school mathematics more closely related to the experiential worlds of children by using more complex and more authentic problem situations in the mathematics lessons. The connection between in- and out-of-school mathematics is not easy to make because the two contexts differ significantly. Just as mathematics practice differs in and out of school, so does mathematics learning. In common teaching practice, the habit of connecting mathematics classroom activities with everyday-life experience is still substantially delegated to word problems. But besides representing the interplay between these two contexts, word problems are often the only means of providing students with a basic sense experience in mathematization and mathematical modelling. Recent studies have documented that the practice of word problem solving in school mathematics promotes in students the exclusion of realistic considerations and a *"suspension of sense-making"* (Schoenfeld, 1991) and rarely reaches the idea of mathematical modelling and mathematization. Primary and secondary-school students, and also student-teachers, tend to ignore relevant and plausible familiar aspects of reality and exclude real-world knowledge from their observation and reasoning. Several studies point to two causes for the abstention from using everyday-life knowledge: a) textual factors relating to the stereotyped nature of the most frequently-used textbook problems; b) presentational or contextual factors associated with practices, environments and expectations related to the classroom culture of mathematical problem solving. Furthermore, it has been noted that the use of stereotyped problems and the accompanying classroom climate relate to teachers' beliefs about the goals of mathematics education.

Finally, in my opinion, the practice of word problem solving is relegated to classroom activities, having meaning and location, in terms of time and space, only within the school. Rarely will students encounter these activities outside of school.

This indicates a difference in views on the function of word problems in mathematics education. The researchers, and probably the drafters of new curricula, relate word problems to problem solving activity and applications. For student-teachers (and probably teachers in general) word problems are nothing other than exercises in the four basic operations, which also have a justification and suitable place within the teaching of mathematics, though certainly not that of fostering a process of providing students with a basic sense experience in *"mathematization"* or *"realistic mathematical modelling"*.

According to Greer, Verschaffel and Mukhopadhyay (2006) the term mathematical modelling is not only used to refer to a process whereby a situation has to be problematized and understood, translated into mathematics, worked out mathematically, translated back into the original situation, evaluated and communicated. Besides this type of modelling, which requires that the student has already at his disposal at least some mathematical models and tools to mathematize, there is another kind of modelling, wherein model-eliciting activities are used as a vehicle for *the development* (rather than the application) of mathematical concepts. This second type of modelling is called 'emergent modelling' (Gravemeijer, 2004). In this contribution the focus will be on the second aspect of modelling.

We deem that if we wish   i) real problems arising from the children's real experiences, so that students may connect reasoning, practices and experiences both in- and out-of-school in a back and forth process,   ii)   situations of realistic mathematical modelling, in problem solving activities, iii) that the students learn mathematics by mathematizing, by overcoming the dichotomy between mathematics as an activity and mathematics as a body of knowledge, we have to change: a) the type of activity aimed at creating interplay between the real world and mathematics with more realistic and less stereotyped problem situations; b) students' beliefs and attitudes towards mathematics, this means changing teachers' conceptions, beliefs and attitudes as well; c) the classroom culture, by establishing also new classroom socio-mathematical norms   (Bonotto, 2006). In this contribution we discuss how these changes can be brought about at primary school level through classroom activities which are more easily related to the experiential world of the student and consistent with a sense-making disposition. In particular we will show, through a study, which is a paradigmatic example, how suitable cultural artefacts and interactive teaching methods can play a fundamental role in bringing students' everyday-life experiences and informal reasoning into play.

In our approach in and out of school mathematics, even with their specific differences, in terms both of practices and learning processes, are not seen as two disjunct and independent entities. Furthermore we think that the conditions that often make out-of-school learning more effective can and must be re-created, at least partially, within classroom activities. Indeed, though there may be some inherent differences between the two contexts, these can be reduced by creating classroom situations that promote learning processes closer to those arising from out-of-school mathematics practices (Bonotto, 2006).

2. OUR APPROACH

Two of important points of views of Realistic Mathematics Education, mostly determined by Freudenthal's view on mathematics (Freudenthal, 1991), are mathematics must be close to children and be relevant to every day life situations and mathematics as human activity. Mathematics education organized as a process of *guided reinvention*, where students can experience a similar process compared to the process by which mathematics was invented. Moreover, the reinvention principle can also be inspired by informal solution procedures. Informal strategies of students can often be interpreted as anticipating more formal procedures. In this case, the reinvention process uses concepts of mathematization as a guide. In agreement with this perspective we believe that the progressive mathematization should lead to algorithms, concepts and notations that are rooted in a learning history which starts with students' informal experientially real knowledge. The idea is not only to motivate students with everyday-life contexts but also to look for contexts that are experientially real for the students and can be used as starting points for progressive mathematization (Gravemeijer, 1999).

Furthermore we stress that besides to foster *the process of bringing the real world into mathematics* by starting from student's everyday-life experience it is necessary to foster *the process of bringing mathematics into reality*. In other words, besides *mathematizing everyday experience* it is necessary '*to everyday' mathematics* (Bonotto, 2006). We believe that this can be implemented in the classroom by encouraging students to analyze '*mathematical facts*' embedded in appropriate '*cultural artefacts*'. In other words, we want to encourage the children to recognize a wide variety of situations as mathematical situations, or more precisely as "mathematisable" situations, since a great deal of mathematics is embedded in everyday life. In this way we can multiply the occasions when students encounter mathematics outside of the school context. The cultural artefacts we introduced into classroom activities, for example, a) supermarket bills to introduce some aspects of multiplicative structure of decimal numbers (Bonotto, 2001a; Bonotto, 2005), b) a ruler to foster children's decimal number understanding (Bonotto, 2001b), c) a cover of a ring binder to introduce the concept of surface area (Bonotto, 2003), d) an informational booklet issued by "Poste Italiane" to estimate and discover area and length dimensions of some envelopes (Bonotto & Ceroni, 2003), are concrete materials which children typically meet in real-life situations.

We have therefore offered the opportunity of making connections between the mathematics incorporated in real-life situations and school mathematics, which although closely related, are governed by different laws and principles. These artefacts are relevant to children; they are meaningful because they are part of their real life experience, offering significant references to concrete, or more concrete, situations. This enables children to keep their reasoning processes meaningful and to monitor their inferences. Roughly speaking *"in the ticket, which is poor in words but rich in implicit meanings, the situation is overturned with respect to the usual buying and selling problem, which is often rich in words but poor in meaningful references"* (Bonotto, 2001a).

The double nature of these artefacts, that of belonging to the world of everyday life and to the world of symbols, to use Freudenthal's apt expression, makes possible the movement from the situations in which it is usually used to the underlying mathematical structure, as well as the reverse process, from the mathematical concepts to the real-world situations; this is in agrees with '*horizontal mathematization*'. An essential property of artefacts, which supports their bilateral influence and offers common bases to culture and discourse, is their being ideal (conceptual) and material. But a different use of these same artefacts, with certain modifications - for instance removing some data present in the artefacts, as for example in Bonotto (2005) - supported the opportunity to favour also '*vertical mathematization*', from concepts to concepts, although only in a weak sense, given the grade level of the students. This occurred when symbols, that is, embedded mathematical facts, became objects to be put in relationship, modified, manipulated, and reflected upon by the children through property noticing, conjecturing, and problem solving. In this way the cultural artefact can be used to introduce new mathematical knowledge through those special learning processes that Freudenthal, 1991 defines '*prospective learning*' or '*anticipatory learning*'.

In this new role these artefacts also may become real "*mathematizing tools*", capable on the one hand of creating new mathematical goals, and on the other of providing pupils and students with a basic sense experience in mathematization which preserves the focus on meaning found in everyday situations. Furthermore we ask children to select other cultural artefacts from their everyday life, to point out the embedded mathematical facts, to look for analogies and differences (for example, different number representations), to generate problems (for example, discover relationships between quantities). So we can present mathematics as a means by which to understand the real world. We deem that in this way we can enable students to become involved with mathematics and to develop a positive attitude towards school mathematics. Besides the use of suitable cultural artefacts discussed above, the teaching/learning environment designed and implemented in our classroom activities is characterized by the application of a variety of complementary, integrated and interactive instructional techniques, and an attempt to establish a new classroom culture also through new socio-mathematical norms.

3. THE STUDY

3.1 Topic

In this quasi-experimental study we decided to exploit as artefact a TV guide from a well-known weekly magazine in order a) to extend students' capacity to calculate from base 10 to base 12, 24 or 60, b) to develop the concept of equivalence between time intervals expressed in different ways (days, hours, minutes), and c) to introduce informally the concept of fractions. To check students' familiarity with TV program guides, the experience was preceded by a phase in which children were asked to bring to class magazines and daily papers they usually use to choose TV programs. It was found that the timetable of television programs, directly or indirectly, is part of the experiential reality of the children involved in the experience. All said that they knew the starting time and duration of their preferred programs, and that they were able to regulate TV viewing with their daily activities.

3.2 Participants

The study was carried out in two third-grade classes (children 8-9 years of age) in a suburb of the city of Padua by the official logic-mathematics teacher, in the presence of a research-teacher. The first class consisted of 20 pupils (10 girls and 10 boys), the second class of 21 (10 girls and 11 boys). In each class there were three children with learning difficulties, and two in the first class and one in the second who displayed demotivated behaviour towards school activities. As a control, two third-grade classes (children 8-9 years old) were chosen from another area of Padua, in keeping with the following criteria: i) the congruence of socio-cultural background, ii) the homogeneous level of performance with the two classes involved in the teaching experiment (as confirmed by the outcome of the pre-test) and finally iii) the use by teachers of a traditional teaching method. The children in the classes involved did not know how to carry out calculus with hours and minutes, however they all knew how to add and subtract in base 10, and remembered from the previous scholastic year that an hour is made up of 60 minutes.

3.3 Procedure

After time to collect, read and comment on the various TV guides gathered by the children, it was decided that all children should work on the same TV guide in order to be able to manage and organize the classes better. The guide included in a weekly supplement of a well-known daily paper was chosen rather than a specialized magazine because of the simpler, compact and ordered structure of the television programs of any one day. This guide also has a section, on the two following pages, dedicated to a review of the films to be televised, where the starting time, duration, but not the finishing time, can be found. Among the details presented is the date of production from which it is possible to calculate the age of the film. Then it was decided to subdivide the teaching experiment into 10 sessions, at weekly intervals, 8 sessions of one hour each and 2 of two hours each, for a total of 12 hours. The first 5 sessions were dedicated to familiarization with the artefact, classification of the various programs according to typology (news, cartoons, films, etc) and to discovering the mathematical facts included, selecting from the many found. The remaining 5 sessions concerned 2 experiences, the first of 3 hours and the other of 4. In the first experience, using the table of television programs, the children were asked to organize their day, and then the week, keeping in mind their activities and commitments, and not exceeding an hour and a half of television a day. The second experience, which took place in 2 two hour sessions, was aimed at reading and interpreting the numerical data in the artefact used - this time the reviews of the two films. The aim also included calculating the duration of the two films in minutes and converting them to hours, and finally establishing a strategy to find the finishing time of the film (see Figure 1 for the requirements of the second experience). The children were then left free to discover other spontaneous scientific dilemmas, for example the age of the film.

Each session of these two experiences was divided into three phases. In the first, each pupil was given an assignment to carry out individually. The children were asked to answer all the questions in writing. In the second phase, the results obtained through personal reflection and elaboration were discussed collectively, sometimes corrected, and then systematized and re-elaborated. The third was aimed at the elaboration of a collective written text comprising the clearer and more convincing explanations emerging from the whole-class discussion.

As far as the control classes were concerned, the class teachers dedicated, within the same time period, exactly 12 hours, to class activities regarding reading and calculation of time duration measured in hours and minutes, according to the modality and techniques normally used in elementary school.

3.4 Data

The research method was both qualitative and quantitative. The qualitative data consisted of students' written work, audio recordings and fields notes of classroom observations and audio recordings of mini-interviews with students. The quantitative data was collected by means of pre- and post-tests, administered to the two experimental classes as well as the other two control classes. The two tests were constructed by the official class teachers, not the research-teacher, by taking some items normally used in the bimonthly tests utilized by the same teachers. Both the pre- and post-tests were organized in such a way as to evaluate the effects of learning on time duration (part 1) and fractions (part 2).

Figure 1.

Film		**Film**	
Courage	***	*The Secret of the Old Forest*	**
Rete 4 / time: 16.00		Channel 5 / time: 0.30	
Producer...		Producer...	
With...		With...	
Review...		Review...	
Comedy	Italy 1956	Fairytale	Italy 1993
Duration 95'	Ó	Duration 134'	Ó

Questions asked: *Make an evaluation of the information presented in the film review, in particular the time the film ends. Write down the procedure you used.*

3.5 Hypotheses

The first general hypothesis was that the children in the teaching experiment class[1] were able to grasp the calculation of hours and minutes and the equivalence between time intervals expressed in different forms (days, hours, minutes) more effectively, compared with the control class, who received a more traditional teaching method. It was also hypothesized that using the clock face, which is divided into half and quarter hours, would allow participants to work out the concepts related to fractions according to "*prospective learning*" (hypothesis II). Furthermore, we hypothesized that, contrary to the practice of word-problem solving documented in the literature, children in this teaching experiment would not ignore the relevant, plausible and familiar aspects of reality, nor would they exclude real-world knowledge from their observations and reasoning (hypothesis III).

Finally children would also exhibit flexibility in their reasoning, by exploring different strategies, often sensitive to the context and quantities involved, in a way that was meaningful and consistent with a sense-making disposition and closer to the procedures emerging from out-of-school mathematics practice; children would also activated problem posing procedures (hypothesis IV).

4. SOME RESULTS

Some early results from the second experience are reported. From the first film review all the children except one, were able to elaborate in their own words the information regarding starting time, channel, year of production, etc (see Figure 1). We note the case of a child, who we will call Emanuele, a repeating student with serious scholastic demotivation and learning difficulties. At the end of the first phase, the written report, he handed in a blank sheet. It was therefore decided to test his knowledge and thought processes by individual interview. We discovered that he knew how to read the data in the artefact and how to correctly work out the equivalence by referring to his preferred interest, football. In fact he knew that the duration of a football match is 90 minutes, and that it corresponds to an hour and a half because he always watches sports programs with his father, the most well-known of which is called *"novantesimo minuto"* (*"ninetieth minute"*). Therefore, 95 minutes for him was equivalent to an hour and half plus 5 minutes. The case of Emanuele therefore confirmed our third hypothesis.

Some significant extracts are presented from written work regarding the finishing time of the film. These show the activation of strategies sensitive to the context and quantities involved and also the emergence of problem posing activities. Claudia: *"I pretended that the film started at exactly 0. I put the 30 minutes to one side. I added 2 hours and that makes 2. I added the 30 minutes and so I got 2 and 30. I added the 14 minutes and so arrived at 2 hours and 44 minutes."*

Claudia tried to simplify the data as much as possible to be able to calculate with greater surety. The explanation was extremely clear, expressed in the language and terminology normally used by children, and for these reasons during the class discussion it led to curiosity, attention, understanding and participation by classmates who were unable to find the finishing time.

Gregorio's protocol included the following:

"1) I found 2 hours and 14 minutes in this way: 60+60=120+14=134. 2) To arrive at 2.44 it was 0.30+2 ore=2.30+14=2.44. 3) To get 8 years we worked out 1993 to arrive at 2001 makes 8 years. We can see 51 minutes of the film "The executors".

It can be seen that in the end that Gregorio faced a spontaneous dilemma with a film whose review was next to the one assigned and whose viewing time partially overlapped. He posed the question *"Once the film "The Old Forest" is finished, how much of the film "The Executors" can I watch?"*.

This shows how the use of an artefact may evoke situations that are in fact experienced, activating the ability to pose and resolve problems.

On the basis of the qualitative results we can say that this experience has reinforced knowledge of the hours in the day and led to calculation in base 60 by means of an informal, non conventional, procedure on the basis of intuition linked to the context or the quantities involved. Among the children's protocols, attempts at formalizing calculation in rows and columns also appeared.

As far as the outcomes of the pre-tests and post-tests are concerned, the errors in the experimental group diminished by 46% overall, while those of the control group remained more or less stationary [hypothesis I]. The two parts of each test are outlined, that is the first part testing reading ability, calculation of hours and the equivalence between time intervals expressed in different forms, and the second part regarding knowledge of the concept of fractions. It emerged that the greater improvement in the experimental group's performance is relative to the abilities tested in the second part of the test, where the concept of fractions was evaluated. There was in fact a 63% reduction in errors in the case of the experimental group, while errors increased for the control group [hypothesis II].

From the results it appears that the teaching experiment had a significant positive effect on achieving learning goals, in particular enhancing and understanding the calculation of hours and minutes and the equivalences between time intervals expressed in different forms, and even more enhancing a first approach to the concept of fractions in a way that is meaningful and consistent with a sense-making disposition. This was not the case in the control group where an increase in errors was found in the second part of the test. It could be supposed that the control group, who received a more traditional type of teaching, may have acquired general algorithmic procedures and formal rules, but these were not well mastered and therefore did not improve performance.

The first two research hypotheses were therefore confirmed. It was also confirmed by the qualitative results that using the TV guide did not activate rigid and general algorithmic procedures but rather specific heuristics, that have an inner consistency and value. The strategies were flexible, local and sensitive to number sizes (hypothesis IV), and were such that children often made reference to parts of the hour (half and quarter hours) to be able to manage calculations better, and in a way that was meaningful and consistent with a sense-making disposition. This aspect made them more sensitive to the concept of fractions according to *prospective learning* and therefore led to the distinct improvement (63%) by the experimental classes in the second part of the post-test. We can say that in our teaching experiments, contrary to the practice of word-problem solving in school mathematics, children did not ignore the relevant, plausible and familiar aspects of reality, nor did they exclude real-world knowledge from their observation and reasoning (hypothesis III).

5. CONCLUSION AND OPEN PROBLEMS

In our view, the positive results obtained in this study, as in our other studies, can be attributed to a combination of closely linked factors: a) the use of suitable cultural artefacts that represent a connection with out-of-school reality or are tied to real-world situations, that allow children good control of inferences and results, and make a connection between symbols and their referents; b) the introduction of particular socio-mathematical norms that played an important role in giving meaning to new mathematical knowledge [prospective learning] or reinforcing previous knowledge [retrospective learning]; c) systematic attention being paid to the nature of the problems and the classroom culture.

In particular by using appropriate cultural artefacts, which students can understand, analyze and interpret, we can present mathematics as a means of interpreting and understanding reality. Teaching students to interpret critically the reality they live in, to understand its codes and messages so as not to be excluded or misled should be an important goal for compulsory education. The computer, as well as other more recent multimedia instruments, has a remarkable social and cultural impact and huge educational potential that perhaps has not yet been fully explored. Obviously, the usefulness and pervasive character of mathematics are merely two of its many facets and can not by themselves capture its very special character, relevance, and cultural value; nonetheless we deem that these two elements can be usefully exploited from the teaching point of view because they can change the common behaviour and attitude held both by teachers and pupils.

For a real possibility to implement this kind of classroom activities, there also needs to be a radical change on the part of teachers. They have to try i) to modify their attitude to mathematics, which is influenced by the way they have learned it, ii) to revise their beliefs about the role of everyday knowledge in mathematical problem solving, iii) to see mathematics incorporated into the real world as a starting point for mathematical activities in the classroom, thus revising their current classroom practice, and iv) to investigate the mathematical ideas and practices of the cultural, ethnic, linguistic communities of their pupils in order to offer them significant references to familiar situations.[2] In agreement with Blum and Niss (1991) we deem that the effective establishment of a learning environment, like the one described here, makes very great demands on the teacher, and therefore requires revision and change in teacher training, both initially and through in-service programs.

NOTES

1. Thanks to the opportunity they had to refer to a concrete reality (the cultural artefact), to explore their strategies and to compare them with those of their schoolmates.

2. *"The main* [regarding rich contexts, author's note] *is that of implementation, which requires a fundamental change in teaching attitudes before it can be solved"* (Freudenthal, 1991).

REFERENCES

Blum, W. and Niss, M. (1991) Applied Mathematical Problem Solving, Modelling, Applications, and Links to Other Subjects - State, Trends and Issues in Mathematics Instruction. *Educational Studies in Mathematics*, 22 (1), 37-68.

Bonotto, C. (2001a) How to connect school mathematics with students' out-of-school knowledge. *Zentralblatt für Didaktik der mathematik*, 3, 75-84.

Bonotto, C. (2001b) From the decimal number as a measure to the decimal number as a mental object. In M.v.d.Heuvel-Panhuizen (ed) *Proceedings of the 25nd PME*. Utrecht: Utrecht University, II, 193-200.

Bonotto, C. (2003) About students' understanding and learning of the concept of surface area. In D. H. Clements and G. Bright (eds) *Learning and Teaching Measurement*, 2003 Yearbook of the NCTM. Reston, Va.: NCTM.

Bonotto, C. (2005) How informal out-of-school mathematics can help students make sense of formal in-school mathematics: the case of multiplying by decimal numbers. *Mathematical Thinking and Learning, 7*(4), 313–344.

Bonotto, C. (2006) How to Replace the Word Problems with Activities of Realistic Mathematical Modelling. In W. Blum, P. Galbraith, M. Niss and H. W. Henn (eds) *Modelling and Applications in Mathematics Education.* New York: Springer.

Bonotto, C., and Ceroni G. (2003) How can the use of suitable cultural artefacts as didactic materials facilitate and make more effective mathematics learning? *CIEAEM 55*, Plock (POLAND), 22 – 28 July 2003, 45-47.

Freudenthal, H. (1991) *Revisiting mathematics education. China lectures.* Dordrecht: Kluwer.

Gravemeijer, K. (1999) How emergent models may foster the constitution of formal mathematics. *Mathematical Thinking and Learning, 1*(2), 155-177.

Gravemeijer, K. (2004) Emergent modelling as a precursor to mathematical modelling. In H.-W. Henn and W. Blum (eds) *ICMI Study 14: Applications and modelling in mathematics education.* Dortmund: Universität Dortmund, 97-102.

Greer, B., Verschaffel, L., and Mukhopadhyay, S. (2006) Modelling for Life: Mathematics and Children's Experience. In W. Blum, P. Galbraith, M. Niss and H. W. Henn (eds) *Modelling and Applications in Mathematics Education.* New York: Springer.

Schoenfeld, A. H. (1991) On mathematics as sense-making: An informal attack on the unfortunate divorce of formal and informal mathematics. In J. F. Voss, D. N. Perkins and J. W. Segal (eds) *Informal reasoning and education.* Hillsdale, NJ: Erlbaum, 311-343.

5.7

STUDENT REASONING WHEN MODELS AND REALITY CONFLICT

Jerry Legé
California State University Fullerton, USA

Abstract–*Upper secondary students were provided data from an experiment which contained a systematic error, all related calculations, and several recommendations for modelling the situation. The students were asked to select one of the given recommendations, or to provide one of their own. The physical context (reality) should have required a direct variation as the model, but data analysis (model) suggested a linear relationship with two parameters. In trying to resolve the internal conflict, a wide range of responses and justifications were offered by participants. Clusters of student responses are characterized, and underlying root issues are suggested as explanations.*

1. INTRODUCTION

For years, advocates have called for incorporating more modelling activity into the classroom because of its positive effect on students' understanding of mathematics. Arguments have included providing a framework for introducing applications (Burghes, 1980), developing problem-solving skills (Blum and Niss, 1991), and promoting conceptual understanding (Lanier, 1999). The kind of thinking that takes place in modelling activity has been associated with a range of other processes, including models for how children learn, especially in mathematics (de Lange, 1987), and the development of meta-cognition among students. Modelling experiences have had the effect of engaging the learner, situating them in the practice of mathematics, and improving their disposition toward more realistic interpretation of word problems (Verschaffel & De Corte, 1997). There has been an inexorable march to develop curricular materials (Blum and Niss, 1991; Usiskin, 1997), and a clearly-articulated rationale for using them. Those arguments range from broad goals such as critical competence (preparing students to be fully functioning members of society) to specific learning outcomes like the acquisition of "*knowledge* of existing models and applications of mathematics..." (Blum and Niss, 1991, p44-45).

Are curricular experiences in modelling a necessary condition for developing a modelling disposition among students? Or can students study mathematics and science as separate disciplines, and synthesize that content knowledge in such a way that it can be activated and adapted when asked to critically examine models for a situation? One way to explore the answer to these questions is to take a group of capable students with no prior experience in modelling, and ask them to build a model using a situation for which the scientific principles and mathematical relationships have been mastered. This paper reports on one such experiment and the responses it generated.

2. BACKGROUND

A comprehensive, college-preparatory school for mathematics and science in the United States was selected to participate in a research study. The school composition had an ethnic and economic diversity which mirrored the community it served. The students were not necessarily gifted or talented in mathematics or science, but were capable and chose to apply to that school because of the concentrated attention in those subject areas. However, the academic success of the school and its students was unquestionably outstanding – top ten ranking in their state, with average scores on standardized tests between the 90th and 80th percentile nationally in mathematics and science. The entire junior class had just completed a course on pre-calculus, and participated in a research study of how the curriculum used at that school shaped their understanding of linearity and linear functions. One aspect of that investigation focused on whether students could clearly distinguish when to use equations of the form $y = mx$ from those with form $y = mx+b$. The task described in the next section was one of the questions used to probe for that understanding, and was administered to half of the students ($n = 75$).

The educational program for schools in that region is defined by various content standards, which serve two main purposes: 1) they prescribe the scope of attention for the standardized tests that students must take, and 2) alignment to these standards forms the criterion for elementary and middle school textbook adoption approval. A review of the content standards for science indicate that by grade 8 (prior to high school), students were to *know* that density is mass per unit volume and *know how* to calculate the density of substances from measurements of mass and volume. The associated narrative in that document even prescribed that density is calculated by dividing the mass of some quantity of material by its volume. The content standards for mathematics indicate that by grade 7, students should be able to "plot the values of quantities whose ratios are always the same...Fit a line to the plot and understand that the slope of the line equals the quantities (sic)" and "solve multistep problems involving rate, average speed, distance and time or a direct variation" (CDE, 1999 p67-68).

In the students' first three years at that high school, all of them completed three additional science courses by the end of their junior year, including a full year of chemistry. Details about the other two courses are not available, but based on the content standards for science, it should have included some physical and earth science but more attention on developing understanding of biological principles. The mathematics curriculum consisted of a sequence of thematic units, usually with an over-riding problem to solve and featuring extensive use of group work, classroom discussion, short-term and long-term problem solving, discovery learning, and attention to developing process skills and higher-order thinking skills, as well as understanding of mathematics content. While none of the thematic units were designed with an emphasis on mathematical modelling, several have contexts in the physical sciences, require students to collect and analyze data, and spiral the mathematical development in increasing complexity.

3. STUDENT TASK

The actual question to which students were asked to respond is the following:

In an attempt to determine the density of a particular gauge of copper wire, the following measurements were made:

Sample	#1	#2	#3	#4	#5	#6
Length (in.)	0.28	0.84	1.15	1.79	2.36	2.81
Mass (g)	0.07	0.11	0.14	0.20	0.25	0.29
Linear Density (g/in)	0.25	0.13	0.12	0.11	0.11	0.10

(Assume the measurements were taken correctly.) The average linear density was found to be around 0.14 g/in, and the regression equation obtained from the data is approximately: $M = 0.089L + 0.040$. It appears to be an excellent fit, but you would think that a piece of wire with length 0" would also weigh 0 g.

Select one of the following recommendations for modelling this situation, and explain why you choose that particular one.
- Go with the regression equation, and restrict the use of the model to the range of data in the table.
- Modify the regression equation somehow so that it "curves" into the origin (0, 0) to match the end behavior.
- Use a direct variation equation of the form $M = k \cdot x$ and begin looking for a source of error to explain why the data doesn't behave correctly.
- Some other explanation (provide the explanation, along with your reasons)

Figure 1. Question 10 from research study.

The intent was to present a conflict between the conceptual understanding that one might apply in modelling the situation (in this case, about proportionality) and the results obtained from doing data analysis on the information. The linear density, average linear density and regression equation calculations were provided to alleviate time constraints and enable students to focus on the model selection. However, students had access to calculators, however, so they could choose to duplicate or verify that work. The given recommendations allowed the participants to 'accept' or 'reject' certain beliefs, including the power of technology to yield the "right" answer, the ability to use mathematics to "fix" aberrant behaviors among data, and the reasonableness of interpolation as a means of prediction. Students were also invited to use their creativity if they were not satisfied with the choices provided, especially if they felt it appropriate to use a different mathematical relationship. The students were not allowed to discuss the problem with their peers, or the person who administered the assessment, thereby denying a powerful social component that was routinely present in their learning environment. In not allowing students to replicate the experiment, the task was made deliberately more abstract in two ways. First, the data was dissociated from the sense-making that would come from obtaining them by direct measurement. If students did not automatically *know* the model that should be applied to this situation, they might need to reconsider those numbers in context in order to reason conceptually about the task. Second, while density was a science subject that the students had encountered, it was less likely that they had studied *linear* density. Also, students had studied proportionality among quantities and direct variation as a function form, but it was unlikely that the situation that they examined while studying these concepts specifically involved linear density. Also, designing the task to take measurements <u>and</u> find the equation which describes the relationship between the data would shift the emphasis from model selection to data analysis; students might report what "is", rather than what "should be".

4. RESPONSES

The good news is that students exist who *can* fuse knowledge about mathematics and science (studied as separate disjoint subjects), apply that synthesis to model a simple situation and critically question the validity of alternative explanations. In the case of these students, the invariant condition was that the mass of the wire should be proportional to the length of the piece under consideration. Several of them stated that condition as assumptions of uniform diameter and constant density. One particularly eloquent response is provided in Figure 2:

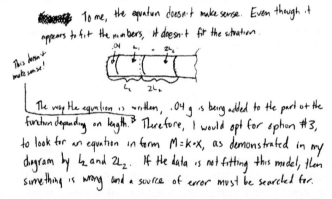

Figure 2. Student response using conceptual understanding.

However, only thirteen students (17%) identified the third recommendation as the best answer, and while the range of explanations was impressive, the rationale was often suspect. Four gave no explanation, one was unintelligible, one argued that errors are always a possibility and their existence should be considered, and one eliminated the other options and provided that as the reason. Two students argued that the linear density calculations were reasonably constant, especially if the first set of data was treated as an outlier. One student argued that there had to be a source of error, since (0, 0) did not satisfy the regression equation. Finally, one student applied Ockham's Razor as a means of resolving the conflict, still convinced that the regression equation should work.

Twenty-three students (31%) selected the first recommendation - that the linear regression equation should be the model used, as long as its domain was restricted to the range of data from which it was obtained. The largest block of students felt that the data determines the equation that is generated, and therefore it seemed reasonable that the equation should represent those numbers. Other arguments that were invoked by multiple students included the fact that interpolation was considered a "safe" process, that more data was required to warrant rejection of the regression equation, and measurement imprecision might explain why the regression equation did not go through the origin. Two students misinterpreted the phrase "assume measurements were taken correctly" to mean that the recordings were accurate, and two students reported a belief that regression always obtained the right answer. On a lighter note, one student argued that models do not have to reflect reality, and another said that this was a school task where the *goal* is to complete the task.

Nineteen students (25%) thought that the second option was the best. The prevailing thought here was that this recommendation captured both the trend in the actual data and the expected end behavior for the model. Secondary arguments included that (0, 0) should be considered a data point (without arguing for redoing the regression calculation with the extra data point included), that this kind of density might produce a graph which is "locally linear", and that polynomial curve-fitting was seen as the goal – it was okay to add more data points and allow the curve which described the pattern to become more complex. One student argued that the availability of calculators made this recommendation possible, but gave no details on how that would be done, and another included (0, 0) with a graph of linear density versus mass, showing a sketch that looked like a probability density function skewed to the right. None of these students considered defining the equations piece-wise to capture the disparate trends in the data, nor did they try other regression models with (0, 0) included in the set of data.

Fourteen students (19%) of the students suggested another recommendation for modelling the situation, with most of the responses reflecting one of two different types of thinking about the problem. First, most of them were willing to resolve the conflict by restricting the domain, either by not allowing $L = 0$, or by allowing extrapolation for large pieces of wire only. Those students suggested arguments like a domain restriction necessarily existed anyway for values of L which were negative, or the relationship involved a ratio which would imply that L couldn't equal 0 anyway! One student claimed that having "no wire" means that it isn't a "piece of wire"; in other words, physical material contained matter, which could be divisible, but the absence of that physical material violated the conditions of the problem situation. The other recurring theme among these students was that the behavior shown in the regularity among the known data points might change as one considers increasingly small pieces of wire. One recommendation was to collect more data points in the interval $0 < L < 0.28$ to validate having the end behavior approach the origin, or to be able to more adequately describe the existing relationship for such small-sized objects. Another student suggested that the precision error would remain constant for the measurements in the data, but the relative size of that error would change as one gets close to zero, which could explain why the "line" needed to "curve" into the origin.

5. CONCLUSIONS

It is possible for *some* students to combine their mathematics and physics knowledge to reason intelligently in a simple yet unfamiliar modelling situation.

However, those students are exceptions when compared to the performance of their classmates, and the conditions at *this* school were more ideal than what exists typically. In fact, the dearth of effective responses and the types of arguments used by the students should suggest that modelling experiences be a necessary component of a vibrant mathematics program. Perhaps because the task was given in a mathematics class, arguments reflect a mathematical bias. Students believe in the power of an equation to describe data, especially when using a "best-fit" one. Domain restrictions resolve mathematical problems like division by zero. Higher-degree polynomials describe complicated patterns in form, even if using them distances you further from reality. There were no clues to suggest that the constant density should be considered, and little evidence that students reasoned conceptually from the fact that the quantities involved should vary proportionally. The only scientific considerations were superficial ones, like interpolation being more accurate than extrapolation, consideration for the effect of measurement imprecision, or the need for more data points.

The suspension of sense-making reflects the school environment, where the goal *is* to complete the task. If any student had reasoned that by starting with the longest piece of wire, one can cut off pieces to create the condition for the other data, that viewpoint would have led to an analysis by finite differences, revealed that the density is constant, and led them to question the calculations that were provided. The blind acceptance of the data as accurately depicting the real situation was exacerbated by the fact that the linear density calculation was provided, which might suggest that the data itself is okay. One might argue that studying statistics, especially if taught well, would instill a "habit of mind" to question information. However, it is unlikely that students would consider three subject areas at the same time when they already have difficulty working in the interface of two domains.

Assuming that learning about existing mathematical models is a valid goal for mathematics education, there are reasons why modelling experiences may be a necessary condition for developing such knowledge. Jablonka (1997) cited several misconceptions that may occur in interpreting mathematical models, including the belief that mathematical exactitude should apply, that models must be reliable, and that the structure of the model must be the structure of reality. In this study, many student responses exhibited similar beliefs that would be offset by such experiences – for example, the equation coming out of a calculator should accurately describe the data upon which it is based, and that curve-fitting as a task should not have a "reality" mirror based on physical laws, natural conditions or common sense.

A second argument may be found in Niss (1999) when he discusses obstacles produced by the process-object duality. These students understood certain processes – some relationships behave in such a way that the amount of one quantity is always proportional to the amount of the other, and that one can create models for these kinds of relationships in a particular form ($y = mx$). Their understanding had not progressed to the point that the form represented *all* situations in which the amount of one quantity is always proportional to the amount of the other, or else there would have been more students who reasoned like the one given in Figure 2. Those students who re-expressed that equation form to say that the ratio y / x stays constant found it more difficult to argue in favor of that model, since it generated a new problem for them involving division by zero. Models may be a situation in which the process-object duality presents an obstacle to learning. In this case, studying proportionality, solving specific problems about direct variation, and expressing and studying functions of the form $y = mx$ may not lead to a deep understanding about a model for a specific situation BEING an equation of that form.

Finally, in discussing the use of contexts in situated teaching as a means for developing understanding of mathematical concepts, DaPueto and Parenti (1999) identified a critical feature as being students' "understanding of, or confidence with, the process by which the *problem is transformed*... (p. 7)". With this set of students in particular, it is likely that they understood the problem transformation, or would have been able to make sense of the situation given time or the opportunity for social discourse. Collectively, it is clear that they lacked the confidence in the process of how the problem was transformed to be able to resolve the inherent conflict that was contained within the problem. And in a manner reminiscent of the conclusions drawn by Lithner (2000), strategy choices are dominated by established experiences – applying methods they know from similar tasks – which can fail when "familiar routines do not work for different reasons (p. 187)."

REFERENCES

Blum, W. and Niss, M. (1991) Applied mathematical problem solving, modelling, applications, and links to other subjects – state, trends, and issues in mathematics instruction. *Educational Studies in Mathematics*, 22, 37-68.

Burghes, D. (1980) Mathematical modelling: A positive direction for the teaching of applications at school. *Educational Studies in Mathematics*, 11, 113-131.

California Department of Education. (1999) *Mathematics framework for California public schools – Kindergarten through grade 12*. Sacramento, CA: CDE.

DaPueto, C., and Parenti, L. (1999) Contributions and obstacles of contexts in the development of mathematical knowledge. *Educational Studies in Mathematics*, 39, 1-21.

Jablonka, E. (1997) What makes a model effective and useful (or not)? In S. Houston, W. Blum, I. Huntley, and N. Neill (eds). *Teaching and learning mathematical modelling – Innovation, investigation and applications*. Chichester: Ellis Horwood, 39-50

Lanier, S. (1999) Students' understanding of linear modelling in a college mathematical modelling course. Ph.D. dissertation thesis, University of Georgia.

Lithner, J. (2000) Mathematical reasoning in task solving. *Educational Studies in Mathematics*, 41, 165-190.

Niss, M. (1999) Aspects of the nature and state of research in mathematics education. *Educational Studies in Mathematics*, 40, 1-24.

Usiskin, Z. (1997) Applications in the secondary school mathematics curriculum: A generation of change. *American Journal of Education*, 106, 62-84.

Verschaffel, L., and De Corte, E. (1997) Teaching realistic mathematical modelling in the elementary school: A teaching experiment with fifth graders. *Journal for Research in Mathematics Education*, 28, 577–601.

5.8

THE CONCEPT OF THE DERIVATIVE IN MODELLING AND APPLICATIONS

Gerrit Roorda, Pauline Vos and Martin Goedhart
University of Groningen, The Netherlands

Abstract–*The question addressed in this paper is how to measure students' transfer skills with respect to the concept of the derivative in modelling and applications. We adapted a framework developed by Zandieh (2000) for analysing students' understanding of the concept of the derivative. Nine grade-11 students made a test and a task-based interview, which consisted of mathematical and application problems. The task-based interview included tasks about the application of the derivative in a physics and economics context. The analysis of interview data showed some difficulties emerging with the framework. With respect to modelling and applications, we found that sometimes students cannot apply knowledge of calculus that they have, and that translations between representations, and especially between a context set in the real world and mathematics, and backwards, cannot be represented in the framework. Our analyses gave directions for further development of a framework for understanding derivatives in relation to applications.*

1. INTRODUCTION

In the Dutch mathematics curriculum for secondary schools, the role of modelling and applications increased in the past fifteen years. When students in grades 10-12 learn the concept of the derivative, most textbooks avoid the use of the formal limit definition (or only mention it on one page). Instead, textbooks provide students with opportunities to learn the concept in different representations (for example, symbolical and graphical) and in applications (for example, physical and economical). Many application problems in mathematics textbooks describe a real situation but often the mathematical model is simultaneously offered by a formula. In terms of the modelling circle described by Blum and Leiß (2006), the real world and the mathematical world are both present in many Dutch textbook exercises. Most exercises are set within the real world and offer a problem context. This context has to be understood (step 1) and structured (step 2). The ensuing step, to mathematise the problem (step 3) into a mathematical model, is already provided for by the textbook authors. What remains is that students have the opportunity to make the last steps in the modelling circle such as 'work mathematically', 'interpret' and 'validate' the model (steps 4, 5 and 6). In tasks involving derivatives, this means that students can use their mathematical knowledge to solve an application problem and compare their answer with the situation in the real world.

The expectation of the designers of the Dutch curriculum was that the use of derivatives in different representations and contexts would give students a better understanding of the concept and would enable them to use their knowledge and skills flexibly. This expectation still has to be supported or refuted by evidence. Therefore, we have embarked on a research, and one of our questions is how to measure students' knowledge and skills with respect to the concept of the derivative in modelling and applications. This question will be explored in this paper.

2. THEORETICAL FRAMEWORK

During the process of learning a certain concept one builds in mind a concept image and a concept definition (Tall & Vinner, 1981). A concept image is the 'total cognitive structure that is associated with a concept' and a concept definition is the 'form of words used to specify that concept'. The difficulty is how to analyse the concept image of students. Zandieh (2000) developed a framework for analysing students' understanding of the concept of the derivative (Figure 1). This framework has two main components: multiple representations or contexts, and layers of object process pairs.

The five *representations or contexts* are (a) symbolic, such as the limit of the difference quotient (b) graphical, such as the slope of the tangent line, (c) verbal, such as the instantaneous rate of change, (d) paradigmatic, physical such as speed or velocity, and (e) other, for other contexts.

The object-process *layers* are: ratio, limit, and function, as can be seen in the symbolic definition of the derivative $f'(x) = \lim_{h \to 0} \dfrac{f(x+h) - f(x)}{h}$. The derivative is a function whose value at any point is defined as the limit of a ratio.

There are fifteen positions in the framework, and at each position the researcher can put an open or a closed circle to indicate the depth of cognition of the student who is interviewed. An open circle means that the student mentions an object within a certain context. A closed circle means that the student also mentions the underlying process.

For example: When a student describes the derivative as the slope of the tangent line, then open circles are placed in the 'graphical' column and the 'ratio' and 'limit' row. The second circle would have been closed if the student also mentions the limit process of average slopes approaching the slope at one point.

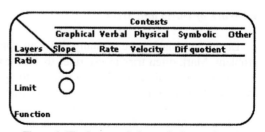

Figure 1. The framework for analysing students' understanding of the derivative (Zandieh, 2000).

The framework is used to describe students' understanding. The emphasis of the framework is on understanding derivatives from a mathematical point of view, but an interesting aspect with respect to modelling, is the column 'Physical'. A student who understands that velocity has to do with derivatives, and uses his or her mathematical knowledge in a physics model, shows competency on a part of the modelling circle (for example, step 5 in the modelling circle: 'interpreting').

Aim of this part of our research is to test if we can use the framework for a larger research project, in which we monitor the cognitive development of a few students for a period of three years.

3. METHOD

We used the framework to assess nine grade 11 students, one month before their final exam from the HAVO (senior general secondary education).

First, we gave them a test with exercises on the derivative. This test gave us an idea of students' knowledge and skills, for example whether they were able to use the power rule, or to calculate at which points two graphs have the same slope (Step 5 in the modelling circle: 'work mathematically'). A few weeks later, we conducted task-based interviews with these nine students (Goldin, 2000), which were video-taped. Three students worked alone during the task-based interview. The other six worked in three pairs. The idea of working in pairs was that a discussion between students of the same level could give us valuable information, because the students use their own 'language' to talk about mathematics. The tasks used during the interviews contained a physics problem, an economics problem and some more general applications.

4. RESULTS AND DISCUSSION

In this section we will focus on the performances of two students, David and Mark, who worked together on the problems during the interview. We selected these two students because they used contrasting methods to solve the tasks although they followed the same mathematics curriculum. We filled in the adapted framework using the work of these two students, and this gave us directions for further development of the framework for describing the learning of modelling, especially with respect to the concept of the derivative in applications.

4.1 Work of David and Mark

David and Mark have a totally different approach, when facing an application problem. Mark often tries to use rules that he learned before. He is not always sure which rules he could use, and also if he is doing the calculations correctly. In some exercises he successfully obtains an answer by following a certain procedure, but in other exercises he starts an erroneous procedure and continues working on his calculations. Mostly, Mark tries to work on the mathematical model but does not take into account the context. His companion, David often tries to understand the context. Before starting to work on the task, he sometimes estimates an answer, he looks at a graph or reflects on the context. When solving a problem, David switches between the mathematical model (graph or formula) and the context.

We will illustrate the differences between David and Mark with a description of their actions during a task from the task-based interview.

When a company has a monopoly position, all customers have to buy their product with that company. If the price increases, the amount of sold products will decrease.

Let $p(q)$ be the price in Euros per unit that the company can charge, if it sells q units. The price function is $p(q) = -0,5q + 12$. The revenue function is

$R(q) = -0,5q^2 + 12q$ and the cost function is $C(q) = 10 + 3q$.

At what production level will the costs and the revenue increase at the same rate?

Description David's actions	**Description Mark's actions**
David starts with a short question about the relation between the three formulas. Then he makes a plot on his graphic calculator. After 2 minutes David says: "*That's funny, when I use the option dy/dx (on the graphic calculator) on this point it is 2.8 and on a point below it is 3. So q is about 8.8.*" After this discovery David continues to think about the relation between the three formulas. When he hears Mark talking about using the derivative, he says: "*No this* (price-function) *is not the derivative* (of the revenue function)."	Mark starts with a remark: "*You have to make them equal.*" He tries to use the 'Solver' on his graphic calculator. After a while he says: "*1.2*". After 2.5 minutes Mark hears a remark from David about the steepness of the graph. Suddenly he begins to calculate the derivatives. Two minutes later Mark has an answer: "*9, I found 9*". Mark tries to convince David: "*The revenue increases at the same rate as the costs, so this must be equal*"(pointing at the calculation).

Figure 2. Exercise of the task-based interview with description of actions.

In this example, David tries to understand the context. He avoids calculations; instead he plots graphs on his graphic calculator. Also, most of the time David tries to investigate the relation between the three formulas. In terms of modelling activities, he tries to understand the mathematisation into a model by reflecting on the context. After finishing his task, David remains interested in the mathematical model. In contrast to David, Mark starts immediately with his calculations. First he commits an error (he calculates the break-even-point, a concept learned in the economy lesson), later he applies the derivative correctly. He needs David's hint about the steepness of the graph. Mark is not really interested in the presented context, nor is he even thinking of validating or interpreting his answers. Unlike David, Mark does not make a connection between the real world and the mathematical world. Another example of the modelling strategies by the two students comes from Task 1 of the task-based interview: a physics exercise about a falling stone. In the exercise a table and a graph are given with the height of the stone, together with a formula for the height of the stone: $h(t) = 90 - 4,9t^2$. The question is: Calculate the velocity at a certain moment and also the velocity when the stone hits the ground.

David's first remark is: "*Velocity has to do with distance and time*". Then he calculates both velocities by using the graph and the difference quotient on a small interval. The interviewer asks: "*Do you know another method to solve the problem?*" David replies: "*I think there is a formula*" and he looks on his graphic calculator where he had stored a list of physics formulae.

When Mark sees the physics exercise he starts with saying: *"It has to do with gravity"*, *"I need formulas"*, *"It has to do with $m \cdot v^2$ "*. Mark needs 8 minutes to think about the formulas from the physics lesson, but he cannot find the right one. Based on the test and the interview, we filled in the framework for David and Mark.

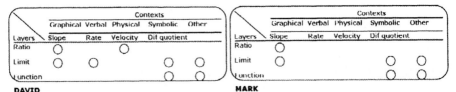

Figure 3. Diagrams for David and Mark.

We placed, for example, open circles in the column 'symbolic' on the rows 'limit' and 'function', because David uses the derivative function to calculate the value of the derivative at one point (limit). David also calculates the steepness of the graph, therefore there is an open circle in column 'graphical' in the row 'limit'. There are only open circles, because David and Mark work with the objects (for example, the derivative function, the slope of the tangent line), but they never mention the underlying processes (for example, the limit process).

4.2 Difficulties with the Framework

The diagrams presented in Figure 3 suggested that David has a better understanding of the concept. The analysis of the interviews revealed some problems with the framework: (1) Students do not automatically connect an understanding of a process in one context with the same process in another context. This cannot be represented in the framework. (2) It happens that students cannot apply their knowledge of calculus. (3) Translations between representations, and between a context and the mathematical model cannot be represented. (4) The framework only displays the overlap with the concept as a notion of the mathematical community. Personal constructions cannot be represented. Below, we will focus on the difficulties (2) and (3), which are primarily related to modelling and application problems.

Applying knowledge: It happens that students cannot apply their knowledge of calculus in new contexts (Selden *et al,* 2000). We observed this phenomenon with David, who does not use derivatives in the economics task. Instead, he tries to understand the situation and he produces graphs on his graphic calculator to find an answer. The phenomenon also occurs when Mark needs much time to calculate the derivatives of the cost and the revenue function. In the other exercise, about the falling stone, again Mark cannot apply his knowledge about derivatives, because he limits himself to physics formulas.

We tried to describe David's and Mark's knowledge using the framework of Zandieh. However, we found that we could not visualise which mathematical knowledge students apply in different contexts. If they apply their knowledge in one context, it is not certain that they will apply the same knowledge in another.

Translation between a context and mathematical model: In many application problems in textbooks the mathematical model is often already offered to the students. In our research it depends on the context and on the student, whether they use the mathematical model or not. David, for example, ignores the mathematical formula in the physics task, but in the economics task he investigates the mathematical model. Mark mostly ignores the context and centres on the mathematical models by applying algorithms and calculations. He never interprets his calculations in terms of the context. In the framework we could not express the step from the context to the mathematical model, and back (the mathematisation and interpretation steps in the modelling circle). Yet, we consider this vital in describing the understanding of derivatives and in describing modelling competencies.

5. CONCLUDING REMARKS

The framework of Zandieh gives a clear representation of students' knowledge of the derivative concept, especially from a mathematical viewpoint. However, an important part of 'understanding derivatives' cannot be represented in this framework. Our analyses gave some clues for further development of a framework, especially with respect to applications of the concept of the derivative. In the first place, we need more columns for different contexts. As shown by David's and Mark's work, it depends on the context which mathematical knowledge is applied by the student. Secondly, we need to visualise translations between representations and between a context and the mathematical model. From a modelling perspective, translations from contexts in the real world to mathematics, and back to the context, are important activities, which will be represented in a forthcoming framework.

REFERENCES

Blum, W. and Leiß, D. (2006) "Filling up" – The Problem of Independence-Preserving Teacher Interventions in Lessons with Demanding Modelling Tasks. In M. Bosch (ed) *Proceedings of the European Society for Research in Mathematics Education CERME 4 (2005)*. IQS FUNDIEMI Business Institute www.fundemi.url.eduT, 1623-1633. ISBN 84 611 3282 3.

Goldin, G.A. (2000) A scientific perspective on structured, task-based interview in mathematics education research. In A.E. Kelly and R.A. Lesh (eds) *Handbook of research design in mathematics and science education*. Mahwah, NJ: Lawrence Erlbaum Associates, 517-545.

Selden, A., Selden, J., Hauk, S. and Mason, A. (2000) Why can't calculus students access their knowledge to solve non-routine problems? In E. Dubinsky, A. Schoenfeld and J. Kaput (eds), *Research in collegiate mathematics education IV*. Providence, RI: American Mathematical Society, 103-127.

Tall, D. and Vinner, S. (1981) Concept image and concept definition in mathematics with particular reference to limits and continuity. *Educational Studies in Mathematics*, 12(2), 151-169.

Zandieh, M. (2000) A theoretical framework for analyzing student understanding of the concept of derivative. In E. Dubinsky, A. Schoenfeld and J. Kaput (eds), *Research in collegiate mathematics education IV*. Providence, RI: American Mathematical Society, 128-153.

5.9

INEQUALITIES AS MODELLING TOOLS IN COMPUTING

Sergei Abramovich
State University of New York, Potsdam, USA

Abstract–*This chapter is a reflection on a technology-enhanced mathematics education course that utilizes context-oriented spreadsheet modelling as a milieu for secondary school teachers' training in the use of inequalities and associated proof techniques. In some cases, a spreadsheet is presented as a generator of applied problems dealing with a computational efficiency of modelling involved and leading to the use of inequalities. In other cases, a context within which computational environments are created is extended to allow for inequalities to be used as problem-solving tools. The chapter illustrates how a traditional pedagogy of utilizing technology for modelling inequalities can be complemented by that of using inequalities as modelling tools in applications to computing.*

1. INTRODUCTION

One of the major changes in curriculum and pedagogy of school mathematics over the last four decades is the focus on application-oriented modelling activities as a useful vehicle for teaching and learning mathematical concepts (Beberman, 1964; Niss, 1989; Blum, 2002). Technology has the potential to enhance the impact of modelling on the construction of meaning of ideas and tools of mathematics by learners (Goos, 1998; Jiang, McClintock & O'Brien, 2003). In North America, as curriculum and didactic changes have been realized in the form of standards (National Council of Teachers of Mathematics, 2000), teachers have come to be increasingly recognized as major agents in the implementation of the standards (Conference Board of the Mathematical Sciences, 2001). This made mathematics teacher education programmes uniquely accountable for providing appropriate environments for the learning of new pedagogy. This chapter shows how the use of inequalities as tools in computing applications can inform spreadsheet modelling in the context of preparation of secondary school teachers.

It has been observed that in the learning environment of a ready-made undergraduate mathematics, prospective teachers are missing context in which problems involving inequalities arise, and they receive little or no training in the use of inequalities. As a result, secondary school students are given almost no support that would enable them, as they continue mathematical studies at the tertiary level, to handle basic techniques of calculus including "epsilon-delta" definitions and convergence tests for infinite series and improper integrals. Therefore, training in the use of inequalities and various methods of proof is important for secondary school students' long-term development. All this begins with the preparation of teachers.

This chapter is a reflection on a technology-enhanced mathematics education course for pre-service and in-service teachers (referred to below as teachers). It shows how one can utilize context-oriented spreadsheet modelling as a milieu for the teachers' training in the use of inequalities and associated proof techniques. In some cases, a spreadsheet will be presented as a generator of applied problems dealing with a computational efficiency of modelling involved and leading to the use of inequalities. In other cases, a context within which computational environments are created will be extended to allow for inequalities to be used as problem-solving tools. In summary, the main goal of this chapter is to show how a traditional pedagogy of utilizing technology for *modelling* inequalities can be complemented by that of *using* inequalities as modelling tools in applications to computing. This is consistent with Lamon's (2003) position "spreadsheets serve the dual purpose of enhancing students' content knowledge and socializing students into the world of mathematical modelling" (p. 81) and "by simultaneously emphasizing tools *and* [italics in the original] their applications in authentic problems" (p. 93) one can gain experience in the usefulness of mathematics. In this chapter, problems on the improvement of computational environments, stage-by-stage solutions of which demonstrate a move from novice practice to expert practice in spreadsheet modelling, are treated as authentic ones, and inequalities are considered as tools used in applications to such problems.

2. THE USE OF INEQUALITIES IN MODELLING LINEAR PROBLEMS

The problem-solving focus of current standards for school mathematics in North America (National Council of Teachers of Mathematics, 2000) has influenced the mathematics curriculum of New York State (New York State Education Department, 1988). From a modelling perspective, many problems found in the curriculum can be presented in the form of Diophantine equations in two or more variables. As an example, consider

Problem 1. *It takes 39 cents in postage to mail a letter. A post office has stamps of denomination 10 cents, 5 cents, and 1 cent. In how many ways can one make this postage out of these three types of stamps?*

In designing a spreadsheet-based environment for the numerical modelling of Problem 1, the following contextual situation arises: Given the cost of postage, determine the maximum number of each type of stamps that might be placed on an envelope. In a decontextualized form, the problem is to find the greatest values of variables x, y, and z which satisfy the Diophantine equation

$$ax+by+cz=n \tag{1}$$

($n=39$, $a=10$, $b=5$, and $c=1$ in the case of Problem 1). As will be shown below, knowing such values of x and y (that is, the largest total for each type of the stamps) enables one to avoid unnecessary computations in generating solutions to equation (1) and, in doing so, to develop competence in spreadsheet modelling. Through resolving such a computationally driven problem, didactically useful activities involving appropriate use of inequalities and corresponding proof techniques can be introduced to the teachers. One such technique is based on reasoning known as proof by contradiction. This type of proof begins with making an assumption contrary to what has to be proved; it then leads to an absurd result enabling one to conclude that the original assumption must have been wrong, since it led to this result. Using this method, the teachers can prove the inequality $x \leq n$. Indeed, assuming $x > n$, yields $n = ax + by + cz > an \geq n$ for all $y \geq 0$ and $z \geq 0$. This contradiction (i.e., the false inequality $n > n$) suggests that x cannot be greater than n, whence $x \leq n$. In much the same way, the inequality $y \leq n$ can be proved. A helpful didactical extension of this formal proof is for teachers to situate its argument in context; that is, to support each algebraic proposition by a situational referent from Problem 1.

An important aspect of teaching to use inequalities as tools in computing applications is to show how these tools can be amended to enhance the efficiency of computations. In other words, by amending inequalities one improves a model (used in search of solutions) to ensure its greater efficiency and, in doing so, develops competence in spreadsheet modelling. With this in mind note, that the inequalities $x \leq n$ and $y \leq n$ can easily be improved. Using a combination of formal (mathematical) and informal (contextual) arguments, one can prove by contradiction the inequality $x \leq n/a$, which is stronger than $x \leq n$ for $a > 1$. Indeed, the formal assumption $x > n/a$ for any $y > 0$ and $z > 0$ yields the following contradictory conclusion: $n = ax + by + cz > a(n/a) = n$. In much the same way the inequality $y \leq n/b$ can be established, both formally and in context.

Furthermore, taking into account that x is an integer variable and using the function INT (described in a spreadsheet environment as a tool that rounds a number down to the nearest integer), yield an even stronger inequality $x \leq INT(n/a)$. By utilizing a measurement model for division, one can build upon the teachers' familiarity with this model and, in doing so, to enhance a contextual representation of the last inequality. Indeed, this model conceptualizes division as repeated subtraction and thus x in equation (1) can be interpreted as a number of ways a is subtracted from n within a set of positive integers. Apparently, the maximum number of such subtractions equals to $INT(n/a)$. In the context of Problem 1 (which has the total of 20 solutions), one cannot use more than three 10-cent stamps in making a 39-cent postage. Similarly, the inequality $y \leq INT(n/b)$ can be proved, interpreted in contextual terms, and used as an upper estimate for the variable y.

Note that the knowledge of lower estimates for variables x, y, and z can be used to further improve computational efficiency of the environment in question. Indeed, the inequalities $y \geq k$, $z \geq m$, where k and m are positive integers less than n enable for the improvement of the x-range found. To this end, one can write $x = (n - by - cz)/a \leq (n - bk - cm)/a < n/a$. Finally, taking into account that x is an integer variable yields the inequality $x \leq INT((n - bk - cm)/a)$ which, as an upper estimate for x, is an improvement over an earlier found estimate $x \leq INT(n/a)$. Likewise, the assumption $x \geq k$, $z \geq m$, $0 < k < n$, $0 < m < n$, and the fact that y is an integer variable can be utilized in refining the y-range found earlier.

In such a way, one can construct a spreadsheet in which a non-trivial information about lower bounds for any two variables is used to improve an upper bound for the third variable. Such a spreadsheet shows that the larger the lower bounds for any two variables, the smaller the upper bound for the third variable. This computationally driven statement can be interpreted in the following contextual terms: the more stamps of one type were used, the fewer stamps of another type could have been used. For example, the so constructed spreadsheet generates 16 solutions to equation (1) with $n=80$, $a=10$, $b=5$, $c=1$, $x\geq3$, $y\geq3$, $z\geq2$. Allowing the variables x, y, and z be any whole number yields 81 solutions. Therefore, the use of inequalities in eliminating possibly extraneous solutions to equation (1) signifies a move from novice practice to expert practice in spreadsheet modelling. Furthermore, the meaning of inequalities can be communicated to teachers in three didactically complementary ways: mathematical, computational, and contextual. More details about this approach can be found elsewhere (Abramovich, 2005).

3. INEQUALITIES AS TOOLS IN REDUCING DIMENSIONALITY OF A MODEL

There are many mathematical problems of both pure and applied nature that afford multiple solutions. The goal of inquiry in that case could be to find their total number rather than solutions themselves. Partitioning problems, bordering two areas of mathematics – combinatorics and number theory, are mostly concerned with the following simple question: "How many"? Indeed, already in the case of Problem 1 there are 20 solutions (all of which could be displayed within a spreadsheet); this number would grow (though not necessarily monotonically) as the cost of postage grows, provided that the stamps' types remain unchanged. For example, one may extend the context of Problem 1 to four types of stamps and, after appropriate decontextualization, to enquire as to how many quadruples (x, y, z, t) satisfy the Diophantine equation

$$ax+by+cz+dt=n \tag{2}$$

Similar to the three-dimensional case, one can use proof by contradiction to establish the inequality $t\leq \text{INT}(n/d)$ which, in turn, enables for the reduction of equation (2) to a family of equations

$$ax+by+cz=n-dt \tag{3}$$

in which integer variable t varies over the set $\{0, 1, 2, \ldots, \text{INT}(n/d)\}$. In such a way, one can use inequalities in order to reduce the computational complexity of a problem; indeed, whereas equation (2) has four variables – more than a spreadsheet can handle concurrently – for each value of $t \in \{0, 1, 2, \ldots, \text{INT}(n/d)\}$, equation (3) turns into equation (1) enabling its direct spreadsheet modelling. Thus one can consider the family of three-variable equations

$$ax+by+cz=n_k \tag{4}$$

where $n_k=n-dk$, $k=0, 1, \ldots, \text{INT}(n/d)$. In finding ranges for variables x and y, one should take into account that these ranges depend on the value of n_k. In other words, for each value of k the following inequalities can be established: $x\leq\text{INT}(n_k/a)$, $y\leq\text{INT}(n_k/b)$. A spreadsheet based on the method described in this section generates 59 solutions to equation (2) for $a=10$, $b=5$, $c=1$, $d=7$, $n=39$. This result can be confirmed by using *Maple* as described in (Abramovich & Brouwer, 2003). A possibility of having an alternative method of verifying the correctness of solution develops one's confidence in the correctness of a model used to find this solution. More details about what may be referred to as the method of inequality-based reduction can be found elsewhere (Abramovich, 2006).

4. USING INEQUALITIES IN MODELLING NON-LINEAR PROBLEMS

Partitioning problems can be associated with (grade-appropriate) non-linear models also. For example, the need to partition a unit fraction into a sum of three like fractions may arise in the context of plane geometry. This brings about the study of new techniques in the use of inequalities as modelling tools. Consider

Problem 2. *How many triangles with integer heights can be circumscribed around a circle of integer radius n?*

Setting x, y, and z to be integer heights of a triangle results in the chain of equations

$$A = n(a + b + c)/2 = ax/2 = by/2 = cz/2 \qquad (5)$$

where a, b, c are the sides of the triangle, A its area. In turn, equations (5) yield the equation

$$1/n = 1/x + 1/y + 1/z. \qquad (6)$$

Equation (6) can be modeled within a spreadsheet in a several ways. One way is to use the above-mentioned method of inequality-based reduction. Another way is to use a method of three-dimensional virtual computing described elsewhere (Abramovich & Brouwer, 2003). Finally, one can utilize the method of level lines combined with appropriate use of inequalities. It is this method that will be discussed below. To this end, equation (6) can be rewritten in the form

$$z = nxy/(xy - n(x + y)) \qquad (7)$$

so that, by using the two-dimensional computational capacity of a spreadsheet, to generate the level lines $z=const$ of integer values; that is, given n, to find those integer pairs (x, y) for which z is an integer also. A computational problem of finding the largest and the smallest values for each of the variables x and y satisfying equation (6) gives rise to interesting activities for teachers in the use of inequalities.

To begin, note that the symmetric nature of equation (6) yields the inequalities $n<x\leq y\leq z$. Next, it might be helpful to start with generating integer values of z defined by equation (7) without regard to the computational efficiency of the environment. In that way, by modelling equation (7) for, say, $n\leq3$, one can get some intuitive ideas as to what estimates for x and y might prove to be helpful in generating solutions for $n>3$. In doing so, one can conjecture, that at most two denominators in the right-hand side of equation (6) may be greater than $3n$, and confirm this formally using proof by contradiction. Indeed, the assumption $3n<x\leq y\leq z$ results in the following false conclusion:
$1/n = 1/x + 1/y + 1/z < 1/(3n) + 1/(3n) + 1/(3n) = 1/n$. This suggests that

$$n+1\leq x\leq 3n \qquad (8)$$

Another computationally-driven conjecture is that at most one of x, y and z may be greater than $2n(n+1)$. In order to prove this conjecture, teachers can be introduced to an *indirect* proof a classic example of which can be found in (Polya, 1945). This kind of argument is a combination of proof by contradiction and *constructive* proof where contradiction is constructed by calculating an appropriate example. To this end, one can choose $k>1$, $y = 2n(n+1)+1$, $z = 2n(n+1) + k$, and then construct a contradictory inequality for such values of k, y and z. Indeed,

$1/n = 1/x + 1/(2n(n + 1) + 1) + 1/(2n(n + 1) + k) < 1/x + 2/(2n(n + 1) + 1) = 1/x$
$+1/(n(n + 1) + 0.5)$.

Therefore, $1/x > 1/n - 1/(n(n + 1) + 0.5) = (n^2 + 0.5)/(n(n^2 + n + 0.5))$,   whence
$x < n + n/(n^2 + 0.5) < n + 1$. This conclusion contradicts to the inequality $x \geq n+1$
yielding $y \leq 2n(n+1)$.

In order to find a lower estimate for y, once again, spreadsheet calculations without regard to their efficiency can be helpful in conjecturing that $y \geq 2n+1$. To prove this conjecture, one can further utilize an indirect proof. To this end, one may assume that for a certain integer $k \geq 1$ there exists an integer m, $k < m < n+1$, such that for $x = n+k$ and $y = n+m$ the value of y in equation (6), contrary to the above computationally-driven conjecture, satisfies the inequality $y < 2n+1$. These values of x and y yield

$1/z = 1/n - 1/(n + k) - 1/(n + m) = (n^2 + 2kn + km)/(n(n + k)(n + m))$ whence
$z = n(1 + n(m - k)/(n^2 + 2kn + km)) = n(1 + (m - k)/(n + 2k + km/n)) < 3n$. This, however, contradicts the inequality $z \geq 3n$ which, in turn, follows from the simple fact that if $x \leq y \leq z < 3n$, then $1/x + 1/y + 1/z > 1/n$. Therefore, the inequalities:

$$2n+1 \leq y \leq 2n(n+1) \tag{9}$$

determine a segment within which the variable y in equation (6) varies.

It should be noted that the ability to amend a model in order to ensure its contextual coherency, that is, to make a connection between mathematics and real world, is an important skill that characterizes one's modelling competence. With this in mind, one should take into account the contextual meaning of the variables involved. Namely, these variables should be related to each other not only to reflect the symmetric nature of equation (6) but also the fact that the sum of any two sides of a triangle is greater than the third side. This proposition (often taken for granted by teachers) is known as the triangle inequality, probably the most basic inequality in geometry. Thus, an earlier assumption $x \leq y \leq z$ considered in combination with geometry-driven relations (5) yields the following inequality

$$1/y + 1/z > 1/x \tag{10}$$

appropriately relating the three heights of a triangle. A qualitative analysis of inequality (10) suggests that in order to maintain a quantitative balance between its sides, x should not be too small and y should not be too large.

The spreadsheet pictured in Figure 1 incorporates inequalities (8)-(10) and it shows that Problem 2 has five solutions for $n=4$. Their distribution within the spreadsheet confirms the above remark about bounds for x and y and suggests that the lower bound for x and the upper bound for y can be improved. Considering jointly equation (6) and inequality (10), one can conjecture that $x > 2n$. The last inequality can, indeed, be obtained from the following formula

$$A = \left[\sqrt{\frac{1}{n}\frac{1}{n}(\frac{1}{x} - \frac{2}{n})(\frac{1}{y} - \frac{2}{n})(\frac{1}{z} - \frac{2}{n})} \right]^{-1} \tag{11}$$

which is a modification of a well-known Heron's formula

$$A = \sqrt{p(p - a)(p - b)(p - c)} \tag{12}$$

for area of triangle in terms of its sides a, b, c, and semi-perimeter p.

Formula (11), however, does not allow one to improve the upper bound for y. A computationally driven conjecture about such an improved bound could be obtained by altering the value of n in cell B2 (Figure 1). In doing so, one can come up with the inequality $y \leq 3.5n$ based on the results of this computational experiment. The inequalities

$$x \geq 2n+1 \text{ and } y \leq 3.5n \tag{13}$$

can then be incorporated in the spreadsheet environment (Figure 2) to further enable its computational efficiency. As an aside, note that equation (6), when considered as a decontextualized Diophantine equation, has 28 solutions; in other words, a unit fraction 1/4 can be represented as a sum of three like fractions in 28 ways (Abramovich, 2005).

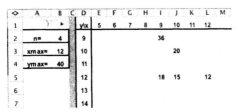

Figure 1. Modelling Problem 2 for $n{=}4$.   Figure 2. Using inequalities (13).

5. HERON'S FORMULA AS A WINDOW ON THE ARITHMETIC MEAN-GEOMETRIC MEAN INEQUALITY

In the context of inequalities, formula (12) can be used not only to support the improvement of an upper bound for variable x. It can serve as an introduction to a famous inequality relating arithmetic and geometric means of three non-negative numbers:

$$(q + r + s)/3 \geq \sqrt[3]{qrs} , \qquad q \geq 0, r \geq 0, s \geq 0 \tag{14}$$

By substituting $q{=}p{-}a$, $r{=}p{-}b$, $s{=}p{-}c$, one can apply inequality (14) – known as the arithmetic mean-geometric mean inequality – to the right-hand side of formula (12) as follows

$$A \leq \sqrt{p(p - a + p - b + p - c)^3 / 27} = \sqrt{p^4 / 27} = p^2 /(3\sqrt{3}) \tag{15}$$

Inequality (15) becomes an equality when $a{=}b{=}c$, a case of an equilateral triangle circumscribed about a circle of radius n. Due to equalities (5), $A{=}pn$; this, in combination with (15), yields $A \geq 3\sqrt{3}n^2$ with equality taking place for an equilateral triangle with height and side equal, respectively, to $3n$ and $2\sqrt{3}n$. In other words, in the context of Problem 2 this triangle has the smallest area. In particular, when $n{=}4$ one gets $(x, y, z){=}(12, 12, 12)$ – a triple associated with cell L5 (Figure 1). In such a way, Heron's formula can be utilized as a window on a tool of great importance for mathematics; by using this tool, one can avoid the application of differential calculus in solving a large number of problems of applied nature, including those concerned with specific metric properties of geometric figures.

6. CONCLUDING REMARKS

In this chapter, it has been demonstrated how a spreadsheet can be utilized as an agent of modelling activities associated with the use of inequalities. Through this agency, inequalities can be introduced to teachers as powerful tools in developing computationally efficient environments for modelling contextually diverse problematic situations associated with partition of integers and their reciprocals. A pedagogical interest in problems of that type stems from the focus of current standards for teaching mathematics on open-ended problem solving in which the multiplicity of correct answers is the rule rather than an exception. Partitioning problems, permeating all levels of K-12 curriculum, by their very nature, are concerned with finding more than one correct answer. Whereas it may not be a requirement for a pre-college classroom to find a complete set of multiple answers, it would be a reasonable expectation for the teachers to do so because some students may want to know how close their efforts are to the complete solution.

The author argues that the role of a secondary mathematics teacher may include the notion of being a resource person for his or her elementary level counterpart. In particular, such a person could be an expert in using technology for modelling problems with multiple answers. Such an expertise can be developed through a course on the applications of spreadsheets and other tools of technology with a focus on modelling contextually diverse problematic situations. That kind of course, as the author's experience indicates, may be of interest to both secondary and elementary pre-teachers and can become a research-like experience for them, as they use technology for both problem solving and problem posing. The educational objectives of the proposed approach include the demonstration of how "basic ideas of number theory and algebraic structures underlie rules for operations on expressions, equations, and inequalities" (Conference Board of the Mathematical Sciences, 2001, p40) as well as providing teachers with opportunities to "examine the crucial role of algebra in use of computer tools like spreadsheets and the ways that [technology] might be useful in exploring algebraic ideas" (p. 41). Assessment methods, being an integral part of the learning process (Burton, 1997), may include independent or group projects for teachers in using inequalities as tools in developing computationally efficient environments for spreadsheet modelling of Diophantine equations in two variables describing contextually diverse situations. Equations that can support such an assessment can be found elsewhere (Abramovich, 2005).

The projects may also include practice in associated proof techniques such as proof by contradiction, proof by construction, and computationally driven proof. The unity of mathematics, context, and computing is conducive to both formal and informal reasoning to be a part of secondary pre-teachers' experiences in mathematical modelling. This, in turn, facilitates the development of competence by the teachers in implementing "instructional programmes [that] enable all students to select and use various types of reasoning and methods of proof" (National Council of Teachers of Mathematics, 2000, p342) and using "increasingly sophisticated technological tools that permit more computationally involved applications" (Conference Board of the Mathematical Sciences, 2001, p37).

REFERENCES

Abramovich, S. (2005) Inequalities and spreadsheet modeling. *Spreadsheets in Education* [e-journal], 2, 1-22. Available at http://www.sie.bond.edu.au .

Abramovich, S. (2006) Spreadsheet modelling as a didactical framework for inequality-based reduction. *International Journal of Mathematical Education in Science and Technology,* 37, 527-541.

Abramovich, S. and Brouwer, P. (2003) Revealing hidden mathematics curriculum to pre-teachers using technology: The case of partitions. *International Journal of Mathematical Education in Science and Technology*, 34, 81-94.

Beberman, M. (1964) Statement of the problem. In D. Friedman (ed) *The Role of Applications in a Secondary School Mathematics Curriculum.* Proceedings of UICSM conference, Monticello, Ill., 14-19 February 1963. Urbana, Ill.: UICSM, 1-13.

Blum, W. (2002) ICMI Study 14: Applications and modelling in mathematics education – Discussion document. *Educational Studies in Mathematics*, 51, 149-171.

Burton, L. (1997) The assessment factor – by whom, for whom, when and why. In S.K. Houston, W. Blum, I.D. Huntley and N.T. Neil (eds) *Teaching and Learning Mathematical Modelling: Innovation, Investigation and Applications.* Chichester: Albion Publishing, 95-107.

Conference Board of the Mathematical Sciences. (2001) *The Mathematical Education of Teachers*. Washington, DC: Mathematical Association of America.

Goos, M. (1998) Technology as a tool for transforming mathematical tasks. In P. Galbraith, W. Blum, G. Booker and I.D. Huntley (eds) *Mathematical Modelling, Teaching and Assessment in a Technology-Rich World.* Chichester: Horwood Publishing, 103-113.

Jiang, Z., McClintock, E. and O'Brien, G. (2003) A mathematical modelling course for preservice secondary school mathematics teachers. In Q.-X. Ye, W. Blum, K. Houston and Q.-Y. Jiang (eds) *Mathematical Modelling in Education and Culture: ICTMA 10.* Chichester: Horwood Publishing, 183-196.

Lamon, S. J. (2003) Powerful modelling tools for high school algebra. In S.J. Lamon, W. A. Parker and K. Houston (eds) *Mathematical Modelling: A Way of Life.* Chichester: Horwood Publishing, 81-94.

Niss, M. (1989) Aim and scope of applications and modelling in mathematics curricula. In W. Blum, J.S. Berry, R. Biehler, I.D. Huntley, G. Kaiser-Messmer and L.Profke (eds) *Applications and Modelling in Learning and Teaching Mathematics.* Chichester: Ellis Horwood, 22-32.

National Council of Teachers of Mathematics. (2000) *Principles and standards for school mathematics.* Reston, VA: Author.

New York State Education Department (1998) *Mathematics Resource Guide with Core Curriculum.* Albany, NY: Author.

Polya, G. (1945) *How To Solve It.* Princeton, NJ: Princeton University Press.

Section 6
The Practice of Modelling

6.1

INTEGRATION OF ENERGY ISSUES IN MATHEMATICS CLASSROOMS

Astrid Brinkmann[1] and Klaus Brinkmann[2]
[1]University of Dortmund, Berufskolleg Iserlohn, Germany
[2]University of Trier, Umwelt-Campus Birkenfeld, Germany

Abstract–*For the young generation, it is indispensable to be concerned itself with the environmental consequences of the extensive usage of fossil fuels and to become familiar with renewable energies. It seems to be particularly suitable to treat future energy issues in mathematics education. However, there is a great lack of appropriate teaching material. Thus, the authors have developed a didactic concept with regard to content and structure of mathematical problems, allowing their direct and broad usage in classroom. On the basis of this concept, the authors have created several series of problems for the secondary mathematics classroom. The aim of this contribution is to present an overview of our project together with some examples of the developed problems.*

1. INTRODUCTION AND MOTIVATION

Especially the young generation would be more affected by the environmental consequences of the extensive usage of fossil fuels. The education of our children should heighten their consciousness of resulting problems. This would be one of the most effective methods of achieving a sustainable change from our presently existing power supply system to a system mainly based on renewable energy conversion. Moreover, for our children, it is essential for them to become familiar with renewable energies, because the decentralised nature of this future kind of energy supply surely requires more personal effort of everyone.

In comparison to the parental education, the schools can provide the possibility of a successful and more easily controlled contribution to this theme; especially, as in mathematics education it is possible to reach all students. In addition this would enhance mathematics education itself, as "the application of mathematics in contexts which have relevance and interest is an important means of developing students' understanding and appreciation of the subject and of those contexts" (National Curriculum Council 1989, ¶F1.4). Such contexts might be environmental issues that are of general interest to everyone. Hudson (1995) states that "it seems quite clear that the consideration of environmental issues is desirable, necessary and also very relevant to the motivation of effective learning in the mathematics classroom". One of the most important environmental impacts is that of energy conversion systems.

However, there is a great lack of mathematical problems suitable for school lessons. Especially, there is a need for mathematical problems concerning environmental issues that are strongly connected with future energy issues.[1] An additional problem is, that the development of such mathematical problems requires the co-operation of experts in future energy matters, with their specialist knowledge, and of mathematics educators, with their pedagogical content knowledge. In such a co-operation, the authors have developed a special didactical concept to open the field of future energy issues for students, as well as for their teachers.

2. DIDACTICAL CONCEPT

The cornerstones of the didactical concept, developed by the authors in order to promote renewable energy issues in mathematics classrooms, are:

- The problems are chosen in such a way, that the mathematical content required to solve them are part of mathematics school curricula.
- Advantageously every problem should concentrate on a special mathematical topic, such that it can be integrated in an existing teaching unit; as project-oriented problems referring to several mathematical topics are seldom picked up by teachers.
- The problems should be of a greater extent than usual, textually explained and posed, in order to enable the students, and also their teachers, to engage more intensively with the subject.
- The problems should not require of teachers special knowledge concerning future energy issues and nor, especially, physical matters. For this reason, all non-mathematical information and explanations concerning the problems' foundations are included in separated text frames.
- In this way, information relating to future energy issues is provided for both teachers and students, helping them to engage properly with the topic.

On the basis of this didactical concept, we have created several series of problems for the secondary mathematics classroom. They deal with topics of rational usage of energy, photovoltaic, thermal solar energy, biomass, traffic, transport, wind energy and hydro power.
(see: http://www.math-edu.de/Anwendungen/anwendungen.html).

3. EXAMPLES OF MATHEMATICAL PROBLEMS

For illustration, here we give short excerpts of some examples of mathematical problems concerning future energy issues. They are suitable for lessons in secondary schools. This paper is focused on the basic structure of the problems, in order to explain the didactical concept and the general principles.

3.1 Example 1 - The Problem of CO_2 Emission

➤ This is an inter-disciplinary problem linked to subject of mathematics as well as to chemistry, physics, biology, geography, and the social sciences. Nevertheless, it may be treated already in lower secondary classrooms.

With respect to mathematics the *conversion of quantities* is practised, knowledge of *rule of three²* and *percentage calculation* is required. The *amount of the annually in Germany produced* CO_2, especially also for the purpose of transport and traffic, is illustrated vividly, so that students become aware of it.

Info:

In Germany, each inhabitant produces, on average annually, nearly 13 t CO_2 (carbon dioxide). Combustion processes (for example from power plants or internal combustion engines in vehicles) are responsible for this emission into the atmosphere.

Assume now, this CO_2 would build up a gaseous layer which remains directly above the ground.

a) What height would this CO_2-layer reach in Germany after one year?

Hints:

- Helpful for your calculation is knowledge from chemistry lessons. There you learn that amounts of material could be measured with the help of the unit 'mole'. *1 mole CO_2 weights 44 g and takes a volume of 22,4 l*, under normal standard conditions (pressure 1013 hPa and temperature 0°C). With these values you can calculate approximately.

- You will find both the surface area and the population of Germany in a lexicon.

Help: Find the answers of the following partial questions in the given order.

i)   How many tons CO_2 are produced in Germany, in total, every year?

ii)   What is the volume in litres of this amount of CO_2? (Note the Hint!)

iii)   How many m³ CO_2 are annually produced in Germany? Express this in km³!

iv)   If the amount of CO_2 produced in Germany covered directly the ground as a lowest gas layer, what would be its height?

Info:

- In Germany, the amount of waste produced every year is nearly 1 t for each habitant (in private households as well as in industry), the average produced amount of CO_2 per habitant is therefore 13 times of this.
- The CO_2, which is produced during combustion processes and emitted into the atmosphere, distributes itself in the air. One part will be absorbed by the plants with the help of the photosynthesis; a much greater part goes into solution in the water of the oceans. However, the potential for CO_2 absorption is limited.
- In the 90's, 20% of the total CO_2-emissions in Germany came solely from internal combustion engines of vehicles (traffic).

b) What height would the CO_2-layer over Germany be, if this layer results only from the annual emissions from individual vehicles? How many km³ CO_2 is this?

3.2 Example 2 - The Problem of Planning Solar Collector Systems

➢   This problem deals with calculations for planning *solar collector systems* by
     using *linear functions*. The understanding and usage of *graphical
     representations* is performed.
Info:

Energy is measured by the unit kWh. An average household in Germany consumes
nearly 16.000 kWh thermal energy per year.
1 l fossil oil provides approximately 10 kWh thermal energy. The combustion of 1 l
oil produces nearly 68 l CO_2.

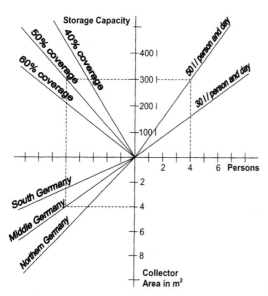

Figure 1. Dimensioning Diagram.

The diagram (Figure 1) provides data for planning a solar collector system for a private household. It shows the dependence of the needed collector area, for houses in different regions of Germany, for the number of persons living in the respective household, for the desired amount of warm water per day and person, as well as the desired coverage of the needed thermal energy by solar thermal energy (in percentages). Example: In a household in middle Germany with 4 persons and a consumption of 50 l warm water per day for each one, it follows that for a reservoir of 300 l and an energy coverage of 50%, a collector area of 4 m² is needed.

a) What collector area would be needed for the household you are living in? Which
   assumptions do you need to make first? What would be the minimal possible
   collector area, what would be the maximal one?
b) On a house in southern Germany, there is installed, a collector area of 6 m² that
   provides 50% of the produced thermal energy. How many persons could be
   supplied with warm water in this household?
c) Describe, by a linear function, the dependence of the storage capacity on the
   number of persons in a private household. Assume, first, a consumption of 50 l
   warm water per day and person, and second, a consumption of 30 l. Compare the
   terms of the two functions, illustrating also their graphical representation.

d) Show, in a graphical representation, the dependence of the collector area on a chosen storage capacity, assuming a thermal energy coverage of 50% for a house in middle Germany.

3.3 Example 3 – Photovoltaic Plant and Series Connected Efficiencies

➢ The aim of this problem is to make students familiar with the *principle of series connected efficiencies*, as they occur in complex energy conversion devices. As an example, an off-grid *photovoltaic plant* for the conversion of solar energy to AC-current as a self-sufficient energy supply is considered. The problem is appropriate for use in a teaching unit on *fractions*.

Figure 2 shows the components of an interconnected energy conversion to build up a self-sufficient electrical energy supply. This kind of supply system is of special interest for developing countries, but also useful for buildings in rural off-grid areas.

Figure 2 sets out the production of electrical energy with solar radiation with the help of a solar generator for off-grid applications. In order to guarantee a gap-free energy supply for times without sufficient solar insulation, a battery as an additional storage is included.

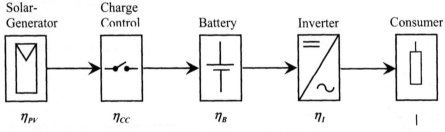

Solar-Generator	Charge Control	Battery	Inverter	Consumer
η_{PV}	η_{CC}	η_B	η_I	

Figure 2. Off-grid Photovoltaic Plant.

Info:

The components of an off-grid photovoltaic (PV) plant are 1) a solar generator, 2) a charge control, 3) an accumulator and optional for AC-applications, 4) an inverter.
The solar generator converts the energy of the solar insulation into electrical energy as a direct current (DC). The electricity is passed to a battery via a charge control. From there it can be converted directly or partly after storage in the battery, to alternating current (AC), such as it is needed by most electrical devices. Unfortunately, it is not possible to use the insulated energy without losses. Every component of the conversion chain produces losses, so only a fraction after each component would be the energy input for the following component. The efficiency η of a component is defined by $\eta = \dfrac{\text{Power going out}}{\text{Power coming in}}$. Power is the energy converted in 1 second. It is measured by the unit W or kW (1 kW = 1000 W).

Assume in the tasks a), b) and c) that the whole electric current is first stored in the battery before it reaches the consumer.

a) Consider that the present insulation on the solar generator would be 20 kW. Calculate the out going power for every component of the chain, if:

Mountain hut with PV plant

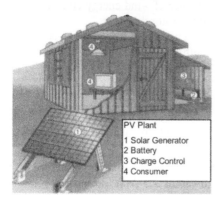

Figure 3. Illustrations of Off-grid Photovoltaic Plants[3].

$$\eta_{PV} = \frac{3}{25}, \ \eta_{CC} = \frac{19}{20}, \ \eta_B = \frac{4}{5} \ \text{and} \ \eta_I = \frac{23}{25}.$$

b) Which is the total system efficiency $\eta_{total} = \dfrac{\text{gained power for the consumer}}{\text{insolated power}}$?

How can you calculate η_{total} by using only the values of η_{PV}, η_{CC}, η_B and η_I? Give a formula for this calculation.

c) Transform the efficiency values given in a) into decimal numbers and as percentages. Check with these numbers your result obtained in a).

d) How do the battery efficiency and the total system efficiency change, if only $1/3$ of the electric power delivered by the charge control, would be stored in the battery and the rest of $2/3$ goes directly to the inverter? What is your conclusion from this?

3.4 Example 4 - Wind Energy Converter and Geometry

➤ This problem deals with *wind energy converters*. It can be treated in lessons of *geometry, especially calculations of circles*. The *conversion of quantities* is practised.

Info:

The nominal power of a wind energy converter depends upon the rotor area A with the diameter D as shown in [4]Figure 4.

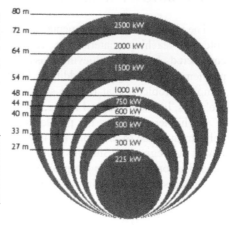

Figure 4. Nominal Power related to the Rotor Area.

a) Interpret the meaning of Figure 4. Show the dependence of the nominal power of wind energy converter on the rotor diameter and, respectively, on the rotor area by graphs in co-ordinate systems.

b) Find the formula, which gives the nominal power of wind energy converter as a function of the rotor area and respectively of the rotor diameter.

c) What is the rotor area that you would expect to need for a wind energy converter with a nominal power of 3 MW? What length should the rotor blades have for this case?

Info:

In Germany, wind energy converter produce in average their nominal power during approximately 2000 hours of a year, with respect to sufficient wind energy conditions.

d) Calculate the average amount of energy in kWh, which would be produced by a wind energy converter with a nominal power of 1.5 MW during one year in Germany.

Info:

An average household in Germany consumes nearly 4000 kWh electrical energy per year.

e) How many average private households in Germany could theoretically be supplied with electrical energy by a 1.5MW wind energy converter? Why do you think, that this could only be a theoretical calculation?

f) What percentage of the year would on average be covered with a nominal power rate supply of a wind energy converter in Germany?

g) Assume the nominal power of a 600kW energy converter would be reached at a wind speed of 15m/s, measured at the hub height. How many km/h is this? How fast are the movements of the tips of the blades, if the rotation speed is 15 revs/min. Give the solution in m/s and km/h respectively. Compare the result with the wind speed.

4. CLASSROOM IMPLEMENTATION

The problems to environmental issues developed by the authors must be seen as potential teaching material. In each case, the students' abilities have to be considered. In lower achieving classes it might be advisable not to present every problem in full length. In addition, lower achievers need a lot of help in solving complex problems that require several calculation steps. The help given in some problems, like in example 1 above, addresses such students.

To higher achievers the problems should be presented without some included help. In the case of example 1 it might be even of benefit not to present the given hint at the beginning. Students would therefore have to find out which quantities are needed in order to solve the problem. The problem would become more open and the students would be more involved in modelling processes.

As the intention of the authors is also an informal one, in order to give more insight in the field of future energy issues, the mathematical models and formulae are mostly given in the problem texts. Students are generally not expected to find out by themselves the often complex contexts; these are already presented thus guaranteeing realistic situation descriptions. The emphasis in the modelling activities rather lies in the demanded argumentation and interpretation processes, with the aim that mathematical solutions lead to a deeper understanding of the issues studied.

Classroom experiences show that students react in different ways to the problem subjects. While some are horrified by finding out, for example, that the worldwide oil reserves are running low already in their life time, others are unmoved by this fact, as twenty or forty years in the future is a time they do not worry about. In school lessons, there are again and again situations in which students drift away in political and social discussions according to the problem contexts. Although desirable, this would sometimes lead to too much time losses for mathematics education itself. Cooperation with teachers of other school subjects would be profitable, if possible.

5. EXPERTS' AND TEACHERS' REACTIONS

The didactical concept described above, and examples of mathematical problems according to this were first published at the '12. Internationales Sonnenforum 2000' in Freiburg (Brinkmann & Brinkmann, 2000). Meanwhile several further presentations followed, within the frameworks of international technical conferences on renewable energies as well as in conferences on didactics on mathematics (Brinkmann & Brinkmann, 2001abc, 2002abc, 2003, 2004) and teacher education events (http://www.math-edu.de/Anwendungen/Energie-Vortraege.htm).

These activities were supported by many positive reactions from experts, as well as from teachers. While experts in future energy matters are familiar with the environmental consequences of the extensive usage of fossil fuels, we found that teachers often underestimate the problems. Thus, teachers were very pleased to get some information and deeper insights in the topic of future energy, and the discussions resulted in a wide consensus about the importance of treating future energy issues in secondary school classrooms.

The didactical concept presented was absolutely convincing. Especially the information frames included in the problems were highly accepted by the teachers. With respect to the implementation of the didactical concept in the presented problems, of course, teachers sometimes had doubts when thinking of their special courses and their own students. Here it was of importance to indicate optional possibilities how to suit the problems according to students abilities (see §4).

6. OUTLOOK AND FINAL REMARKS

Meanwhile, the collection of worked out problems has been edited in German language as a special text book for mathematics school education (Brinkmann & Brinkmann, 2005). However, to be used in school lessons the acceptance and supporting promotion of experts as well as politicians and educators is very important.

In order to integrate future energy issues in curricula of public schools, several initiatives have already been started in Germany, supported and co-operated by the 'Deutsche Gesellschaft für Sonnenergie e.V. (DGS)', the German section of the ISES (International Solar Energy Society). The wide spectrum of respective activities going on has been presented amongst others also on the 3rd Solar Didactica within the Solar-Energy World Exposition, in Berlin, 2001. This event took place under the patronage of the German Minister for Education and Research, E. Bulmahn, expressing thus the great interest and importance accorded by politicians.

Across Europe there is a project, named "SolarSchools Forum", with the aim to integrate future energy issues in curricula of schools. In the frame of this project the German society for solar energy DGS points out the teaching material the authors created (http://www.dgs.de/747.0.html). Most of this material is only available in German language yet. It would be of great importance to make these materials accessible also in other languages.

Although the education in the field of future energy issues is of general interest, the project that we present in this paper seems to be the only greater activity focusing especially on *mathematics* lessons. Thus it would be desirable to develop further materials for mathematics lessons.

NOTES

1. One exception is Böer's (1999) work on the topic of parabolic sun collectors.
2. that is, calculation with proportional quantities.
3. Source: Energieplus e. V., Umweltministerium M ecklenburg-Vorpommern
4. Source: Danish Wind Industry Association 2002
   www.windpower.org/de/core.htm .

REFERENCES

Böer, H. (1999) Konzentrierende Kollektorsysteme. *MUED-Schriftenreihe Unterrichtsprojekte.*

Brinkmann, A. and Brinkmann, K. (2000) Möglichkeiten zur Integration des Themas Regenerative Energien in einen fachübergreifenden Mathematikunterricht. In DGS München, International Solar Energy Society – German Section (eds) *12. Internationales Sonnenforum 2000 in Freiburg, 5-7 July 2000.* München: Solar Promotion GmbH Verlag.

Brinkmann, A. and Brinkmann, K. (2001a) Future Energy Issues in the Secondary Mathematics Classroom. In M. Tzekaki (ed) *Proceedings of the 5th Panhellenic Conference with International Participation on Didactics of Mathematics and Informatics in Education. Oct. 12 - 14, 2001 in Thessaloniki, Greece,* 589-599.

Brinkmann, A. and Brinkmann, K. (2001b) Aufgaben für einen fachübergreifenden Mathematikunterricht zum Thema Photovoltaische Solarenergie. Problems for Applied School Mathematics Concerning the Topic of Photovoltaic Solar Energy. In: OTTI Energie-Kolleg (eds) *16. Symposium Photovoltaische Solarenergie, 14.-16. March 2001 in Kloster Banz, Staffelstein.* Regensburg: OTTI, 114-118.

Brinkmann, A. and Brinkmann, K. (2001c) Electric Vehicles as a Topic for Applied School Mathematics. *The 18th International Electric, Fuel Cell and Hybrid Vehicle Symposium and Exhibition EVS 18 – The World's Largest Event for Electric Vehicles - Proceedings. October 20-24, 2001 in Berlin, Germany.*

Brinkmann, A. and Brinkmann, K. (2002a) Biomass for Future Energy as a Topic in Secondary Mathematics Classrooms. *Proceedings of the 12th European Conference and Technology Exhibition on Biomass for Energy, Industry and Climate Protection. 17 – 21 June 2002 in Amsterdam, The Netherlands.*

Brinkmann, A. and Brinkmann, K. (2002b) Promoting Renewable Energy Issues in Secondary Mathematics Classrooms. In A.A.M. Sayigh (ed) *Renewable Energy. Renewables: World's Best Energy Option. Proceedings of the World Renewable Energy Congress VII. 29 June – 5 July 2002 in Cologne, Germany.* Amsterdam: Pergamon Elsevier Science Ltd.

Brinkmann, A. & Brinkmann, K. (2002c) Wind Energy in Secondary Mathematics Classrooms. *Proceedings of the 1st World Wind Energy Conference and Exhibition. 04 – 08 July 2002 in Berlin, Germany.*

Brinkmann, A. and Brinkmann, K. (2003) Integration der Themen „rationelle Energienutzung" und „regenerative Energien" in einen fachübergreifenden Mathematikunterricht. Begründung – Didaktisches Konzept – Aufgabensammlung. In H.-W. Henn (ed) *Beiträge zum Mathematikunterricht 2003.* Hildesheim, Berlin: Franzbecker, 145-148.

Brinkmann, A. and Brinkmann, K. (2004) Solar Thermal Energy as a Topic in Secondary Mathematics Classrooms. *Proceedings of EUROSUN 2004, 14. Internationales Sonnenforum – The 5th ISES Europe Solar Conference, 20–23 June 2004 in Freiburg, Germany.* Vol. 3. Freiburg: PSE GmbH, 636–645.

Brinkmann, A. and Brinkmann, K. (2005) *Mathematikaufgaben zum Themenbereich Rationelle Energienutzung und Erneuerbare Energien.* Hildesheim, Berlin: Franzbecker.

Danish Wind Industry Association (2002) *http://www.windpower/dk* , http://www.windpower.org/de/core.htm.

Hudson, B. (1995) Environmental issues in the secondary mathematics classroom. *Zentralblatt für Didaktik der Mathematik*, 27, 95/1, 13-18.

National Curriculum Council (1989) *Mathematics Non-Statutory Guidance.* York: National Curriculum Council.

6.2

MODELS OF ECOLOGY IN TEACHING ENGINEERING MATHEMATICS

Norbert Gruenwald[1], Gabriele Sauerbier[1],
Tatyana Zverkova[2] and Sergiy Klymchuk[3]
[1]Wismar University of Technology, Business and Design,
Germany
[2]Odessa National University, Ukraine
[3]Auckland University of Technology, New Zealand

Abstract–*This paper describes a recent joint project of three universities from Germany, Ukraine and New Zealand. The project is based on innovative pedagogical strategies in teaching applications in first-year university engineering mathematics. The aim of the project is to develop a set of real ecology models suitable for first-year engineering students in order to involve students in solving real practical problems from the first year of their study, motivate them by showing the relevance of the mathematics that they learn in their course and encourage them to pay attention to environmental issues. Two models that are presented in the paper were given as a project to the first-year engineering students in their mathematics course in the Wismar University, Germany and the Auckland University of Technology, New Zealand. A questionnaire was given to the students to find out their difficulties with the project and their attitudes towards using it as a part of assessment. The students' responses to the questionnaire and comments are presented and analysed in the paper.*

1. INTRODUCTION

There are many papers devoted to investigating first-year undergraduate students' competency in different steps of the mathematical modelling process. We mention some recent research. A measure of attainment, for stages within a modelling cycle, has been developed by (Haines & Crouch, 2001). Crouch and Haines (2004) expanded their study to compare undergraduates (novices) and engineering research students (experts). They suggested a three level classification of the developmental processes which the learner passes in moving from novice behaviour to that of an expert. One of the conclusions of that research was that "students are weak in linking mathematical world and the real world, thus supporting a view that students need much stronger experiences in building real-world mathematical world connections" (Crouch & Haines, 2004). These results are consistent with the findings from a study of 500 students from 14 universities in Australia, Finland, France, New Zealand, Russia, South Africa, Spain, Ukraine and the UK (Klymchuk & Zverkova, 2001). The study indicated that the students found it difficult to move from the real world to the mathematical world because of lack of practice in application tasks. An investigation of undergraduate students' working styles in a mathematical modelling activity has been done by Maull and Berry (2001) and a further study by Nyman and Berry (2002) addresses the development of transferable skills in undergraduate mathematics students through mathematical modelling. Some relationships between students' mathematical competencies and their skills in modelling were considered by Galbraith and Haines (1998) and by Gruenwald and Schott (2000). Kadijevich pointed out at an important aspect of doing even simple mathematical modelling activity by new coming undergraduate students: "Although through solving such ... [simple modelling] ... tasks students will not realise the examined nature of modelling, it is certain that mathematical knowledge will become alive for them and that they will begin to perceive mathematics as a human enterprise, which improves our lives" (Kadijevich, 1999). In this paper we will consider first-year engineering students' feedback on the usage of real ecology models developed by professional mathematicians working in industry in assessment. Although the students were given almost complete models, they had to go through some stages of the mathematical modelling process. In particular, in Model 1 they had to finish the formulation step and set up the (differential) equation describing the model, in Model 2 they were asked to write an extensive interpretation, in both models they had to find the solution analytically and in Model 2 also with the help of a computer program.

2. STUDY

2.1 Description of the Models

Our intentions in giving such models to the students as a project were:
- to involve the students in solving real practical problems from the first year of their study
- to motivate them by showing relevance of mathematics they learn in their course
- to encourage them to pay attention to environmental issues.

The models have the following features:
- Each model is ecological. Environmental issues are getting more and more important for many human activities worldwide. This, on the first glance non-traditional area of application for engineering students, will help them to broaden their vision and prepare them to take responsibility in future because nearly every engineering activity relates to the environment. Chances are that most engineering graduates will be dealing with those issues in one way or another at their work place. Environmental issues cannot be overestimated. We face discussions on pollution, limited natural resources, climate change and other important ecological issues every day. "In the course of the 20th century, the relationship between mankind and nature underwent fundamental change. Human activity impacted life on Earth to a much greater extent than ever before. Population growth and the constant development and application of new

technologies are the reasons for this impact. Both result in growing demand for resources, which in turn means that the Earth's climatic equilibrium is in danger of being knocked out of balance. Global ecological stability is at risk."
Helmholtz Association, Germany
http://www.helmholtz.de/en/Research_Fields/Earth_and_Environment.html .

- Each model is adjusted to the region where the students study. For example, in case of New Zealand, Water Quality Control in *Taupo* Lake and Trout Fishing in *Rotorua* Lake are considered in the models. We assume that this psychological strategy will help students to relate to the models in a *personal* and an *emotional* way and increase their motivation and enthusiasm.
- Each model is developed by professional mathematicians working in industry and is based on a real practical problem.
- Each model is adapted and presented in a way understandable by first-year engineering students.
- Each model is a little bit beyond the scope of the mathematics course the students study. So they need to learn on their own some new concepts. For example, a model can be based on separable differential equations that students study but further investigation of equilibrium solutions and stability that are not covered in the course is required. We assume that this discovery learning strategy can help the students to enhance their investigation and research skills.
- Each model has questions that require analytical solutions and in some cases applications of programs such as Omnigraph or Matlab or Maple.
- Each model is mathematised to a large extent. So the students do not go through the first few steps of mathematical modelling process (collecting the data, making assumptions, formulating the mathematical model, etc.). But, apart from solving the given mathematical models, they do practice in other important steps of mathematical modelling process including interpreting the solutions and communicating the findings through writing the report.

2.2 Examples of the Models

Below are two models that were giving to the students after they learnt the differential equation topic in their Calculus course. The models can be easily adjusted to any country and region by selecting a nearby lake, putting its name in the title of the model and entering the corresponding values of the parameters into the model.

1. Model of Water Quality Control in Taupo Lake (adapted from Biswas, 1976)

Polluted water enters Taupo Lake from a recently built factory at a constant rate N. A mathematical model of concentration of pollution has been developed under certain assumptions including the following:
- the upper levels of water are mixed in all directions
- change in mass of pollution is equal to the difference between the mass of the entering pollution and the mass of the pollution which is being decomposed
- the rate of decomposition of pollution is constant

- decomposition of pollution takes place due to biological, chemical and physical processes and/or exchange with the deeper levels of water.

Under the assumptions the equation of the balance of mass of pollution on any interval of time Δt can be written in the form:

$$V\Delta C_N = N\Delta t - Q\, C_N\, \Delta t - KV\, C_N\, \Delta t, \tag{1}$$

where V – volume of the upper levels (constant), Q – water consumption rate (constant), $C_N = C_N(t)$ - concentration of pollution at time t, K – decomposition rate (constant).

Questions:

1. Set up a differential equation for the concentration of pollution from equation (1) by dividing both sides of equation (1) by Δt and taking a limit when $\Delta t \to 0$.
2. Solve that differential equation to find the concentration of pollution as a function of time provided that the initial concentration of pollution was zero.
3. Determine the equilibrium concentration of pollution C_{Ne} (that is the concentration when $t \to \infty$ or $\dfrac{dC_N}{dt} = 0$).
4. Determine time needed to reach p portion ($p = C_N\,(t)/C_{Ne}$) of the equilibrium concentration. Will it take more time to reach p portion of the equilibrium concentration in case when there is no decomposition of pollution?
5. Set up a differential equation for the concentration of pollution from the differential equation in question 1 in case when the initial amount of pollution entered the lake was C_o and after that pollution is not coming to the lake anymore (that is $N = 0$).
6. Solve the differential equation from question 5.
7. Determine time needed to reach the $(1-p)$ portion ($C(t)/C_o = 1 - p$) of the initial concentration C_o.

2. Model of Trout Fishing in Rotorua Lake (adapted from Arnold, 2004)

The differential equation

$$dx/dt = (1 - x)x - c \tag{2}$$

describes utilization of the fish population, where $x(t)$ is the portion of the maximum possible amount of fish in the lake (when there was no fishing) and t is time. The constant parameter $c > 0$ is the proportion of the fish population allowed for fishing. It characterises the allowed rate of the fishing-ground (intensity) and is called a quota. Choosing the values of the parameter c is an important factor of the control of the fish population.

Questions:
1. Solve equation (2). Separating the variables and completing the perfect square of the quadratic you will get 3 different solutions for different values of parameter c:

    a) $0 < c < 1/4$ b) $c = 1/4$ c) $c > 1/4$.

In cases a) and c) you need to choose and apply two appropriate formulae from the Table of Integrals of your main textbook or other Calculus textbooks. You can leave the solutions in an implicit form, that is not making $x(t)$ the subject.

2. Find the number of *equilibrium* solutions of equation (2) depending on the value of parameter c, that is, solve the equation:
$$dx/dt = 0 \quad \text{or} \quad -x^2 + x - c = 0.$$

3. Draw direction fields and integral curves using Omnigraph or Matlab or Mathematica or Maple or any other program for the following values of parameter c: a) 1/6; b) 1/4; c) 1/3.
    In each case a), b) and c) draw 10 integral curves for the following values of the initial condition: $x(0) = 0.1; 0.2; 0.3; 0.4; 0.5; 0.6; 0.7; 0.8; 0.9; 1$. For each equilibrium solution state whether it is *stable* or *unstable*.

4. Interpret the equilibrium solutions by giving recommendations about permitted quotas for fishing.

2.3 The Questionnaire

After completing the project the students from the Wismar University, Germany and the Auckland University of Technology, New Zealand were given the following questionnaire.

Question 1. Do you find the project to be practical?
    a) Yes   Please give the reasons:
    b) No   Please give the reasons:
Question 2. Do you find the project to be relevant and useful for your future career?
    a) Yes   Please give the reasons:
    b) No   Please give the reasons:
Question 3. What were the hardest things for you to do in the project? Why?

3. THE RESULTS OF THE STUDY

The results in both universities were so similar that we decided to combine them. The total number of students who completed the project was 147 in both countries. Participation in the study was voluntary. The number of students who answered the anonymous questionnaire was 63 so the response rate was 43%.

3.1 A brief Statistics and Common Students' Comments

Question 1. Practical?   Yes – 48%
Selected comments:
  •   the models describe the real world
  •   a good way of increasing students interest in the subject
  •   it was so helpful for my other subjects
  •   I didn't realize modelling is used for fishing quotas. It also helped me realize the effects of sneaky illegal fishing (which most of us have done)

Question 1. Practical?   No – 52%

Selected comments:
- it is not possible to calculate the nature
- it did give a practical situation but you barely think about that at all when doing the assignment

Question 2. Relevant for your career?   Yes – 35%
Selected comments:
- mathematics is the base needed to go into the Engineering World, so it will help a lot
- in engineering, we will be dealing with these kind of situations
- we are more motivated to solve such real problems than working with dry examples
- everything you learn is bound to be beneficial at some point

Question 2. Relevant for your career?   No – 65%
Selected comments:
- I don't see how it relates to mechanical or electrical engineering (most common comment)
- I don't compute formulas, I have to calculate beams…

Question 3. Hardest things?
- The assignment was difficult - 77% (this question actually was not asked)
- Understanding the questions – 32%
  "Nothing really, probably need more time to sit down and read the question"
- Using computer program – 24%
- Interpretation of the solutions – 17%

3.2 Students' Challenges

Below we describe most common questions asked by students while doing the project:
- The most common question was related to Model 1 where the students were asked to "set up a differential equation for the concentration of pollution from equation (1) by dividing both sides of equation (1) by Δt and taking a limit when $\Delta t \to 0$ ".

The most common question was: what is $\lim\limits_{\Delta t \to 0} \dfrac{\Delta C_N}{\Delta t}$?

It was a big surprise for us. Many students could not see the derivative there and for that reason could not set up the differential equation. They learnt the definition of the derivative at school and revised it earlier in the course. They did some applications in a familiar context, for example when defining the instantaneous velocity: $v = \lim\limits_{\Delta t \to 0} \dfrac{\Delta s}{\Delta t}$.

But facing an unfamiliar context and notations they could not make a connection between the topic they studied earlier in the course and the current topic on differential equations. It looks as though those students learn the concepts in isolation. This students' challenge was related to the "formulation" phase of the modelling cycle – the hardest phase for non-mathematics major students according to the findings of Klymchuk and Zverkova (2001). This was in spite of detailed guidance given in that question.

- Another common question was about a simple technique: how to complete the perfect square in the quadratic $x^2 - x + c$? Many students forgot this technique from their intermediate school years.
- The third most common question was:

  which of the formulae $\int \dfrac{dx}{x^2 + a^2}$ or $\int \dfrac{dx}{x^2 - a^2}$

  is the right formula for the integral $\int \dfrac{dx}{(x - \frac{1}{2})^2 + c - \frac{1}{4}}$?

  Those formulae were not covered in the course so the students had to choose the correct formula from the table of integrals in their textbook and apply the substitution method.
- Next common question was about how to use the computer program to draw direction fields and integral curves in Model 2. The students from the Wismar University had weekly laboratories where they studied Matlab. Still they needed much help in producing the diagram below (Figure1).

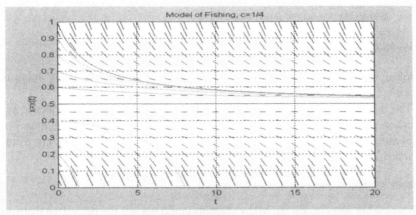

Figure 1. Direction fields of equation (2) when c = ¼.

The students from the Auckland University of Technology did not study any software as part of their mathematics course so they were advised to use a user-friendly graph drawing package Omnigraph. A brief introduction of the package was given in the lecture and the students were able to produce a similar diagram using Omnigraph. The above 3 students' challenges were related to the "solving" phase of the modelling cycle – the first two to the pure mathematics techniques and the third to the use of technology.

- Finally, many students had problems with the interpretation stage. The concepts of the equilibrium solutions and stability were not part of the course, so many students were challenged to produce the interpretation like this:"When $c = 1/4$ there is one unstable equilibrium solution. Fishing with such a quota is optimal when the initial proportion of the fish population is sufficiently large ($x > 1/2$) and mathematically is possible for as long as we need. However, the quota $c = 1/4$ is not allowed, since any small decrease of the proportion of the fish population in the equilibrium leads to the situation when the fish population would be completely fished out in a finite period of time."

The last students' challenge was related to the "interpretation" phase of the modelling cycle. Although only 17% of the students reported that they had difficulties with interpreting in fact there were many more of them who had such problems. This is because the students filled the questionnaire before they received their marked projects back so they thought that their interpretations were reasonable.

4. CONCLUSIONS

After consulting the students in the project, analysing its results and the students' comments and answers to the questionnaire we made the following conclusions:

- Our assumption that the engineering students will relate to *important ecology* problems from *their region* in *personal* and *emotional* way was too optimistic. One reason could be that both universities are located in ecologically good regions and countries. Another reason might be too brief and superficial introduction of the project by their lecturers.
- The majority of the students (77%) had another name for *discovery learning strategy* – "too hard".
- Many students are not used to learn on their own.
- Many students cannot select essential information for solving a particular question.
- Many students can't apply their knowledge from one part of the course to another (for example, the concept of the derivative for setting up the differential equation).

At the same time we should not be too critical to the students – after all they are in their first year at university. We as lecturers have learnt good lessons from the project. For example, we should use every opportunity to show connections between different parts of the course, we should use a variety of different notations for relationships between variables. But the main lesson we have learnt was the importance of a proper detailed introduction of the project to the students and a discussion of the results. A vital part in both the introduction and the discussion of the results would be creating a good learning environment, in particular explaining the benefits for the students from doing the project:

1. Improving independent learning
2. Enhancing modelling skills (regardless of the context!)
3. Developing problem solving skills on real problems
4. Acquiring useful computer skills

5. Boosting confidence

Those valuable skills will definitely pay-off the students in the future – in other courses and at their work places.

REFERENCES

Arnold V.I. (2004) *"Hard" and "Soft" Mathematical Models*. MCCME Press, Moscow (in Russian).

Biswas, A.K. (ed) (1976) *Systems Approach to Water Management*. McGraw Hill, Inc.

Crouch, R. and Haines, C. (2004) Mathematical modelling: transitions between the real world and the mathematical world. *International Journal of Mathematics Education in Science and Technology*, 35 (2), 197-206.

Galbraith, P. and Haines, C. (1998) Some mathematical characteristics of students entering applied mathematics courses. In J.F. Matos et al. (eds) *Teaching and Learning Mathematical Modelling*. Chichester: Albion Publishing, 77-92.

Gruenwald, N. and Schott, D. (2000) World mathematical year 2000: Challenges in revolutionasing mathematical teaching in engineering education under complicated societal conditions. *Global Journal of Engineering Education,* 4 (3), 235-243.

Haines, C. and Crouch, R. (2001) Recognising constructs within mathematical modelling. *Teaching Mathematics and its Applications,* 20 (3), 129-138.

Kadijevich, D. (1999) What may be neglected by an application-centred approach to mathematics education? *Nordisk Matematikkdidatikk*, 1, 29-39.

Klymchuk, S.S. and Zverkova T.S. (2001) Role of mathematical modelling and applications in university service courses: An across countries study. In J.F. Matos et al. (eds) *Modelling, Applications and Mathematics Education – Trends and Issues*. Ellis Horwood, 227-235.

Maull, W. and Berry, J. (2001) An investigation of students working styles in a mathematical modelling activity'. *Teaching Mathematics and its Applications,* 20 (2), 78-88.

Nyman, A. and Berry, J. (2002) Developing transferable skills in undergraduate mathematics students through mathematical modelling. *Teaching Mathematics and its Applications,* 21 (1), 29-45.

6.3

MODELLING AS AN INTEGRATED PART OF THE CLASS ON CALCULUS

Adolf Johannes Riede
Ruprecht-Karls-Universität, Heidelberg, Germany

Abstract–*This paper reports on the practice of teacher education in modern teaching and learning methods including mathematical modelling. The report may encourage teachers and their educators and trainers to overcome frequently-cited obstacles to learning mathematics.*

1. INTRODUCTION

During recent decades, many teachers have reported on the difficulties of enhancing interest, motivation and joy in mathematics and finally of achieving understanding and competency in using, applying and teaching mathematics. On the other hand, teachers and researchers have found new insights into the learning process and have developed new learning and teaching methods and tested them successfully in practice. See, for example, the references of this article. During a didactical exercise course at the University of Heidelberg on calculus in the classroom for future high school teachers H. Eichhorn (Dietrich-Bonhoeffer-Hochschule, Weinheim) and the author used the new methods. We report on the organisation, the tuition and the students' performance. It is shown how students can rediscover and practise mathematical notions and theorems by transitions from real-world situations.

2. LEARNING AND TEACHING METHODS AND ORGANIZATION

Eichhorn gave an introductory talk where he started from actual educational discussions, didactical research, technological developments, and basic experiences with mathematics as a subject and with its teaching. Since then, in the German state of Baden-Württemberg (See http://www.ls-bw.de/allg/lp/bpgykurs.pdf, p189ff or http://www.rzuser.uni-heidelberg.de/~jx8/NeueMeth.html) many of the points have been accepted as standards. Here, the author promotes the following additional point: instead of considering the lack of understanding as a sign of missing ability, accept this lack of understanding as a challenge for didactical improvements.

In the following, the participants of the course are called students, in contrast to high school students, who are called pupils. Instead of a traditional talk, every participating student undertook the task of playing the part of the teacher and holding a modern lesson with the other participants in the roles of the pupils- that is, they had to train a pupil-centred and teacher-guided class. The speakers had to study one to two articles. To help them to read with comprehension, they received from us a list of questions, suggestions and exercises to be solved by them. They had to design some simple exercises for the pupils to include them in the studying process during the talk, and to initiate discussions. Furthermore, they had to organise a group exercise for the end of their presentation. Fourteen days before the presentation, a first written version of their presentation had to be handed in. Tuition was available throughout, and from this point on, the participant received intensive tuition, often by e-mail. At the end the final version of the written paper was to be handed in.

3. GROWTH, EULER NUMBER AND EXPONENTIAL FUNCTION

Already at the very beginning of the introduction of the Euler-number the students can rediscover the sequences $(1+\frac{bt}{n})^n$ and $(1+\frac{1}{n})^n$ by modelling a real-world problem, namely growth under uniformly distributed and time-invariant conditions. For short we shall speak of *uniform growth*. Such conditions can be obtained in a laboratory, where temperature, light and nutriments can be kept constant and unvarying from one point of the container to another. Furthermore, it can be assumed that for some time there is no spatial restriction to the growth. Let $x(t)$ denote the biomass at time t measured in suitable units.

Discrete models: It is confirmed by experiments that the biomass grows within a time interval of length Δt by a fixed factor $q > 1$ where q is independent of t and of the actual value $x(t)$ of the biomass.

$$x(t+\Delta t) = q \cdot x(t) \tag{1}$$

The growth in the interval $[0, n \cdot \Delta t]$ can be considered as n steps of growth in subintervals of length Δt. Hence the biomass is n times multiplied by q:

$$x(n \cdot \Delta t) = q^n \cdot x(0) \tag{2}$$

Continuous models: q will depend on Δt. For a larger Δt q will be larger, too. For $\Delta t = 0$ we must have $q = 1$. Thus as a simple model of the function $q = q(\Delta t)$ we can try that q is a linear function of Δt: $q = 1 + b \cdot \Delta t$ where b is a constant.

For arbitrary real t put $\Delta t = \frac{t}{n}$ and get $q = 1 + \frac{bt}{n}$. (2) suggests to try the models:

$$x(t) = (1+\frac{bt}{n})^n \cdot x(0) \quad \text{and} \quad x(t) = \lim_{n \to \infty} (1+\frac{bt}{n})^n \cdot x(0) \tag{3}$$

Thus, the sequences $(1+\frac{bt}{n})^n$ and $(1+\frac{1}{n})^n$ for $b = t = 1$ do not drop from the sky but can be introduced in the context of a growth problem. For completeness we note that the procedure is heuristic since b depends on Δt, that is, for $\Delta t = t/n$ on n.

A further continuous model: Denote $q_n = q(1/n)$ and put $a := q_1$. Then (2) implies for $n = \Delta t = 1$ $x(1) = a \cdot x(0)$ and for $\Delta t = 1/n$ $x(1) = (q_n)^n x(0)$. Hence $q_n = a^{1/n}$ as we can assume $x(0) \neq 0$. Doing m steps of length $1/n$ we find:

$$x(m/n) = q_n^m x(0) = (a^{1/n})^m x(0) = a^{m/n} x(0) \tag{4}$$

Thus at an early stage of the class on calculus it can be shown that an exponential function models continuous uniform growth at least for rational $t = m/n$, namely:

$$x(t) = a^t \cdot x(0) \qquad (5)$$

Now the question is in the air: Can a^t be given a meaning for not rational t such that $x(t)$ is determined by formula (5) for all real numbers t ? Thus by modelling uniform growth, interest and motivation can be evoked for the definition of the exponential function for all real t.

4. MOUNTAIN PROFILE AND THE SPEED OF A SKIER

A participant had to design a worksheet for the situation of a skier in the mountains.

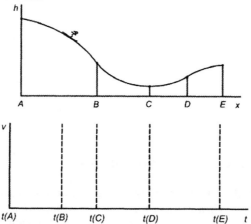

Describe qualitatively the mountain profile and the skier's speed in each of the regions (open intervals)]A,B[,]B,C[,]C,D[and]D,E[:

    a.   In striking everyday speech.
    b.   In precise mathematical speech.
    c.   By a mathematical formula.

Sketch the speed function.

$t(A),t(B),t(C),t(D)$ are the times when the skier is at $x=A$, $x=B$, $x=C$, $x=D$.

]A,B[	]B,C[	]C,D[	]D,E[
Mountain profile			
a.	a.	a.	a.
b.	b.	b.	b.
c.	c.	c.	c.
Speed function			
a.	a.	a.	a.
b.	b.	b.	b.
c.	c.	c.	c.

Table 1. Worksheet for the skier in the mountains.

This exercise was inspired by Schlögelhofer (2000). As help the student got a picture where the mountain was already subdivided into regions marked by the open intervals $]A,B[$, $]B,C[$, $]C,D[$ and $]D,E[$ on the x-axis. The worksheet should ask for a description of the profile of the mountain and of the speed in these regions. See Table 1. The mathematical background was the concepts of monotonicity and convexity as well as the mathematical criteria for them by means of the first and second derivatives of a function. Here they had to look at the height function h and the speed function v. For simplicity it should be assumed that the first two derivatives are defined and different from 0 in the respective open intervals. For example:

Definition: A real function h is *strongly concave in an interval* $]A,B[$
$\Leftrightarrow \forall x, \tilde{x} \in]A,B[$ the straight line from $(x, h(x))$ to $(\tilde{x}, h(\tilde{x}))$, except the endpoints, sits below the graph of h restricted to $]A,B[$.

Criterion: Under the above assumption h is strongly concave in $]A,B[$
$\Leftrightarrow \forall x \in]A,B[\ h''(x) < 0$.

The form of the worksheet should encourage pupils to make the transition from everyday speech to a mathematical description and vice versa. Necessary assumptions were to be stated. The student should find some hints for the pupils to get started and detailed guidance. Furthermore she should create a prototype of a solution and criteria for assessment. As a group exercise the other participants should solve the worksheet. As detailed guidance the solution for the mountain profile in the region $]A,B[$ might be given: a. permanently steeper downwards, b. strong decreasing and strong concave, c. $h' < 0$ and $h'' < 0$. A funny relation of the form of the mountain and the graph of the speed function can be observed. The performance of the groups was shown on transparencies, explained by the groups and discussed.

5. CIRCUMFERENCE, AREA AND VOLUME

Blum and Kirsch (1996) emphasize the importance of basic visualisations of mathematical notions. Their concept was practised within two of their exercises.

Exercise: Why is the changing rate of the area a of a disc with respect to changing radius r the circumference c? Argue without using the formulas which represent a and c as functions of r.

This result follows from the inequalities $\Delta r \cdot c(r_0) < \Delta a < \Delta r \cdot c(r_1)$ for $\Delta r = r_1 - r_0 > 0$, division by Δr, the Intermediate Value Theorem applied to the function $c(r)$ and the intermediate value $\Delta a / \Delta r$ and taking the limit $\Delta r \to 0$.

During the meeting we asked us how the starting inequalities could be made clear. On her own initiative the student tried to make them plausible in her final written report. Her ideas were to cut the circular area between the circles of radius r_0 and r_1 and bend it till it looks like a trapezium and then compare with the area $\Delta r \cdot c(r_1)$ of a rectangle with edges of length Δr and $c(r_1)$ to get one of the inequalities. Although there is the problem of controlling the area during this deformation the idea of deforming the surface is good and leads to a precise proof:

Follow-up: We can regard $\Delta r \cdot c(r_0)$ as the area of a cylinder of height Δr and cross section a circle of radius r_0. We can deform this cylinder by rotating each straight line on the cylinder by the angle $\alpha = 90°$ as in Figure 1. To understand this deformation better imagine an umbrella with flexible cloth and hold the umbrella downwards. Then the deformation is like opening the umbrella.

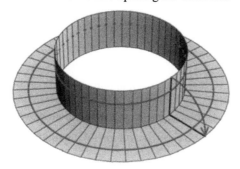

Figure 1. Deformation of a cylinder into a ring.

On the straight lines the length is preserved and the cylinder circles -except the lowest- get longer and circles and straight lines are perpendicular to each other. This makes plausible that area is increased by this deformation, hence the second inequality holds. Similarly the first can be deduced. Elementary arguments from differential geometry of surfaces in 3-space show that these ideas lead immediately to a proof.

Someone could ask: Why do we struggle for a proof without using the formulas for the functions $c(r)$ and $a(r)$. In the sense of Blum and Kirsch (1996) the answer is that we want give associations of the contents of a term outside the world of calculus. On the other hand they pose an exercise where it is difficult to find the appropriate formulas and indeed it may not be possible to find some. Surprisingly for us these exercises about these mathematically more sophisticated objects of the kitchen turned out easier than the above analogue exercise on the disc:

Exercise: Let V be the volume of a drink that is filled in a glass up to the height h. Which is the interpretation of dV / dh ?

The student demonstrated by filling a glass and gave some figures and a sketch proof that the derivative is the cross sectional area. We assumed that the glass had a circular cross section whose radius is *any* function increasing with h. Using the inequalities $a(h_0) \cdot \Delta h < \Delta V < a(h_1) \cdot \Delta h$, which in the course had been regarded as being obvious, a formal proof was given analogously to the previous exercise.

Follow-up: What is obvious or trivial for teachers or authors is often not obvious and not at all trivial for students or readers of a mathematical book. It is a pedagogical experience that in many cases this issue can be solved by reflecting and saying or writing why you consider it obvious. Often this contemplation leads to an explanation that makes the fact obvious to students and readers as well. Here the inequality follows from the monotony property of volume. In everyday language: A bigger glass has a bigger volume. See Figure 2. ΔV is the volume of a part of the

Figure 2. Cylinders inside and outside of a part of a glass.

given glass. $a(h_0) \cdot \Delta h$ is the volume of a cylindrical glass, which is smaller than the part of the glass, and $a(h_1) \cdot \Delta h$ is the volume of a cylindrical glass, which is bigger than the part of the glass.

6. THE FUNDAMENTAL THEOREM: ALTERNATIVE APPROACHES

We used the articles: Blum and Kirsch (1996), Kirsch (1996), Hughes-Hallet et al. (1998), Riede (1993). To prompt the student who was assigned to this topic we mentioned two statements, which seemed to contradict each other because the purely mathematical content of two deductions of the Second Fundamental Theorem seemed to be equal. Hughes-Hallett et al. (1998) remark to their deduction (p168): "The argument we have given makes the Fundamental Theorem plausible. However, it is not a mathematical proof." On the other hand Blum and Kirsch write: "This argumentation is - after specification of the assumptions - a valid proof."
All sums in this section are taken from $i = 1$ to $i = n$. Here are some statements of the student's report:
- In Hughes-Hallet's Figure 3.9 and 3.10 (p. 153) non-monotonic functions should be used in order that left and right sums are not mistaken for upper or lower sums. The
imprecise "about to be equal" (p. 167) can be improved by replacing

$$G(b) - G(a) \approx \sum G'(x_{i-1})\Delta x \quad \text{by} \quad G(b) - G(a) = \sum G'(x_{i-1})\Delta x + \varepsilon \Delta x . \tag{6}$$

- Most of all in R's deduction, importance is attached to mathematical completeness. It needs the Mean Value Theorem of differential calculus. The starting statement presupposes a good preparation.
- He appreciates that in Kirsch's article the students deduce for themselves the Fundamental Theorem with the help of questions and exercises. By Kirsch's and Blum's deduction the pupils achieve a good familiarity with the theorem and its meanings in real-world situations and its geometrical interpretation. He reports that a difficulty being encountered by Kirsch's students can be avoided if one does not identify the notions line and length of a line or surface and area of a surface.
Amplification after having read the student's report: For $f \geq 0$ regard the surface S between the interval $[a,x]$ and the graph of f. Then the quantities

$F_a(x) = \int_a^x f(t)dt$, the area of S and the length of the line from $(x,0)$ to $(x, F_a(x))$ are different but they are equal as far as their absolute measures are concerned.

-Here you may find the student's summary: There should be many examples for illustration. Basic comprehension and meanings in real-world situations should take centre stage. A precise proof is indispensable.

Remark: The above use of an ε is heuristic because ε depends on i. Nevertheless it points in the direction of a precise proof. Unfortunately this proof needs much "epsilontics" for an exact determination of the limit $\Delta x \to 0$.

Follow-up: **From average speed to the Second Fundamental Theorem**

Reading the student's report the question arose: What did the student mean by "good preparation"? In the author's opinion the explanations given in the mathematical proof were sufficient. However, close examination reveals that in contrast to the modern methods a deductive procedure had been chosen. Here is an updated procedure that avoids a deductive procedure and engages the pupils via questions. In the class this can be dealt with at the very beginning of integral calculus. Prerequisite is differential calculus including the Mean Value Theorem. Then no further preparation ought to be necessary.

Question 1: The path covered from time a to time b divided by the difference in time $\frac{s(b)-s(a)}{b-a}$ is usually called the *average speed* in the interval $[a,b]$. Why is it justified to speak of an average?

Answer: Besides the usual explanations there is one that connects to the notion of the arithmetic mean. Decompose the interval by dividing points $a = t_0 < t_1 < t_2 < ... < t_n = b$ into n equally spaced subintervals of length $\Delta t = t_i - t_{i-1} = (b-a)/n$ for $i = 1,2,...,n$. Then the average speed in the whole interval is the *arithmetic mean* of the average speeds in all subintervals. To include them in the procedure the pupils may prove this as an exercise. Note that no mathematical formula is given and the pupils do not have to insert something. Instead they have to find the formula by translating the statement and verifying the formula. The result is the following equation:

$$\frac{s(b)-s(a)}{b-a} = \frac{1}{n}\sum \frac{s(t_i)-s(t_{i-1})}{\Delta t} \qquad (7)$$

Question 2: Can this equality lead to a further meaning if you simplify it or if you substitute for an expression something else?

Answer: The summands on the right side of the equation look like the left hand side of the Mean Value Theorem of the differential calculus; thus apply the Mean Value Theorem. Remember $v(t) := s'(t)$ denotes the speed. In order to get an equation without divisors multiply the equation by the denominator $b - a = n \cdot \Delta t$. We arrive at the simplified equation: $s(b) - s(a) = \sum v(c_i)\Delta t$ for suitable $c_i \in [t_{i-1}, t_i]$. Now it is the teacher's part to interpret this equation and inform the pupils that they have rediscovered a so-called *Riemann sum* on the right hand side. Have in mind that up to this point no theory of integration was needed. Now the teacher can give as a little digression a simplified definition of the integral. A rather more general definition of a Riemann sum than occurs here should be given: *A Riemann sum* for the speed function v in the interval $[a,b]$ is a sum of the form $\sum v(c_i)\Delta t$ where c_i is *any* value in the ith interval. A graphical explanation should follow. Define the integral as usual as the limit of such sums, namely:

$$\int_a^b v(t)dt := \lim_{\Delta t \to 0} \sum v(c_i)\Delta t = \lim_{n \to \infty} \sum v(c_i)\frac{b-a}{n} \qquad (8)$$

and cite that for continuous functions such limits exist and are all the same. This should be sufficient for a first encounter with the integral.

After this digression let us come back to the equation with the special Riemann sum where the c_i were chosen according to the Mean Value Theorem. In this case the question of convergence is quite easy. The sequence of *Riemann sums* is a constant sequence and converges to its constant value $s(b) - s(a)$. No "epsilontics" are needed. You get the second Fundamental Theorem in a real-world situation:

$$s(b) - s(a) = \int_a^b s'(t)dt = \int_a^b v(t)dt \qquad (9)$$

The length of the path from time a to time b is the integral of the speed function. The same procedure shows that the total change of *any* (continuously differentiable) function f from a to b is the integral of its local rate of change $f'(t)$. Replacing $x = b$ and regarding x as variable leads to the other interpretation of the Fundamental Theorem: Integration is recovering a function from its derivative. A real life situation where the path has to be recovered from the speed function occurs when you have to recover the driven route from the tachograph, which gives the graph of the speed function as real data (Hussmann, 2001).

8. THE CABLE REEL

At the end of the course we dealt with modelling problems that were not from the very beginning related to any concept of calculus. With reference to Förster's and Herget's article (2002) we dealt with the question: How long is the cable on the cable reel? To get a first estimation of the length their students studied the circle model where the cable is cut into appropriate pieces and wound as circles onto the reel. It turned out by calculation that 14.6 circles can be put on the reel in one layer, which means in reality 14 circles.

As a more realistic model they used helixes. In their exciting article Förster and Herget reported how a feature of the circle model is used in the helix model too, which led to a completely unrealistic model. In our course the audience was asked to carry out the exercise to search for more points that were taken from the circle model but which were different in the helix model. One point is the following: If in the helix model there is room for the cable to progress 0.6 of a cable diameter towards the side limit of the cable reel then you can wind by a further winding angle $\alpha = 0.6 \cdot 2\pi$ since the progression to the side limit is proportional to the winding angle. At first glance you might think that you can do 14.6 windings in the helix model but indeed the cable already approaches a distance to the side limit of 0.6 of a cable diameter after 13 windings. So you can do only 13.6 windings. This is visualized in Figure 3, which is drawn for 3 instead of 14 circles. Instead of 792 m for 14 windings we can estimate only a length of 769 m, which is 2.9% less. Thus the example of the cable reel pays attention to the dangers when you use results for a simplified situation in a more realistic model.

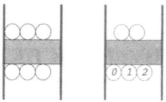

Figure 3. Cross section of the first layer in the circle and helix model.

8. CONCLUSION

We found out that none of our students was acquainted with any of the new ideas and had learned mathematical modelling at school. During the semester, most students approved emphatically of the modern methods as was evident from the lively meetings of the course and as was documented in the students' written reports of 10 to 30 pages. Thus the message of this article is addressed to all teachers and educators of teachers who want to enhance interest, motivation, joy and success in mathematics. The message is: The new ideas do work and they help in an essential way. Use the results of research and practical investigations as they are documented in a broad field of literature and apply it in practice. To initiate this an effective organization and an intensive individual tuition system is necessary.

ACKNOWLEDGEMENT

The author thanks Hanspeter Eichhorn for the good teamwork during several courses on high school teacher education and training and J. Roy Dennett, formerly Department of Mathematics, University of Hull for advice in preparing this chapter.

REFERENCES

Blum, W. and Kirsch, A. (1996) Die beiden Hauptsätze der Differential- und Integral- rechnung. In G. Malle (ed) *mathematik lehren 78*, Seelze: E. Friedrich, 60 – 65.

Hussmann, S. (2001) Mathematik konstruktiv und eigentätig entwickeln. In G. Malle (ed) *mathematik lehren 109*, Seelze: E. Friedrich, 51-56.

Kirsch, A. (1996) Der Hauptsatz – anschaulich? In G. Malle (ed) *mathematik lehren 78*, Seelze: E. Friedrich, 55-59.

Förster, F. and Herget, W. (2002) Die Kabeltrommel. In G. Malle (ed) *mathematik lehren 113*, Seelze: E. Friedrich, 48 – 52.

Hughes-Hallet, D. et al. (1998) Calculus. New York: John Wiley, p153 and pp167-168.

Riede, A. (1993) *Mathematik für Biologen*. Wiesbaden: vieweg, 90-91 and 119- 120.

Schlögelhofer, F. (2000) Vom Fotograph zum Funktionsgraph. In G. Malle (ed) *mathematik lehren 103*, Seelze: E. Friedrich, 16-17.

6.4

CASE STUDY: LEAK DETECTION IN A PIPELINE

Andrei Kolyshkin
Riga Technical University, Latvia

Abstract—*A mathematical model for leak detection in pipelines is analysed in the present paper. The model is used as a case study for a course in mathematical modelling at undergraduate or graduate level.*

1. INTRODUCTION

It is recognized nowadays that a course in mathematical modelling should be one of the key elements in undergraduate and graduate engineering curriculum. The interdisciplinary nature of mathematical modelling brings together mathematicians, engineers and people in other fields. These specialists share their knowledge and expertise in order to analyze industrial processes, improve the performance of existing devices, and develop new products and services. However, there is no definite answer to the question: "How should mathematical modelling be taught?"

One of the possibilities could be case study approach. It is widely used in business schools all over the world. The attractiveness of case studies lies in analyzing real-world problems which is crucial to any business programme, for example, MBA. Case studies can also be successfully used in engineering programmes to teach mathematical modelling at undergraduate and graduate level. A collection of case studies based on problems in some British companies is presented in (Burghes et al., 1996) at elementary undergraduate level. Several textbooks (MacCluer, 2000; Friedman & Littmann, 1996; Svobodny 1998) discuss case studies and projects from industry at upper undergraduate and graduate level. The above mentioned recent publications relating to mathematical modelling clearly show the shift in pedagogy from standard teaching, where mathematical methods are discussed first and then used to solve problems, to modern teaching, where real-world problems are discussed first and mathematics is used as a necessary tool to solve these problems.

In order to be really useful for engineering students, a case study should satisfy the following requirements:

1. The case study should be based on a real-world problem.
2. Relatively simple mathematical models (that can be understood with basic knowledge of calculus and linear algebra) should be used in the analysis.
3. Several mathematical methods should be used in order to solve the case.

4. Interpretation of the solution should be based on the comparison of the model results with the actual data.

In the present paper, a case study approach is used to analyze the important practical problem: leak detection in pipelines.

Leakage from water supply systems is an important problem, which may lead to substantial environmental damage and economic loss. It is shown by Rao and Sridharan (1996) that in 35 of the 38 analyzed cities in the Asian and Pacific region, unaccounted for water loss reaches 36% on the average. In particular, the values for Singapore and Dhaka are 8% and 62%, respectively. Previous studies (Makar & Chagnon 1999) indicated that in some North American cities unaccounted for water losses can reach 25%, despite the fact that regular leak detection programmes are conducted. Therefore it is necessary to develop effective leak detection methods in order to minimize economic and environmental losses.

The steady-state network problem is analyzed in Pudar and Liggett (1992) where the analysis is based on the assumption that friction coefficients in all pipes are known. Leak location is determined by minimizing the difference between measured and calculated pressures. In this case large errors can occur if friction is not accurately known. An inverse transient analysis model for pipe networks in Liggett and Chen (1994) is based on the time-dependent pressure history. A leak detection method based on the time history analysis of pressure data in a pipeline is suggested by Wang et al. (2002). The main idea of the method is the representation of the damping rates for different harmonic components as a sum of the steady-state friction damping factor (R) and the leak-induced damping factor (R_{nL}). It is suggested in (Wang et al., 2002) that the ratio of leak-induced damping rates can be used to determine the location of the leak. The solution of one-dimensional linearized equation for the pressure perturbation in the presence of the leak is obtained (Wang et al., 2002) under several implicit assumptions. A simple formula for leak detection is suggested on the basis of the obtained solution. Reasonable agreement of the proposed formula with experimental data is found. However, the mathematical analysis of Wang et al. (2002) cannot be fully justified since the system of ordinary differential equations described by them is not uncoupled.

The present paper is organized as follows. Analytical solution of one-dimensional linearized equation for the pressure perturbation in the presence of a leak is discussed. It is suggested to use this problem as a case study in any course of mathematical modelling at upper undergraduate and graduate level.

All the four conditions specified above are satisfied in the case study considered in the paper. In particular, the following mathematical methods are needed to solve the case: the method of separation of variables, Fourier series and delta-function, series summation, Laplace transform and the residue theorem, matrix eigenvalues, Newton's method for the numerical solution of nonlinear equations. Suggestions and recommendations for possible teaching strategies are presented.

2. ONE-DIMENSIONAL LINEARIZED MODEL

Consider one-dimensional water-hammer equations which take into account the elasticity effects, compressibility of the pipe and the presence of a leak (Wang et al., 2002):

$$\frac{\partial H_*}{\partial t_*} + \frac{Q_*}{A_*}\frac{\partial H_*}{\partial x_*} + \frac{a^2}{gA_*}\frac{\partial Q_*}{\partial x_*} + \frac{a^2}{gA_*}Q_*^L\delta(x_* - x_*^L) = 0, \tag{1}$$

$$\frac{\partial H_*}{\partial x_*} + \frac{1}{gA_*}\frac{\partial Q_*}{\partial t_*} + \frac{Q_*}{gA_*^2}\frac{\partial Q_*}{\partial x_*} + \frac{fQ_*^2}{2D_*gA_*^2} - \frac{Q_*Q_*^L}{gA_*^2}\delta(x_* - x_*^L) = 0, \tag{2}$$

where H_* is the piezometric head (the pressure which is equivalent to the height H_* of water column), Q_* is the flow rate, a is the wave speed in the fluid, g is the acceleration due to gravity, A_* is the cross-sectional area of the pipe, x_* and t_* are the longitudinal coordinate and time, respectively, D_* is the diameter of the pipe, Q_*^L and x_*^L are the total discharge at the leak and the location of the leak, respectively, and $\delta(x_*)$ is the Dirac delta-function. The leak discharge has the form

$$Q_*^L = C_d A_*^L \sqrt{2g\Delta H_*^L}, \tag{3}$$

where C_d is leak discharge coefficient, A_*^L is the area of the leak and $\Delta H_*^L = H_*^L - z_*^L$ is the pressure head at the leak where H_*^L is the piezometric head in the pipe and z_*^L is pipe elevation at the leak.

It should be emphasized at this stage by the instructor that the formulated model is still too complicated for the analysis by analytical methods. The major source of difficulty is nonlinearity. Therefore linearization of the governing system (1) – (3) is performed as described below.

Suppose that H_* and Q_* can be written in the form $H_* = H_*^0 + h_*$, $Q_* = Q_*^0 + q_*$ where H_*^0 and Q_*^0 are the steady-state head and flow values, and h_* and q_* are small unsteady perturbations. Linearizing equations (1) – (3) in the neighborhood of H_*^0, Q_*^0 and eliminating q_* we obtain the following dimensionless equation

$$\frac{\partial^2 h}{\partial x^2} = \frac{\partial^2 h}{\partial t^2} + [2R + F_L\delta(x - x_L)]\frac{\partial h}{\partial t} - 2RF_L\delta(x - x_L)h \tag{4}$$

where h is the dimensionless head deviation, t and x are dimensionless time and longitudinal coordinates, $R = fLQ_0 /(2aD_*A_*)$, L is the length of the pipe, f is the steady-state friction factor, x_L is the dimensionless location of the leak and

$$F_L = \frac{C_d A_*^L}{A_*}\frac{a}{\sqrt{2gH_0}} \tag{5}$$

The variables h_*, x_*, t_* and q_* are made dimensionless by means of the scales H_0, L, L/a and Q_0, respectively, where H_0 and Q_0 are the characteristic piezometric head and flow rate, respectively. All the variables without asterisks are dimensionless. The boundary and initial conditions are

$$h(0,t) = 0, \quad h(1,t) = 0 \tag{6}$$

$$h(x,0) = f(x), \quad \frac{\partial h}{\partial t}(x,0) = g(x). \tag{7}$$

The instructor should discuss problem (4) – (7) with students in detail. It is recommended at this stage to pay students' attention to the fact that (4) is a linear partial differential equation and, therefore, can be solved by the method of separation of variables. This method is used below to solve (4) – (7) for the following two cases: 1) without a leak and 2) with a leak.

3. SOLUTION FOR THE CASE WITHOUT A LEAK

Since $F_L = 0$ (no leak), the solution to (4) is

$$h(x,t) = \sum_{n=1}^{\infty} h_n(t)\sin n\pi x, \tag{8}$$

where

$$h_n(t) = \exp[-Rt]\left(f_n \cos b_n t + \frac{g_n + Rf_n}{b_n}\sin b_n t \right). \tag{9}$$

The constants f_n and g_n in (9) are the Fourier sine coefficients of the functions $f(x)$ and $g(x)$, respectively. The roots of the characteristic polynomial

$$s^2 + 2Rs + n^2\pi^2 = 0 \tag{10}$$

are

$$s_{0n} = -R \pm ib_n, \tag{11}$$

where $b_n = \sqrt{n^2\pi^2 - R^2}$

Note that the equation for $h_n(t)$ is the second order ordinary differential equation with constant coefficients. The instructor should remind the students how to solve such equations using the roots of the characteristic polynomial.

4. SOLUTION FOR THE CASE WITH A LEAK

In this section the solution method is briefly described. We apply the Laplace transform to (4) and seek the solution to the resulting equation in the form

$$\overline{h}(x,s) = \sum_{n=1}^{\infty} \overline{h}_n(s)\sin n\pi x \tag{12}$$

where $\overline{h}(x,s)$ is the Laplace transform of h, and s is the parameter of the Laplace transform.

A few comments are necessary at this stage. First, the instructor should clearly demonstrate the application of the method of Laplace transform to the solution of (4). Second, the instructor should remind the students how to work with the delta-function. Finally, it should be made clear that the obtained expression for $\overline{h}_n(s)$ is not a solution of the original problem in the transformed space since it contains an unknown constant C.

The role of the instructor at this stage is to demonstrate that such an expression is rather inconvenient for the subsequent use and it would be difficult to invert the Laplace transform in this case. In order to obtain the solution in the form suitable for further analysis one needs to use series summation formulas (Antimirov et al, 1998) and the residue theorem.

The Laplace transform of the solution has the form

$$\overline{h}(x,s) = \overline{G}(x,s) - \frac{F_L \overline{G}(x_L,s)(s-2R)\sinh(\alpha - \alpha x_L)\sinh \alpha x}{\alpha \sinh \alpha + F_L(s-2R)\sinh(\alpha - \alpha x_L)\sinh \alpha x_L}. \tag{13}$$

It follows from the residue theorem that the complex damping rates for the case with the leak are the (complex) roots of the equation

$$\psi(s) = 0, \tag{14}$$

where

$$\psi(s) = \alpha \sinh \alpha + F_L(s-2R)\sinh(\alpha - \alpha x_L)\sinh \alpha x_L. \tag{15}$$

The instructor should emphasize at this stage that using (13) one can obtain the solution of the original problem by means of the residue theorem. However, in practice (Wang et al. 2002)) we are interested in the damping rates of different harmonic components which are obtained as the difference ε_n between the rates with and without the leak. This means that ε_n is the difference between the roots of equations (10) and (14). Suppose that s_{0n} and s_{1n} are the roots of equations (10) and (14), respectively. Then

$$\psi(s_{1n}) = 0 \tag{16}$$

and

$$s_{0n}^2 + 2Rs_{0n} = -n^2\pi^2. \tag{17}$$

Note that for small F_L the difference $\varepsilon_n = s_{1n} - s_{0n}$ is also expected to be small. Expanding the function $\psi(s_{1n})$ in Taylor series around $s = s_{0n}$, taking into account only linear terms and assuming that F_L is small we obtain

$$\varepsilon_n = \frac{s_{0n} - 2R}{s_{0n} + R} F_L \sin^2 n\pi x_L. \tag{18}$$

It should be underlined by the instructor that the procedure described above is essentially a linearization procedure which is used quite often in applications. Calculating ε_n^r, the real part of ε_n, we obtain

$$\varepsilon_n^r = F_L \sin^2 n\pi x_L.$$ (19)

Formula (19) can be rewritten in the form

$$\frac{\varepsilon_m^r}{\varepsilon_n^r} = \frac{\sin^2 m\pi x_L}{\sin^2 n\pi x_L}.$$ (20)

Note that (20) is used by Wang et al. (2002) for leak detection. However, our analysis shows their formulas (23) and (24) are not valid for arbitrary values of F_L and x_L. These formulas should be considered as approximations to the true values of ε_n^r for small values of F_L only.

5. SOLUTION OF THE INVERSE PROBLEM

In this section results of the numerical experiments are discussed in order to obtain the region of validity of (20) for leak detection purposes. For fixed values of R, F_L and x_L, the first three complex roots of equation (14) are calculated. The value of R is fixed at $R = 0.0742$. This value is equal to the value used in (Wang et al. 2002) and corresponds to a Reynolds number of 396000. The roots of (14) are calculated as follows. In order to get a good initial guess for the root, we used the sine Fourier transform (8) to solve equation (4). The corresponding system of ordinary differential equations for the functions $h_n(t)$ is then truncated (three terms in (8) are used) and the eigenvalues of the corresponding constant matrix are chosen as the initial guesses for the roots. This is a good opportunity for the instructor to remind the students some basic facts from linear algebra (matrix eigenvalues and their calculation). Newton's method is then used to find better approximations to the roots. An example calculation is shown in Table 1 where the roots are accurate to 6 decimal places.

Table 1. Roots of equation (14) for $F_L = 0.05$, R = 0.0742 various values of x_L.

x_L	0.15	0.25	0.35
s_{11}	$-0.084527 + 3.140131i$	$-0.099260 + 3.139255i$	$-0.113973 + 3.138139i$
s_{12}	$-0.106971 + 6.282132i$	$-0.124232 + 6.280973i$	$-0.106921 + 6.281354i$
s_{13}	$-0.123018 + 9.423596i$	$-0.099188 + 9.423583i$	$-0.075428 + 9.424460i$

The roots of (14) are also calculated for different values of F_L. Note that in real pipeline systems the value of F_L (defined by (5)) is usually small. For this reason the calculations are restricted to the interval $0 < F_L \leq 1$. For each fixed pair (x_L, F_L) we calculate the difference $\varepsilon_n^r = s_{1n}^r - s_{0n}^r$ (here s_{1n}^r and s_{0n}^r are the real parts of the roots s_{1n} and s_{0n}, respectively) and we compare the ratios

$\varepsilon_2^r / \varepsilon_1^r$ and $\varepsilon_3^r / \varepsilon_1^r$

to the ratios $\sin^2 2\pi x_L / \sin^2 \pi x_L$ and $\sin^2 3\pi x_L / \sin^2 \pi x_L$, respectively.
The absolute percentage difference

$$\Delta_n = \frac{\varepsilon_n^r / \varepsilon_1^r - \sin^2 n\pi x_L / \sin^2 \pi x_L}{\varepsilon_n^r / \varepsilon_1^r} \quad \text{for } n = 2 \text{ and } n = 3 \text{ is shown in Table 2.}$$

Table 2. The percentage difference Δ_n for $n = 2$ and $n = 3$.

$F_L = 0.01$					$F_L = 0.05$			
x_L	0.15	0.25	0.35	0.45	0.15	0.25	0.35	0.45
Δ_2	0.017	0.040	0.042	0.026	0.073	0.175	0.211	0.145
Δ_3	0.035	0.057	0.034	0.020	0.127	0.287	0.186	0.100

$F_L = 0.1$					$F_L = 0.5$			
x_L	0.15	0.25	0.35	0.45	0.15	0.25	0.35	0.45
Δ_2	0.118	0.287	0.430	0.325	0.493	1.167	2.579	3.093
Δ_3	0.134	0.576	0.416	0.198	4.319	3.064	4.019	0.908

As can be seen from Table 2, the relative error Δ_n in using formula (20) for the values of F_L in the interval $0 < F_L \leq 0.5$ is very small. Thus, our analysis supports the use of the approach suggested in (Wang et al. 2002) for small values of F_L. The results presented by Wang et al. (2002) compare well with experimental data and it is concluded by them that formula (24) can be used for reliable leak detection in a straight long pipeline. The analysis presented above clearly demonstrates the limitations of the model and restricts the applicability of (20) to small values of F_L.

6. CONCLUSIONS

In this section several recommendations to instructors using the case study are presented. First, problem (1) – (3) is rather complicated and it is unlikely that students can solve the case independently without the help from the instructor. The instructor should provide guidelines for the analysis. For example, it would be a very good idea to discuss the linearization procedure in detail. In fact, this gives an opportunity to the instructor to pay students' attention to the importance of linearized models in applied mathematics.

Second, linearized problem (4) – (7) also needs a careful discussion in class. Students may be familiar with equations similar to (4) for the case $F_L = 0$. It is advisable at this stage to review the method of separation of variables with proper references to the textbook. The instructor may even ask students to solve similar problems at home or at least analyze several examples of the method of separation of variables in class. Then students may be asked to present the solution of problem (4) – (7) without leak working in groups.

Third, basic steps of the solution with leak should be clearly identified by the instructor, that is, the technique of the Laplace transform, basic properties of the delta function, methods of complex analysis such as series summation and the residue theorem. It is suggested to discuss all these methods with students and then ask them to find the Laplace transform of the solution working in groups.

Fourth, the instructor should underline that the obtained solution is still complicated enough and needs to be simplified. Students should review Taylor series and basics of numerical analysis (methods of solution of nonlinear equations, in particular, Newton's method). The instructor then should ask students to reproduce results in Table 1 and Table 2. Finally, students should analyze the limitations of the model and discuss its domain of applicability.

REFERENCES

Antimirov, M.Ya., Kolyskin, A.A. and Vaillancourt, R. (1998) *Complex Variables*. Academic Press.

Burghes, D., Galbraith, P., Price, N. and Sherlock, A. (1996) *Mathematical Modelling*. Prentice Hall.

Friedman, A. and Littman, W. (1996) *Industrial Mathematics: A Course in Solving Real-world Problems*. SIAM.

Liggett, J.A. and Chen, L.C. (1994) Inverse transient analysis in pipe networks. *Journal of Hydraulic Engineering*, 120, 934-955.

MacCluer, C.R. (2000) *Industrial Mathematics: Modelling in Industry, Science and Government*. Prentice Hall.

Makar, J. and Chagnon, N. (1999) Inspecting systems for leaks, pits, and corrosion. *Journal of the American Water Works Association*, 91, 36-46.

Pudar, R.S. and Liggett, J.A. (1992) Leaks in pipe networks. *Journal of Hydraulic Engineering*, 118, 1031-1046.

Rao, P.V. and Sridharan, K. (1996) Inverse transient analysis in pipe networks. *Journal of Hydraulic Engineering*, 122, 278-288.

Svobodny, T.P. (1998) *Mathematical Modelling for Industry and Engineering*. Prentice Hall.

Wang, X.J., Lambert, M.F., Simpson, A.R., Liggett, J.A. and Vitkovsky, J.P. (2002) Leak detection in pipelines using the damping of fluid transients. *Journal of Hydraulic Engineering*, 128, 697-711.

6.5

DISCRETE AND CONTINUOUS MODELS OF LIZARD POPULATIONS

Michael Jones and Arup Mukherjee[1]
Montclair State University, New Jersey, USA

Abstract–*Difference and differential equations are used to model the mating strategies of side-blotched lizards. The lizard population is divided into three sub-classes based on mating strategies. The discrete model using difference equations assumes that the lizards are playing a "rock-paper-scissors game" and results in long-term behaviour contradicting observations. The same evolution can be modelled using a system of ordinary differential equations similar to predator-prey systems and results in a long-term behaviour that matches the observations. The lizard project is assigned as a capstone project in a second year undergraduate course with single variable calculus as a prerequisite. The students use Excel and Maple to study the long-term behaviour of the various models, justify their observations using analytic solutions when appropriate, explore and learn the process of checking model validity through comparison with observed data, and learn verbal and written communication skills.*

1. INTRODUCTION

Mathematical modelling can be described as an iterative process where successive models of often increasingly more sophisticated mathematics are compared to real-world data. This conjures the image of converging to a suitable mathematical model, as described by Klamkin (1980) and Pollak (1959), and the process is quite different from the student experience of solving homework problems. Not only is the iterative process absent, but homework problems are often narrow in scope, it being clear what mathematics is necessary to solve the problem. In this paper, we consider a project that requires students to develop successive models and compare the outcomes to observed data. The project serves as a capstone for a beginning second-year college course that introduces students to mathematical modelling and serves as a transition to the upper-level courses. In this article, we place the project into the context of the course, describe the requisite mathematics, and discuss its pedagogical impact.

The project requires students to consider three ways to model the effect of mating strategies on the evolution of lizard populations. Sinervo and Lively (1996) use game theory to model and, subsequently, to describe the evolution of side-blotched lizard populations on the mounds of Los Banos Grandes in Merced County, California. Specifically, the lizards' mating strategies are compared to an evolutionary version of "rock-paper-scissors" game or Roshambo. The lizards exhibit three types of mating strategies that are perfectly correlated with the lizards' coloration. When one of the three-colored lizards becomes dominant in the population, then one of the other colored lizard's strategy becomes more successful from an evolutionary standpoint. For example, in the "rock-paper-scissors" analogy, when more players use rock, then paper is more successful. Just as in "rock-paper-scissors", one strategy defeats a second strategy, but is defeated by a third strategy.

Students model the evolutionary behaviour using both difference and differential equations. Using difference equations, the evolutionary process can be viewed as a Markov chain. The model predicts that all sub-classes of lizards attain an equilibrium value in the long run. Students reject this model as it contradicts the oscillatory behaviour observed in real life. Students are encouraged to consider continuous models. A first naïve continuous model is obtained from the discrete model through limits, but also leads to spurious results. By extending a two-species explosion/extinction model coupled with interaction terms to a three-species case, students develop a system of ordinary differential equations that results in long-term behaviour that more accurately represents the observed data. The project demonstrates how the modelling process evolves and why it is often necessary to add complexity to a model to achieve more realistic results.

In §2, we place the lizard project into the context of the course. The lizard problem is described in detail in §3, including the three mathematical models describing the effects of mating strategies on species evolution. The mathematics highlights the modelling process in which the students consider different models and compare the outcomes with what is observed in nature. In §4, we discuss the impact that the lizard population project has had on the course and the course has had on both the programme and students.

2. PEDAGOGICAL CONTEXT OF THE LIZARD PROJECT

The lizard problem is assigned as a capstone project in a second year undergraduate course with single variable calculus as a prerequisite. The course is designed to transition students to higher-level applied mathematics courses, taking the place of what is often termed a "proofs" course. Rather than focus on pure mathematics as the content, students learn how to analytically solve elementary difference and differential equations in an applied setting where the equations studied are motivated by real-life examples. At the beginning of the course, students use Microsoft Excel to explore how changes in initial conditions and parameters affect the long-term behaviour of sequence data generated by difference equations. This leads to conjectures that are proved; techniques of proof are introduced in this setting. Systems of linear difference equations and elementary linear algebra are also introduced and motivated through applications.

Differential equations are introduced because of the limitations of difference equations. Students transition from discrete to continuous models through in-class examples and Interdisciplinary Lively Application Projects (Arney, 1998), a collection of extended problems that tie mathematics to partner disciplines. Although students learn how to solve elementary differential equations analytically, more emphasis is placed on using technology to analyze numerical and graphical representations of the solutions. Jones and Mukherjee (2005) provide an overview of the course, comparing it to traditional proofs courses, as well as motivate its development. The text by Arney, Giordano, and Robertson (2001) was used as a primary text. Much of the same material also appears in (Fulford et al., 1997).

Students work in 2-3 person groups on two projects during the semester with the capstone project assigned as a third project, given *in lieu* of a final exam. Students have the last two weeks of the term to work on the capstone project; they are given two class meetings of one hour and fifteen minutes to work on the capstone project. In comparison, students are given one class meeting to work on the earlier projects. By providing time in class to work on the projects, we assure that the groups understand the problem and have a firm foundation in which to complete the project. All of the projects require students to prepare written reports and to present their results orally. The demands of the students' reports and presentations increase throughout the term. Students give 25-minute talks on the capstone project.

Specific details about the mating strategies of the lizards are given in the next section. The descriptions are used to create one discrete and two continuous models of the effect of the mating strategies on the evolution of the lizard population. We provide the mathematical content of the models and emulate the students' work by discussing whether or not the outputs of the models accurately reflect the observed data. As the students would do in their written report and their oral presentation, we include a brief comparison of the models.

3. LIZARD POPULATIONS: DISCRETE AND CONTINUOUS MODELS

In this section we provide mathematical models for the mating strategies of side-blotched lizards (*Uta stansbriana*) living on the mounds of Los Banos Grandes in Merced County, California. Male side-blotched lizards can have three distinct throat colorations and each color based sub-population uses a different mating strategy which is inherited by their offspring who have the same throat coloration. Female side-blotched lizards are slightly smaller than male lizards and exhibit yellow throat colorations. The three male sub-populations can be classified as: non-territorial yellow throated males who can impersonate female behaviour and are called "sneakers", high testosterone orange throated males who are territorial and aggressive, and mate-guarding blue throated males who are not fooled by the sneakers. The mating strategy of each sub-population has strengths that allow it to win against another sub-class, while suffering from a weakness when compared to the third sub-class. In particular, blue-throated males are not fooled by the yellow-throated male sneakers, but can be overpowered by orange-throated males. Because the orange-throated males patrol a larger territory, they are susceptible to the yellow-throated sneakers' mating strategy. This coupled interaction between the sub-populations leads to an evolutionary stable system where the percentages of the three sub-populations fluctuate around 1/3. A short article describing the rock-scissors-paper mating strategies appeared in *The Economist* (1996) and provides a good introduction to the problem. For a more detailed discussion on the mating strategies of side-blotched lizards, consider Sinervo and Lively (1996).

3.1 Discrete Model: "Rock-paper-scissors"

"Rock-paper-scissors" or Roshambo is a two player game with no clear choice among the three possible options. In the absence of information about an opponent's strategy, the equilibrium strategy is to randomize between rock, paper, and scissors, assigning a probability of 1/3 to each. The mating strategies of side-blotched male lizards can be modeled as a discrete dynamical system paralleling a game of Roshambo considered as an evolutionary process. When Roshambo is viewed as a game, rock defeats scissors, ties with rock, and loses to paper resulting in payoffs of 1 (win), 0 (tie), and -1 (loss). Accordingly, scissors defeats paper (+1), ties with scissors (0), and loses to rock (-1) and, paper defeats rock (+1), ties with paper (0), and loses to scissors (-1). For our purposes, we consider the payoffs as describing the success of the strategies on the population of the next generation. An equivalent representation is 2 for a win, 1 for a tie, and 0 for a loss.

If $y(n)$, $o(n)$, and $b(n)$ denote the percent of the male lizard population with yellow, orange, and blue throat colorations, then the population at time $n+1$ depends on the population distribution at time n via the matrix equation

$$\begin{bmatrix} o(n+1) \\ b(n+1) \\ y(n+1) \end{bmatrix} = \begin{bmatrix} 1/3 & 2/3 & 0 \\ 0 & 1/3 & 2/3 \\ 2/3 & 0 & 1/3 \end{bmatrix} \begin{bmatrix} o(n) \\ b(n) \\ y(n) \end{bmatrix} \quad or \ S(n+1) = T\,S(n) \qquad (1)$$

where $S(n)$ is the generation dependent 3-vector of unknowns and T is the 3x3 matrix describing the evolution of the population proportions of side-blotched lizard sub-populations from one generation to the next. We normalize the payoffs of 2, 1, and 0 to 2/3, 1/3, and 0 so the next generation percentage populations sum to 1. A generalized discrete model of "rock-paper-scissors" from an evolutionary viewpoint appears in (Weibull, 1995).

Although not the type of analysis the students do, realize that the matrix T is doubly stochastic and ergodic (a stochastic matrix has non-negative entries with row sums of 1; a matrix is doubly stochastic if both the matrix and its transpose are stochastic; a doubly stochastic matrix is ergodic if the digraph of its non-zero entries is strongly connected and aperiodic). It is well known from the theory of Markov chains that if a square matrix A of order m is doubly stochastic then the m-vector with entries $1/m$ is a fixed point or equilibrium vector of the matrix. Moreover, if the matrix A is ergodic, the equilibrium vector is attracting, see, for example Isaacson and Madsen (1985).

Based on what they have learned during the course of the term, students recognize that the discrete dynamical system (1) is first order, homogeneous and linear. They recall that (1) has an analytical solution of the form

$$S(k) = c_1 \left(\lambda_1 \right)^k v_1 + c_2 \left(\lambda_2 \right)^k v_2 + c_3 \left(\lambda_3 \right)^k v_3 \qquad (2)$$

where λ_i, v_i, $i = 1,2,3$ are the eigenvalue, eigenvector pairs of the matrix T. The matrix T has a pair of purely imaginary eigenvalues with norm less than 1 and a real eigenvalue of 1. The eigenvector corresponding to the eigenvalue 1 is the equilibrium vector. As part of their report for the lizard project, students study the long-term behaviour of system (1) using two different approaches: by iterating the system to numerically study the long term behaviour and by taking the limit of the analytic solution as k goes to infinity. They quickly discover that each sub-population of color-throated side-blotched lizard approaches the attracting equilibrium value 1/3 independent of the initial population distribution. In addition, they discover the same behaviour even when one of the sub-populations is assumed to be initially zero!

3.2 Naïve Continuous Model

A naïve approach to overcome the shortcomings of the discrete model in §3.1 is to use a continuous model obtained by approximating differences by difference quotients. In particular, equation (1) can be rewritten as

$$S(n+1) - S(n) = \begin{bmatrix} o(n+1) - o(n) \\ b(n+1) - b(n) \\ y(n+1) - y(n) \end{bmatrix} = \begin{bmatrix} -2/3 & 2/3 & 0 \\ 0 & -2/3 & 2/3 \\ 2/3 & 0 & -2/3 \end{bmatrix} \begin{bmatrix} o(n) \\ b(n) \\ y(n) \end{bmatrix}. \tag{3}$$

Considering each of the variables as continuous functions of time and making crude approximations of the form $o(n+1) - o(n) \cong o'(t)$ for each variable, we arrive at a naïve continuous model for the evolution of lizard sub-populations given by

$$\frac{d}{dt} S(t) = \begin{bmatrix} o'(t) \\ b'(t) \\ y'(t) \end{bmatrix} = \begin{bmatrix} -2/3 & 2/3 & 0 \\ 0 & -2/3 & 2/3 \\ 2/3 & 0 & -2/3 \end{bmatrix} \begin{bmatrix} o(t) \\ b(t) \\ y(t) \end{bmatrix}. \tag{4}$$

Realize that $o'(t) + b'(t) + y'(t) = 0$ indicates that the population is constant.

Equation (4) is a system of first order, homogeneous, ordinary differential equations. During the course of the term leading up to the final lizard project, students' learn how to write analytic solutions to (4) using the eigenpairs of the matrix and appropriate exponential functions. In addition, students in the transition course have extensive experience in using Maple to visualize and analyze solutions to (4). In the context of the lizard problem, they quickly realize that this particular transition from discrete to continuous did not result in a more realistic model. Each component of the solution still approaches the value 1/3 as time evolves and some components can even take on negative values—a contradiction to the basic assumption that each component represents the proportion of a lizard sub-population.

3.3 Better Continuous Model

3.3.1 Transition from Single to Multiple Species
When multiple species (*for example*, plants growing together in the same plot) use similar resources (*for example*, light, water, nutrients) they are said to be in inter-specific competition. This kind of competition typically results in reduced fecundity or survivorship. The competition is called density dependent if the

competing effects are directly proportional to the numbers present. The Lotka-Volterra model for inter-specific competition extends the logistic equation to the case of two species and thus incorporates the density-dependent effects of competition. A typical Lotka-Volterra model is given by the system of differential equations

$$N_1'(t) = \alpha_1 N_1(t)\left(1 - N_1(t)/C_1 - \lambda_{12} N_2(t)/C_1\right), \qquad (5)$$

$$\text{and} \quad N_2'(t) = \alpha_2 N_2(t)\left(1 - N_2(t)/C_2 - \lambda_{21} N_1(t)/C_2\right). \qquad (6)$$

In the absence of the competition terms, each population satisfies the logistic model. Moreover, N_2 individuals of species 2 have the same effect on the growth rate of species 1 as $\lambda_{12} N_2$ individuals of species 1 would have. This system of non-linear, homogeneous, ordinary differential equations can be studied using the built-in tools in Maple. Towards the end of the transitions course, students explore the Lotka-Volterra model through a carefully prepared Maple worksheet. Maple has the capability to produce direction field plots for this system while particular solution curves are obtained using the available numerical solvers in Maple. The explosion-extinction or doomsday vs. extinction model, where females rely solely on chance encounters to meet males for reproductive purposes, yields a population model with mathematical structure similar to the logistic model.

3.3.2 A Non-linear model for lizard populations

Based on the knowledge acquired from the transition outlined in sections 3.3.1 and in-class discussions about standard nonlinear population models like the logistic and explosion-extinction models, the students are guided to understand why the non-linear system of ordinary differential equations

$$y'(t) = y(t)^2 - y(t) + 2y(t)o(t) \qquad (7)$$

$$b'(t) = b(t)^2 - b(t) + 2b(t)y(t) \qquad (8)$$

$$o'(t) = o(t)^2 - o(t) + 2o(t)b(t) \qquad (9)$$

is a good model for the proportion of sub-populations of colored lizards. Notice that each sub-population satisfies the extinction-explosion model with a threshold of 1 in the absence of interactions. Thus, the sub-populations are not self-sustaining (unless they start exactly at the equilibrium value of 1) while the interaction terms have a positive effect on the growth rates of each sub-population. It is expected that this model will replicate the observed behaviour of the side-blotched lizards.

Numerical solution of this system shows that the long-term behaviour is independent of the initial conditions in the sense that the lizard sub-populations oscillate with fixed amplitude and frequency. However, the amplitude and center of the oscillations are determined by the initial distribution of the lizard sub-populations. Students compare this model's outcome to the data and realize that this model more closely models the population dynamics of side-blotched lizards. Slight perturbations of the above model result in different behaviour where the equilibrium population is either attracting or repelling; see the generalized "rock-paper-scissors" systems of differential equations models in (Weibull, 1995). For our model, the equilibrium point is a center.

3.4 Extensions of this material for higher-level classes

The system of non-linear ordinary differential equations presented in the previous section is a natural gateway to introduce students to many other topics in differential equations. For example, non-linear autonomous systems in two variables like the Lotka-Volterra model are typically analyzed using either a graphical approach where different regions of the phase plane are identified through the properties of the functions appearing on the right hand sides of the equations or an analytical approach where the system is linearized around the equilibrium solutions, and the eigenvalues and eigenvectors of the Jacobian Matrix are used to identify the nature of solutions near the equilibrium. The graphical approach does not lend itself to generalization when more species are present, while the analytical approach is perfectly suited for this extension. Since the proportion of lizard sub-populations satisfy $y(t) + o(t) + b(t) = 1$ for all times, one of the variables can be eliminated to reduce the system to two-dimensions. In particular, eliminating $o(t)$ yields

$$y'(t) = y(t)^2 - y(t) + 2y(t)\left[1 - b(t) - y(t)\right] \qquad (10)$$

$$b'(t) = b(t)^2 - b(t) + 2b(t)y(t) \qquad (11)$$

which has fixed points at $(0,0)$, $(1,0)$, $(0,1)$, and $(1/3,1/3)$. Although linearization of non-linear systems is presented in a junior/senior level ordinary differential equations course, it can also be the topic of an independent study for an advanced student in the transitions course. The fixed point at $(1/3,1/3)$ is a stable node. The amplitude of the stable oscillations of the lizard sub-populations depend on the initial proportion distribution. As far as we know, the exact nature of this relationship is not known.

Another possibility is to study the effect of varying an "interaction strength parameter." In particular this means solving a system of the form

$$y'(t) = y(t)^2 - y(t) + 2\lambda y(t)\left[1 - b(t) - y(t)\right] \qquad (12)$$

$$b'(t) = b(t)^2 - b(t) + 2\lambda b(t)y(t) \qquad (13)$$

for various values of the interaction strength parameter λ. It follows from the linearization analysis that when the interaction parameter is less than 1, the fixed point at $(1/3,1/3)$ shifts down and to the right in the phase plane and the fixed amplitude stable node becomes a decaying amplitude stable spiral where the variables decay to the fixed point in the long run. On the other hand, when the parameter is greater than 1, the fixed point shifts up and to the left in the phase plane and the fixed amplitude stable node becomes an unstable spiral so that the variables oscillate and grow without bound in the long run.

4. PEDAGOGICAL RESULTS AND IMPACT

Throughout the course, the students learn to model different phenomena using either discrete or continuous models. The complexity of the models increases as the term progresses. In each case, students were encouraged to investigate the models by changing initial conditions and by perturbing the model parameters. This exploration leads the students to conjectures that they often prove.

Because lizard mating strategies can be modeled by both discrete and continuous methods, it provides a good capstone project. Students appreciate an opportunity to place all of the topics introduced in the course into context through the analysis of a single problem. Such a problem helps prevent the students from compartmentalizing the mathematics that they learn to be used only in very narrow situations. By having the students use more than one mathematical model and set of tools, they see the value of multiple approaches and understand that the specifics of a model can affect the outcome (in our case, the long-term behaviour of the lizard population).

The course ran as a pilot from Spring 2004 through Spring 2005. Jones, Mukherjee and Weinstein (2004) report on evaluating the course, including the analysis of survey data on the students' impression of the course and its effect on their education. The course is in the process of becoming a required course for mathematics majors at Montclair State University. The department has voted to make the course required for mathematics majors, as part of changes to the programme that include students taking an additional two mathematics courses to graduate. This brings our total units up to the level of neighboring institutions.

A goal of the course is to excite students about applied mathematics early in the students' collegiate career. The lizard project requires students to use different types of mathematics, evaluate the models using Excel and Maple, and communicate their results in both oral and written formats. In short, the project simulates how applied mathematics is done and how applied mathematicians work and communicate. We expected to see an increase in the number of students who major in applied mathematics. Although two students changed their majors from mathematics education to applied mathematics, many of the other students were already majoring in applied mathematics. One of the two students who changed his major received full funding to attend a graduate programme in applied mathematics in fall 2005; the other doesn't graduate until spring 2007. She has already decided to go to graduate school.

We have seen an increase in the number of students who work on research projects with faculty and go to graduate school. From the first class of 18 students in spring 2004, four students paired up with faculty to perform research in Fall 2004; three students' works have resulted in article submissions to peer-reviewed journals. All of these students have presented their work at local conferences and some have presented their work at national conferences. One student was among the winners of the annual Undergraduate Poster Competition as part of the Joint Mathematics Meetings of the American Mathematical Society and the Mathematical Association of America.

ACKNOWLEDGEMENT AND NOTE

1. The authors were supported by National Science Foundation DUE Grant 0310753. Any opinions, findings, and conclusions or recommendations expressed in this material are those of the author(s) and do not necessarily reflect the views of the National Science Foundation.

REFERENCES

Arney, D.C. (1998) *Interdisciplinary Lively Application Projects.* Washington DC: Mathematical Association of America.

Arney, D.C., Giordano, F.R. and Robertson J. R. (2001) *Mathematical Models with Discrete Dynamical Systems.* McGraw-Hill Higher Education.

Fulford, G. Forrester, P. and Jones, A. (1997) *Modelling with Differential Equations and Difference Equations.* New York: Cambridge University Press.

Isaacson, D.L. and Madsen, R.W. (1985) *Markov Chains: Theory and Applications.* Malabar FL: Robert E. Kreiger Publishing Company Inc.

The Economist It's only a game, (1996), 339, 7961, p78.

Jones, M.A. and Mukherjee, A. (2005) A proofs course that transitions students to advanced, applied mathematics courses. In Dick Maher (ed) *Innovative Approaches to Undergraduate Mathematics Courses Beyond Calculus, MAA Notes,* 67. Washington DC: Mathematical Association of America, 39-52.

Jones, M.A.., Mukherjee, A. and Weinstein, G. (2004) A second year transitions course: Pedagogy, projects, and evaluation. In *Proceedings of the Hawaii International Conference on Statistics, Mathematics, and Related Fields.*

Klamkin, M.S. (1980) Mathematical Modelling: Die Cutting for a Fresnel Lens. *Mathematical Modelling,* 1, 63-70.

Pollak, H.O. (1959) Mathematical research in the communications industry. *Pi Mu Epsilon Journal,* 2, 494-496.

Sinervo, B. and Lively, C.M. (1996) The rock-scissors-paper game and the evolution of alternative male strategies. *Nature,* 340, 240-246.

Weibull, J.W. (1995) *Evolutionary Game Theory.* Cambridge MA: MIT Press.

6.6

MODELLING AND PROBLEM SOLVING IN BILLIARDS

Burkhard Alpers
University of Aalen, Germany

Abstract–*In this article we discuss several models for the billiards scenario and identify a typology of problems. For these, several examples are provided. Moreover, we discuss how technological tools like a Dynamic Geometry System and a billiards machine can be used in the problem solving process. We also report on using billiards problems in mathematical application projects.*

1. INTRODUCTION

In so-called "Mathematical Billiards" balls are modelled as points and the motion of such points on different table geometries is investigated (Drexler & Gander, 1998). This model can be found frequently in the didactical literature. It is used to treat reflection geometry and reflection groups (Stowasser, 1976; Shultz & Shiflett 1988), closed trajectories and irrational numbers (Barabash, 2003), envelope curves as well as chaotic behaviour (Bettinaglio & Lehmann, 1998). Although mathematical billiards is a suitable starting point to get to interesting mathematical problems, it lacks authenticity, since the real billiards scenario, including several balls, is not modelled. In order to overcome this problem, this contribution addresses the following questions:

- How can billiards be modelled more realistically (but not too in too complex a fashion) and which interesting learning opportunities and problems emerge?
- How can technological tools like Dynamic Geometry Systems (DGS), and a real billiards machine support modelling and problem solving processes?

Modelling the real billiards ball motion is quite complex (Marlow, 1995). Differentiating between sliding and rolling and inclusion of spin would exceed the capabilities of most high school students. Therefore, we suggest a sequence of models which is simpler, yet sufficiently realistic, for describing the behaviour of a billiards machine we constructed in our laboratory. For these models, we set up a typology of problems consisting of basically four types. Examples of these types will be presented where, even in the simplest model, interesting geometric questions emerge. Although most of the problems we identified could also be tackled using pencil and paper, technological tools can be very useful for getting and testing ideas and hypotheses as well as for motivating model propagation. We investigate the potential and limitations of a DGS (Cinderella), and a real billiards machine. Finally, we report on using the billiards scenario and the tools in mathematical application projects for mechanical engineering students.

2. SEQUENCE OF MODELS

We use three levels of modelling:
- Mathematical billiards with balls and impacts
- Physical billiards including friction
- Physical billiards including friction and loss of (kinetic) energy on collisions

We do not consider any spin in our models. This is certainly a restriction with respect to modelling the human billiards player but not with respect to the billiards machine we describe below.

2.1 Mathematical Billiards with Balls and Impacts and Exchange of Velocities

In this model we consider balls which have a certain two-dimensional velocity. Since friction is not included in the model, the balls move with constant velocity until there is a collision with another ball or with a cushion. Such a collision or impact is called centrical if the centres of mass of the two bodies involved lie on the normal line through the point where the bodies touch each other at collision time. With billiards balls, this is always the case. Moreover, an impact is called centrically straight ("head-on collision") if the velocity vectors of the two balls lie on this normal line as shown in Figure 1. Otherwise an impact is called oblique (cf. Figure 2).

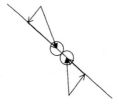

Figure 1. Head-on Collision.   Figure 2. Oblique Impact.

We first consider a centrically straight collision of two bodies. The principles of conservation of momentum and of energy provide equations from which one can easily compute the velocities $\overline{V}_1$ and $\overline{V}_2$ after collision (m_1 and m_2 are the masses involved, v_1 and v_2 are the respective velocities before collision)

$$\overline{V}_1 = \frac{2 \cdot m_2 v_2 + (m_1 - m_2) \cdot v_1}{m_1 + m_2}, \quad \overline{V}_2 = \frac{2 \cdot m_1 v_1 + (m_2 - m_1) \cdot v_2}{m_1 + m_2} \tag{1}$$

Two situations are interesting here:

When a ball hits another ball, we have equality of masses and we get $\overline{V}_1 = v_2, \overline{V}_2 = v_1$. This means that the relative velocity of the two masses changes its sign but the absolute value is preserved: $\overline{V}_1 - \overline{V}_2 = v_2 - v_1$.

When a ball with mass m_1 hits a cushion, we have $m_2 \gg m_1, v_2 = 0$, so we can neglect m_1 and get $\bar{v}_1 = -v_1$, $\bar{v}_2 = 0$ (we could also take limits for $m_2 \to \infty$).

When the collision is not centrically straight but oblique, then the velocity vectors are simply decomposed into normal and tangential components as shown in Figure 2. For the normal components the considerations of the centrically straight impact can be applied whereas the tangential components are not influenced by the collision and hence do not change.

Since there is no friction in the model, the balls can move infinitely long. So, it obviously does not model physical reality and hence we still call this model "mathematical billiards". Nevertheless, working with this model can already give answers to those practical questions which are purely geometrical in nature.

2.2 Physical Billiards Including Friction

In our next level of modelling, friction is included. We assume that the balls are always rolling (that is, there is no sliding phase) and friction is modelled by a constant force that is proportional to the gravitational force with proportionality factor μ_r. Therefore, the distance over time function is given by:

$$s(t) = v_0 t - \frac{1}{2}\mu_r g t^2 \text{ (until } \dot{s}(t) = 0) \tag{2}$$

This model is already quite realistic but experiments with the billiards machine show that it can be quite coarse and lead to results where a desired collision is not achieved.

2.3 Physical Billiards Including Friction and Loss of Kinetic Energy

In this model, we assume that when two balls collide centrically straight, conservation of momentum is still given but the absolute value of the relative velocity is no longer preserved, that is, we no longer have $\bar{v}_1 - \bar{v}_2 = v_2 - v_1$ but $\bar{v}_1 - \bar{v}_2 = k \cdot (v_2 - v_1)$ with $0 \le k \le 1$. Here, k is called the coefficient of restitution. This is equivalent to a loss of kinetic energy. Solving the equations for the velocities after collision leads to

$$\bar{v}_1 = \frac{(1+k) \cdot m_2 v_2 + (m_1 - km_2) \cdot v_1}{m_1 + m_2}, \bar{v}_2 = \frac{(1+k) \cdot m_1 v_1 + (m_2 - km_1) \cdot v_2}{m_1 + m_2} \tag{3}$$

There are three interesting special cases:
- If two balls collide, then the masses are equal and can be cancelled.
- If a ball collides with a cushion, then $m_2 \gg m_1, v_2 = 0$, and hence:
  $$\bar{v}_1 = -k \cdot v_1$$
- If the cue of the billiards machine collides with a resting ball, then $m_2 \gg m_1, v_2 = 0$, and hence $\bar{v}_2 = (1+k) \cdot v_1$.

The coefficients of restitution are different for collisions between balls, collisions between a ball and a cushion, and collisions between the "cue" of the billiard machine and a ball. They can be derived from measurements.

3. TYPES OF TASKS IN BILLARDS

We assume that the basic underlying physical laws are already known, that is, conservation of momentum and energy as well as motion with constant velocity or acceleration. The modelling task in billiards then consists of applying these concepts to the billiards scenario. For the first and second model, the modelling task is quite simple. The only challenge is to proceed from the one-dimensional to the two-dimensional situation by splitting up the velocity vector into two components. This is an important principle when considering multi-dimensional situations and hence has a value in itself. When getting to the third model, one has to introduce the coefficient of restitution which models the loss of relative velocity and hence implicitly the loss of kinetic energy. The motivation for advancing to the third model might come from observing the motion of real balls in the billiards machine (section 5) which shows that the law of reflection used in the second model is (more or less) violated in reality.

Once the above models have been set up, they can be used in solving a variety of problems. This problem solving activity using algebraic and geometric mathematical knowledge forms the main activity in our billiards scenario. We identified a typology of problems consisting of basically four types where up to three balls are involved:

- Given a start configuration and an impact, what happens?
- Given a start configuration, how can a carambolage be achieved?
- Construction of configurations and impacts with given properties
- Computation of system coefficients from real data

Within each type one can start with simple problems involving just one ball and one can use the results when working on more complicated problems. There are a lot of possible variations, so that students can be asked to vary tasks or to create their own tasks. We just give an example for each of the above types.

Example 1: Given the configuration depicted in Figure 3 (that is, the positions of the balls and the velocity vector of ball1), what happens?

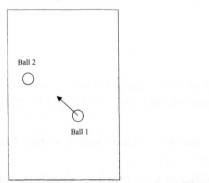

Figure 3. Impact computation.

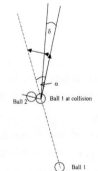

Figure 4. Impact with loss of energy.

This task can be considered in all model types: Always, the question of whether or not ball 1 hits ball 2 comes up. This can be solved geometrically by drawing a circle with radius two times the ball radius around the centre of ball 2 and intersecting it with the line through the centre of ball 1 going in the direction of the velocity vector. The intersection point (and hence the position of ball 1 at collision time) can also be computed algebraically by solving the corresponding system of equations (quadratic and linear). The behaviour after collision depends on the type of model under consideration. In both "mathematical billiards" and "physical billiards including friction" the path of ball 1 after collision is orthogonal to the path of ball 2. In the physical model including loss of kinetic energy, we have a deviation from orthogonality since ball 1 still has a non-zero component in the direction of the motion of ball 2 as is shown in Figure 4 (resulting in a deviation of δ where δ can be computed from $\tan\delta = \frac{1}{2}(1-k)\tan\alpha$).

Example 2: Given the configuration depicted in Figure 5 (that is, the positions of the balls), compute a velocity vector for ball 1 such that it impacts ball 2 centrically straight?

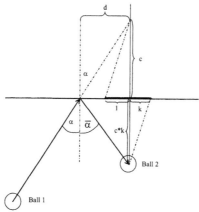

Figure 5. Collision with cushion and loss of energy.

In the first and second model this task can simply be solved by reflecting ball 2 at the cushion (or more precisely: at a line having a distance of a ball radius from the cushion). Similar tasks can be solved for several cushions by composing reflections. In the third model, the situation is more interesting and a possible solution is depicted in Figure 5. It can be solved algebraically using the equation $\tan(\alpha) = k \cdot \tan(\bar{\alpha})$ or geometrically performing a composition of reflection and dilation. This can be done by using intercept theorems as is shown in Figure 5 (any multiplication can be performed geometrically this way). Connecting the image of ball 2 (that is, its centre) with the centre of ball 1 provides the desired direction of the impact on ball 1.

Variations of this task include a consideration of the interval of angles such that the second ball is hit at all (the width of this interval can later be used as a quality criterion when several solutions are possible for a certain task in three cushion billiards).

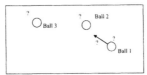

Figure 6. Setting up a configuration with certain properties.

Example 3: *Find a configuration (that is, the positions of the balls and a velocity vector for ball 1), such that ball 1 hits ball 2 and would hit ball 3 afterwards if the latter had not been moved away by ball 2 in the meantime (see for example, Figure 6).*

The solution of this task requires the variation of positions and of the velocity vector in order to achieve the desired effect. Performing the "experiments" with pencil and paper is quite tedious such that a simulation would be the right tool to use.

Example 4: *Given the (centre) positions of a ball rolling on the table before and after hitting a cushion, determine the coefficient of restitution for the collision between a ball and a cushion.*

For performing this task, one can compute lines of regression for the path before and after collision and compute k from the angles (for the formula cf. the model description above).

The learning goals for students working on these tasks can be outlined as follows:

- Students should recognise that basic geometrical concepts and operations like line, circle, reflection etc. are useful in solving practical problems. They should be able to use the concepts and construction methods (orthogonal line, tangent, reflection, dilation etc) to determine positions and directions. They should also recognise that and how geometric theorems like the theorem of Thales can be used for constructive purposes.
- Students should be able to set up corresponding algebraic models and set up and solve (using technology) the algebraic equations. This includes the equations for lines and circles, equations in triangles (rule of sine/cosine, theorem of Pythagoras), two-dimensional coordinate geometry and vectors.
- Students should learn how to tackle a complex problem by starting with special cases and simplified situations and reducing the more complex cases using the results obtained earlier.

These goals are important in secondary as well as tertiary education.

4. BILLARDS IN A DYNAMIC GEOMETRY SYSTEM (DGS)

In our scenario, the DGS can be used as a constructive tool for setting up problem situations and for finding positions and velocity directions. When setting up the problem situation (table, initial positions of balls), interesting questions already come up. It is a quite natural requirement to perform the table construction in such a way that it can be easily adapted to different table sizes. Therefore, it is not sufficient to draw a simple rectangle using four suitable points in a mesh. If one then moves one point, the rectangle property is destroyed. So, the first challenge for students consists of finding a construction using parallel and orthogonal lines such that the rectangle property is invariant when making the table larger. This is an open task with different possible solutions. Such a "design for modification" is also required later on when ball positions and velocity directions are to be varied.

If an exact solution for a problem is not known yet, the method of "soft construction" (Healy 2000) can be applied in order to get ideas. Here, one requirement for an exact construction is left out and the dragging mode of the DGS is used to find a position or angle such that the remaining condition is also fulfilled. This configuration is then investigated in order to detect an additional property which could be used for exact construction ("robust construction" in terms of Healy). Moving a soft construction to achieve an additional property is well-known from classical problems concerning constructible numbers (for example, trisection of an angle).

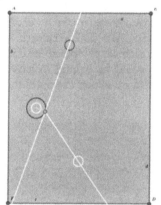 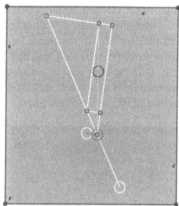

Figure.7 Carambolage construction.   Figure 8. Construction with loss of energy.

Figure 7 shows a soft construction for a carambolage in the model without loss of kinetic energy. The line through the lowest ball can be rotated and by doing this one can achieve that the second line goes through the centre of the third ball. Students then have to retrieve properties from this desired configuration in order to do a "robust construction". In this case, one might see that the centre of the first ball, when colliding with the second one, also lies on the Thales circle over the line segment between the centres of balls 2 and 3. Students can always make use of the Thales theorem in such a constructive way when a point with unknown position is the angular point of a right angle where points on the legs are known. Figure 8 shows a soft construction for the same problem in the model including loss of energy. A construction using the theorem of intersecting lines is used to guarantee the relation between $\tan(\delta)$ and $\tan(\alpha)$.

Soft constructions can be considered as blue prints for real mechanisms, that is, aiming devices. This way, one gets from modelling reality to realizing models. The usage of DGS for constructing and animating planar mechanisms can also be found in mechanism theory and design (Corves, 2004), and we actually realised a Cinderella construction as a coupler device as will be shown in the last section.

When using a DGS for solving problems in the billiards environment, there are also some limitations. One can only model potential motion lines (or paths). There might be other balls getting in the way or balls might stop because of friction. So, there is no dynamic modelling over several collisions and final resting positions cannot be constructed. Moreover, the construction does not give any feedback on the real motion ("real" with respect to one of the models outlined above) as a simulation does. Hence, one cannot make experiments to get initial ideas on how solutions might look like and one cannot check results.

5. BILLIARDS MACHINE

The real billiards machine is depicted in Figure 9. Balls can be positioned on the table using the 3-axes-machine and a sucking device. A cylinder that is attached to the vertical axis is used as a substitute cue. This way, impacts are always centrical, English shots cannot be performed. Since this is left out in all of our models, this is not an important restriction but rather desired. The cylinder can be programmed to move in a certain direction with a given velocity. Hence it can be used to test constructions performed on paper or with a DGS. We had a camera installed temporarily in order to record real motion from which we computed the coefficient of friction and the coefficients of restitution. It would be helpful to have such a device permanently installed combined with a software which computes the centre coordinates of balls in single pictures.

Figure 9. Billiards Machine.   Figure 10. Aiming Device.

We see several kinds of usage within the billiards learning scenario:
1.   check validity of a model, give rise to model propagation
2.   retrieve data for coefficients from camera pictures
3.   test and adapt coefficients by trials
4.   retrieve data on realistic restrictions (for example, maximum velocity).

Real objects can show the difference between model and reality and give rise to model progression. Seeing in the real billiards machine, that the law of reflection is violated gives rise to get from model stage 2 to stage 3.

6. DISCUSSION AND CONCLUSIONS

We consider two kinds of reality ("real" billiards, billiards machine) and a sequence of models (mathematical billiards, physical billiards with friction, physical billiards with friction and loss of energy). The billiards machine provides a subset of the "real human billiards" in that it only allows centrically straight shots in the plane. The sequence of models we describe includes more and more features necessary to model the reality of the billiards machine. Thus, it provides a pathway to proceed from problem solving in a simple model to that in a more complex, realistic model.

As yet, we have used the billiards scenario in mathematical application projects as well as in diploma theses for mechanical engineering students (Alpers, 2002). In the projects, students worked with the billiards machine reality. They have learnt about the underlying impact situation in their technical mechanics lectures. They recall their knowledge to set up the models presented in this paper, so the modelling part rather consists of understanding already existing models. The main task then is to solve problems within the models and check with reality how well the respective model fits. Solving a problem in a simpler model often gives hints on solving it in more complex models (making changes instead of starting from scratch). Moreover, it is a good engineering principle to keep a model as simple as possible.

One project dealt with the situation described in example 2. Students had to set up a CAS worksheet for computing centrically straight shots via a cushion in the different models. They then realised the shots on the billiards machine which showed that the model without friction produced shots without collision whereas the model including loss of energy led to a shot that was "nearly" centrically straight. Another student group had the task to investigate the model assumption in the most advanced model that there is a constant coefficient of restitution for ball-cushion collisions. They first had to understand the respective model and then to think up a measurement scenario for computing the coefficient k. They performed several trials with the billiards machine using different "ingoing" angles and measured the "outgoing angles" from which k can be computed. These resulted in values varying around 0.7 (from 0.6 to 0.8) showing that the modelling assumption is reasonable but the possible variation must also be taken into account. In yet another project, position data retrieved from a highspeed camera was used to compute the angles using regression lines. Moreover, for the configuration shown in Figure 8 (carambolage in the model including loss of energy), an aiming device was constructed that is depicted in Figure 10. For this, the respective DGS construction was used as a blue print for the device. The students had to first understand the construction and then convert it into a coupling structure with swivel and translational joints. The DGS turned out to be an excellent support for understanding the motion of the joints making up the device. We got the overall impression that the billiards scenario provides a wealth of project tasks for a variety of levels of difficulty and time frames.

REFERENCES

Alpers, B. (2002) Mathematical application projects for mechanical engineers – concept, guidelines and examples. In M. Borovcnik and H. Kautschitsch (eds) *Technology in Mathematics Teaching* Vienna: öbv&hpt, 393-396.

Barabash, M. (2003) Cycloids, Billiards, Lissajou: Using the Computer to visualize irrational numbers, and what can this be good for. *International Journal of Computers for Mathematical Learning*, 8, 333-356.

Bettinaglio, M. and Lehmann, F. (1998) *Mathematisches Billard - zwei Vorschläge zu projektartigem Unterricht*. Zürich: ETH, Bericht No. 98-08.

Corves, B. (2004) Computer-aided Lectures and Exercises: Graphical Analysis and Synthesis in Mechanism Theory. Proceedings of the 11th World Congress in Mechanism and Machine Science, Tianjin, China.

Drexler, M. and Gander, M.J. (1998) Circular Billiard. *SIAM Review* 40, 315-323.

Healy, L. (2000) Identifying and explaining geometrical relationship: Interactions with robust and soft Cabri constructions. *Proceedings of PME 24, Hiroshima 1*, 103-117.

Marlow, W.C. (1995) The Physics of Pocket Billiards. Palm Beach Gardens: Marlow Advanced Systems Technologies.

Shultz, H.S. and Shiflett, R.C. (1988) Mathematical billiards. *The Mathematical Gazette* 72, 95-97.

Stowasser, R. (1976) Küstenschiffahrt, Landmessen, Billard - drei Problemfelder der Geometrie. *Mathematikunterricht* 3, 24-52.

6.7

THE LOTTERY OF CASANOVA

Hans-Wolfgang Henn and Andreas Büchter
University of Dortmund, Germany

Abstract–*This paper presents the results and experiences of a project we did at Dortmund University with students in our teacher education department. The open-ended task to analyse the "Lottery of Casanova" provided a productive learning experience and leads naturally to various modelling approaches. These different approaches emerge through different perspectives of the situation. The context opens up a large potential for the design of modern mathematics teaching, oriented at current didactical concepts starting from phenomena and leading to the formation of mathematical theories.*

1. AUTHENTIC STOCHASTIC MODELLING

The Lottery of Genoa or "Lottery of Casanova", as we call it, which will be described, has been used within our project, "authentic stochastic modelling". Our goal was the development of learning environments for stochastic modelling for prospective primary and lower-secondary mathematics teachers. Our theoretical background was Wittmann's constructivistic approach to mathematics education as a "design science" (Wittmann, 1995). The project was embedded in a lecture series "Introduction to Stochastics" and "Didactics of Stochastics", which is usually taken by students in two consecutive semesters. Freudenthal (1983) characterizes stochastics as a classic example of realistic mathematics having many associations. Although some "simple mathematics" can be sufficient for stochastics modelling, it is often not clear how this simple mathematics has to be used. Typical situations are:

- Different solution strategies seem to be equally plausible, but lead to different results.
- Different solution strategies lead to the same result, but it is not obvious, why?

Working with such "paradoxa" is a suitable method to develop "Grundvorstellungen" (basic ideas) of stochastic concepts and modelling competence. We will illustrate this by using the "Lottery of Casanova" which was part of this project. Known worldwide and recently introduced in Germany, the Keno lottery has more or less the same structure as the historic Genoa lottery. The mathematical treatment with both lotteries is a paradigmatic example of authentic stochastic modelling. The open-ended task, to analyse the lotteries, provides a productive learning experience and leads naturally to various modelling approaches. These different approaches emerge through different perspectives of the situation. The context opens up a large potential for the design of modern mathematics teaching, oriented at current didactical concepts, starting from phenomena and leading to the formation of mathematical theories. The analysis of the structurally related Keno lottery proves to be a suitable tool to check the learning achievement of students in assessment situations.

2. THE LOTTERY OF GENOA

The origin of the Lottery of Genoa lies in the election of senators, which was introduced in 1575 following a coup d'ètat (Krätz & Merlin 1995, p65). Five citizens from a list of 90 were given the senator status to complement the existing council. Resulting from various bids on which citizens would rise to senator status, a lottery developed in Italy until the year 1643. This Lottery of Genoa was introduced to many European countries. In 1758, the Venetian adventurer Casanova (1725 – 1798) introduced a lottery similar to the model of the Lottery of Genoa in France. In this lottery five winning numbers were drawn out of ninety. The participants could put one number, two numbers (Ambe) or three (Terne) on their coupon. One drawn winning number was reimbursed with a multiple of fifteen of the stakes, two with a multiple of 270, and three winning numbers with a multiple of 5,200 of the stakes (Childs 1961). Table 1 gives an overview of the different options.

Options	Single number	Ambe	Terne
Numbers chosen	1	2	3
Drawing	5 out of 90 balls are drawn		
pay-out	15 times	270 times	5,200 times

Table 1. Options of the Lottery of Genoa.

3. DIFFERENT SOLUTION STRATEGIES GIVE DIFFERENT RESULTS

The lecture started with the fundamentals of descriptive statistics. Afterwards, beginning with the theme 'probability' a short introduction to the Lottery of Genoa was given, students were then asked to analyze this lottery. In small laboratory course groups students had to choose questions to work on. One question could be chosen by more than one group. All student groups started working on their chosen questions by trying to determine winning probabilities for the three lottery options. The most popular approach, presented here for "*Terne*", was the following:

$$P(\text{Terne}) = \frac{\binom{5}{3} \cdot \binom{85}{2}}{\binom{90}{5}} \approx 0.08123\%$$

For the denominator the students started – close to the situation – from the number of possibilities to draw five out of 90 numbers, oriented on the drawing process. For the numerator they calculated with the number of chosen and the number of drawn numbers. The argumentation usually went as follows: "*Three numbers have to appear within five numbers, therefore two are left over which have to be among the other 85 numbers*". Reflecting on this approach, students discussed how this approach is different to a fictional game option where five numbers are chosen of which three are then among the drawn numbers. For this fictive option the same formula holds. Obviously, it is more difficult to choose three numbers, which are then all drawn, compared to choose five numbers, out of which three are then drawn. After that, the students tried to save the above approach by the procedure:

$$P(\text{Terne}) = \frac{\binom{5}{3}}{\binom{90}{5}} \approx 0.00002\%$$

Further reflection on this attempted rescue lead to more objections. If this approach was correct, the chance to choose three winning numbers would be the same as the chance to choose two winning numbers:

$$P(\text{Terne}) = \frac{\binom{5}{2}}{\binom{90}{5}} = \frac{\binom{5}{3}}{\binom{90}{5}} \approx 0.00002\%$$

On the basis of these considerations the students realized that this approach to model the Lottery of Genoa led to a dead end.

4. DIFFERENT STRATEGIES LEAD TO THE SAME RESULT

To come out of the dead end the students started to discuss the (to them) better known German lottery "6 out of 49", which led to further solution strategies in the small working groups. Again, we look at the *"Terne"*.

Approach A: $P(\text{Terne}) = \dfrac{\binom{3}{3} \cdot \binom{87}{2}}{\binom{90}{5}} = \dfrac{1}{11,748} \approx 0.00851\% \cdot$

To calculate this winning probability the drawing of the numbers is – *close to the situation* – modelled with the Laplace approach as a stochastic experiment. Five numbers are drawn out of 90. The denominator is the number of possible results. Taking into consideration that three of the numbers drawn have to coincide with the three chosen numbers and that the other two drawn numbers have to come from the 87 not chosen numbers, gives the term for the numerator. Another interpretation of the same model was: Choosing colours three balls from an urn holding 90 balls as personal winning balls. What now is the probability to get these three balls by drawing five balls without replacement?

Approach B: $P(\text{Terne}) = \dfrac{\binom{5}{3}}{\binom{90}{3}} = \dfrac{1}{11,748} \approx 0.00851\% \cdot$

The second Laplace approach for modelling the Lottery of Genoa looks at the possible and the favourable chosen numbers. *In spite of the chronology* of the lottery the assumption is made that the drawn numbers are fixed and the process of doing the lottery is taken as the stochastic experiment. Three out of 90 numbers are chosen; therefore the denominator counts the number of possible choices. For the numerator the three chosen numbers need to be among the five drawn numbers. In the alternative way of thinking this means: The later drawing colours five balls from 90 as winning balls. How big is the probability by choosing three different numbers (model of drawing without replacement) to get three out of the five coloured balls?

Approach C: $P(\text{Terne}) = \dfrac{5}{90} \cdot \dfrac{4}{89} \cdot \dfrac{3}{88} = \dfrac{1}{11,748} \approx 0.00851\%$.

A third, also discussed, modelling approach works directly with the alternative interpretation. Again – *against the chronology* of the lottery – the model is based on an urn with five "winning balls". The process of doing the lottery, here the drawing of three balls without replacement, is calculated: For putting the first number on one`s coupon the chance to gain a winning ball results from the ratio of five winning balls to 90 balls altogether. In the same way the chances for the second and third number are found. The resulting probability for winning is the product of the three fractions according to the path rule.

A look at the three reduced fractions for the three approaches (A, B, C) and the winning percentages shows that the numerical check of all three approaches results in the same chance. The correspondence of the solutions can be seen as the mutual validation of the three solution strategies. Students, who prefer one of the approaches and have no insight in the other approaches, can be convinced that the other approaches also work. They can be motivated to try to understand the other approaches, too. The question "why" becomes interesting and leads to a structural validation of the different solution approaches.

5. WHAT DOES THE STATE EARN?

Following the introduction of the random variable concept the Lottery of Genoa was discussed again. Emphasis was given to the question of most interest for the organizer of the lottery: "How big is the expected profit?" For an intuitively apparent idea, which will later become the expected value, the stakes are chosen to be 1 € per bet. The winning probability for the single-number-bet is $^1/_{18}$. For a "normal" course there is exactly one win of 15 € per 18 players in the long run. That means, 18 times the state receives 1 €, one time the state will have to pay out 15 €. If 18 players have placed a single-number-bet

$$EP(\text{single number}) = 18 \cdot 1 \, € - 1 \cdot 15 \, € = 3 \, €$$

will be the expected profit for the state. The abbreviation "EP" stands for "expected profit". Analogous considerations for the other two possibilities to bet result in the profit expectations

$$EP(\text{Ambe}) = 801 \cdot 1 \, € - 2 \cdot 270 \, € = 261 \, €,$$
$$EP(\text{Terne}) = 11,748 \cdot 1 \, € - 1 \cdot 5,200 \, € = 6,548 \, €.$$

Now it would be incorrect to deduct that the bid "Terne" is the most profitable for the state, because the analysis is based on different numbers of participating gamblers. The necessary remedy is obvious: The comparison must be based on the relative profit per gambler

$$\text{ep(single number)} = \frac{\text{EP(single number)}}{18} \approx 0.17\,\text{€},$$

$$\text{ep(Ambe)} = \frac{\text{EP(Ambe)}}{801} \approx 0.33\,\text{€},$$

$$\text{ep(Terne)} = \frac{\text{EP(Terne)}}{11,748} \approx 0.56\,\text{€}.$$

From the point of view of the state the option "Terne" is indeed the most profitable, if everything runs smoothly. This is also the option which will – because of the large possible profit – attract most gamblers. But the game could not run smoothly and it might happen that about 10% of all gamblers have chosen the winning numbers. As the profit margin is a fixed number (5,200 times the stakes), the bank might collapse. To analyse the question "How big is the expected profit for the state?" sensibly further considerations are necessary. If we are interested in the profit which the state can make as additional income, further information have to be available or further assumptions have to be made. These normative actions consist in the decision how many gamblers can participate in the lottery, how big the stakes are, how many of them choose which option of the lottery and which spread of winners is to be expected for the three different options. Suitable methods for these issues can be developed easily. For this, we analyse in detail the relative profit per gambler for the single number bid from the viewpoint of the state's profit and loss probability:

If the state accepts a single number bid, then it has to pay 14 € with the probability $^1/_{18}$ and with the probability $^{17}/_{18}$ it wins 1 €. The single number bid can be described as a random variable Z_1, measured in €, that can take the values -14 and 1 with a probability of $P(-14) = ^1/_{18}$ and $P(1) = ^{17}/_{18}$. The expected profit of the state per gambler, measured in €, can then be described by

$$E(Z_1) := \text{ep(single number)} = \frac{\text{EP(single number)}}{18} = \frac{17 \cdot 1 - 1 \cdot 14}{18}$$

$$= \frac{17}{18} \cdot 1 + \frac{1}{18} \cdot (-14) = P(Z_1 = 1) \cdot 1 + P(Z_1 = -14) \cdot (-14) = \frac{1}{6} \approx 0.17.$$

What we have found is the concept of the *expected value E(Z)* of a random variable Z. We compare this approach with the calculation of the arithmetical mean of a sample using relative frequencies: If 1,000 gamblers bid on one number and 50 of them have won the balance of the state, again in €, will be

$$\text{Average profit per gambler} = \frac{950 \cdot 1 - 50 \cdot 14}{1,000} = \frac{950}{1,000} \cdot 1 + \frac{50}{1,000} \cdot (-14) = 0.25.$$

Substituting the relative frequencies in the formula for the arithmetic mean of a sample by the probability of a random variable, leads to the expected value of the random variable. Analogously, the expected values for the two other options are:

$$\text{Ambe: } E(Z_2) = \frac{261}{801} \approx 0.33; \quad \text{Terne: } E(Z_3) = \frac{6,548}{11,748} \approx 0.56.$$

The positive expected value of the Lottery of Genoa obviously does not guarantee a profit for the state. If, by chance, among 18 gamblers there are two winners, the state looses $16 \cdot 1$ € $- 2 \cdot 14$ € $= -12$ €. A sensible analysis calls for a prediction of the spread to be expected. Our naïve approach for the expected value has lead to a formula which is analogue to the arithmetic mean. Accordingly, **variance V(Z)** and **standard deviation** σ_Z are introduced. These definitions, again, are analogue to the definitions of empirical variance and standard deviation of samples, where probabilities take the place of relative frequencies.

$$V(Z_1) := (1 - E(Z_1))^2 \cdot P(Z_1 = 1) + (-14 - E(Z_1))^2 \cdot P(Z_1 = -14) \approx 11.81 \quad \text{and}$$

$$\sigma_{Z_1} := \sqrt{V(Z_1)} \approx 3.44 .$$

The corresponding values for the other two options are

Ambe: $\sigma_{Z_2} \approx 13.47$; Terne: $\sigma_{Z_3} \approx 47.97$.

Compared with the expected value the value of the standard deviation is rather large and points at the risk for the state.

In order to come to more precise statements about the "Lottery of Casanova", the topic was re-introduced for a third time at the end of the lecture course in relation to the binomial and normal distributions. Some of the students proposed the following, again shown using the option Terne: One bet is looked upon as a Bernoulli experiment with the possible results 0 = "the state wins" and 1 = "the gambler wins". The corresponding random variable therefore possesses the values P(X=0) = $\frac{11,747}{11,748}$ and P(X=1) = $\frac{1}{11,748}$ =: p . We consider a lottery where n gamblers want to play Terne. Together with the sensible model assumption that all gamblers bet independently from each other, the situation can then be described as a random variable Z following a binomial distribution with parameters p and n. The state receives 1 Euro from each gambler and pays 5,200 Euro to each winner. If m is the number of winners a profit for the state is only possible if $n - 5{,}200{\cdot}m \geq 0$. For a given n there should not be more than $m^* := \mathrm{trunc}(\dfrac{n}{5{,}200})$ winners. The risk for the state is that there are more than m* winners. For that the probability is given by

$$P(Z \geq m^*) = \sum_{i=m^*+1}^{n} \binom{n}{i} \cdot p^i \cdot (1-p)^{n-i}$$

Trying to evaluate this formula for concrete values of n using MAPLE or any other CAS, the limits of the computer are reached fast (by only calculating the sum from 0 to m*, too.). So, the normal distribution has to be used as an approximation for the binomial distribution. Using the expected value $\mu = n{\cdot}p$ and the standard deviation $\sigma = \sqrt{n{\cdot}p{\cdot}(1-p)}$ of the binomial distribution, the following formula has to be evaluated

$$P(Z \geq m^*) \approx \int_{m^*}^{\infty} \frac{1}{\sqrt{2\pi}.\sigma}.e^{-\frac{1}{2}\left(\frac{x-\mu}{\sigma}\right)^2} dx$$

This is a much easier task for MAPLE. For example, the risk probability for n = 100,000 is approximately 0.00041, for n = 1,000,000 this probability is 0 (calculated with a precision of 20 digits).

6. THE KENO LOTTERY

The Keno Lottery, introduced in 2004 in some German federal states is a good real life example for mathematics teaching. The lottery has several aspects which are of interest for teaching. It is the first virtual lottery in Germany. Winning numbers are drawn by a computer, which has been developed especially for this purpose by the *Fraunhofer Institut für Rechnerarchitektur und Softwaretechnik* – a fact that motivates to reflect on the generation of "random numbers" by computers.

The Keno Lottery is played mostly via the Internet and aims at adolescents and young adults. Keno is probably the world's oldest lottery. More than 2,000 years ago it was already played in China as *White Doves Game*. Since quite some time Keno is very popular in Anglo-Saxon countries.

From a stochastic point of view, Keno's structure is similar to the Lottery of Genoa. 20 out of 70 numbers are drawn. Gamblers have to decide to play one of the options Kenotype 2 to Kenotype 10 and then choose two to ten numbers, respectively. Figure 1 shows a lottery coupon of the German Keno Lottery which allows the gambler 5 different bets. The gambler has to mark with a cross the chosen Kenotype ("Anzahl getippter Zahlen") for every bet and then to mark the desired numbers with a cross, for example 5 numbers when the Kenotyp 5 is chosen. As stakes gamblers can choose (with a cross in "Einsatz") 1 €, 2 €, 5 € or 10 €. "plus 5" is an additional game, a cross in "Anzahl der Ziehungen" indicates the duration (in weeks) of validity of the coupon.

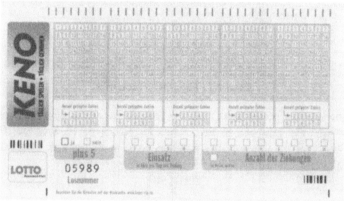

Figure 1. The German Keno Lottery.

In each case the number of numbers chosen and numbers drawn are different. Contrary to the Lottery of Genoa the gambler does not only win if all chosen numbers are drawn. In the options with eight, nine or ten numbers to choose (Kenotype 8, 9, and 10), the gambler also wins if none of his or her numbers are selected. Therefore, Keno is the first lottery where unlucky fellows win. Altogether there are 36 prize categories. And, obviously, there is a lot to calculate. On the reverse side of the Keno Coupon (Figure 2) all winning classes are described. The first column shows the Kenotyp and the second the number of correctly chosen numbers. The columns 3 to 6 show the profit for the stakes 1 €, 2 €, 5 €, and 10 €.

As for the Lottery of Genoa the Keno Lottery has fixed winning ratios, between one and 100,000 times the stakes. Theoretically the lottery can turn out a loss for the provider. To avoid this, the two highest winning categories have an additional provision. It states that in Kenotype 10 for 10 correctly predicted numbers a win of 100,000 times the stakes is only given to five winners. If more than five gamblers win in this category the maximum sum is divided among all those winners. Similar rules apply to Kenotype 9 for 9 correctly predicted numbers, with a maximum number of ten winners. The actuality of this lottery motivates students to deal with it in detail.

For an analysis of the winning situation described in Figure 2 we define the event   **m|n:** "a hit rate of m correctly predicted numbers in Kenotype n"

Figure 2. The prize categories.

Now the winning possibilities from the perspective of the numbers drawn can be deduced easily. For example, to have 8 hits for Kenotype 10, 8 of the 20 winning numbers and 2 of the non-winning numbers have to be marked. This leads to the probability

$$P(8|10) = \frac{\binom{20}{8}\binom{50}{2}}{\binom{70}{10}} \approx 0.00039$$

For the same Kenotype 10, the probability to have chosen none of the winning numbers is only

$$P(0|10) = \frac{\binom{50}{10}}{\binom{70}{10}} \approx 0.0259.$$

Generally, the probability to have m hits in Kenotype n is:

$$P(m \mid n) = \frac{\binom{20}{m} \cdot \binom{50}{n-m}}{\binom{70}{n}}.$$

The largest single winning probability is reached for 2 hits in Kenotype 3 with
$$P(2|3) \approx 0.174.$$
The probability to win anything at all in Kenotype n is given by

$$P(\text{Win} \mid \text{Type n}) = \sum_{\substack{m=0 \\ \text{m leads to profit}}}^{n} P(m \mid n).$$

This winning probability takes its maximum for Kenotype 4 with $P(\text{Win} \mid \text{Type 4}) \approx$ 0.321 and its minimum for Kenotype 2 with $P(\text{Win} \mid \text{Type 2}) \approx 0.079$.

As for the "Lottery of Casanova" an interesting task for Keno is to analyze the state's risk for each of the game options.

7. CONCLUSION

This paper on making use of the "Lottery of Casanova" in university teacher education shows some aspects of problem contexts which have a positive effect for productive learning environments. To design a teaching-learning environment according to current conceptions in mathematics education contexts including these aspects have to be developed. Searching for and developing such contexts is therefore essential for the further development of mathematics education. But a *good context* alone is not sufficient. It has to be used for good teaching, which is not always easy in a concrete teaching-learning situation. It is important that enough time is planned for individual learning and the reflection on this processes, as well as for different solution strategies and their discussion.

REFERENCES

Büchter, A. and Henn, H.-W. (2004) Stochastische Modellbildung aus unterschiedlichen Perspektiven – Von der Genueser Lotterie über Urnenaufgaben zur Keno Lotterie. *Stochastik in der Schule*, 24 (3), 28-41.

Childs, J.R. (1961) *Casanova: A Biography Based on New Documents*. London: George Allen and Unwin Ltd.

Freudenthal, H. (1983) *Didactical phenomenology of mathematical structures*. Dordrecht: Reidel.

Krätz, O. and Merlin, H. (1995) *Casanova. Liebhaber der Wissenschaften*. München: Callwey.

Wittmann, E.Ch. (1995) Mathematics education as a "design science". *Educational Studies in Mathematics*, 29, 355-374.

6.8

MODEL TRANSITIONS IN THE REAL WORLD: THE CATWALK PROBLEM

Thomas Lingefjärd and Mikael Holmquist
Gothenburg University, Sweden

Abstract– *In our courses, we use examples from real-world phenomena as a forceful argument why modelling could be useful, valuable, interesting, and lead to a deeper competence in mathematics. This study examines whether a "real-world phenomena problem" makes any difference to the way students approach a problem.*

1. INTRODUCTION

The usefulness and value of mathematical modelling is often discussed, at least at Swedish universities. Although the concept of mathematical modelling is embedded in the national curriculum for secondary education in Sweden, most prospective teachers are not educated in the process of mathematical modelling. There is consequently always a strong demand for valid reasons as to why one should teach mathematical modelling and in what way.

Our purpose with the mathematical modelling course we gave was to illustrate the rich area of real-world phenomena to a class of prospective mathematics and physical training teachers in the teacher programme at Gothenburg University, and to encourage them to teach mathematical modelling as future teachers. The group of students consisted of two women and eight men. Since all of the students except one were athletes and interested in physical training as a school subject, we decided to select a problem about physical movement. In the spring of 2002, Bob Speiser presented the so-called "Catwalk" problem in a research seminar) at Gothenburg University. The problem is about how to determine the movement of a specific cat, a cat moving in a particular way by changing from a walk to a gallop. The problem was presented by photographs of the cat and by encouraging the students to actually do the catwalk in order better to understand the vast difference between a cat's change from a walk to a gallop. Since many of them were active athletes, they showed a natural interest in the cat's moving. To handle the transition from the real world to the model and back again, some of the students decided to use a corridor lane and made marks on the floor, mirroring the walk of the cat, but enlarged by a factor of 10 since humans simply are too large to do the catwalk in scale of one-to-one.

2. MODEL TRANSITION

Model transition is a central part of the modelling process but it can also be seen as an important activity in the sense of learning how to use mathematical concepts and methods. Another important activity is to learn how and when a mathematics tool (for example, a calculator or software) should be used, and to value the results from this mathematics tool. Since we are educating prospective mathematics teachers for the Swedish school system where such tools are common, this is an important and identified goal for their programme.

The transition from real-world phenomena to a mathematical model and back again, is in this case expressed by the students written reports of the catwalk problem. This in turn involves another hidden circumstance, which the students in general have weak experience and knowledge of, namely the process of writing mathematics.

For many different subjects, in school as well as at universities, the importance of fostering the students to write as part of the learning experience is natural. For the last 10 or 15 years there has been a growing concern about how to create and implement relevant writing assignments also throughout the mathematics curricula. Different reasons such as an increasingly advanced and available technology among learners of mathematics, a growing interest among teachers of mathematics at all levels to learn more about what and how their students learn, and a likewise growing certainty among researchers in mathematics education that assessing knowledge in mathematics is much more than just a written test at the end of a course, have all contributed to a movement about the use of writing assignments in mathematics.

Many authors have discussed the need and importance of students' ability to communicate mathematics verbally and literally (NCTM, 2000; Meier & Rishel, 1998; Cowen, 1991). Another continuing ongoing discussion regarding the need for students to take responsibility for their own learning often connect this aim with empowering the students to read, write and discuss mathematics intelligently (Lingefjärd & Kilpatrick 1998; Cobb, 1986). The communicative aspect of learning mathematics is so strong and coherent that the Principles and Standards for school mathematics (NCTM, 2000) have it as a strand for all grades. It concludes that:

Writing is a valuable way of reflecting on and solidifying what one knows, and several kinds of exercises can serve this purpose. For example, teachers can ask students to write down what they have learned about a particular topic or to put together a study guide for a student who was absent and needs to know what is important about the topic. A student who has done a major project or worked on a substantial long-range problem can be asked to compare some of their early work with later work and explain how the later work reflects greater understanding. In these ways, teachers can help students develop skills in mathematical communication that will serve them well both inside and outside the classroom. Using these skills will in turn help students to develop deeper understandings of the mathematical ideas about which they speak, hear, read, and write. (NCTM, 2000 p352). This is also the case in the national syllabus for both the compulsory school and the gymnasium in Sweden.

Develop their ability to understand, carry out and use logical reasoning, draw conclusions and generalize, as well as orally and in writing explain and provide the arguments for their thinking, (Skolverket, 2005, p16). An important part of solving problems is designing and using mathematical models and in different ways communicating mathematical ideas and processes of thinking. Both in everyday and vocational life, there is an increasing need to understand the meaning of and be able to communicate on issues with a mathematical content. (Skolverket, 2000, p62)

Whenever we try to communicate mathematical conjectures, statements and/or mathematical ideas, we realize that the language of mathematics is specific and precise. More than with most other sciences, in the writing of mathematics "what you see is what you get". There are in general few hidden meanings, little should be left unsaid, and there is not too much room for interpretation or context. When students write about the solution of a mathematical modelling problem, they are both writers in the formal language of mathematics, at the same time as they are asked to explicitly argue for their choice of model, explain their interpretation of the results, and explore the conclusions of their modelling activity. All together this complexity often causes students to be reluctant to the idea of writing up a report about what has taken place when they have solved a modelling problem. We strongly argue that the students will learn more about mathematics when they write, as well as they, unsurprisingly, also learn how to write. To respond to student writing is doubtless one of the most challenging part of teaching writing in a mathematical environment. It takes a tremendous amount of time as well as it demands a substantial deal of intellectual activity, it also affects to a large extent how students feel about their ability to write.

The quality of students' writing (as defined by the teachers and researchers involved) developed over time points to the possibility that, with extensive experience and feedback from teachers and from peers, students come to learn the features of the genre that will be valued by their teachers. (Morgan, 1998 pp41-42)

Whenever one wants to examine student's performance in any subject, there is evidently a need for understandable and functional criteria for the evaluation of the achievement. At the first level of examining a written assignment in mathematics, one should ask if (Lingefjärd & Holmquist, 2001 p207):

- The mathematical content – is it correct in terms of notation, figures, diagrams, and conclusions?
- The report – is it written in a language and style that is structured, clear and distinct?
- The problem – is it solved, generalized, explored, and investigated to the limit of all available resources?

Nevertheless if we are using a checklist or not, we are obliged to inform our students what criteria we are using when we are grading their written assignments. This ought to be followed by clear and distinct comments on the papers.

To decide criteria for assessing students' achievements in a written assignment is naturally influenced by the assignment itself. What criteria that will be possible to use, is in a way already decided when the task is formulated. When we are designing mathematical assignments, we must ask the following kind of questions:

- What mathematical concept or knowledge am I trying to evaluate with this assignment?
- Is the assigned time length for the assignment appropriate?
- Are the students prepared for this assignment?

- What relevant topics do I value, and which of them is relevant here?
- Does my assignment indicate my priorities to the student?
- Is technology an option or a requirement to accomplish the assignment?

All of this was in our minds when we chose to have the Catwalk problem as one out of three examination items in our mathematical modelling course.

3. THE CATWALK PROBLEM

We decided to use the catwalk problem in a modelling course in the spring of 2004. One aspect of this problem is that it tests how much calculus students actually know or understands. Speiser and his colleagues have tried the problem with college students as well as high school students. We decided that it could also be used with students who have taken courses in calculus at university, and designed a study inspired by, and partly similar to, the one Speiser and his colleagues reported on. Basic calculus is a way to study change and motion. In the catwalk problem, one challenge is to build connections between local rates of change and total changes, based on real-world data. The problem was originally designed to expose some of the complexity inherent in the use of mathematics to examine motion. Work on this problem by college calculus students has been reported in three papers by Speiser and Walter (1994a, 1994b, 1996).

The problem is illustrated by a series of photographs that there is not space enough to reproduce here in full. The photographs consist of 24 frames of a single cat, entitled *Cat in Walk Changing to a Gallop*. Edward Muybridge made the photos in 1880, by using 24 cameras that were activated successively at intervals of 0.031 second. They show the cat against a background grid, composed of lines spaced 5 centimeters apart. Every tenth line is darker. The 24 photographs show the cat over a total time of action of 0.71 second. We gave our students copies of the photos, the information described above, and asked them to construct one or two mathematical models describing how the cat moved over that time period. They were specifically asked to answer the following two questions: How fast is the cat moving in Frame 10? How fast is the cat moving in Frame 20? Figure 1 illustrates two consecutive frames out of the 24.

Figure1. Catwalk.

4. STUDENTS WORK AND ASSESSMENT

The students were all prospective teachers at Gothenburg University, preparing to become teachers of mathematics as a complimentary subject to physical training or music in the grades 7 to 9 or 10 to 12 (Swedish compulsory school or gymnasium). All the students, two women and eight men, in the group took the modelling course at the end of their teacher education programme, meaning that they took the course during their seventh or eighth term out of total 9. Earlier in the mathematics education part of the programme, the students had taken courses in number theory, Euclidean geometry, linear algebra, real analysis, probability, statistics, and discrete mathematics. The catwalk problem was presented as part of the final examination during the spring 2004. Technology was presented as an option to accomplish the assignment.

Student 4: In this case, we can compare with the speed of a human. If I, for instance, will run 100 meters, it maybe takes 16 sec and if we divide 100 meters with 16, we get a speed of 6.25 meter/sec equal to 625 cm per sec. I think that the whole sequence of the 24 frames represents the start of the cat's movement forward.

For the students the first thing was to try to understand and to get an idea of the problem situation, the context. At this stage they were in the beginning of the transitional process and many discussions followed both among the students and between the students and the teachers. As a result the students were asked to try to imitate the Catwalk in order to better understand the explosiveness of a cat moving from walk to gallop. Figure 2 & 3 illustrates the measuring of a walk according to scale, and a student trying to illustrate the movement of a cat.

Figures 2 and 3. Students trying to do the Catwalk.

Some of the students also made use of the illustration when they were in the concluding part of their writing.

Student 10: I calculated the speed of the cat in frame 10 to 1.24397 m/s and the speed in frame 20 to 3.24043 m/s. To verify the relevance in our result, Lisa and I performed the corresponding movement for a human. We assumed that a human is about 10 times larger than a cat. Therefore we enlarged the x-values and the y-values 10 times. This resulted in a movement of 13 meter at 7.1 seconds. We labeled the whole distance by notes in a corridor, and I tried to do the catwalk. Not all of the students were interested in really doing the catwalk and instead they chose to go more straightforward to the measurements. For some of them it ended up with a report which lacked information about what was behind the tables of measured values. Despite the fact that they perhaps got acceptable values for the speed of the cat there were no clear indication as to why the student was performing her or his calculations. As part of the assessment procedure that lead to comments from the teacher. One important part of the assessment work was to find out to what extent the student's arguments support the conjectures and conclusions they present in their reports. As part of the assignment they were asked to explicitly argue for their choice of model. When arguments like that should be communicated, an essential part is the language, how the students use the linguistic possibilities to intertwine the formal mathematical language with a descriptive non-formal mother tongue. In this context, writing can serve as tool for the teacher to try to understand students understanding. It is important to acknowledge that this linguistic transition back and fort is hard for the students to handle, they often position the mathematical language to be in control at the same time as they allow the descriptive language to weaken the precision in their mathematical formulations.

Student 2: I like to clarify that I know that one can fit any polynomial to any set of points. Even if one has a set of 51 data points, one can fit a polynomial of degree 50 to that set. So a polynomial can be adjusted to fit perfectly to any graph, even the steepest.

Student 4: When I look at the cat's movement in the frames, I interpret that the movement after frame 11 straightens out and can be seen as a fairly straight line. This might mean that the cat relatively quick reach his or hers top speed.

All students in the group expressed a clear ambition to find arguments for their choice of mathematical model, but the students were different in the way they argued for the connection between their mathematical model and the physical situation it was selected to describe and represent. This argument constitutes a central part of students writing and makes it possible for the teacher to get an insight into the students' ability to deal with the transition between model and reality.

Student 2: In the program Curve Expert there are a variety of different functions one may use and I have chose to analyze my values with help of a 10-degree function. By testing my way trough the different alternatives in Curve Expert, I found that the 10-degree function is the best choice, according to standard error deviation and correlation coefficient. The values from a 10-degree function were the best and closest I could get. Interpretations are also an important part of the students' transitional reasoning, the ability to value and understand the mathematical model and to translate the mathematical expression in light of the real situation. Many of the students seem to simplify this process into an analysis of the graphical situation created by the function they picked in the curve fitting tool as a way to describe the cat's movement.

Student 8: The diagram also gives a clear picture of the cat's movement during the whole period, where you clearly see how the cat's movement changes during three different phases. During the first part, the cat restrains its speed somewhat and almost stands still at the end. Maybe the cat saw something and prepared to be ready for action? Thereafter starts a huge acceleration during a short time. One explanation might be that the cat chase something, is forced to change direction with the loss of speed as a result, and then speeds up again to catch the prey. The hunt is over. Interpretations may also lead into conjectures and conclusions regarding the validity of a mathematical model and several of the students demonstrated that kind of reasoning.

Student 6: I want the cat's initial speed to be zero, which forced me to construct my own second degree model. My model is not quite accurate since the speed increase to levels far above what is reasonable for a cat. Nevertheless, I consider it useful for the limited time the frames describe, and therefore I consider it useful as a mathematical model for frame 1-24.

When the students were at the end of their writing they reached the question of how to verify the results. We found it interesting to notify how several of the students searched for basis and facts about a cat's movement scheme in order to justify and validate the results from their mathematical model.

Student 3: After a frantic search, I found a source who told me that cats can run up to 50 km/h. That makes my calculations look more realistic, since the cat probably not has reached its highest speed during the 24 frames.

Student 9: A car that drives slowly (for instance at "Watch out – playing kids" signs) has a speed of about 30 km/h. A bicycle in a slow but steady pace will go about half that speed, equal to approximately 15 km/h. A cat in acceleration runs probably at about the same speed as this bicycle.

Apparently, it also turned out that several students became so interested in investigations and what we would call "sophisticated mathematical models", that the most straightforward solution was omitted by these students. Merely by plotting the measured values by hand or by computer and thereafter using a simple ruler, a close look could give a good hint about the expected range for the values asked for in the problem. Figure 4 illustrates this simple and unsophisticated method.

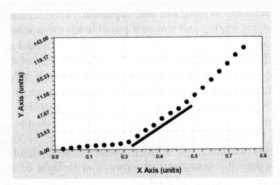

Figure 4. A straightforward approach to the Catwalk problem.

The evaluation was not only about verification of calculated values and figures. In this part of the writing process, the students also had the possibility to formulate themselves around how they valued their fulfilled assignment.

Student 7: At the same time the cat's position of its head was hard to determine, partly because of the dark background on the photos. Since I have picked the cat's nose as my measuring point, this turn of the cat's head might play a major role for my results. This means that I have to take into account that the cat might have turned its head several times, thereby yielding that my measuring of distance is altogether wrong.

In this part of the writing process some of the students demonstrated an evaluation of their own mathematical modelling competencies.

Student 1: Finally I like to express that it should be observed that the eventual sources of error in my modelling are: My ability to measure accurate with a ruler.

the photo quality in the frames; my choice of measuring point on the cat; my overall mathematical ability.

This statement indicates that mathematical writing may lead students to become more involved in the assessment procedure, a self-assessment activity.

5. RESULTS AND CONCLUSIONS

To work with students' writing in mathematics in a course in mathematical modelling raise a lot of questions, not in the least regarding the assessment procedure. The assessment part of a mathematical modelling course requires time and effort and demands an ongoing communication between the involved teachers in the course. Beyond that we also seek to identify what ideas and methods of assessing student's investigative work that are relevant and applicable.

In our opinion, it is crucial that before and during the actual work is carried out; one formulates the criteria for assessing and grading and communicates these criteria to the students. Our experience is that this kind of openness regarding how we view our assessing procedure creates a constructive dialogue around the assessment process. When we are talking about the writing up part of the assignment, we sometimes ends up describing what constitutes appropriate writing in this specific context. Fundamentally, it also touches upon how we as university teachers of mathematics view the role the written report plays in the students learning process.

An alternative position views "investigations" as a means of assessing part of the mathematics curriculum. This alternative position is associated with a different view of the relationship of the written report of the investigative work to the problem solving process itself. (Morgan, 1998 p29)

Naturally it is also important how the students view their investigative work in and around an assignment like this and their writing about it. Several of the students expressed opinions about the task quite spontaneously.

Student 2: I found the assignment fun and challenging. From a mathematical view, it wasn't so hard but there were several different factors that play a significant role when one is trying to get the best result.

Student 5: Finally I like to add that this was a different but interesting assignment to work with. A task that probably asked for more thinking and analyzing than just pure mathematical knowledge.

It is important to bear in mind that these quotations are from students who have studied a considerable amount of mathematics at the university level. Our interpretation is that the students in this group expressed their opinion that in their work with *The Catwalk problem* they really had to try out their mathematical knowledge in reality. This was an unusual situation for these students, where the modelling process raised demands on their ability to handle the transition from the real world to the model and back again, to communicate the content both orally and in writing, and to explore the conclusions of their modelling activity.

We believe this to be a good example that gives the opportunity to learn mathematics from another point of view. We don't consider this as the only right and true way, more as a complement to traditional ways to approach a mathematical content and to show how mathematics can be used in practice. Our work in this mathematical modelling class helped the students to develop deeper understandings of the mathematics they used and gave them more insight in the process of writing mathematically. Our ultimate goal is to make the students well prepared to become professional teachers of mathematics and thereby ready to help their own students to develop their ability to communicate on issues with a mathematical content. As university teachers, we have above all acquired both new and deepened experiences in assessing students' investigative work with modelling transitions in the real world.

REFERENCES

Cobb, P. (1986) Contexts, goals, beliefs, and learning mathematics. *For the Learning of Mathematics,* 6(2), 2-9.

Cowen, C. (1991) Teaching and testing mathematics reading. *American Mathematical Monthly*, 98(1), 50-53.

Lingefjärd, T. and Holmquist, M. (2001) Mathematical modelling and technology in teacher education – Visions and reality. In J. F. Matos, W. Blum, S. K. Houston, and S. P. Carreira (eds.), *Modelling and mathematics education. ICTMA 9: Applications in science and technology.* Chichester: Horwood, 205-215.

Lingefjärd, T. and Kilpatrick, J. (1998) Authority and responsibility when learning mathematics in a technology-enhanced environment. In D. Johnson and D. Tinsley (eds.), *Secondary school mathematics in the world of communication technologies: Learning, teaching and the curriculum.* London: Chapman & Hall, 233-236.

Meier, J. and Rishel, T. (1998) *Writing in the teaching and learning of mathematics.* Washington DC: Mathematical Association of America.

Morgan, C. (1998) *Writing Mathematically. The discourse of investigation.* London: Falmer.

NCTM (2000) *Principles and standards for school mathematics.* Reston, VA: National Council of Teachers of Mathematics.

Skolverket. (2000) *National Science Programme. Programme goal, structure and syllabuses.* Stockholm: Fritzes.

Skolverket. (2005) *Compulsory school Syllabuses.* Retrieved June 22, 2005, from http://www3.skolverket.se/ki03/front.aspx?sprak=EN&ar=0405&infotyp=23&skolform=11&id=3873&extraId=2087

Speiser, R. and Walter C. (1994a) Constructing the Derivative in First-Semester Calculus, *Proceedings International Group for the Psychology of Mathematics Education, North American Chapter XVI.* Baton Rouge, LA., 116-122.

Speiser, R. and Walter C. (1994b). Catwalk: First-Semester Calculus. *Journal of Mathematical Behavior, 13*,135-152.

Speiser, R. and Walter C. (1996). Second Catwalk: Narrative, Context, and Embodiment. *Journal of Mathematical Behavior, 15*, 351-371.

6.9

FRACTAL IMAGE COMPRESSION

Francesco Leonetti
University of L'Aquila, Italy

Abstract–*Fractal image compression gives us more models for images. It also uses simple mathematical tools: scalings, translations and rotations. Thus, fractal image compression is suitable when showing that mathematics is very useful.*

1. INTRODUCTION

We consider an image **i** on a black and white screen: each pixel of the screen is on or off. Changing the status of one pixel will change the image **i** on the screen. Every pixel is identified by its cartesian coordinates **(x,y)** thus we set

$$\mathbf{f\_i\,(x,y)} = \begin{cases} \mathbf{1} & \text{if the pixel } \mathbf{(x,y)} \text{ is ON} \\ \mathbf{0} & \text{if the pixel } \mathbf{(x,y)} \text{ is OFF} \end{cases} \qquad (1)$$

Such a function **f_i** is uniquely associated to the image **i**. Thus, functions are the first natural model for images. Fourier analysis allows us to approximate a function **f** by means of a finite linear combination of sine and cosine; the Fourier procedure gives us the coefficients of such a linear combination. JPEG (Joint Photographic Experts Group) compression of images relies on the previous ideas. This discussion gives a first mathematical model for images; it also shows the importance of teaching advanced calculus to students in Computer Science or Information Technology. However, we do not think that Fourier analysis is suitable for high school pupils. Luckily, fractal image compression uses simpler mathematical tools: scalings, translations, and rotations. This kind of image processing is useful to give new motivation for studying linear algebra. Moreover, fractal compression provides some further mathematical models for images: self-similar pictures are identified with contractive mappings; a non self-similar image is a "list of pairs of squares". In §2 self-similar fractals and their compression are discussed; here, the question arises: Are standard geometric pictures (like squares, triangles, circles) self-similar? The answer sheds light on old geometric arguments. Non self-similar sets are dealt with in §3, introducing weaker property of local type. Such a local property is the key to perform fractal image compression on computers. A basic version is shown in §4, which can be achieved simply by means of a sheet of paper and a pencil. Such an activity is useful when dealing with students with a weak background in mathematics: it makes them confident and they do not give up.

2. SELF-SIMILARITY AND COMPRESSION

Fractal image compression dates back to the last decade, (Barnsley and Hurd, 1993; Fisher, 1995). It is based on the approximation of fractals by means of an iterative procedure that relies on the

Fixed Point Theorem:
let X be a complete metric space and G be a distance reducing map from X into X itself; then there exists a unique x^* in X such that $G(x^*)=x^*$; x^* is called fixed point; moreover, for every a in X, let us consider the sequence $x_1=a$, $x_2=G(x_1)$, $x_3=G(x_2)$, ..., $x_{n+1}=G(x_n)$; it results that x_n converges to the unique fixed point x^*.

In our framework, X is the collection of all pictures, x^* is the fractal picture that we want to approximate and G is built by means of scalings, rotations and translations, (Hutchinson 1981, 2000). The previous theorem guarantees that, for a fixed G, the fractal picture x^* is the unique picture enjoying $G(x^*)= x^*$. So, the fractal x^* is uniquely determined by G. Then, when recording x^*, it is enaugh to record G: this procedure will save storage (compression) provided the amount of parameters that we need when recording G is less than the number of points that we need when drawing x^*. How can we recover x^* from G? The second part of the theorem helps us: we take a picture a as we like, we transform a by means of G, the resulting picture $G(a)$ is transformed again by means of G and so on; after some steps we arrive close to x^*. Let us consider the following example. We take the ½ scaling $S_{1/2}$ thus

$$S_{1/2}(x,y) = (x/2, y/2); \qquad (2)$$

now we consider the translation $T_{u,v}$ so that

$$T_{u,v}(x,y) = (x+u, y+v). \qquad (3)$$

We compose $T_{40,0}$ with $S_{1/2}$ and we set $V_2 = T_{40,0} \circ S_{1/2}$. Let V_3 be the composition of $T_{20,(\sqrt{3})20}$ with $S_{1/2}$ thus $V_3 = T_{20,(\sqrt{3})20} \circ S_{1/2}$. Eventually, set $V_1 = S_{1/2}$ thus $V_1 = T_{0,0} \circ S_{1/2}$. Now we define the mapping G acting on any picture a as follows:

$$G(a) = V_1(a) \cup V_2(a) \cup V_3(a) \qquad (4)$$

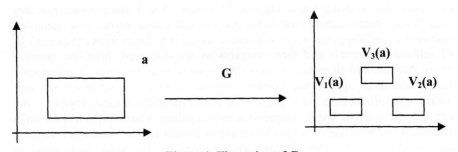

Figure 1. The action of G.

We consider the set **X** of all the pictures; we endow **X** with the Hausdorff distance: thus **X** turns out to be a complete metric space and **G** : **X** → **X** turns out to be a distance reducing mapping. Fixed point theorem guarantees that there exists one and only one picture **x*** that is not changed under the action of **G**. The same theorem tells us how to find this picture **x***: it is enaugh to take a picture **a** as we like, we transform **a** by means of **G**, the resulting picture **G(a)** is transformed again by means of **G** and so on; after some steps we arrive close to **x***. In the case of the above mentioned **G**, an approximation of **x***is shown in Figure 2: we recognize Sierpinski's triangle, (Diallo, 1998; Masini, 1997). Sierpinski's triangle **x*** has been drawn inside a square screen with 80X80 points (pixels); thus **x*** needs 6400 parameters: at each point of the screen we have to know whether the pixel is on or off. On the other hand, since Sierpinski's triangle **x*** is the unique fixed point of **G**, it needs only the parameters of **G**: let us count them. **G** has been built by means of **V₁**, **V₂**, **V₃**. For every $V_i = T_{u,v}$ o S_r we need the scaling parameter **r** and the two translation parameters **u** , **v** : this makes three parameters for every **V$_i$** , and 9 parameters for **G**. Thus, Sierpinski's triangle **x***, as the unique fixed point of **G**, needs only 9 parameters. This second way is cheaper and it gives rise to the compression ratio 6400/9=711. Does the previous procedure work for other pictures? The key ingredient is the fixed point property with respect to some transformation similar to the previous **G**. Let us write this property for Sierpinski's triangle: **x*** = **G(x*)** = **V₁** (**x***) ∪ **V₂** (**x***) ∪ **V₃** (**x***). Let us write explicitly the equality between the left and the right hand side:

$$x^* = V_1 (x^*) \cup V_2 (x^*) \cup V_3 (x^*) \tag{5}$$

Equality **(5)** tells us that Sierpinski's triangle **x*** can be decomposed into 3 reduced copies of itself, as Figure 2 shows.

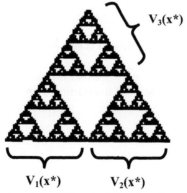

Figure 2. Approximation of Sierpinski's triangle and its decomposition.

Equality **(5)** says that Sierpinski's triangle **x*** is self-similar. Let us consider a picture **i**; we would like to decompose **i** into k reduced copies of itself:

$$i = W_1 (i) \cup W_2 (i) \cup \ldots \cup W_k (i). \tag{6}$$

If we are able to do that, we say that **i** is self-similar. In such a case, if k is not large enaugh, then we are able to compress **i** by means of the parameters of $\mathbf{W_1}$, $\mathbf{W_2}$, ... , $\mathbf{W_k}$. We will recover **i** by means of the fixed point theorem: we consider the operator **H** acting as follows

$$\mathbf{H(a)} = \mathbf{W_1\,(a)} \cup \mathbf{W_2\,(a)} \cup\ ...\ \cup\ \mathbf{W_k(a)}; \qquad (7)$$

we start from some picture **a**, we transform **a** by means of **H** thus creating the sequence $\mathbf{x_1}$=**a**, $\mathbf{x_2}$= **H** $(\mathbf{x_1})$, $\mathbf{x_3}$= **H** $(\mathbf{x_2})$, ...; the fixed point theorem guarantees that the previous sequence will approximate the picture **i** since **i** is fixed for **H**:

$$\mathbf{i} = \mathbf{W_1\,(i)} \cup \mathbf{W_2\,(i)} \cup\ ...\ \cup\ \mathbf{W_k\,(i)} = \mathbf{H\ (i)}. \qquad (8)$$

The previous discussion says that contractive mappings are a new mathematical model for self-similar images. The above mappings **G** and **H** are called Iterated Function System, IFS for short, (Barnsley and Hurd, 1993). Now we ask ourselves if the square **i** is self-similar. Let us half the four sides and let us join the middle points: we get four smaller squares, each of them is a ½ reduced copy of the starting square **i**, as Figure 3 shows.

 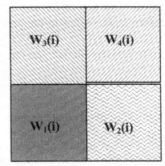

Figure 3. Square **i** is self-similar.

A similar procedure works for triangles: we half the three sides, we join the middle points and we get four reduced copies of the starting triangle, as shown in Figure 4.

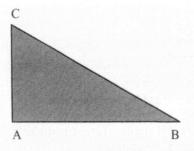 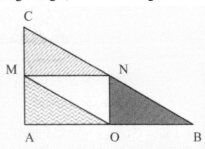

Figure 4. Self-similarity for triangles.

Are we sure that the four smaller triangles in Figure 4 are similar to the starting one? The positive answer relies on the following

Theorem about triangles.
Let ABC be a triangle. Let M, N be the middle points of sides CA and CB, as in Figure 4. Then MN is parallel to the third side AB; moreover, the lenght of MN is half of the lenght of AB.

We apply three times the previous theorem and we get that the four smaller triangles in Figure 4 are equal; moreover, they are ½ reduced copies of the starting one. Now we ask ourselves whether circles are selfsimilar or not: the answer is negative, as next theorem says.

Theorem
Any circle **k** is not self-similar.

Proof. We argue by contraddiction. Let us assume that there exist distance reducing similitudes $V_1,\ldots, V_n$ such that

$$k = V_1(k) \cup \ldots \cup V_n(k). \qquad (9)$$

The previous equality gives

$$k = (V_1(k) \cap k) \cup \ldots \cup (V_n(k) \cap k) \qquad (10)$$

Let us recall that, for every V_j there exists $0 < s_j < 1$ such that

$$d(V_j(x), V_j(y)) = s_j\ d(x, y) \qquad (11)$$

for every **x,y** in the plane, where $d(a, b)$ is the standard distance between two points **a** and **b** in the plane. Since **k** is a circle, its center will be some point x° and its radius will be some $r > 0$. We keep in mind that V_j is a similitude, thus $V_j(k)$ is contained in the circle with radius $s_j\ r$, centered at $V_j(x^\circ)$. Since $s_j\ r < r$, then the two circles have different radii, thus they share no more than two points. This means that the right hand side in **(10)** has at most **2n** points: this is the desired contraddiction, since **k** has infinitely many points. This ends the proof. □

Let us remark that the key tool in the previous argument is the following: two circles, with different radii, share no more than two points, as Figure 5 shows.

Figure 5. Intersection of two different circles.

This well known property, that we learned when attending high school, gives rise to a non self-similar set!

3. LOCAL SELF-SIMILARITY

In the previous section we have seen that circles are not self-similar. We consider the "L" picture $i = \alpha \cup \beta$ shown in Figure 6: it can be proved that i is not self-similar, (Manetta, 2006). However, such a picture i can be decomposed into four pieces

$$i = a \cup b \cup c \cup d \tag{12}$$

The first piece a is the ½ reduced copy of the vertical piece α; the same holds true for b; c is the the ½ reduced copy of the horizontal piece β and the same holds for d.

Figure 6. The "L" picture $i = \alpha \cup \beta$.

This means that i is local self-similar: i can be decomposed into a finite number of pieces such that every piece is a reduced copy of another piece of the starting picture i. In the next section we show how we can decompose any image i into a fixed number of pieces in such a way that every piece is approximately a reduced copy of the starting picture i. Such a procedure is known as fractal image compression, (Barnsley and Hurd, 1993).

4. COMPRESSION AND RECOVERY

We consider the image i represented as dotted lines inside the 80x80 grey, as in Figure 7.

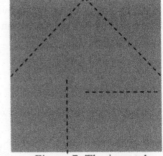

Figure 7. The image i.

Let us make two copies of the picture **i** as in Figure 8: the left one has been subdivided by means of small 20X20 squares, the right one has been decomposed by means of larger 40X40 squares.

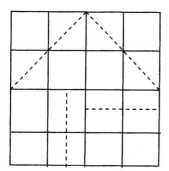
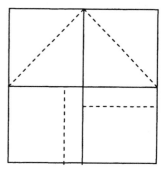

Figure 8. The two copies of the image **i**.

Small squares are called D_1, D_2, . . . , D_{16} and larger squares are named R_1, R_2, R_3, R_4. In Figure 9 we show the two grids with their squares.

D_1	D_2	D_3	D_4
D_5	D_6	D_7	D_8
D_9	10	11	12
13	14	15	16

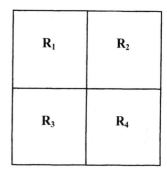

Figure 9. The two grids.

Let us consider the piece of **i** inside each D_j; it will be compared with the part of **i** inside R_1, R_2, R_3, R_4. Let us consider the piece of **i** inside D_1: there is no **i** inside D_1 thus we go on. Let us consider the piece of **i** in D_2: we compare it with $i \cap R_1$, $i \cap R_2$, $i \cap R_3$, $i \cap R_4$ and we see that $i \cap D_2$ is the reduced copy of $i \cap R_1$ thus we link up D_2 with R_1. Let us consider the piece of **i** in D_3: we compare it with $i \cap R_1$, $i \cap R_2$, $i \cap R_3$, $i \cap R_4$ and we see that $i \cap D_3$ is the reduced copy of $i \cap R_2$ thus we link up D_3 with R_2. There is no **i** inside D_4 thus we go on. We see that $i \cap D_5$ is the reduced copy of $i \cap R_1$ thus we link up D_5 with R_1. There is no **i** inside D_6 and no **i** in D_7 then we proceed. We see that $i \cap D_8$ is the reduced copy of $i \cap R_2$ thus we link up D_8 with R_2. There is no **i** inside D_9 and we go on. Let us consider the piece of **i** in D_{10}: we compare it with $i \cap R_1$, $i \cap R_2$, $i \cap R_3$, $i \cap R_4$ and we see that the "closest" one is $i \cap R_3$ thus we link up D_{10} with R_3. Let us consider the piece of **i** in D_{11}: we

compare it with $i \cap R_1, i \cap R_2, i \cap R_3, i \cap R_4$ and we see that the "closest" one is $i \cap R_4$ thus we link up D_{11} with R_4. In the same way we link up D_{12} with R_4.
There is no i inside D_{13} thus we go on. The argument we used for D_{10} can be applied to D_{14} thus we link up D_{14} with R_3. There is no i inside D_{15} and D_{16}. Eventually we have the following links:
D_2 with R_1, D_3 with R_2, D_5 with R_1, D_8 with R_2, D_{10} with R_3, D_{11} with R_4, D_{12} with R_4, D_{14} with R_3.

We store the previous 8 links: thus we record only 8 parameters instead of 6400: this is the compression for i. This means that we can use another mathematical model for the image i: a list of pairs of squares. The previous comparisons were made at glance: in order to automatically compare $i \cap D_j$ with $i \cap R_k$ we have to carry $i \cap R_k$ into D_j: in order to do that, we need the scaling $S_{1/2}$ and a suitable translation $T_{u,v}$; the right choice of u and v will give $D_j = T_{u,v} \; o \; S_{1/2} \; (R_k)$. When trying to compress by means of a computer, we cannot glance any more! On the contrary, we have to specify a procedure that evaluate the distance between $i \cap D_j$ and $T_{u,v} \; oS_{1/2} \; (i \cap R_k)$ for $k=1,...,4$: thus Hausdorff distance, that we mentioned before, comes into play again! Let us come back to our links between $D_1,..., D_{16}$ and $R_1,..., R_4$. In order to recover our picture i from the 8 links that we stored, let $R_{k(j)}$ be the 40x40 square linked to the smaller 20x20 square D_j. By means of the ½ scalling $S_{1/2}$ and a suitable translation $T_{u,v}$ we are able to map $R_{k(j)}$ onto D_j: $D_j = T_{u,v} \; o \; S_{1/2} \; (R_{k(j)})$. Let us set $W_j = T_{u,v} \; o \; S_{1/2}$ thus $D_j = W_j \; (R_{k(j)})$. By means of our links we are able to build an operator L acting on any picture a as follows

$$L(a) = W_2 \; (a \cap R_1) \cup W_3 \; (a \cap R_2) \cup W_5 \; (a \cap R_1) \cup W_8 \; (a \cap R_2) \cup$$
$$W_{10} \; (a \cap R_3) \cup W_{11} \; (a \cap R_4) \cup W_{12} \; (a \cap R_4) \cup W_{14} \; (a \cap R_3) \qquad (13)$$

Let us briefly describe how L acts; the picture $L(a)$ is obtained by gluying all the pieces $W_2 \; (a \cap R_1)$, $W_3 \; (a \cap R_2)$, ..., $W_{14} \; (a \cap R_3)$ where $W_2 \; (a \cap R_1)$ is obtained as follows: we take the part of a inside R_1, we shrink it by ½ and we shift it into D_2. The same argument applies to $W_3 \; (a \cap R_2)$: we take the part of a inside R_2, we shrink it by ½ and we shift it into D_3. In general, the piece $W_j \; (a \cap R_k)$ is obtained as follows: we take the part of a inside R_k, we shrink it by ½ and we shift it into D_j.
Let us consider the square a; in order to get $L(a)$ we have to subdivide a by means of the 40x40 grid; then we take the empty 20x20 grid and we have to fill it by means of the operator L: the square D_2 is filled by means of the part of a inside R_1, the square D_3 is filled by means of the part of a in R_2 and so on. We proceed in this way until we get $L(a)$. We now let L act on $L(a)$ and we get $L(L(a))$. If L acts on $L(L(a))$, we get $L(L(L(a)))$ as in Figure 10.

Note that $L(L(L(a)))$ is a "good" approximation of our image i. For the previous picture i we stored 8 links, thus 8 informations, instead of 80X80=6400. For more complex images, the previous procedure will get 16 links, one link for every D_j, thus we will get the compression ratio 6400/16=400. For pictures with more details, we need to consider different grids: we use 8X8 small squares $D_1,..., D_{100}$, and 16X16 larger squares $R_1,..., R_{25}$; in this way we get 100 links, thus 100 informations to be stored, instead of 80X80=6400, (Medoro, 2002); this gives rise to the compression ratio of 6400/100=64. Let us come back to the picture i: the previous discussion about its compression and reconstruction can be done in the classroom; after that, students have first to compress their initials, then to recover them, using millimeter paper, looking at initials as 100X100 images, using 10X10 small squares and 20X20 larger squares. This activity can be performed even without a computer, by means of a pencil and a sheet of paper; it has been tested during my teaching in 2005, students, with a weak background in mathematics, did not give up; on the contrary, the previous activity made them confident.

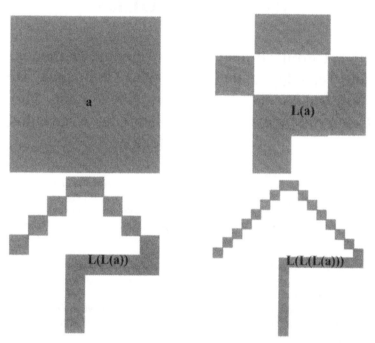

Figure 10. The square **a** and its transformations **L(a), L(L(a)), L(L(L(a)))**.

REFERENCES

Barnsley M. and Hurd L.(1993) *Fractal image compression*. A K Peters.

Diallo D. (1998) Implementazione di acuni frattali. Graduate Thesis, University of L'Aquila.

Fisher Y. (1995) *Fractal image compression: theory and application to digital images*. Springer Verlag, New York.

Hutchinson J.(1981) Fractals and self-similarity. *Indiana University Mathematics Journal*, 30, 713-747.

Hutchinson J. (2000) Deterministic and random fractals. In T.R.J. Bossomaier and D. G. Green (eds) *Complex systems*. Cambridge: CUP, 127-166.

Manetta M. (2006) Autosimilarita' e trattamento delle immagini. Graduate Thesis, University of L'Aquila.

Masini P. (1997) Costruzione di frattali. Graduate Thesis, University of L'Aquila.

Medoro C. (2002) Frattali e applicazioni. Graduate Thesis, University of L'Aquila.

6.10

MODELLING HEAT FLOW
IN WORK ROLLS

Leticia Corral[1], Rafael Colás[2], Antonino Hernández[3]
[1]Instituto Tecnológico de Cuauhtémoc, Chihuahua, México
[2]UANL, San Nicolás de los Garza, Nuevo León, México
[3]Centro de Investigación en Materiales Avanzados,
Chihuahua, México

Abstract–*The continuing quest for explanations that resolve real-world problems leads to present a two-dimensional model for work rolls used for hot rolling of steel strip in a continuous mill. This model is developed in a novel form using analytic and numerical methods by considering the roll as a semi-infinite solid. Due to the excessive thermal cycles to which the roll is subjected during the processing, the industry requires that the research predicting its thermal behaviour is robust. Therefore, this current research was developed and the heat flow at a steady-state through the two-dimensional model computed in a way that is valued in the industry, and meets its requirements. This paper contributes to the state of art of the modelling and simulation of industrial processes of the real world, as well as in significant learning through the pedagogic practice, and its assessment. This work presents a novel insight for the development of future researches in the field of materials science.*

1. INTRODUCTION

The linking of industry and education has turned out to be a key factor that contributes to the benefit of both areas. The present work involves the study of an industrial process and its background. This study case of modelling is appropriate because it represents one of the research lines of modelling and simulation of industrial processes that the Advanced Materials Research Centre (CIMAV) offers to students.

In this work, the heat transfer phenomenon is represented by means of an analytical-numeric mathematical model, which can be simulated to automate a process or other application. In this way, it contributes to the art of the modelling and simulation of industrial processes.

In any production line in industry, there is a necessity of having clear and concise information of the productive process for its control, in order to satisfy the requirements that are demanded from the product.

The development of the computation systems, and access to larger groups of researchers, has allowed, among other things, the development of simulators of industrial processes with the purpose of reproducing the process or parts of it under operational conditions, and a prediction tool has also been developed. With this tool it is possible to isolate the effect of the operation parameters, separating the variables that act in direct form from the casual ones, optimizing the process and making it possible to simulate new operation conditions, with the purpose of obtaining better or new characteristics in the product (mechanical, thermal, electric, etc.) requested by the market. That is to say, the simulator offers a function of parameters that, under a combination of them, gives the characteristics of the product.

This provides the additional benefit of knowing the limits of the production plant; it allows the prediction of the feasibility of the generation of new products with the machinery. This, together with an analysis of future markets, allows the appropriate investment in the modernization of the plant in the sectors that are really more necessary, anticipating the necessities of new products or new norms of quality.

The publication of works on hot rolling begins in 1970 with Holander (1970); today efforts are made for the creation of new models in the search of a better representation of reality. In particular, the purpose of the present work is to carry out a contribution in the simulation of the thermal changes that happen in the work roll during hot rolling of steel strip; this is in the steady state in the production line, subject to the operation conditions, as distinct from works (Yuen, 1884; Patula, 1981; Haubitzer, 1974). The steady state is reached when the changes in the operation factors are not significant; if they are, this state becomes a dynamic steady state in the production line.

The contribution of the mathematical model presented here is to model the complex temperature profile of the work roll, taking into account the variable temperature of the strip and obtaining a sensitive simulator to changes under operation conditions in plant. The applications background to the problem presented in this work is based on the mission that the CIMAV has to prepare researchers in the different areas, and considers important aspects that have to do with educational aspects that are described next. The starting point is the form in which the individual learns and which has much to do with the significance this learning has for the student.

Changes need to be made in traditional teaching for a significant learning, that is to say, to change assumptions that the student goes to the classroom to receive information and he receives accumulative learning, where the knowledge that is already possessed is not questioned, further, it is accepted without doubting. It is necessary to move towards changes in the teaching in which the student goes out of the classroom with questions, without answers and/or procedures to find them (to build, to knock down the above-mentioned). In this way, the student analyzes processes as specific forms of construction of the knowledge. The teacher should make sure of the learning achieved by the student (appropriateness of contents, development of pertinent abilities, enjoyment to solve problems), since the forms in that each student learns are endless.

The learning process, is based in theories whose paradigm represents a change in the teaching of the modelling providing the "significant learning" in the construction of the individual's own knowledge. In this respect, Jean Piaget's theory by means of the triad: assimilation-accommodation-equilibration (Piaget, 1990; Piaget & Garcia, 1982; Beth & Piaget, 1980), changes in the psychology of the learning, the factors that contribute in the process of preparing students (influence, system educational), and the existence of changes in the teaching, among others, converts the teacher in "facilitator ".

This allows the teacher to shorten the time needed in favorable scenarios forward for the development of a certain research. The designs of didactic situations starting from resolution of problems according to the concept that, to know how to solve problems, a student should try to not only solve many problems, but a great variety of the same ones (Borasi, 1986). This is included within the subjects that are imparted for each student, in support of his research line.

The methodology used during the setting the scene of a course has as base the resolution of problems considering a problem since two points of view: a) search of an action and achievement of an objective (Polya, 1945), and b) situation (group or individual) which has to find a way to solve it (Krulik & Rudnik, 1980).

During the phase of resolution of problems (the context of the problem, the formulation of the problem, the group of solutions, complexity of the behavior in the resolution of problems, the focus in the resolution of problems and the method), the student goes by stages of acceptance, blockage and exploration (Borasi, 1986; Schoenfeld, 1985; De Guzmán, 1991; Beagazgoitia et al., 1997; Callejo, 1994; Mason et al., 1982).

During the setting of the scene of learning situations, three elements are taken into account: a) knowledge: the student responds to a certain situation, b) student: learns for construction, as consequence of experiences, and, c) teacher: it should provide an enough quantity of rich experiences in significance (Piaget, 1990).

2. INDUSTRIAL PROCESS

In the production process, the steel strip is conformed to high and low temperatures by means of rolls that are in contact with the strip being deformed. Most of the deformation is carried out hot (see hot rolling in Figure 1), to take advantage of the smallest mechanical resistance as well as the greatest ductility in the worked material.

A hot rolling train is constituted by intermediate stations that are denoted mills, in which the rolls are housed (see continuous mill in Figure 1). Accordingly they are either in contact or not in contact with the steel (Mikhailus et al., 1990), the rolls can be working or back-up; the first have the function of deforming the strip, and the second search to avoid the flexion of the first.

The temperature of the work rolls has a tendency to increase during the rolling, so a cooling system is required to avoid excessive heating. Figure 1 shows the process of the obtaining of hot steel strip. First, the iron mineral and its collimation are obtained, and then it is transported to the direct reduction plant (HyL process). This reduces the iron mineral by means of a gas reducer giving spongy iron. The spongy iron is founded with scrap in an electric oven and it is emptied in slab mould, letting the steel solidify. The slabs, the raw material of the process are made in this way. The slab is transported to the reheating graves, in which they remain until they reach the appropriate temperature (depending on the type of steel that is rolled); they then go to the hot rolling mill. The hot rolling begins with roughdresses, in which two reversible mills are used to reduce the thickness of the slab (about 0.46 m or 18 inches thick), transforming it to a strip of thickness 0.0279 m (1.1 inches). During the process the width of the strip is controlled and the oxide formed is removed from the surface of the strip.

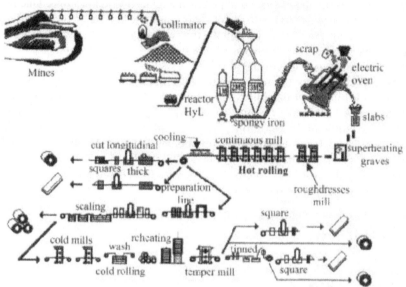

Figure 1. The process of obtaining of steel strip, starting from the extraction from the mineral to the finished product.

Next tips are cut off the strip as it exits the roughdresses mill and the oxide formed on the surface of the strip is removed again (scaling), with jets of water before the strip enters the continuous mill. The continuous mill is formed of six unidirectional type-four mills (so called because of each mill has two back-up rolls and two work rolls), known as finishers mills, because they complete the hot rolling process to obtain the steel strip (see Figure 2). Finally, the steel strip is transported to the cooling table, to be rolled up and transported to a patio for cutting or to be prepared for cold rolling.

Figure 2. Diagram of the continuous mill for hot rolling of steel strip.

The diameters of the work rolls and back up rolls are 18.75 and 37 inches, respectively, and both are 48 inches long. The mill is moved by three 4000 hp motors continuously from mill M1 to mill M3, and three 2300 hp motors from mill M4 to mill M6, each one of them working from 250 to 500 rpm (Figure 2).

3. THE TWO-DIMENSIONAL MODEL

A work roll is considered to be an infinitely long, semi-infinite solid cylinder. This geometry is often quite useful in practical engineering applications. One reason for this is in finite geometries where, for short times, the heating or cooling effects at the surface do not extend very far into the material. In this work, step changes in the surface temperature of a solid work roll are made at small enough times so that the centre temperature is still at its initial value.

The model leads to a Fourier series solution that normalizes the results of the heat flow allowing great flexibility. From this a database is generated that is able to reproduce the production. This space-time model calculates the complex temperature profile of the work roll. Thus, the heat flow transferred to the work roll by various means is evaluated as described below. The work roll was divided for into eight angular sectors of heat transfer following Peréz (1994) and Peréz et al. (2004). The heat flow of the strip to the work roll can be expressed as:

$$f = \beta_s H_s \left[T_s^* - (T - T_a) \right] \tag{1}$$

While the heat flow of the back-up roll to the work roll can be expressed as:

$$f = \beta_b H_b \left[T_b^* - (T - T_a) \right] \tag{2}$$

where: $\beta_s = 1 + \sqrt{\dfrac{(\rho C_p K)_r}{(\rho C_p K)_s}}$, $\beta_b = 1 + \sqrt{\dfrac{(\rho C_p K)_r}{(\rho C_p K)_b}}$

$$T_s^* = \frac{T_s^0 - T_a + (\beta_s - 1)(T_{2\pi} - T_a)}{\beta_s}, \quad T_b^* = \frac{T_b^0 - T_a + (\beta_b - 1)(T_{\theta_b} - T_a)}{\beta_b}$$

$$H_{(\theta_s)} = \beta_s H_s, \quad H_{(\theta_b)} = \beta_b H_b, \quad H_{(\theta_w)} = H_w$$

$$\phi_{(\theta_s)} = T_s^* / T_s^* = 1, \quad \phi_{(\theta_b)} = T_b^* / T_s^*, \quad \phi_{(\theta_w)} = T_w^* / T_s^* = 0$$

The thermal flow settles down toward the roll subject to the boundary conditions or in function of the angular position, as:

$$f = \sqrt{(\rho C_p K \omega)_r} \, T_s^* h_{(\theta)} \left[\phi_{(\theta)} - \lambda_{(\theta)} \right] \tag{3}$$

where: $h_{(\theta)} = \dfrac{H_{(\theta)}}{\sqrt{(\rho C_p K \omega)_r}}$

The normalized temperature in the work roll, for a harmonic source of heat prescribed, it is given by:

$$\lambda = \bar{\lambda} + \sum_{n=1} \frac{\exp(-\alpha\sqrt{n})}{\sqrt{n}} \left[A_n \cos(n\theta_{(t)} - \alpha\sqrt{n}) + B_n \sin(n\theta_{(t)} - \alpha\sqrt{n}) \right] \tag{4}$$

where: $\lambda = \dfrac{T - T_a}{T_s^*}$, $\alpha = r\sqrt{\left(\dfrac{\omega}{2\kappa}\right)_r}$, $\omega_r = 2\pi / t_r$, $\theta_{(t)} = 2\pi t / t_r$

The expression of thermal flow in the work roll obtained starting from the temperature λ is:

$$f = \sqrt{\left(\rho C_p K \omega\right)_r} T_s^* \sum_{n=1} \left[A_n \cos\left(n\theta_{(t)}\right) + B_n \sin\left(n\theta_{(t)}\right)\right] \qquad (5)$$

where: H_s and H_b are the heat transfer coefficients in the zone of contact between the work roll and strip, and between the work roll and back-up roll, respectively, $H_{(\theta)}$ is the heat transfer coefficient in the different angular sectors defined for: the contact between the work roll and strip ($H_{(\theta_s)}$), the contact between the work roll and back-up roll ($H_{(\theta_b)}$), the contact between the work roll and jets of water -free and forced convection- ($H_{(\theta_w)}$), H_w is the heat transfer coefficient of water, $h_{(\theta)}$ is the normalized angular heat transfer coefficient in function of the boundary conditions, β_s and β_b are constants which depend on the physical properties of the work roll and strip and of the work roll and back-up roll, respectively, T_s^0 and T_b^0 are the temperatures in the strip surface before of the contact with the work roll and in the back-up roll surface before of the contact with the work roll, respectively, $T_{2\pi}$ is the temperature in the surface of the work roll before of the contact with the strip, T_a is the ambient temperature, T is the temperature in the surface of the work roll, λ are the normalized temperature in the work roll, $\bar{\lambda}$, A_n and B_n are Fourier coefficients, T_s^* and T_b^* are the temperatures in the surface of the work roll immediately after the contact with the strip and in the surface of the back-up roll immediately after the contact with the work roll, respectively, T_{θ_b} is the temperature of the back-up roll during the contact with the work roll, ρ_r, ρ_s, and ρ_b are the density of the work roll, strip and back-up roll, respectively, C_{p_r}, C_{p_s}, and C_{p_b} are the specific heat of the work roll, strip and back-up roll, respectively, K_r, K_s, and K_b are the thermal conductivity of the work roll, strip and back-up roll, respectively, ω_r is the angular velocity of the work roll, $\phi_{(\theta)}$ is the normalized temperature to the means to which transfers: zone of contact between work roll and strip ($\phi_{(\theta_s)}$), zone of contact between the work roll and back-up roll ($\phi_{(\theta_b)}$), and zone of free or forced convection ($\phi_{(\theta_w)}$), T_w^* is the temperature in the surface of the work roll during contact with jets of water, α is a constant which depends on the physical properties and angular velocity of the work roll, r is the radial coordinate, κ_r is the thermal diffusivity of the work roll, t_r is the rotation period, s, and t is the temporal coordinate, $\theta_{(t)}$ is the angular position in function of the time t in $\left(0, t_r\right]$.

4. RESULTS AND DISCUSSIONS

The model described here simulates in appropriate form the heat flow subjected to a work roll in a hot mill in the industry. The model also simulates the thermo-elastic response of strains, stresses and dilatation (Corral et al., 2004). These theoretical models of real-world phenomena provided insights into these phenomena. The process and search for explanations to resolve problems and the theory on the

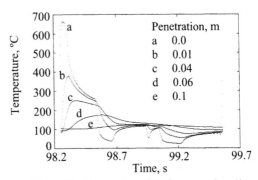

Figure 3. Thermal profile for a work roll during one cycle.

construction of the knowledge (Piaget, 1990; Borasi, 1986; Polya, 1945; Krulik & Rudnik, 1980; Schoenfeld, 1985; Freudenthal, 1991; García, 1997) in this investigation line are basic points that contributed to the achievement of the objectives.

The results of the modelling and simulation of heat flow, given by the theoretical model during one cycle is important in the surface of the roll (Corral et al., 2004; Perez, 1994; Guerrero et al., 1999; Perez et al., 2004) (see Figure 3). We can observe that, as it penetrates from the surface to the centre of the roll, the temperature falls, because the information is not transmitted to the interior by means such as: thermophysical properties (as thermal diffusivity), and free and forced convection effects during the process.

Figure 4 shows the thermal effects during the series of complete production in hot rolling (Figure 4: a, c, e), with different numbers of strips during the sequences of the process and the times of shutdown (Figure 4: b, d, f). In Figure 4 (a to e), we can see that when it is penetrating to the interior of the roll, the temperature falls throughout. On the other hand, in Figure 4 (f), it is observed that the lines corresponding to the surface and at a depth of 0.01 m into the interior of the roll, begins to vary until they cross, when this section begins to reach the same temperature, whilst other lines have the same behavior to that sections before (f).

Figures 3 and 4 coincide with the experimental data obtained in the hot rolling industry. The model considers the process of strip deformation for finite length and variable temperature so that the thermal response of the work roll, subject to the variation of the operation parameters and physical properties of the roll, is known; this helps in knowing which properties of the work rolls are desirable.

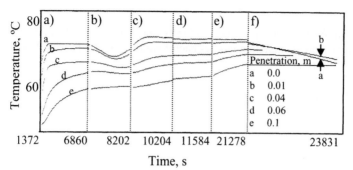

Figure 4 Profile of average temperature in the work roll during:
a) the Series I, with 79 strips; b) their first stop (28 minutes); c)
the Series II, with 24 strips; d) their second stop (20 minutes); e)
the Series III, with 136 strips; f) before being removed of the
mill (4 minutes).

5. CONCLUSIONS

Two types of implications emerge from this research. First, learning of industrial
applications within aspects of the construction of the knowledge, and the second, the
implications within the industry for hot rolling steel strip. With regard to the first
implication, having developed, analyzed and assessed several problems of physical
phenomenon during the present research, it allows us to conclude that, this problem
has the scope to be a useful model for educational purposes, since this teaching type
leads to a significant learning. This leads to further research in materials science.
Nowadays, the mathematical model has had an unprecedented success in the
industry of hot rolling work rolls, specifically, in a continuous hot strip mill. In this
case, the theoretical model agrees with the experimental data.

The above-mentioned, completes the expectations required by the industry
guided to give maintenance to the work rolls, besides opening the horizon to new
research lines open to the Research Centre which carried out this work,. This
reinforces in important form the research through industrial applications. On the
other hand, as for the second implication, several interesting observations can be
made from the model. First, the phase shift is a function of depth. Thus, there are
positions under the work roll's surface that are increasing in temperature at the
surface. Second, the amplitude of the oscillations decreases with depth as it can be
seen in the graphs. Thus the thermal effects for some few turns do not penetrate as
deeply into the work roll as they do during the whole process. The repercussion in
the industry of the results above mentioned, it is linking intimately to its contribution
to the art state in the modelling and simulation in hot rolling process, and they have
to do with the success reached as for the service and selective maintenance in the
area of hot mills.

REFERENCES

Beagazgoitia, A., Castañeda, F., Fernández, S. and Peral, J.C. (1997). La resolución de problemas en las matemáticas del bachillerato. Universidad del País Vasco: Servicio Editorial.

Beth, E.W. and Piaget, J. (1980) Epistemologìa Matemática y Psicología: relaciones entre la lógica formal y el pensamiento real. Barcelona: Editorial Crítica, Grijalbo.

Borasi, R. (1986) On the nature of problems. *Educational Studies of Mathematics,* 17, 125-141.

Callejo, M. (1994). *Un club matemático para la diversidad.* Nancea.

Corral, R.L., Colás, R. and Pérez A. (2004) Modelling the thermal and thermoelastic responses of work rolls used for hot rolling steel strip. *Journal of Materials Processing Technology* 153-154, 886-893.

De Guzmán, M. (1991). *Para pensar mejor.* Labor.

Freudenthal, H. (1991) *Revisiting Mathematics Education.* Dordrecht: Kluwer

García, R. (1997) *La Epistemología Genética y la Ciencia Contemporánea.* Barcelona: Gedisa.

Guerrero, M.P., Flores, C.R., Pérez A. and Colás, R. (1999) Modelling heat transfer in hot rolling work rolls. *Journal of Materials Processing Technology*, 94, 52-59.

Haubitzer, W. (1974) Steady State Temperature Distribution in Rolls. *Architectir fuer das Eisenhuttenwesen,* 46, 635-638.

Hollander, F. (1970) Mathematical models in metallurgical process development. *Iron Steel Insitute,* 123, 46-74.

Krulik, S. and Rudnik, J. (1980) *Problem Solving.* London: Alleyn & Bacon Inc.

Mason, J., Burton, L. and Stacey, K. (1982) *Thinking Mathematically.* New York: Addison-Wesley.

Mikhailus, A.S., Shatik, Y.S. and Moiseenko, I.I. (1990) *Steel in the USSR*, 20.

Patula, E.J. (1981) *ASME Journal of Heat Transfer*, 103, 36-41.

Pérez, A. (1994) Simulación de la temperatura en estado estable-dinámico para un rodillo de trabajo de un molino continuo de laminación en caliente. PhD Thesis, Universidad Autónoma de Nuevo León, México.

Pérez, A., R.L. Corral, Fuentes, R. and Colás, R. (2004) Computer simulation of the thermal behaviour of a work roll during hot rolling of steel strip. *Journal of Materials Processing Technology* 153-154, 894-899.

Piaget, J. (1990) La equilibraciòn de las estructuras cognitivas. Problema central del desarrollo. Madrid: Siglo XXI de España.

Piaget, J. and García, R. (1982) Psicogénesis e Historia de la Ciencia. México: Siglo XXI.

Polya, G. (1945) *How to solve it.* Princeton: Princeton University Press.

Schoenfeld, A. (1985) *Mathematical Problem Solving.* New York: Academic Press.

Yuen, W.Y.D. (1984) *ASME Journal of Heat Transfer*, 106, 578-585.

6.11

APPLICATIONS OF MODELLING IN ENGINEERING AND TECHNOLOGY

[1]Sanowar Khan, Kenneth Grattan and Ludwik Finkelstein
City University, London, UK

Abstract–*The use of mathematical models in the design, investigation and prediction of sensors, actuators and other devices constitutes one of the major advances in engineering design. Efficient numerical solutions are now available for a wide range of mathematical problems that are impossible to tackle by analytical methods. Based on a number of industrial case studies, this paper will describe the use of numerical modelling techniques, such as the finite element technique, for the design and analysis of sensors, actuators and other electrical devices. In doing so, it will focus on the scope and importance of physical modelling at a sub-system level which ultimately contributes to modelling activities at a global systems level. It highlights the fact that because of its importance in design and analysis, the topic of mathematical modelling and the underpinning techniques are of wide pedagogical interest.*

1. INTRODUCTION

With the availability of powerful desktop computers, significant progress is being made in the application of numerical techniques such as the finite element method for modelling, computer aided design (CAD), performance prediction and validation of instrument transducers and sensors. Mathematical modelling is a key enabling tool and a means by which the functioning of systems and sub-systems can be predicted from a description of the system's physical principles, geometric features and material properties.

A model of a system is the description of the system in a formal language, such that relations between symbols in statements in the language imply and are implied by relations between the objects and attributes of the system and its components (Finkelstein & Grattan, 1994). In other words, a model can be looked upon as the representation of a physical process and possesses the essential attributes of that physical process. Models are extensively used in design and in this context modelling is the study of the mechanisms inside a system, and through using basic laws and relationships, a model is inferred. In terms of representation schemes, there could be linguistic, pictorial and mathematical models (Abdullah et al., 1994). This paper focuses on mathematical models in which physical sub-systems are described as a set of mathematical relations (for example, equations, discrete data, etc.) representing the physical processes, properties and behaviour of the sub-systems.

2. DESIGN AND ANALYSIS BY MODELLING

Figure 1 shows a simplified heuristic procedure for mathematical modelling and design of physical sub-systems in engineering and technology. It is a simple procedure yet it could be identified as the core activity in modelling and design of many sensors, actuators and devices, especially at a sub-system level.

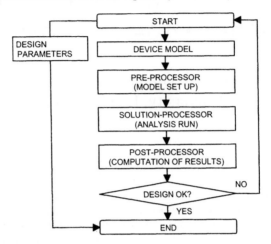

Figure 1. Heuristic procedure for mathematical modelling and design of physical sub-systems (for example, sensors, actuators and devices).

The procedure comprises three main processes - the process of defining and setting up appropriate mathematical models (pre-processor), the process of solving the defining equations (solution-processor), and the process of calculating the necessary output parameters from the solution of these equations (post-processor). This procedure is repeated until a satisfactory design is obtained. In many areas of science and technology, sub-systems that are implemented using standard information technology, are modelled using the techniques of signals and information systems science. The relation between a sub-system's physical embodiment and its function can be represented by idealised lumped parameter models (Finkelstein & Watts, 1983). Such models are based on the relation between the input power flow to a sub-system, or system, and the output flow.

While idealised models like these are useful in the representation and analysis of concepts, detailed analysis and design require models which relate the detailed geometric and material properties of the object modelled to its functional behaviour. Engineering objects are characterized by complex geometries and distributed properties. The physical laws governing their behaviour are represented by partial differential equations, which are often non-linear and transcendental. Analytical solutions of such realistic models are generally not feasible. However, various numerical methods, such as finite element method (FEM), boundary element method (BEM), hybrid FEM-BEM, etc. have made possible the formulation and analysis of such models. The following Sections show the application of the widely used FEM (Binns et al., 1992) based on a number of case studies in engineering exemplified in modelling and design of sensors, actuators and devices.

3. MATHEMATICAL MODELLING IN ACADEMIC CURRICULA

Because of its importance in design and analysis, the topic of mathematical modelling and the underpinning techniques are of wide pedagogical interest. Hence it is included in the curricula of a wide variety of degree courses both at the undergraduate (UG) and postgraduate (PG) levels. The knowledge and understanding of the underlying concepts of mathematical modelling can be considered of generic nature. In the School of Engineering and Mathematical Sciences at City University, London the topic of mathematical modelling and design is extensively covered in relevant UG and PG curricula. One of the examples is the Systems Engineering and Design module taught on a number of BEng/MEng (Hons) degree programmes. The aim is to provide basic principles of physical and systems modelling and structured design, introduce fundamental aspects of linear and non-linear dynamics, and the qualitative aspects of systems behaviour. The module is divided into four distinct but interrelated parts comprising physical modelling, systems modelling, systems and advanced dynamics, and control analysis. It is, in part, underpinned by a Numerical Computing and Statistics module which provides essential numerical techniques for modelling. In addition to subject specific learning outcomes, together these two modules provide generic knowledge and transferable skills in modelling and design that transcend subject boundaries.

At a physical sub-system level the implementation and subsequent realisation of the principles of mathematical modelling techniques to solve real-life design and analysis problems require considerable expertise and the bridging of the gap between the physical reality and its representation by models. This paper addresses this issue by describing the successful application of a finite element (FE) modelling technique for solving real-life industrial problems of design and analysis of physical sub-systems.

4. MATHEMATICAL MODELS IN ENGINEERING

As mentioned earlier, the use of mathematical models in the design and investigation of sensors, actuators and other devices constitutes one of the major advances in engineering design. The performance of such physical sub-systems is often represented by partial differential equations which constitute their main mathematical model. For capacitive devices (for example, capacitive sensors) this is adequately represented by the following Laplace's (1) or Poisson's equation (2) giving the electrostatic field distribution in the three-dimensional problem domain $\Omega(x, y, z)$:

$$\nabla \cdot \varepsilon \nabla \Phi = 0 \text{ in } \Omega(x, y, z), \qquad (1) \qquad \nabla \cdot \varepsilon \nabla \Phi = -\rho \text{ in } \Omega(x, y, z), \qquad (2)$$

where $\Phi = \Phi(x, y, z)$ is the electrostatic potential and $\varepsilon = \varepsilon(x, y, z)$ – dielectric permittivity distribution in the problem domain $\Omega(x, y, z)$. Under given boundary conditions, the numerical FE solution of the above equations gives the unknown potential $\Phi(x, y, z)$ in the problem domain. This can then be used to calculate the field intensity and flux density vectors E and $D = \varepsilon E$, and subsequently, such global quantities of interest as capacitance C, charge Q and electrostatic field energy E_e. The capacitance C is calculated either from the field energy or from electric charge:

$$E_e = \frac{1}{2}CV^2, \qquad\qquad (3) \qquad C = \frac{Q}{V}, \qquad\qquad (4)$$

where, V is the known potential difference. The capacitance in (4) is calculated by integrating vector D over appropriate electrode surfaces using the Gauss's law (Khan & Abdullah, 1993):

$$Q = \oint_S D_n ds = \oint_S D\cos\theta\, ds = \oint_S D\cdot n\, ds = \oint_S D\cdot ds \qquad\qquad (5)$$

Equation (5) says that the integral of the normal component, $D_n = D\cos\theta = D.n$ of vector D over any closed surface s enclosing a capacitive electrode is equal to the total charge Q distributed on it. If the surface s is chosen very close to the capacitive electrode, the calculation of Q becomes easier to perform as there is no need to find D_n since electric field lines are always normal to a conducting surface. This means:

$$Q \approx \oint_S D\, ds \qquad\qquad (6)$$

In comparison, the modelling of EM actuators pose significant problems because of the complexity of interrelated physical processes involved. The EM part of the system is represented by electric and magnetic circuits with self-inductance, resistance and reluctance which are subject to variations, in general, due to eddy currents, saturation conditions, motional electromotive force, demagnetisation and hysteresis. The mechanical part is represented by friction, damping, elasticity and inertia as well as external forces. Thus the nonlinear and transient EM, thermal, and motional problems to be solved in EM actuators pose substantial challenges.

In general, the mathematical model of an EM actuator can be adequately represented by the following differential equations: an electrical circuit equation for the excitation coil and control circuitry (7), a nonlinear magnetic field equation (Poisson's equation) for the flux producing the magnetic force (8), a mechanical equation for this force, load, friction, acceleration, speed, etc. (9), and a nonlinear thermal diffusion equation for the heat produced by electrical power losses (10):

$$u(t) = iR + N\frac{d\Psi(i,z)}{dt} \qquad (7) \quad \text{curl}(v\,\text{curl}\,A) = J - \sigma\frac{\partial A}{\partial t} + \sigma V \times (\text{curl}\,A) \quad (8)$$

$$F_m(i,z) = m\frac{d^2 z}{dt^2} + B\frac{dz}{dt} + Kz + F_e \quad (9) \quad \rho C\frac{\partial T}{\partial t} - \nabla\cdot[k(T)\nabla T] = q^B \qquad (10)$$

In the above equations $u(t)$, i and $\Psi(i, z)$, and z are the applied voltage, coil current, flux linkage with the coil, and the displacement of the moving part respectively, R and N are the coil resistance and the number of turns in the coil, J, A, V are the coil current density, magnetic vector potential, and the velocity; m, B, K, F_m and F_e are the mass of the moving part, damping coefficient, spring constant, magnetic force and the load force respectively; T, and q^B – temperature and internal rate of heat generated per unit volume. The material parameters v, σ, ρ, C and k are magnetic reluctivity ($v = 1/\mu$, μ– permeability), electric conductivity, density, specific heat and thermal conductivity respectively. In general, the above equations are nonlinear and inseparable. The current produced by (7) creates the magnetic field given by (8) and produces the magnetic force which causes the displacement, speed and acceleration of the actuator obtained from (9). The current also generates the heat and the resulting temperature distribution is given by (10).

At the sub-system level, the modelling of such actuators is mainly based on the modelling and computation of the two- or three-dimensional nonlinear magnetic field distribution. This often involves the steady-state and transient solutions of nonlinear Poisson's equation (8). The results are used for performance prediction and design optimisation. The thermal modelling involves the development of thermal models and the numerical solution of the steady-state and/or transient heat transfer equations given by (10) above. Following the solution of (8), the global quantities of interest such as inductance L and force F_m are calculated from the EM field energy E_m:

$$E_m = \frac{1}{2} Li^2 \Rightarrow L = \frac{2E_m}{i^2} \quad (11) \qquad\qquad F = \frac{\partial E_m}{\partial x}\bigg|_{i=\text{const.}} \qquad (12)$$

Besides the above virtual work method of calculating force F, there exist two other methods for calculating the magnetic force – Maxwell stress tensor method and the magnetizing current method.

5. INDUSTRIAL CASE STUDIES

5.1 High-speed Long-lifetime Electromagnetic Valve Actuators

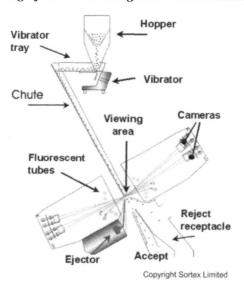

Copyright Sortex Limited

Figure 2. Schematic diagram of an optical sorting machine showing
the optical and ejector sub-systems.

Electromagnetic (EM) actuators are used as pneumatic ejector valves in high-speed optical food sorting machines (Bee & Honeywood, 2002). Operating under continuous duty cycles at 150-300 Hz, they produce a large force (8-15 N). Combined with a stroke length of 0.05-0.1 mm, 'on' and 'off' times of 0.2 ms and 0.46 ms respectively, and a requirement for multibillion cycle operation (in excess of 5 billion cycles) without maintenance, these actuators operate at the limit of what can be achieved by solenoid-based EM actuator technology.

Figure 2 shows the schematic of an optical system implemented in the current generation sorting machines. The unwanted food product is removed (or ejected) by 'firing' a high-pressure (200-550 kPa) air jet controlled by an EM ejector valve (on/off variety). The ejectors are arranged either singly or in a linear array (ejector bank) across the product stream. Once the optical system has identified an object with a defect, the corresponding ejector is triggered electronically to release an air jet that ejects the 'defect' item from the main product stream.

At the sub-system level the mathematical model of the above ejector valve is adequately given by equations (7)-(10). The core modelling activity here involves the modelling and computation of magnetic field distribution in the valve given by (8). The results are used for design optimisation and for investigating the effects of various design parameters on the output performance.

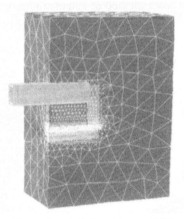

Figure 3. Typical three-dimensional finite element (FE) model of the EM valve actuator (1/4 of the full model shown).

Figure 3 shows the typical FE model of an ejector valve consisting of an excitation coil wound around a c-shape magnetic core that attracts or releases a movable valve plate depending on the excitation state of the coil. In the FE method the direct solution of the governing field equation is replaced by the minimisation of an energy functional corresponding to this equation. For this, the problem domain is divided (discretised) into a mesh consisting of a finite number of 'elements' (for example, triangles in two dimensions) and the solution is sought at the vertices of these elements. The three-dimensional FE model in Figure 3 consists of about 250k tetrahedral elements. For all modelling purposes commercial software package Opera-3d was used. The three-dimensional models refined the two dimensional FE models developed for the above ejector valve actuators (Khan et al., 2005). They were used to design and investigate the next generation commercial ejector valves.

5.2 High Voltage Power Cables

Skin and proximity effects in high voltage power cables increase their per-unit-length a.c. (alternating current) resistance (R_{ac}). It is a key cable parameter that determines its current carrying capacity (ampacity) for which it operates within specified thermal limits. The skin effect in a conductor carrying a.c. is the redistribution of current in the conductor produced by time-varying magnetic field of the conductor itself. The proximity effect is the redistribution of current in the conductor produced by time-varying magnetic field of a neighboring conductor carrying current of the same frequency. In multiconductor systems (for example, high voltage power cables) both skin and proximity effects exist at the same time. By increasing the a.c.-d.c. resistance ratio (R_{ac}/R_{dc}) these effects may considerably increase the power loss in cables. Hence there is a need for detailed analysis of these effects by modelling and computation of magnetic field distribution in order to minimize these effects. In the past attempts had been made to calculate skin and proximity effects by analytical (Woodruff, 1938; Murgatroyd, 1989) and experimental techniques. The analytical techniques are applicable to only conductor systems with simple topology and linear materials (Khan et al., 2000). The experimental techniques are time consuming and extremely expensive. In comparison, the numerical approach based on FE modelling is fast, accurate and applicable to any conductor system.

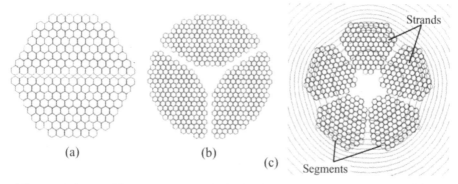

Figure 4. FE modelling of multistrand multisegment power cables: (a) 2-segment, (b) 3-segment, (c) 5-segment (with magnetic field distribution) Milliken type cables.

Results from FE modelling were used to investigate the effects of various geometric (for example, number of segments N_s, diameter of conductors D, number of conductors (strands) per segment N_c, total number of conductors per cable $N_{ct}=N_s \times N_c$, shape, size and arrangement of segments, etc.) and material parameters on R_{ac}/R_{dc} ratio of multisegment Milliken type power cables. It has been shown that by appropriately selecting these parameters this ratio can be reduced considerably.

5.3 Multilayer Printed Circuit Boards

In today's ever increasing use of information technology and electronic devices the need for robust, reliable and long lifetime printed circuit boards (PCB) cannot be over emphasized. This is especially true for highly specialized multilayer PCBs designed and manufactured to function under demanding operating conditions (for example, extreme temperature, shock, vibration, etc.). With "correct-by-design" product development strategy adopted by electronics industry, the demand for accurate modelling and simulation at the design stage of highly integrated PCBs has increased considerably. This mainly involves modelling and analysis of thermal and structural behaviours of the circuit board (board-level modelling) as well as the semiconductor components (packages) mounted on it (component-level modelling).

Figure 5. Typical multilayer printed circuit board (PCB) with surface mounted on-board semiconductor devices (components).

Modern PCBs are multilayered and densely populated (Figure 5). Higher performance, decreasing costs and short product design cycles have resulted in the downsizing of packages and reduction of board-level functionality to a single component. They are composed of multiple interconnected layers (number varying from 2 to more than 20) including substrate layers, typically made of fibreglass and electrical conducting layers, usually made of copper. To route signals vertically between layers, vias are used which are made of copper plated holes. Within conducting layers signals are carried by either irregular shaped planes or linear traces.

The critical mechanical and thermal issues in PCBs which affect their reliability and life expectancy are tackled by numerical modelling and analysis. The life expectancy of individual on-board packages is very much linked to the life expectancy of the board as a whole. In modern PCBs the board level power dissipation can be up to 400 W. In considering the heat flow paths from a package mounted on a PCB it is important to note that only a proportion of the heat is transferred directly from the package surfaces to the ambient air by convection and radiation. Other major paths for this heat flow are by conduction to the PCB via the package pin junctions (leads/balls) and by convection/radiation from its free bottom A PCB itself transports heat by convection and radiation to the ambient air. Thus, the board acts as a heat sink for packages mounted on its surface. Hence the circuit board plays a vital role as a heat dissipator and its thermal performance strongly affects the temperature of solder joints of the package leads.

In addition to external mechanical load (shock, vibration), the cyclic thermal load and non-uniform distribution of temperature in the three-dimensional volume of a PCB give rise to cyclic mechanical stress, especially in the pin junctions. This often leads to cracks in solder joints leading to its fatigue failure. Solder joints act as both the mechanical and electrical connection to the board. The prediction of this thermo-mechanical fatigue life of pin junctions is essential for the calculation of package life expectancy and, hence the life expectancy and reliability of a PCB.

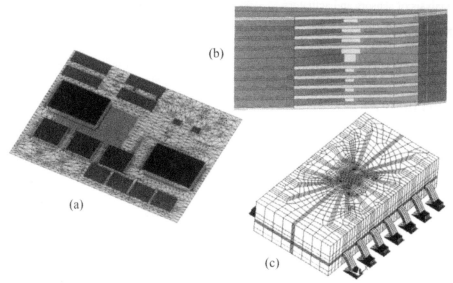

Figure 6. three-dimensional finite element modelling of multilayer PCBs: (a) typical board-level FE model, (b) model of multiple layers including signal and dielectric layers, and vias, (c) component-level model of a SOP-14 package (small online package).

Thus the FE modelling of PCBs involves the numerical solution of thermal, mechanical (structural) and coupled thermo-mechanical problems. The complexity of model realization and its solution is compounded by highly populated three-dimensional wing assemblies, complex topology, use of 'exotic' materials for PCB layers, thermal nonlinearities, etc. To tackle this, the FE modelling and analysis of PCBs is performed in a number following stages: board-level thermal modelling, board-level structural modelling, component-level thermal and thermo-mechanical modelling. Following the board-level modelling, detail component-level modelling is usually performed for a single selected package. This critical package is selected by considering the areas of high temperature and/or maximum mechanical stress.

The three-dimensional FE modelling methodologies described above were used to develop an integrated modelling tool (called PCB-FEA) for finite element analysis of multilayer printed circuit boards (Neumaier et al. 2001, Neumaier et al., 2004).

6. CONCLUSIONS

It is clear from the case studies presented here that mathematical modelling is a key enabling tool in the analysis and design of sensors, actuators and devices. Recent advances in computer based modelling, for example in computational electromagnetics, have added powerfully to the capability of design and analysis, especially at the physical sub-system level. With the availability of powerful computer hardware and software tools significant progress is being made in the application of advanced numerical techniques such as the FEM for modelling and analysis of sensors and actuators.

NOTES

1. Measurement and Instrumentation Centre, School of Engineering and Mathematical Sciences, City University, London.

REFERENCES

Abdullah, F., Finkelstein, L., Khan, S.H. and Hill, J.W. (1994) Modelling in measurement and instrumentation – an overview. *Measurement*, 14, 41-54.

Bee, S.C. and Honeywood, M.J. (2002) Colour Sorting for the Bulk Food Industry. In D.B. MacDougall (ed) *Colour in Food: Improving Quality*. Woodhead Publishing Limited, Chapter 6.

Binns, K.J., Lawrenson, P.J. and Trowbridge, C. (1992) *The Analytical and Numerical Solution of Electric and Magnetic Fields*. Chichester: John Wiley & Sons.

Finkelstein, L. (1994) (eds) *Concise Encyclopedia of Measurement and Instrumentation*. Oxford: Pergamon Press.

Finkelstein, L. and Watts, R.D. (1983) Fundamentals of Transducers-Description by Mathematical Models. In P. H. Sydenham (ed) *Handbook of Measurement Science*, Chichester: Wiley, 747-795.

Khan, S.H. and Abdullah, F. (1993) Finite element modelling of multielectrode capacitive systems for flow imaging. *IEE Proceedings-G*, 140, 3, 216-222.

Khan, S.H., Cai, M., Grattan, K.T.V., Kajan, K., Honeywood, M. and Mills, S. (2005) Design and investigation of high-speed, large-force and long-lifetime electromagnetic actuators by finite element modelling. *Journal of Physics: Conference Series*, 15, 300-305.

Khan, S.H., Grattan, K.T.V. and Attwood, J.R. (2000) Calculation of skin and proximity effects in solid conductors – comparison of analytical and numerical techniques. In P. Di Barba and A. Savini (eds) *Studies in Applied Electromagnetics and Mechanics*. The Netherlands: IOS Press, 18, 165-168.

Murgatroyd, P.N. (1989) Calculation of proximity losses in multistranded conductor bunches. *IEE Proceedings-A*, 136, 3, 115-120.

Neumaier, K., Haller, D., Khan, S.H. and Grattan, K.T.V. (2001) three-dimensional solid modelling of multilayer printed circuit boards. In *Proceedings of the CAT Engineering 2001 Conference*, Stuttgart, 269-283.

Neumaier, K., Haller, D., Khan, S.H. and Grattan, K.T.V. (2004) Automated modelling and thermal FEA simulation of printed wiring assemblies. In *Proceedings of the 20th IEEE Semiconductor and Thermal Measurement and Management Symposium*, SEMI-THERM XX, 271-278.

Woodruff, L.F. (1938) *Principles of Electric Power Transmission*. New York: John Wiley & Sons Inc.

Section 7
Behaviours in Engineering and Applications

7.1

MATHEMATICS IN ARCHITECTURE EDUCATION

Igor Verner and Sarah Maor
Technion – Israel Institute of Technology

Abstract–*This paper considers a Mathematical Aspects in Architectural Design course in a college of architecture, which focuses on experiential learning activities in the design studio. The design process is tackled from three "geometrical complexity" directions: tessellations, curve surfaces, and subdividing space by solids. Mathematical needs in architecture design and relevant learning methods were selected from interviews with practising architects and educational literature. The assessment criteria for the portfolio focused on the project contents, design solutions and mathematics applications. Results of the course follow-up revealed a variety of mathematically-defined complex geometrical shapes applied in students' design projects. The increased interest, challenge, motivation, and positive attitude towards application of mathematics in architectural design were demonstrated. We conclude that continuous mathematics studies in architecture education are required, in order to assimilate mathematical concepts and turn them into a practical tool in architectural design.*

1. INTRODUCTION

Recent research has emphasized the value of mathematical thinking in architecture, particularly in geometrical analysis, formal description of architectural concepts and symbols, and engineering aspects of design. The studies call for accommodating authentic mathematical learning in architecture education (Lewis 1998, Gilbert 1999, Unwin 1997, Burt 1996). The two general approaches to teaching mathematics in context are Realistic Mathematics Education (RME) and Mathematics as a Service Subject (MSS). In RME, the mathematics curriculum integrates various context problems which are experientially real to the student (Gravemeijer and Doormen, 1999), while the MSS approach considers mathematics as part of professional education and focuses on mathematical skills needed for professional practice (Houson et al., 1988 p8).

Our study utilizes the RME and MSS approaches to developing an applications-motivated mathematics curriculum for colleges of architecture. At the first stage, we developed a first-year mathematics course, based on the RME approach (Verner & Maor, 2003). The two-year follow-up indicated the positive effect of integrating applications on motivation, understanding, creativity and interest in mathematics. However, from the analysis of 52 graduation design projects of students who studied the first year RME-based mathematics course, we found that the students did not apply mathematical knowledge acquired in the course in their architectural design projects. This situation motivated us at the second stage of the study to develop a Mathematical Aspects in Architectural Design (MAAD) course based on the MSS approach. The MAAD course is given in the second college year. It relies on the first year mathematics course and offers mathematical learning as part of hands-on practice in architecture design studio. The course focuses on the analysis of different types of geometrical forms used in architectural design. This paper presents the concept of the MAAD course, its architectural design assignments, mathematical models and learning activities. The paper also reports students' achievements and attitudes towards the course.

2. GEOMETRICAL FORMS

In developing the course subjects we considered the principles of classification of geometrical forms in architecture education. Salingaros (2000) and Consiglieri and Consiglieri (2003) formulated the principles of geometrical complexity of architectural forms and proposed studying them from elementary geometrical objects to their complex compositions. Grounded on these principles, our study dealt with three directions of geometrical complexity in architectural design. The course contents for each direction were selected from architectural education literature or recommended by the architects, as presented below:

Arranging regular shapes to cover the plain (tessellations). Boles and Newman (1990) developed a curriculum which studied plain tessellations arranged by basic geometrical shapes with focus on proportions and symmetry. Applications of Fibonacci numbers and golden section in designing tessellations were emphasized. Frederickson (1997) studied geometrical dissections of figures into pieces and their rearranging to form other figures, using two methods: examining a shape as an element of the module, and examining a vertex as a connection of elements. Ranucci (1974) studied mathematical ideas and procedures of tessellation design implemented in Escher's artworks.

Bending bars and flat plates to form curve lines and surfaces (deformations). Hanaor (1998) developed a course "The Theory of Structures" which focused on "the close link between form and structure, between geometry and the flow of forces in the structure". He pointed that distributed loads on bars and surfaces effect deformations which can be described by different mathematical functions. The reciprocal connection between form and construction characterizes Gaudi's approach to architectural design (Alsina & Gomes-Serrano, 2002). Gaudi systematically applied mechanical modelling to create geometrical forms and to examine their properties. He also created three dimensional surfaces such as paraboloids, helicoids and conoids by moving generator profiles "in a dynamic manner" (Alsina, 2002).

Intersecting solids (constructions). Burt (1996) examined integrating and subdividing space by different types of polyhedral elements. He emphasized that this design method can provide efficient architectural solutions. Alsina (2002 pp119-126) considered the design of complex three-dimensional forms by intersecting various geometrical forms. He analyzed the use of these forms in Gaudi's creatures in order to achieve functional purposes such as light effects or symbolic expressions.

3. CASE STUDY FRAMEWORK

The MAAD course has been implemented in one of Israeli colleges as part of the architecture program which certifies graduates as practical architects. The course consists of three parts corresponding to the three directions of geometrical complexity introduced in the previous section of this paper. Each part of the course includes the following components: mathematical concepts and methods with connections to architecture, practice in solving mathematical problems, and design projects. The 56-hours course outline is presented in Table 1.

The first column includes the three course subjects (tessellations, curved surfaces, and solids intersections). The second column details the above mentioned components for each of the course subjects. The third column contains instructional goals which directed mathematical learning in each of the subjects. The fourth column describes learning activities towards achieving the objectives.

In all three subjects, the mathematics concepts were introduced in the form of student seminars. Each of the seminars was given by a number of students and included definitions of mathematical concepts and their applications in architecture. Hands-on activities and discussions were encouraged. The practice sections of the course focused on specific mathematical skills related to the subjects. The projects were performed as individual assignments, while the course meetings were dedicated to project guidance and studio discussions. Emphasis was put on the mathematical aspects of the project.

The tessellation project assignment. Design a tessellation of a floor surface of 34×55 m^2 by means of identical rectangular modules. The module should be a periodic combination of various geometrical figures. Define proportions and dimensions of the figures using golden section ratio and Fibonacci numbers. Develop a concept of the designed module choosing one of the following metaphoric subjects: a temple, kinder garden, political message, harmony with nature, and musical impression.

The curved surfaces project assignment. Design a plan of a gas station. Start from a zero level plan including access roads, parking, pumps, car wash, coffee shop, and an office. Design a top covering for the pumps area, or the roof of the coffee shop and office building. Find a design solution answering the stability, constructive efficiency, complexity and aesthetics criteria. The project stages are: (1) Identifying the project data, (2) Developing an architectural programme, (3) Defining design factors relevant to the project, (4) Generating alternative solutions, (5) Analyzing alternatives and selecting the solution, (6) Producing drawings, calculations, and a physical model.

The solids intersections project assignments. Select a known public building which was designed with solids intersections. The project requirements are: (1) Seminar on the building design process, (2) Mathematical analysis of the solids intersection, (3) Building a model which accurately presents solids intersection in the building, (4) Designing an additional functional module in the solids intersection area.

Table 1. The MAAD course outline.

Course subjects	Components	Instructional goals	Learning activities
1. Tessellations (16 hours)	A. Mathematical concepts of tessellations design	Understanding harmonic dimensions and their use in art, music, and architecture	Seminar presentations on golden section, Fibonacci sequence, logarithmic spiral, and applications.
	B. Practice in solving mathematical problems related to tessellations	Acquiring basic skills in analysis of proportions, symmetry, and drawing tessellations	Drawing logarithmic spirals and tessellation fragments, analyzing basic geometrical figures and their combinations
	C. Tessellation design project	Acquiring experience in tessellation design, modularity, differentiation, proportion and harmony	Designing a flat tessellation of a given floor surface as a combination of modules inspired by a certain metaphoric subject
2. Curved surfaces (20 hours)	A. Algebraic surfaces used in constructions	Identifying types of mathematical surfaces utilized in roof design of existing gas stations	Seminar presentations on roof surfaces such as the Sarger surface, hypar, and ellipsoid.
	B. Practice in drawing algebraic surfaces	Developing skills of drawing surfaces through calculating parameters and coordinates	Exercises of drawing algebraic surfaces (ellipsoid, cylinder, hypar, etc.) and calculating their volumes and areas.
	C. Gas station design project	Application of calculus to defining complex roof shapes for large span solutions	Designing and building a model of a gas station roof which answers the stability, and constructive efficiency criteria.
3. Solids Intersections (20 hours)	A. Algebraic solids and intersections	Defining algebraic solids, their features, parameters and intersections	Seminar presentations on solid intersections (cylinder, oblique cone, pyramid, sphere, and ellipsoid) in constructions.
	B. Practice in calculating solids intersections parameters	Developing skills of mathematical analysis of solids and intersections	Drawing algebraic solids, their sections and involutes. Calculating volumes, surface areas.
	C. Solids and intersections project	Acquiring the skill of analysis of composed structures using analytic geometry	Decomposing an existing architectural structure into solids and analyzing intersections. Building a model of the structure

4. METHODOLOGY

The new MAAD course was implemented in the college in the 2003-2004 academic year by one of the authors (Maor) and attended by 26 second-year students. The goal of the course's follow-up was to examine mathematical learning in the architectural design studio. The study focused on the following questions: (1)

What are the features of mathematical learning in the studio environment? (2) What is the effect of the proposed environment on learning mathematical concepts and methods? The study used qualitative and quantitative methods, which analyzed architecture design experience and observed learning behaviour within the context of design studio. Data on the features of mathematical learning were gathered from interviews with experts. Two experienced practicing architects considered their professional activities and relevant mathematical concepts throughout the design stages. From these considerations we derived ideas of the design activities in the three course projects, of the design education features, and of mathematical needs in design. Relevant teaching methods using context, visualization, heuristic and intuitive reasoning, algorithmic analysis, and reflection were selected in architectural design and mathematics education literature. Through integrating them with the ideas given by the architects we developed the concepts of learning activities in the course.

Data on the effect of the environment were collected by means of design project portfolios, attitude questionnaire and interviews. The design assessment criteria were based on the existing practice of studio evaluation and referred to the three following aspects: concept, planning/ detailing, and representation/expression. The mathematics assessment criteria were: perception of mathematical problems, solving applied problems, precision in drawing geometrical objects, accuracy of calculations and parametric solutions. The post-course questionnaire asked students to list the mathematical concepts studied in the course, give their opinion about its importance, and evaluate the learning subjects and methods. The in-depth interview with one of the students in the end of the course focused on his experience of applying mathematics in design before and in the course.

5. ANALYSIS OF RESULTS

5.1 Interviews and Literature

From the architectural design literature and interviews given by practicing architects we identified relevant mathematical activities throughout the design stages: concept design, data collection and analysis, design alternatives development, design criteria formulation, design solution selection, models and drawings producing and presentation, solution examining and revising. We selected design criteria and related mathematical concepts which include the following: aesthetics, geometrical form, space division, proportions, functionality, culture, environment, symbolism, climate, geology, topography, construction rules and processes, flow of public, energy, materials, stability, durability, building limitations, efficiency, modularity, and accuracy. As noted by the architects, when developing structural forms they use mathematical concepts such as reflection, symmetry, function, fractals, topological features, chaos, proportion, equality, identity, scale, algebraic surfaces, surface area, dimensions, volume, and polyhedron. The literature sources mentioned in the geometrical forms section of the paper present numerous applications of these mathematical concepts in architecture design. The architects recommended to teach the mathematics concepts through architectural design projects which deal with curved surfaces, transformations, large structures and spans in airports, stadiums etc. They emphasized the importance of mathematical methods for obtaining accurate design solutions. They suggested introducing mathematics activities in the architectural design studio and focusing the activities on inquiry and learning discovery, design experiments, and critical discussions. The architects proposed to teach geometrical objects in the order of their geometrical complexity, from a point to a line, plane, surface, and three dimensional volume.

5.2 Design Project Portfolios

Each of the students in the MAAD course performed three projects and reported them in project portfolios. Our focus in evaluating student projects was on the correlation between students' achievements in design and in mathematics. Tables 2 and Table 3 present mean project assessment grades in the three course projects, following the design and mathematics criteria mentioned in the Tables.

Table 2. Design assessment grades.

Project 1	Subject expression	Conception & application	Geometrical variety	Graphic representation	Design grade
Mean	83.1	89.0	77.4	64.6	74.7

Project 2	Efficiency	Aesthetics	Functionality	Program quality	Design grade
Mean	76.9	85.6	73.5	87.5	78.9

Project 3	Structure selection	Geometrical form	Model aesthetics	Additional module	Design grade
Mean	90.0	68.0	76.3	77.5	77.0

Table 3. Mathematics assessment grades.

Project 1	Problems perception	Calculus application	Modularity application	Proportion harmony	Calculations	Drawings precision	Parametric solutions	Math grade
Mean	72.1	56.2	78.4	87.5	80.2	79.4	58.1	73.1

Project 2	Dimensions calculation	Surface parameters	Roof calculation	Building a model	Model precision	Model analysis	Geometrical complexity	Math grade
Mean	76.9	68.8	81.3	76.3	81.9	81.9	85.0	78.9

Project 3	Dimensions calculation	Use of parameters	Intersection calculation	Building a model	Model precision	Intersection analysis	Geometrical complexity	Math grade
Mean	76.9	68.8	81.3	76.3	81.9	81.9	85.0	78.9

Tables 2 and 3 reveal the following features:
1. The mean grades of the three projects are similar in design (74.7-78.9) and in mathematics (73.1-78.9). The project 1 grades were slightly lower than of projects 2 and 3. A possible reason is that in the project the students dealt with tessellations design for the first time.

2. Mean grades for the use of parameters in the three projects were lower than for other mathematics criteria. These difficulties originated from the students' mathematical background.

3. Close correlation between the individual design and mathematics grades was found in project 1 ($\rho = 0.665$) and in project 2 ($\rho = 0.698$). This result indicates the tight integration of design and mathematical aspects of these course projects. In project 3 the correlation between the grades was lower ($\rho = 0.398$). A possible explanation is that project 3 does not include a design component, but focuses on analyzing existing structures.

5.3 Project Examples

Students' project portfolios included concept explanations, module drawings and allocation plans, physical models, and descriptions of design and mathematical solutions. Here we present examples taken from the three projects performed in the course. The example are a tessellation solution drawing, a gas station roof model, and an existing building model (Figures 2A, 2B, 2C). The tessellation drawing presented in Figure 2A shows one student's solution based on logarithmic spirals. The student combined drawing spirals with the golden section procedure and from

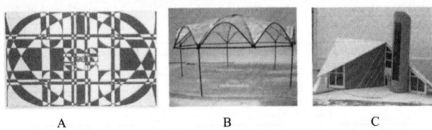

| A | B | C |

Figure 2. Examples of the three course projects: A. Tessellation design; B. Gas station roof design; C. Solids intersection in an existing structure.

her own curiosity defined their analytical parameters by measurements and calculations. In the gas station roof example (Figure 2B) the student applied knowledge of algebraic surfaces and selected a solution based on the Sarger segment, which was suitable to cover the pumps area. This solution was precisely calculated and implemented in the physical model. When building the model, she acquired experience of dealing with efficiency and construction stability factors. When selecting a public building (a museum in the north of Israel) the student recognized the solids intersection part of the structure (cylinder-pyramid). She measured building's dimensions and extracted geometrical data from the architecture design plans and was able to construct the precise model of the building (Figure 2C).

5.4 Attitude Questionnaire

The post-course questionnaire included four open questions. The students' responses were categorized through context analysis and calculating frequencies. The first question evaluated students' opinions about the importance of the three course subjects (design of tessellations, curve surfaces, and solids' intersections) for their architectural studies. Answers to this question indicated that the majority of the students (93%) noted the high importance of the three course subjects. The students emphasized the course contribution to deeper thinking on the tessellation subject, understanding mathematical concepts applied in architecture, designing geometrical forms and connecting mathematics and architecture.

Table 3. Mathematical concepts learned in the course.

Mathematics concepts	Frequency (%)
Proportions, sequences, logarithmic spirals, polygons, symmetry, harmonic division, algebraic surfaces and line intersections (polyhedral, cylindrical, spherical, elliptic, and conic)	90-100
Cartesian and polar coordinates, circles and arcs, exponential and logarithmic functions	70-89
Similarity of triangles, irrational numbers, geometrical dimensions	50-69
Fractals, trigonometric functions, derivatives	30-49
Parabolas, limits, radians, tangents, equations and inequalities, differentiability, vectors and matrices	Less than 30

The second question asked each student to list mathematics concepts learned in the course. Their responses show that the students in the course were exposed to a variety of mathematics concepts learned in class or on a need-to-know basis (Table 3). The third question related to the impact of the course on students' attitudes toward mathematics. Responding to this question, most students (72%) said that the course changed their attitude towards mathematics. Almost all these students (68%) expressed that the course roused their interest in mathematics and demonstrated its relevance to architecture. Some students recognized the challenge and looked for new applications of mathematics in architecture on their own. The fourth question asked students to evaluate instruction in the course. Here, 64% of the students think that studio based instruction increased their motivation to learn mathematics, for 84% it stirred their interest, curiosity and was a challenge. This method enhanced students' creativity (60%) and opened a window to mathematics (68%).

6. CONCLUSIONS

Our study shows the positive change of students' ability to apply mathematics to design of architectural structures as a result of integrating mathematics and architecture design curricula. To achieve this integration, we developed, implemented and evaluated the second year MAAD course. The course is based on the Mathematics as a Service Subject approach, offering mathematical learning with hands-on practice in an architecture design studio. The three directions of complexity in geometrical objects for architectural design in the course are: (1) arranging regular shapes to cover the plain (tessellations); (2) bending bars and flat plates to form curved lines and surfaces (deformations); (3) integrating and subdividing space by solids (constructions).

In the course follow-up study we used qualitative (ethnographic) methods, which observed learning behaviour within the context of the design studio using observations and interviews, attitude questionnaire, and project portfolios. Our observations showed that the students expressed curiosity and motivation to the project experience, and interest to deepen in mathematical subjects and their use. Assessment of students' activities in the projects indicated that the majority of them refreshed and practically applied their background mathematical knowledge. The correlation between design and mathematics grades showed the tight integration of the two subjects in the projects. Analysis of the attitude questionnaires revealed the students' high positive evaluation of the course. The majority of the students noted that the course roused their interest in mathematics and demonstrated its relevance to architecture. The studio method encouraged students' creativity. Some of the students recognized the challenge and even looked for new applications of mathematics in architecture on their own.

REFERENCES

Alsina, K. and Gomes-Serrano J. (2002) Gaudian Geometry. In D. Giralt-Miracle (ed) *Gaudi. Exploring Form: Space, Geometry, Structure and Construction*. Barcelona: Lunwerg Editores, 26-45.

Alsina, K. (2002) Conoids. In D. Giralt-Miracle (Ed) *Gaudi. Exploring Form: Space, Geometry, Structure and Construction*. Barcelona: Lunwerg Editors, 88-95.

Alsina, K. (2002) Geometrical Assemblies. In D. Giralt-Miracle (Ed) *Gaudi. Exploring Form: Space, Geometry, Structure and Construction*. Barcelona: Lunwerg Editors, 118-125.

Boles, M. and Newman, R. (1990) *Universal Patterns. The Golden Relationship: Art, Math & Nature*. Massachusetts: Pythagorean Press.

Burt, M. (1996) *The Periodic Table of the Polyhedral Universe*. Haifa: Technion.

Consiglieri, L. and Consiglieri, V. (2003) A Proposed Two-semester Program for Mathematics in the Architecture Curriculum. *Nexus Network Journal*, 5(1), http://www.nexusjournal.com/Didactics_v5n1-Consiglieri.html .

Frederickson, G. (1997) *Dissections: Plane and Fancy*. Cambridge: CUP.

Gilbert, H. (1999) Architect – Engineer Relationships: Overlappings and Interactions, *Architectural Science Review*, 42, 107-110.

Hanaor, A. (1998) *Principles of Structures*. Oxford: Blackwell Science.

Howson, A., Kahane, J., Lauginie, P. and Turckheim, E. (1988) In: A. Howson et al. (eds) *Mathematics as a Service Subject*. Cambridge: CUP.

Gravemeijer, K. and Doormen, M. (1999) Context Problems in Realistic Mathematics Education: a Calculus Course as an Example. *Educational Studies in Mathematics*, 39, 112-129.

Lewis, R. K. (1998) *Architect? A Candid Guide to the Profession*. MIT Press.

Ranucci, E. (1974) Master of tessellations: M.C. Escher, *Mathematics Teacher*, 4, 299-306.

Salingaros, N. (2000) Complexity and Urban Coherence. *Journal of Urban Design*, 5, 291-316.

Unwin, S. (1997) *Analyzing Architecture*. London: Routledge.

Verner, I. and Maor, S. (2001) Integrating Design Problems in Mathematics Curriculum: An Architecture College Case Study, *International Journal of Mathematical Education in Science and Technology*, 32(6), 817-828.

7.2

MODELLING IN ENGINEERING: ADVANTAGES AND DIFFICULTIES

Maria Salett Biembengut and Nelson Hein
Universidade R Blumenau, Brazil

Abstract–*This article presents the results of a research whose empirical data were obtained from the use of mathematical modelling as a teaching method for Differential Integral Calculus in a Civil Engineering course. The objectives of the research were to evaluate students' learning of mathematics, analyze their competence and ability in using the model and verify the principal advantages and difficulties in establishing modelling as a teaching methodology in a regular degree course.*

1. THE MATHEMATICS SCENARIO IN ENGINEERING COURSES

Mathematics has made up a significant part of Engineering School curricula ever since the precursor courses of Artillery and Fortifications. Since the 19th century for example, differential and integral calculus and geometry have become indispensable tools for engineers. However, the teaching of mathematics that has been employed in engineering courses in most Brazilian universities reveals itself to be inadequate and poorly adjusted. Among the principal causes for this, Biembengut (1997) and Briguentti (1994) point out that:
- The disciplines of Calculus, Algebra and Geometry are dealt with by teahers as if they were hermetically sealed, without any connection between them;
- Most Engineering professionals affirm that the mathematics learned in their courses is insufficient for a practising engineer;
- Differential Calculus makes up part of the requirements of many courses and is taught in accordance with the course in question.

Consequently, the following can be highlighted:
- Students do not see any use for this, and furthermore, they remain unclear about the future application of what they are taught;
- The mathematics developed in the first four semesters ends up forgotten by students just when they need to use it in their professional disciplines;
- Algebra and Calculus record the highest indices of withdrawal and failure;
- Concepts like functions, limits, derivatives and integrals are taught in the same way for all courses (Chemistry, Biology, Economics, Mathematics etc.), the only difference occurring in the emphasis given by the teacher.

Throughout Brazil, in many Engineering Courses, the subjects of Differential and Integral Calculus, Linear Algebra, Analytic Geometry and Numerical Calculus correspond to roughly 480 classroom hours (or 15% of the Course total). Emphasis is given to techniques and not to applications, without any connection to Engineering while neglecting some fundamental concepts in favour of rules and techniques. In general, teaching does not facilitate the ability of the student to interpret and solve problems. This is because only the techniques of solving problems are passed on to the student without his understanding either the genesis of problem solving or the required application. As a consequence, when these students, future professionals, find themselves with a problem or situation to evaluate, analyze or resolve that requires a decision about better performance, they not recognize the tools that they have already acquired during their education. Another aspect is that teachers in charge of the mathematics program for a course look for textbooks that are, in general, better suited to themselves, that is, the same that were used in their own education, and rewrite them for use in their respective classes (Biembengut, 1997).

In order to be able to promote knowledge, teachers cannot limit themselves to the mere transposition of the curriculum. It is fundamental that they reorganize the contents and methods of teaching in such a way that permits articulation between mathematics theory and the activities of engineers, so that the method of teaching allows them to emphasize the physical principles of Engineering and the mathematical theory that describes them (Carmadella, 1995). Under these terms, the mathematics teacher in an Engineering course needs to incorporate some basic knowledge of the course in question into his teaching practices and relate those strictly with mathematical language in order to assist the future engineer.

It is because of this scenario that Mathematical Modelling has come to be defended as a teaching and research method for course in Higher Education, in particular, those in Engineering. Mathematical Modelling can become an interdisciplinary activity that would be sufficient for verifying the areas in Engineering where mathematics is necessary.

2. MATHEMATICAL MODELLING AS A TEACHING METHODOLOGY

Due to the existing structure of Engineering courses (semesters, credits system, disciplines, hours) and the difficulty the mathematics teacher has in a short space of time of acquainting himself with an engineering problem which requires instrumental mathematics, some re-direction of the Modelling method is necessary for teaching mathematics (Differential and Integral Calculus, Linear Algebra). This method is guided by the *development of programmatic content*, beginning with mathematical models applied to various areas of knowledge and, concomitantly, to the *orientation of students towards research-modelling* (Biembengut, 1997; 2004). The following is a synthesis of the method.

2.1 Development of Programmatic Content

In order to develop programmatic content, the teacher chooses one or more mathematical models applied to the Engineering in question, adapting them to the teaching. This adaptation should contain a brief exposition of the theme, some questions that instigate the students to respond and a process that leads to the formulation of questions in such a way as to raise the programmatic content. Once the content is raised, the teacher develops it, presenting the theory along with analogous examples, returning to the original question at the end of the process.

In synthesis, the stages of the process to be followed by the teacher in the development of his classes for the course in question are:

(1)   Present a synthesis of the *theme* relative to Engineering
(2)   Propose some questions about the *theme*, encouraging students to respond
(3)   Select the question that is most appropriate for developing the programme's mathematical content
(4)   Formulate questions in such a way as to raise the mathematical content necessary for resolving them
(5)   Present the programmatic mathematical content
(6)   Propose analogous examples so that the content is not restricted to the model, as soon as the necessary content is sufficiently developed for answering or resolving this stage of the work
(7)   Solicit students to make research into the subject, if deemed necessary
(8)   Return to the question that began the process, presenting a solution
(9)   Guide the students in analysing their results and its validity.

A mathematical model can be used to develop each mathematical topic of the program, or a single one for the semester, trimester, etc. In the case of a single one being used for the entire period of study, it must be wide enough in scope to develop the program at the same time, as well as interesting enough to motivate the students. The important thing is that this model be 'tuned in' to the knowledge and expectations of the students (Biembengut and Hein, 2005).

2.2 Orientation of Students Towards Research - Modelling

The objective of this work is to create conditions in which students learn to make mathematical models, increasing and perfecting their knowledge. A group of students, under the supervision of the teacher, must be responsible for the choice and direction of their own work. It falls to the teacher to create conditions where they can develop this autonomy.

3. EXPERIMENTAL WORK

The research objectives were to evaluate students' learning of mathematics, analyze their competence and ability in using the model and verify the principal advantages and difficulties in establishing modelling as a teaching methodology for Differential and Integral Calculus in a regular degree course. The experiment was made in the four disciplines comprising the subject, during 4 semesters (two years) in the Civil Engineering Course at the Universidade Regional de Blumenau (FURB). It began with CDI I class of 50 students and was followed with the same for the other CDI disciplines (II, III and IV). To illustrate this process, we present one of the efforts made in the Calculus I that used a mathematical model as a guide for developing the subject and the directions the students used for doing the modelling.

3.1 Development of Programmatic Content

In order to develop programmatic content for Differential and Integral Calculus I in Civil Engineering, the *theme* of Public Health and Sanitation was used, in particular. This theme is part of the program used in the discipline Sanitation I (given in the 7th semester). Data concerning the theme were obtained from the work of Imhoff and Imhoff (1986) and the Water and Sewage Treatment Service of the city of Blumenau, SC, Brazil, and duly modelled and adapted in order to develop the programmatic content for Calculus during the semester period (75 classroom hours).

(1) Present a synthesis of the *theme* relative to Engineering;

In order to interest the students in the theme, a brief exposition:

Water pollution is one of the main problems faced by cities, a problem caused by industries and members of the community throwing rejected or domestic waste materials, mainly chemical products, into lakes, rivers, water supplies and the sea. This pollution destroys the environment, eliminates flora and fauna, provokes flooding and leads to endemic diseases, thus compromising the quality of life of the very population that produces it. The concentration of substances that exist in the sewage system depends on the amount of water consumed by inhabitant in one day and their eating habits, the existence or non-existence of rainwater, and other factors. In accordance with local, particular conditions the level of impurity in effluents varies during the course of the day, as does discharge. Generally speaking, maximum discharge occurs near mid-day, at which time the pollution level also reaches its maximum levels. Solving this problem requires action on two fronts: employing the abilities of pollution specialists for correcting the existent problems and for treating waste water, and Education.

(2) Propose some questions about the *theme*, encouraging students to respond;

After exposition of the *theme*, the teacher (the author of this proposal) encouraged the students to raise questions. Among the questions proposed, the following were selected: *What is the daily run-off? What is the daily quantity of sewage? What are the treatment processes for sewage? How is the treatment of effluents effected?*

(3) Select the question that is most appropriate for developing the programme;

We begin by establishing the relationship between *time x discharge* and between *time x sediment producing solids*, using the data below from IMHOFF and IMHOFF (1986).

Square: Variation of discharge and average amount of sediment producing solids during 24 h

Time (h)	discharge (m³/h)	sediment (g/m3)
0	208	3.5
2	180	2.5
4	150	1.5
6	180	2.0
8	240	5
10	360	9.5
12	500	12.2
14	530	11
16	480	9
18	350	7
20	310	5
22	250	4

(4) Formulate questions in such a way as to raise the mathematical content necessary for resolving them.

The data in the tables show that discharge and sediment producing solids, in an effluent, vary during the day, that is, both discharge and sediment producing solids depend on the time. *What is the daily run-off? What is the daily quantity of sewage?*

(5) Present the programmatic mathematical content

Starting with discharge analysis (m^3/h) of the effluents in a 24h

(i) Partial resolution: *What are the laws that govern these two functions?*

Since discharge of an effluent is continuous in the time period, we can determine a 'law' that governs this function. Since we make us of only a finite number of points, 12 to be exact, it becomes necessary to use some 'tools' in order to pass on to the process called 'discrete to continuous'.

The data is connected to questions allow us to conceptualize Real Function: domain, image, graphic representation and types of function. By means of a computational program, students made various adjustments (straight, parabolic, cubic, exponential, logarithmic and trigonometric), opting in the end for the one that provided the best coefficient of correlation. Through a polynomial of degree 4, they obtained a good approximation for both functions.

$$V(t) = 0.0251t^4 - 1.3763t^3 + 22.5958t^2 - 98.4173t + 241.2295 \qquad (a)$$

where $V(t)$ is the rate of discharge at each instant.

$$S(t) = 8.8098(10)^{-4}t^4 - 0.0457^5 t^3 + 0.7134^6 t^2 - 2.97721t + 4.30827 \text{ (b)}$$

where $S(t)$ is the rate of sediment producing solids at each instant

(ii) Partial resolution

The rate of discharge and the amount of sediment producing solids vary over time. For example, discharge and the amount of sediment producing solids increases between 06:00 and 14:00, reducing between 14:00 and 04:00 (of the following day). That is to say, the rates of discharge and the amount of sediment producing solids depend on time *(t)*. The functions found allow us to perform an analysis on the tendency of discharge or sediment producing solids at a given time. For example:

What are the discharge and sediment producing solids near mid-day?

$$\lim_{t \to 12} 8.8098(10)^{-4}t^4 - 0.0457^5 t^3 + 0.7134^6 t^2 - 2.9772t + 4.30827 \cong 10.532 .$$

The hour-to-hour discharge can be determined by calculating the average, for example: the discharge rate between 08:00 and 18:00 is $11m^3/h$.

(iii) Partial resolution

However, the rate varies during the day at each moment, thus: *What is the discharge rate at each moment?*

Through the data presented, the maximum discharge and the maximum rate of entry of sediment producing solids occur from 12:00 to 14:00. If we derive the functions that govern the situations proposed and make them equal to zero, we can obtain the moments of greatest and least discharge and entry of sediment producing solids, that is, the critical points of the functions.

The above questions allow us to represent the concepts of Limit, Derivation, maximum, minimum and inflection points and n- extreme derivative.

(iv) Partial resolution

The function S(t) presents the 'behaviour' of the rate (grams per cubic meter – g/m^3) of sediment producing solids that enter into an effluent at each moment, but:

What is the daily amount of sediment producing solids?

Since the entry of sediment producing solids is practically continuous along the course of 24 hours, in order to calculate the accumulated quantity it is necessary to add (integrate) the amounts that enter at each moment. In these examples, V(t) and S(t) are derived functions (functions established by means of the average rates). In order to find the quantities owed requires application of an Anti-derivative, or a Defined Integral.

At this moment, the students were presented with the concept, definition and properties of the fundamental theorem of Calculus.

$$\int_0^{24} S(t)dt = \int_0^{24} (8.8098^{(10)^{-4}} t^4 - 0.0457^5 t^3 + 0.7134^6 t^2 - 2.97721t + 4.30827)dt$$

where S(t) is the rate of sediment producing solids at each instant of time.

Although good results were obtained, the rate of discharge and the entry of sediment producing solids at each moment of time indicate periodic functions. In this sense, trigonometric regression can be employed.

Each of these questions was developed (or partially modelled) together with the students in the classroom, according to the 9 stages of the Modelling method described above. By the end of the semester, topics in Differential and Integral Calculus (functions, limits, derivatives and integrals in **R**), along with others belonging to Numerical Calculus, such as theory of errors, minimum square root method and the roots of functions had been dealt with completely. Numerical Calculus topics did not cut into the hours allotted to the discipline as computers were used to solve exercises, reducing the time necessary for accomplishing this.

Although the 'modelled' questions allowed students to obtain satisfactory results, it is not possible to affirm that the mathematical models obtained are valid since they did not incorporate relevant variables such as time of year, regional climactic conditions and culture of the community, among others. However, their value as guides in the discipline of Differential and Integral Calculus I was fully satisfactory. It is worth noting that analogous examples (6th stage) and exercises were proposed at various times, as were tests. Also, students made research into the subject (7th stage) and returned to the question that began the process to analyze it (8th and 9th stages).

3.2 Orientation of Students Towards Research - Modelling

In accordance with the method described above, students were encouraged to form groups, in order for them to design a Modelling project on some engineering theme, to be done together with the classes. As Differential and Integral Calculus I entails 75 hours, 15 were given over to this work, sub-divided into 5 meetings within the classroom, in order to allow the teacher to supervise the students in the design of their projects.

The first meeting took place only after presentation of part of the program by means of the modelling guide, that is, after 15 hours. This was done because it was understood to be the minimum time needed for the student to understand the process and still be able to choose a theme that he/she intended to develop. At the 1st meeting the students formed themselves into 20 groups.

Among the themes chosen by the students were: Impermeability; Chlorination; Foundations, Evaluation of Loads on Building Floors; Urban Planning; Metallic Structures; Cement Resistance; Study of Soils; Standards for Railroads; Tensile Strength; Structural Analysis; Use of Additives in Concrete; Porosity; Water Supply; Metallic Supports; Septic Tanks; Ceramic Materials; Corrosion; Plastics and Resistance of Steel. Each of these themes/subjects is an area of Civil Engineering.

After this meeting, a week was given for the students to design a synthesis about the theme, along with ten questions. This synthesis allowed the teacher to acquaint herself with the subject and to select, as a suggestion, three questions for each group, questions for which the mathematics necessary for solution made up part of the program of the discipline. At the 2nd meeting, the students, under supervision, applied the Function and Method of Minimum Squares, also using graphic calculators; in the 3rd and 4th meetings, students performed analysis on the model questioned and later, validation though use of Limits, Derivative, Differentials and Integrals. A seminar took place at the 5th meeting, with presentation of the work of each group. Most of the works lived up to the expectations of the students themselves.

3.3 Some Considerations

During 4 semesters of lecture, various forms and instruments were used for gathering and organizing the data evaluated in the research, such as: individual written tests in the traditional form (solving problems oriented towards engineering, explanation of mathematical theory and application of techniques for solving problems); Modelling projects made in groups; observations about the attitudes of students in the classroom and during orientation of their work and interviews with 12 students that participated in an experimental project (students from the four levels of the Calculus program).

At the beginning of the semester, this class of CDI was made up of 50 students. Of these, 60% passed, 8% failed and 32% gave up. Of those that quit, most had already taken the discipline at least once. They thus gave up less than one month into the semester, seeing that the process required more study and research, which does not occur in traditional teaching. Of the 30 students who passed, 25 continued in this class. Since it is treated as a regular course, there were about 50 students in CDI II, meaning that 25 of these students did not know the method (accustomed, that is, to traditional teaching). Thus in the following CDI's (III and IV), there were both drop-outs and entrants in the class without knowledge of the method. By the end of CDI IV, 20 students had concluded the discipline. Of these, 12 students (24%) of the 50 that had started in CDI I participated in the entire experimental method. These students started a modelling project in the first semester and concluded it at the end of the fourth CDI discipline. They displayed a positive attitude in relation to the method adopted.

It is worth remembering that each of the CDI disciplines (II, III and IV) used an engineering *theme* to develop programmatic mathematical content (prepared previously) and, together with their classes, the students designed a Modelling project. Students that entered in CDI II, or III, or IV, integrated themselves into a group from the initial semester or chose other themes they wished to model. These students also had much more difficulty in making a model.

4. FINAL CONSIDERATIONS

It is understood that, for a student in the first stages of a Course, it is necessary to know that mathematical theory can and should be adapted to the area of Engineering. In this sense, for the majority of students in this class in particular, Mathematical Modelling, besides enabling the learning of mathematical content integrated into the area in question, also facilitated conditions for learning how to formulate, solve and take decisions about Engineering problems.

Modelling is a dynamic process that can contribute above all to the improvement of knowledge in both the student and teacher (Blum & Niss, 1991). The results from the experimental works that were made show that when compared to the traditional model of teaching Calculus, Mathematical Modelling as a teaching and research method in Engineering provoked the following: improved grasp of concepts, through making mathematical content emerge from mathematical models in Engineering; stimulating participation and creativity, due to the questioning process, research and design of models; highlighting the importance of mathematics to the student, not only in application (as a tool) but also the theory that sustains it; drawing the beginning student closer to specific Engineering disciplines, properly so-called; enabling the mathematics teacher to acquaint herself or himself with Engineering problems, contributing to his or her continued training; making interaction possible between teachers of introductory disciplines and those of professional disciplines, and allowing a tightening of the relationship between new technologies, leading to periodic revision of questions relative to the curriculum.

All of these led to greater interest on the part of students in regard to applicability, stimulating classroom participation and increasing the number of research projects presented periodically in the form of seminars, besides leading to an increase in the general average of grades in written examinations, which resulted in a reasonable reduction in the number of withdrawals and failures. Despite these results in CDI, some facts may be highlighted: resistance on the part of the students in regard to the teaching proposal, since the students, accustomed as they are to the "conventional forms" of teaching, see modelling as a more difficult road since it is a process that demands research, creativity and reasoning; the non-continuity of the semester system, made it difficult for some students to realize a long term research project; the absence of interaction among other teachers of basic disciplines (Algebra/Mathematics, Physics, Chemistry) and other specific disciplines. There was no joint planning among teachers that permitted mutual informing about the needs of specific disciplines and technological innovations. The results of this research were significant, principally for understanding that modelling is a good path for those that want to study, but is not a medicine for those that are still not sure about what path to follow in their own lives.

The Modelling Method proposed above seeks to offer conditions for promoting the teaching of mathematics that, besides theory and practice, provides the future engineer with the ability to discern and discuss aspects related to his area, thus becoming an autonomous and critical professional. This entails knowing mathematics that arms the beginning student with the possibility of acquainting himself with questions that he will have to deal with in the future, promoting union between disciplines of the course and providing a more integrated approach for the curriculum. To summarize, the Method proposes reaction and interaction between the teaching staff and students, involving both in the continuous and necessary production of knowledge - sharing together the experiences acquired (Berry, 1987).

Upon teaching mathematics integrated into an area relative to engineering, the teachers contribute, above all, to the 'energizing' of the students, all future engineers, to the chances offered by critique and questioning. These critiques and questionings strengthen the degree to which the students are able to design their mathematical model, by means of becoming aware of alternative ways and of understanding what is really significant. The mathematical content that can lead to better training of future engineers is that which will provide sustenance for perfecting new technologies. This shows that Engineering curricula, mathematics in particular, cannot remain static. It will only become dynamic by interconnecting with other areas, making the course point out to what is more relevant. This will allow for the tailoring and programming of Engineering Courses that are capable of attending to the needs of society, as well as orienting the students towards assuming a critical posture in light of growing technological domination.

REFERENCES

Berry, J. et al. (eds) (1987) *Mathematical Modelling Courses*. Chichester: Ellis Horwood.

Biembengut, M.S. (2004) *Modelagem Matemática & Implicações no Ensino e na Aprendizagem de Matemática*. 2nd edition. Blumenau-SC: Edifurb.

Biembengut, M.S. and Hein, N. (2005) *Modelagem Matemática no Ensino*. 4th edition. São Paulo – SP: Contexto.

Biembengut, M.S. (1997) *Qualidade no Ensino de Matemática da Engenharia*. Doctoral Thesis, UFSC, Florianópolis - SC.

Blum, W. and Niss, M. (1991) *Applied Mathematical Problem Solving, Modelling, Applications, and Links to Other Subjects - State, Trends and Issues in Mathematics Instruction*. Educational Studies in Mathematics 22 (1), 37-68.

Briguenti, I. (1994). O Ensino na Escola Politécnica da USP: Fundamento para o ensino de Engenharia, EPUSP: São Paulo.

Camardella, A. (1995) *Detectando problemas no ensino de Engenharia. Revista de Ensino de Engenharia- ABENGE, 11*. Niteroi-RJ.

Imhoff, K and Imhoff, K. R. (1986) *Manual de tratamento de águas residuarias , tradução de Max Lothar Hess.* São Paulo: Edgard Blucher.

7.3

MODELLING: DIFFICULTIES FOR NOVICE ENGINEERING STUDENTS

Marta Anaya, María Inés Cavallaro and
María Cristina Domínguez
University of Buenos Aires, Argentina

Abstract–*The purpose of this work is to study the different responses and reactions of novice students when a real problematic situation is presented with different degrees of specification in the given information. The main objective is to achieve a deeper understanding of the difficulties in the modelling process. The research was developed in two stages and it was focused on the structuring of the real model (strategies, assumptions, feasibility, self-evaluation), the mathematical model and mathematising processes, and the incidence of working conditions.*

1. INTRODUCTION

Mathematical models in engineering are usually developed in order to study real-world problems. In fact, the abilities, which are emphasised in model eliciting activities, are similar to the abilities required in the practice of engineering. These activities become relevant in a pedagogical sense as well. However, the characteristics and difficulties of these sorts of students at different stages of the modelling process are still issues of interest to be investigated.

With the aim of studying the characteristics and difficulties, of novice engineering students modelling activities, when different styles of guidance are provided, a real-problem situation was presented to four groups of students. The study was conducted in two stages. During the first one, students responded to a questionnaire related to the resolution of the problematic situation, each group with different degrees of specification in the given information and stressing different styles of working. During the second stage, students interpreted the process involved in the situation contextually.

2. THEORETICAL CONSIDERATIONS

Concepts and notions such as: *mathematical modelling process, problem, real-world situation, mathematising and mathematical model* are used according to the definitions stated in the ICMI Study 14 Discussion Document (2000).

The role of creativity in the formulation of strategies for (mathematical) modelling has been interpreted following the ideas about creativity in mathematics in Ervynk (1992).

The role of intuition in the formulation of strategies and the interpretation of the elaborated model was analysed following Fischbein's theories.

In line with these ideas, the theory of intuitive rules developed by Stavy and Tirosh (2000) proposes that students' responses may be often determined by irrelevant external features of the tasks rather than by underlying concepts.

The use and availability of mathematical knowledge in the process of modelling different kinds of mathematical knowledge was studied according to Hiebert and Lefevre (1986) who propose a classification in conceptual and procedural knowledge.

3. METHOD

Population:

The study was conducted with 140 engineering students which constituted four entire classes of an average of 35 students each (ages 19-20). All of them had already taken the same first courses in General Chemistry, Physics, Calculus and a brief introductory course to ordinary differential equations.

Procedure:

The study was conducted in two stages.

a) In a first stage, each class, constituting a group, was given a different design of the same problem with different degrees of details and mathematical specifications.

The four groups G1, G2, G3 and G4 worked with a general initial situation related to different versions of a well-known inverse problem presented by: *E. Aboufadel y S.J. Tavener (Newsletter for the C-ODE-E Winter 1996, p11)*

> *"A tank is buried in the ground. It is filled with salty water and it is suspected to be leaking, polluting nearby groundwater. It is not possible to measure directly neither the volume of salty water into the tank nor the amount of salt. Yet it is possible to measure the concentration of the solution any time and it is also possible to inject and to remove liquid into the tank. "*

The four groups had to answer finally to the following questions:

> *Q1. How would you find out whether the tank is leaking?*
> *Q2. Which would be your assumptions to solve this question?*
> *Q3. How general is your method. Point out possible weaknesses of your method.*

G1: It consisted of 35 students who were only given the initial situation and the questions and they worked on them spontaneously on the spot for two hours.

G2: It consisted of 36 students who were first requested to find out the volume in a simplified case (No Leak) and after this, they were guided to construct a more complex model (Leakage) with data related to a process of injection and removal of liquid. The translation to the mathematical model was neither given nor suggested. They worked on the spot for two hours.

G3: It consisted of 38 students who received the same information as students in G2, but guidance was added to construct the mathematical model of the situation using a differential equation. The work was carried out on the spot for two hours.

G4: It consisted of 32 students who received the same information and guidance as students in G3. However, they worked at home for three weeks in groups of four students each, naturally made up by them. They were allowed to use software to find the parameters and to draw the graphics in order to write an individual report accounting for their findings.

The students' performance in the first questionnaire was analysed according to:

I) Modelling processes:

I.1) *Structuring the real model.* Strategies that the students developed, the context of resolution, the knowledge they invoked according to the guidance they received to formulate the real model. Self-evaluation of the model (comparing G1 and G2)

I.2) *Mathematical modelling and mathematising* process according to the amount of information with which each group had been guided, the mathematical level they handled to solve the tasks, and how they interpreted the model (comparing G2 and G3)

II) Impact of different working conditions (analysed in G3 and G4)
These differences related to guidance and details in the information and the working conditions are summarised in Table I:

Activity	G1	G2	G3	G4
Guidance for Real Model	-	x	x	x
Guidance for Math Model	-	-	x	x
Required Presence when solving	x	x	x	-

Table 1. Activities Planned for each group.

b) In a second stage, another questionnaire was given to each student of G2, G3 and G4 requiring a brief analysis of the evolution of the concentration during the process of injections and extractions of solution. The aim of these questions was to gain an insight of the students' understandings and interpretation of the global process and the relation to their mathematical model. G1 was not included because these students were not suggested a dilution process.

4. RESULTS AND DISCUSSION

FIRST STAGE: Modelling Processes and Different Working Conditions

I) Modelling processes.

I.1) *Structuring the real model* . (G1 and G2 students)
I.1.1) *Strategies*: Students' performance are summarised in Table 2.

Strategies for Leakage Detection		G1 n=35	G2 n= 36
Strategies	Workable proposals	20%	0%
	Incomplete	16%	17%
	No responses	10%	28%
	Non suitable methods	54%	55%

Table 2. Groups G1 and G2 performances in formulating strategies.

Group G1: Within the 90% of the students that attempted a strategy to detect the leak, four different types of strategies worked out in different contexts were observed:

1)*Direct measurement of volume* (filling the tank or emptying it through different methods: 17% (Unworkable method, previously stated as not possible).

2) *Measuring and relating pressure and temperature*: (12%). (using previous knowledge)

3) *Measuring pollution of* ground layers (close to / far from the tank) with different assumptions (11%).

4) *Measurement of the concentration of the solution in the tank* considering different assumptions. (50%) For example:

"I could compute the initial salt concentration, then I would add a known amount of water and measure the salt concentration again, I would repeat this procedure and I would be able to know the change in the proportion of concentrations. But I don't see how to find out the variation of volume. I can't find the solution." (M)

In order to elaborate the strategy, the students seemed to focus only in part of the brief information given, neglecting the remainder. This selection might be based on the associations between the student available knowledge and this given information.

Only 14% of G1 students attempted to use mathematics, and they were not successful.

Group G2: 67% of these students considered the measuring of salt concentration as a strategy for solving the problem. Though 17% of the students successfully elaborated a method to find out the volume of liquid into the tank (adding fresh water and measuring concentrations), none of them could elaborate a workable strategy to detect the leak.

Difficulties in the modelling process:

Mental processes involved in modelling activities that were observed during this stage:

Understanding: The richer the student cognitive system is in concepts and meaningful relationships, the richer they are in resources to design modelling strategies.

Choice of the available knowledge: This is related to a balance of effectiveness and efficiency in order to gain feasibility. This selectivity lies on the modeller's knowledge cognitive structure and on the modeller's experiences.

Creativity: It is required for the design of strategies (and Mathematical Models) for this design may demand the student to create new ideas and to relate old ideas in a new way, extending the context in a way that is different from the context that was known before.

Difficulties in the former processes were observed in G1 and G2 students:

❏ *Difficulties related to the relational understanding* of the situation to be modelled, including difficulties in the *identification* and *discrimination* of variables and unknowns.

❏ *Difficulties related to creativity* in establishing *associations* and *relationships between pieces of knowledge that eventually might not have been related up to that moment.*

❏ *Difficulties related to the choice of the available knowledge and the use of the given information.*

I.1.2) Self- evaluation of the model:

Only 14% of the students of G1 explained that the strength of their method was "generality". The rest of the students focused in weaknesses. 34.28% of G1 students and 11% of G2 students stated the given assumptions as weaknesses. For example: *"the conditions of the problem might not be fulfilled"(D)*

Evaluation of the Model Weaknesses	Based on the students' own assumptions	43%	34%
	Based on the lack of knowledge	25%	6%
	Based on general difficulties (*not exact, troublesome*)	9%	11%
	No Response	23%	50%

Table 3. Students' evaluation of their model.

While G3 Students were very far from evaluating the mathematical model, some students of group G4 (11%) commented about the resolution method but not about the model:

"The results obtained are not exact because we had to estimate some unknown data. For example we had to use the command FindRoot of Mathematica. These tentative data were randomly chosen but our teacher helped us." (ML)

The following difficulties were observed in G2 and G4 denoting the students' evaluation of their own model:

❏   *Seldom use of the MM to confirm or predict results of the real model.* Observed in the leakage detection in G2 and G4.

❏   *Analysis of the real situation may be contradictory with the mathematical results.*

I.2)Mathematical modelling and mathematising (G2 and G3 students)

In the mathematising stage, 60% of **G2 students** in the Leak case used proportionality, linear and non linear algebraic equations to describe the volume. The rates of change of salt and concentration were rarely associated to the corresponding derivative (only 6%), and if so, no ordinary differential equation (ODE) was stated to describe the process. The knowledge of mathematics that they handled was not the one that was studied at university, but a previous one based in proportionality. 32% of the students failed to identify or take the inflow and removal of fluid and /or the leakage into account. There was a tendency to use as information what in fact was an unknown quantity, and to neglect information. For instance, 22% assumed that the volume is measurable at any time. Other students assumed that the parameters calculated under a NL model were appropriate for the L model.

Group G3: The mathematising process in this group was guided so that the students could obtain a first order linear differential equation for the amount of salt. However, 80% of G3 students who attempted to solve the ODE showed difficulties concerning its resolution. These difficulties were related mainly to the recognition of the ODE type (60% failed), and to procedural difficulties related to integrals, logarithms and functions, only 5% solved the differential equation correctly. G3 students did not obtain or either tried to arrive at a conclusion.

Time was an important factor influencing performance: ODE was a subject recently taught and they had to solve the equation on the spot. They also had to deal with limitations in time and resources as well as with difficulties in the previous mathematical knowledge (concepts and procedures). For example, the salt concentration, which is an unknown function of time C(t), was taken in many calculations (25%) as a constant.

I.2.1) Difficulties related to the mathematical model:
❏ *The knowledge of mathematics that is spontaneously used is more elementary than the knowledge that is required by the problem.* Observed in G2
❏ *Mathematics procedural difficulties.* Observed in G2 , G3.
❏ *Difficulties with the mathematical representation* and interpretation of the contextual concept of variation (derivative as rate of change). Observed in G2.
❏ *Difficulties with units.* Observed in G2, G3
❏ *Persistence of the linear model even if not appropriate.* Observed in G2
❏ *Implicit misleading assumptions.* Observed in G2

Mathematising and operating within the mathematical model involve: an isomorphic correspondence between concepts and relationships in both contexts: the real context and the mathematical context. They also involve a correct choice of concepts and relationships of the available knowledge to state the mathematical model and a correct interaction between conceptual and procedural knowledge. All these factors failed in these students.

II) Different working conditions

Group G4: Students solved the differential equation successfully and estimated parameters using a software program, presenting graphics of the concentration and quantity of salt, but they obtained poor conclusions from them. For these students the main concern was to show the way they had solved the equation with the software. Only 15% of G4 students went back to the original situation although some of them misunderstood the concept of modelling a situation, as for example: *"The value of the initial amount of salt is different in both cases, so there is a leak" (GF)*
53% of the students did not explain their calculation, except for the commands they had used.

Comparing G4 with G3, it was observed that while G3 Students were very far from evaluating the mathematical model, some students of group G4 (11%) commented about the resolution method but did not mention the model:

"The results obtained are not exact because we had to estimate some unknown data. For example we had to use the command FindRoot of Mathematica. These tentative data were randomly chosen but our teacher helped us." (ML)

Some students spontaneously observed coincidences with the real model, while other students were surprised with their results. This fact shows that their mathematising was far away from the real model. Only 5% was able to spontaneously make a correct analysis of the process supported in their plots. Their responses followed intuitive models of thinking and intuitive heuristic rules

SECOND STAGE: Interpretation of the Process

The required qualitative analysis of the dilution process (in the second questionnaire) was aimed to evaluate whether the students considered the model capable of predicting the behaviour of its variables. The results of this stage are summarised in Tables 4, 5.

Concentration (C)		Leak (L) %			No Leak (NL) %		
		G2 n =36	G3 n=38	G4 n=32	G2 n =36	G3 n=38	G4 n=32
Increases		11%	0%	0%	22%	0%	0%
Decreases		**61%**	**37.5%**	**100%**	**52%**	**50%**	**37.5%**
Reasons for decreasing	Volume	0	0%	50%			
	Dilution	39%	25%	0%			
	Both	**22%**	**12.5%**	**12.5%**			
	No reason	0%	0%	37,5%			
Constant		17%	12,5	0%	20%	12.5%	62.5%
No response		11%	50%	0%	6%	37.5%	0%

Table 4. Concentration Evolution in each case L&NL.
(Correct responses are in bold).

Concentration (C) in (L) & (NL)	G2 n=36	G3 n=38	G4 n=32
Decreases (L) &constant (NL)	6%	0%	62,5%
Decreases (L)&Decreases (NL)	**50%**	**37.5%**	**37.5%**
Constant (L) & constant (NL)	10%	12,5%	0
Increases (L) & constant (NL)	11%	0%	0
Increases (L) & Increases (NL)	11%	0	0
No response (one or more items)	12%	50%	0

Table 5. Concentration Evolution in cases L&NL.
(Correct responses are in bold).

The different percentages of correct responses in cases L and NL show dissociation between the real model and the mathematical model, especially in G4 students, who had solved the differential equations and plotted the salt concentration as a function of time. All of them observed that concentration decreased in the leaking case, but associated this change to the variation of volume, which had been used by them in the differential equation.

A possible explanation of these interpretations is that many of these responses might have been framed within intuitive models of reasoning. In this case the intuitive rules were observed to influence the responses of the three groups to different extents. For example: 20% for G2, 12.5% for G3, and 62.5% for G4 considered that concentration remains constant because "*the volume remains constant*". These responses are clearly in line with the intuitive rule Same (vol)-Same (C).

Other students (G2) believed that the concentration would increase because "*we are adding salt*", forgetting about the dilution effect and reasoning with the rule More(S)-More (C).

The students also showed a difficulty in handling all the variables simultaneously, which reflects on their responses in the leaking case.

The following **difficulties** were observed in the students' evaluation of their own model:

❑ Seldom use of the MM to confirm or predict results of the real model (G2 and G4).

❑ The analysis of the real situation may be contradictory with the mathematical results.

5. FURTHER REMARKS AND CONCLUSIONS

The process of "modelling" seemed blurred in groups G3 and G4. The explicit demands of mathematising and solving in the case of G3, including additional use of computing software and numerical estimations in the case of G4, seem to have shifted the students' attention: they transferred mathematising and solving to the foreground, and the real model that was suggested, to the background. On the other hand, G1 students focused on the real model, shifting mathematics to the background, and as G2 students' mathematising process was neither complete nor successful, their responses were framed within the real model with a limited use of mathematical knowledge.

The students also showed a difficulty in handling different information simultaneously.

In any case there is a dissociation between the real model and the associated mathematics. This factor may cause Mathematics to be perceived like a tool but, at the same time, the conditions under which this tool may be used are neglected.

There is a need to investigate the interpretation and the analysis of the mathematical results. These interpretations should be contrasted with the intuitions that the student might have.

The entire process of Mathematical modelling seems to be slow and time consuming. It comes to be fruitful as long as the different stages are interrelated.

Structuring the real model proved to be an overriding stage within the whole Mathematical Modelling process. It is a multidisciplinary stage, in which the desired students' achievements have to be very clear, for specialised knowledge is required. The knowledge about the subject and the interest that the problem may generate in *students (and teachers)* may come to be an encouraging challenge or an obstacle for the attitude and favourable disposition for developing a proposed modelling activity.

The Mathematical Model is not always used to strengthen the analysis of the real model. For example, its predictive aspect is not always recognised.

The students' strategies showed basically unworkable methods and /or neither explanatory nor predictive methods as was evidenced in group G1.

The students' responses about the dilution process showed that the structuring stage requires a holistic view of all parts and of the relationships between these parts. In addition, the responses to the qualitative questionnaire revealed that the awareness of the mathematical model (the ODE describing the real situation), even if constructed by the student, does not mean that they have grasped the real model. Moreover, the responses of the students of G1 and G2 that had chosen the suggested strategy might have possibly developed a mathematical model had they received suitable assistance.

The guidance of students, even when attending particular issues should focus on the general processes of relational understanding, associations of concepts and relationships, and interpretation. Research is needed about appropriate assistance at each stage.

REFERENCES

Blum, W. et al. (2002) ICMI Study 14: Applications and modelling in mathematics education – Discussion document. *Zentralblatt für Didaktik der Mathematik,* 34, 5, 229-239.

Dubinsky, E. (1991) Reflective Abstraction in Advanced Mathematical Thinking. In D.Tall (ed) *Advanced Mathematical Thinking.* Dordrecht: Kluwer Academic Publishers, 95-123.

Ervynck, G. (1991) Mathematical Creativity. In D.Tall (ed) *Advanced Mathematical Thinking.* Dordrecht: Kluwer Academic Publishers, 42-53.

Fischbein, E. (1987) *Intuition in Science and mathematics: an educational approach.* Dordrecht: D.Reidel Publishing Company.

Hiebert, J. and Lefevre, P. (1986) Conceptual and Procedural Knowledge in Mathematics: an Introductory Analysis. In *Conceptual and Procedural Knowledge: The Case Of Mathematics.* Hillsdale NJ: Lawrence Erlbaum Associates, 1-23.

Skemp, R., (1977) Relational understanding and instrumental understanding. *Mathematics Teaching,* 77, 20-26.

Stavy, R. and Tirosh, D. (2000) *How students (Mis)Understand Science and Mathematics: Intuitive rules.* New York: Teachers College Press.

7.4

INTEGRATION OF APPLICATIONS IN THE TECHNION CALCULUS COURSE

Shuki Aroshas, Igor Verner and Abraham Berman
Technion – Israel Institute of Technology

Abstract–*This paper considers an education design experiment in which applications were integrated in the Multivariable Calculus course. Its purpose was to emphasise the connections between the mathematics course and the science and engineering disciplines, and examine the effect of applications on students' understanding of calculus concepts, course achievements, and learning motivation. In the experiment the conventional curriculum was extended by optional application. The experiment tested several supplemental instruction frameworks and focused on the applied problem solving skills, developing teaching methods and materials, and testing the course outcomes. Data analysis indicated high positive evaluation of supplementary applications classes, their contribution to understanding calculus concepts, and significant positive effect on students' achievements and motivation.*

1. INTRODUCTION

Science and Engineering students in the course of their academic studies and further carriers continue applying calculus as a professional tool. In higher education, the ability to apply mathematics has been recognized as one of the main learning outcomes required from graduates of the engineering programs (Engineering Criteria, 2000). The mission of mathematics education is not only to impart the knowledge of mathematical rules, theorems and procedures, but to develop the ability to put mathematical knowledge and skills to functional use in a multitude of contexts (PISA, 2000). Applications and modelling are a central theme in mathematics education and mathematics education research. "Very many questions and problems, concerning human learning and the teaching of mathematics affect, and are affected by relations between mathematics and the real world" (Henn & Blum, 2004 p3).

There is an intensive debate on applications and modelling in mathematics curricula for engineering students. Many educators believe that applied mathematics skills can be developed in undergraduate mathematics courses, particularly in calculus. Kumar and Jalkio (2001) proposed a conceptual framework for teaching mathematics from the application point of view. It concerns the mathematical skills required by the engineering disciplines, mathematics courses for developing these skills, and relevant applied problems. A number of mathematics with applications textbooks have recently been published (Stewart, 2001; Harvard Consortium, 2005), which implemented the following principles: (1) Topics are presented geometrically, numerically, analytically, and verbally. (2) Formal definitions and methods evolve from the investigation of practical problems. (3) The real-world problems are open-ended and may have more than one solution.

This paper reports a study of applications-integrated Multivariable Calculus course at the Technion. In the study we developed and tested different methods of integrating applications in the calculus course without affecting its mathematical level and scope. The study examined the effect of learning applications on students' achievements, understanding calculus concepts and attitudes towards the course. We developed applied problems which related to Technion students' majoring subjects and their real-world situations. By experiments, we found that applied problem solving involved the students in the mathematical modelling cycle. In this cycle a new mathematical concept is introduced from the need to describe a real life situation and represent it mathematically, and then the theoretical solution is interpreted in reality. The modelling cycles helped the students to better understand the mathematical concepts.

2. EDUCATIONAL CONCEPTS

Cognitive psychologists noted that instruction should refer to individual characteristics of learners. The educational approach which coordinates student's abilities and teaching methods is Aptitude Treatment Interaction (ATI) (Cronbach & Webb, 1975; Snow, 1991; Cramer, 1989). ATI offers a variety of instructional methods and gives students opportunities to choose those which fit their learning styles. It emphasizes team-based inquiries and project assignments in which the students can select their preferred learning strategies. The ATI indicated that integrating different instructional methods provided more students with opportunities of successful and motivated learning. It gave rise to substantial examination of different learning styles and approaches to address them in curriculum and instruction. The two central aspects of the multiplicity of intelligence are cognitive performances and learning styles. Sternberg (1985) shifted the focus from formal learning processes to those emerged when studying and solving real practical problems. His theory supports the integration of applications in science education. Gardner (1983) pointed out a number of different intelligences which characterize different learning styles. He aspired to build learning processes which implement multiple intelligences in order to fit learning methods to different students and provide harmonic development of their abilities. The Gardner theory directs us to integrate various teaching methods and emphasizes the value of experiments and applications. Wilson (1993) pointed that learning and knowing are integrally and inherently situated in the every day world of human activity and that the context in which something is learned is a very important part of learning. The theory of situated cognition supports meaningful learning and enriches the learning process by providing practical experiences of real life situations (Choi & Hannafin, 1995). Educational studies indicated that situated cognition is a way to "address difficulties students have in retention and generalization" (Gersten & Baker, 1998).

3. EDUCATION DESIGN EXPERIMENT METHODOLOGY

The term Education Design Experiment (EDE) denotes a certain type of educational research methodology, which utilizes the similarity between educational and engineering process (Cobb et al., 2003). The EDE objective is to engineer (create, test, and revise) a new learning environment and examine (implement, evaluate, and conceptualize) educational processes emerging in the environment. In this section we consider the features of design experiments and their relevance to our study. The education design experiment removes certain biases of the traditional educational study:

- The primary motivation for the experiment is to improve the educational process and examine efficiency of the proposed approach.
- The environment, instruction, and learning are studied as a whole.
- The investigated factors and research methods are goal-oriented and can be changed during the experiment.
- The EDE explicitly delimits the range in which its results are significant.

The EDE is an adequate framework for research in which the learning environment is developed along with the educational process and its characteristics largely determine the efficiency of learning. The design experiment can be conducted at individual, class, and educational program levels. Following (Cobb et al., 2003), the EDE consists of three stages: preparation, experiment, and retrospective analysis. Preparations for a design experiment include the following activities:

1. Identifying learning behaviours, which can indicate the contribution of the proposed learning method and research instruments to be used for measuring the outcomes.
2. Determining the concepts and subjects, in which the proposed approach can lead to efficient learning. This can possibly affect developing new frameworks that match the approach.
3. Ascertaining students' prior knowledge and attitudes and specifying their prospective progress in the learning subject.
4. Formulating the experiment's outline.

At the experiment stage, the primary objective is "to improve the initial design by testing and revising conjectures informed by ongoing analysis of both the student's reasoning and the learning environment" (Cobb et al., 2003). At the retrospective analysis stage, the central challenge is to generate a coherent conceptual framework that accounts the effects observed in the course of the experiment and gives directions for dissemination and future development of the proposed educational approach. The reasons for grounding our study on the EDE methodology are the following: (1) Our research motivation was to test the effectiveness of teaching mathematics with applications as a means of improving learning in the Calculus course; (2) Dependency on the existing course constraints in relation to time, curriculum, assessment, involvement, and student access limited our opportunities of using conventional research frameworks; (3) We developed, tested and revised the learning environment and methods of teaching applications in conjunction with teaching the course; (4) The study strived to draw conclusions based on real experience.

These limitations and factors influenced the action research experiment framework of our study.

4. STUDY FRAMEWORK

Conventionally, the Multivariable Calculus course weekly schedule included four lecture hours and two hours of classes. With development of the Technion Mathematics Web tutoring system the classes were reduced to one hour a week. In our study the course schedule was extended by supplementary applications classes. The applications classes were given voluntarily in conjunction with teaching a conventional Multivariable Calculus class.

The study included three teaching experiments which examined different forms of supplementary applications classes. In the first (pilot) experiment Supplementary Applications Classes (SACs) were given by one of the authors (Aroshas) in the fall semester 2002-2003. The optional classes were designed for 30 students, but in practice about 75 students participated regularly. The students came from different science and engineering faculties. The SACs were coordinated with the main calculus class so that theoretical concepts and their use in real problems were studied in the same week. Condensed subject matter descriptions were given when needed. The classes were followed up by a pilot study. In conjunction, we developed and tested applied problems and teaching methods in class. To characterize the applied problem solving skills we asked a number of experts from science and engineering faculties at the Technion for their opinion regarding the need for learning applications in the mathematics courses, and how important it was for their students. We also accepted their advice on how to present and analyze applied problems in the course.

The second (central) teaching experiment was conducted in the spring semester of 2003-2004 and designed in the following way. Two groups (experimental and control) of Multivariable Calculus students were created with the same number of students (N=33). Both groups attended the same Calculus lectures with no emphasis on applications, but their classes were different. Throughout the course the control group participated in two weekly hours of conventional classes (without applications), while the experimental group participated in one weekly hour of conventional class and one of SACs. There were no significant differences between the experimental and control groups with regard to mean grades in Calculus 1 and in the pre-course test, interest in studying calculus with applications. Both groups represent the same engineering faculties. In this experiment the study focused on testing the effect of teaching calculus with applications through its comparison with the conventional teaching approach. The comparison related to learning achievements, understanding calculus concepts, and students motivation.

In the third (additional) teaching experiment two supplementary 2-hours SACs were given in the fall semester 2004-2005 and attended by more than 50 students from the multivariable calculus class. The goal of the classes was to introduce the calculus concepts of Lagrange multipliers and multiple integrals prior to their formal study in the lecture. In the sessions the concepts were recreated from the practical need and through the analysis of applied problems. In the follow-up we examined to what extend the SACs helped students to understand the concepts taught in the lectures.

5. RESEARCH QUESTIONS, INSTRUMENTS, AND APPLIED PROBLEMS

The objective of our study was to develop methods and materials for teaching multivariable calculus with applications and examine the effect of applications on students' understanding calculus concepts, course achievements, and learning motivation. The research questions were as follows:

1. What is the effect of the applications-based mathematical instruction on learning achievements, understanding calculus concepts, and motivation in the course?
2. What are the possible ways of integrating applications in the Multivariable Calculus course while keeping its existing constraints and limitations?

5.1 Instruments

In the first teaching experiment we used pre-course and post-course questionnaires. Four questions of the pre-course questionnaire were repeated in the post-course one. They tested student opinions on the following aspects: (1) Anticipated effect of integrating engineering and science problems on understanding the calculus concepts; (2) Interest in solving calculus problems from the area of specialization; (3) Viewing the calculus capabilities as a condition to succeed in the area of specialization; (4) Interest in attending the applications motivated course in addition to the conventional calculus class. The post-course questionnaire also inquired student opinions about the contribution of the three teaching methods used in the course: demonstrating mathematical problems of science and technology, constructing and solving mathematical problems in context, visualization through computer simulations.

The central teaching experiment applied pre-course and post-course questionnaires and tests. The pre-course questionnaire examined students' attitudes and learning styles, and collected personal information. This information and a One Variable Calculus applications test were used for creating experimental and control groups. The two post-course questionnaires were also conducted in both groups. The first one tested students' opinions about the value of tutoring sessions and their preferences in learning with applications. The second one was an understanding test which examined students' possible misconceptions of Multivariable Calculus concepts. The midterm exam and the final course exam grades of students in the control and experimental groups were also collected.

In the third teaching experiment we used an attitude questionnaire which was conducted after each of the supplementary applications classes and the following lecture. In the questionnaire the students evaluated the contribution of the classes to their understanding of the calculus concepts given in the lecture.

5.2 Creating a Pool of Applied Problems

Through the literature search we found sources of applied calculus problems from various science and engineering domains which fit the scope and level of the course such as (Stewart, 2001). Other interdisciplinary texts and our professional experience served as sources of ideas for developing applied problems following the criteria proposed in (Alsina, 1998). The problems were discussed with highly experienced lecturers of Calculus, Science, and Engineering. Following are examples of problems given in the course:

Problem 1 (Application of the double integral)
In a theatre, the average waiting times in a ticket line is 10 minutes, and the average waiting time in a popcorn line is 5 minutes. Assuming that the waiting time in the two lines are independent of each other, and their density functions are exponential

$$f(t) = 0.1 \times e^{-t/10}, \ g(t) = 0.2 \times e^{-t/5} \text{ for } t \geq 0 \text{ and } f(t) = g(t) = 0 \text{ for } t < 0, \text{ then}$$

calculate the probability that a person coming to the theatre will buy a ticket and popcorn in less than 20 minutes.

Problem 2 (Application of the line integral)
The change of state of 1 mol mass of an ideal gas in a thermodynamic process from the initial state $A(V_1, P_1)$ to the final state $B(V_2, P_2)$ is described by the function

$$T = \frac{P \cdot V}{R}$$. Here P is pressure, V is volume, and T is Kelvin temperature of the

gas, R is the universal gas constant. The process is presented on the coordinate plain (V, P) by a path AB. The increase in internal energy of the gas in the process

is given by the line integral $U = \int \frac{C_P}{R} \cdot P \cdot dV + \frac{C_V}{R} \cdot V \cdot dP$, where heat capacities

C_P, C_V are constants of the gas.

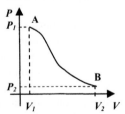

A) Is the field $\left(\frac{C_P}{R} \cdot P, \frac{C_V}{R} \cdot V \right)$ continuous on the path AB?

B) What condition provides that the field is conservative?
C) Supposing that $C_P \neq C_V$, is the change of the gas state along the pass AB from point A to point B and back to A accompanied by heat transfer between the gas and the environment?

5.3 Analysis of Results

First teaching experiment
The pre-course and post-course questionnaires included four similar questions related to the following aspects: (1) Anticipated effect of integrating engineering and science problems on understanding the calculus concepts; (2) Interest in solving calculus problems from the area of specialization; (3) Viewing the calculus capabilities as a condition to succeed in the area of specialization; (4) Interest in attending the applications motivated course in addition to the conventional calculus class. An additional question of the post-course questionnaire related to the

contribution of the three teaching methods used in the course to understanding calculus concepts. The methods were: demonstrating mathematical problems of science and technology, constructing and solving mathematical problems in context, visualization through computer simulations. The pre-test findings were as follows:

- The absolute majority of students pointed their high level of expectations from integrating applied problems on understanding the calculus and interest to solve problems from the area of specialization.
- A majority of the students (70.7%) recognized the connection between success rates in the first-year mathematics and the majoring subjects.

The post-test indicated: (1) The students did not change their opinion about the effect of integrating applied problems, and continued to be interested in solving problems related to the area of their specialization (as given by the t-test). This shows that the course met the students' expectations. (2) There was an insignificant increase in the average evaluation of the importance of the calculus capabilities for success in the area of specialization (as given by the t-test). (3) All the students reported that they would recommend attending the applications course to their classmates. (4) All three teaching methods significantly contributed to the understanding of calculus concepts (as revealed by F-test). The contribution of visualization through computer simulations was the highest.

Second teaching experiment
The pre-course questionnaire was used to divide calculus students into two "equal" groups: the experimental group and the control group. As a result, there were no significant differences between the control and experimental groups in One Variable Calculus grades, interest in learning mathematics with applications, results of the one-variable calculus applications test, in representation of different faculties and in learning styles. Results of the comparison between the experimental and control groups at the end of the course are summarized in Table 1.

Table 1. Mean grades of course exams and SACs contributions.

Item	Control group	Experim. group
Midterm exam grade in Calculus 2	57.2	66.1
Final exam grade in Calculus 2	62.1	68.4
Contribution to understanding calculus concepts	3.96	4.66
Contribution to stimulating interest in calculus	3.52	4.03
Contribution to studying other disciplines	3.15	4.12
Contribution to presenting practical aspects	4.12	4.51
Contribution to broadening horizons	3.78	4.15
Contribution to exercising course material	3.66	4.60
Contribution to interest in application-driven learning	3.51	4.42

The groups were compared with regard to their mean grades in the midterm and final course exams, and their answers to the post-course questionnaire. The exams did not include applied problems. The mean grades are presented in the first two rows of the table. The next rows show mean grades given by the students in the post-course questionnaire for contributions of the course. These grades were given in the five-point Likert scale with 1 indicating no contribution and 5 indicating high contribution. The features indicated by the table are as follows: (1) The mean exam grades of the experimental group were 8.9 points higher than that of the control group in the midterm exam and 6.3 points higher in the final exam. The differences between the groups in both exams were found significant. (2) The experimental group gave a significantly higher evaluation for the contribution of the course in relation to all the aspects mentioned in the table. (3) The two groups are of the same opinion about the need of addressing different learning styles and integrating applied problems in the course.

It is worth mentioning, that in the teacher evaluation survey of the Technion Centre for Promotion of Teaching, the mean grades given by the control and experimental groups to the calculus teacher were similar (4.85 and 4.86 out of 5). This fact supports the view that the advantage of the experimental group was a result of studying calculus with applications. The calculus understanding test consisted of 14 theoretical and applied questions related to the following concepts: equipotential lines, directional derivative, gradient, tangential plain, Lagrange multipliers, and extremum. The test was validated by two experienced practicing lecturers of the Technion Multivariable Calculus course. Significant advantage of the experimental group over the control group in the percentage of correct answers to the test questions was indicated. For some of the questions correct answers were given by 70-80% of students in the experimental group vs. 25-35% in the control group. Typical student reflections to a question on the contribution of studying applications for understanding calculus concepts were as follows: *"Through applications I grasped the complex calculus concepts"*; *"The impact of a one-hour application session is the same as of a regular (two-hour) tutorial"*; *"It is a pity that applied problems were not given in the first Calculus course".*

Third teaching experiment
The attitude questionnaire conducted after each of the two supplementary applications classes asked to evaluate its contribution in the following aspects: understanding a lecture, following the lecturer's explanations, geometric interpretation of the concepts, linking new and formerly studied concepts, and development of problem solving skills. The answers were very positive. More than 90% of the students mentioned high positive contribution of the SACs in all the abovementioned aspects. In response to a question on preferred teaching methods, an absolute majority of the students supported teaching calculus with applications through practical examples, discussions of possible applications, and visualization of calculus concepts.

6. CONCLUSIONS

Our study integrated applications in the Multivariable Calculus course at the Technion and tested their effect on students' understanding of calculus concepts, course achievements, and learning motivation. The education design experiment methodology was used to develop, implement, and evaluate   supplementary applications classes which extended the conventional curriculum, meeting the constraints and limitations of the existing course. The study included three teaching experiments throughout the course which examined different forms of supplementary applications classes. The first (pilot) experiment combined teaching applications classes with developing and testing applied problems and teaching methods. The second experiment tested the effect of teaching calculus with applications vs. the conventional teaching approach, through the comparison of experimental and control groups. The third experiment tested the opportunity of introducing calculus concepts from the practical need and through the analysis of applied problems prior to their formal study in the lectures. Results of all three experiments were very positive. Learning achievements of students who attended supplementary applications classes were significantly higher than of other students. Through applications the students explore mathematical modelling cycles, and this practice contributed to better and easier understanding of calculus concepts, higher learning motivation and interest in the subject. The absolute majority of the students involved in the study supported integrating applications in the Multivariable Calculus course, recommended to continue teaching the applications course in the future, and even extend it.

REFERENCES

Alsina, C. (1998) Neither a Microscope Nor a Telescope, Just a Mathscope. In P. Galbraith, W. Blum, G. Booker and I. Huntley (eds) *Mathematical Modelling. Teaching and Assessment in a Technology-Rich World*, Chichester: Horwood, 3-10.

Cobb, P., Confrey, J., diSessa, A., Lehrer, R. and Shauble, L. (2003) Design Experiments in Educational Research. *Educational Researcher*, 32(1), 9-13.

Cramer, K. (1989) Cognitive Restructuring Ability, Teacher Guidance, and Perceptual Distracter Tasks: An Aptitude-Treatment Interaction Study. *Journal for Research in Mathematics Education*, 20(1), 103-110.

Cronbach, L. and Webb, N. (1975) Between-class and within-class effects in a reported aptitude treatment interaction: reanalysis of a study by G. L. Anderson. *Journal of Educational Psychology*, 67(6), 717-724.

Engineering Criteria 2000, Third Edition. In *Criteria for Accrediting Programs in Engineering in the United States*. Baltimore, MD: The Accreditation Board of Engineering and Technology (ABET), 32-34.

Gardner, H. (1983) *Frames of Mind: The Theory of Multiple Intelligences*. New York: Basic Books.

Gravemeijer, K. and Doorman, M. (1999) Context Problems in Realistic Mathematics Education: A Calculus Course as an Example. *Educational Studies in Mathematics*, 39, 111-129.

Greeno, J. G. (1998) The Situativity of Knowing, Learning, and Research. *American Psychologist*, 53(1), 5-26.

Harvard Consortium (2005)
http://hea.wiley.com/WileyCDA/Section.rdr?id=100324.

Kumar, S. and Jalkio, J. (2001) Teaching Mathematics from an Applications Perspective, *Journal of Engineering Education,* 24, 275-279.

Lave, J. (1988) *Cognition in Practice: Mind, Mathematics, and Culture in Everyday Life.* Cambridge: CUP.

Snow, R. (1991) Aptitude-Treatment Interaction as a Framework for Research on Individual Differences in Psychotherapy. *Journal of Consulting and Clinical Psychology*, 59(2), 205-216.

Sternberg, R. (1985) Human intelligence: the model is the message, *Science*, 230, 111-118.

Stewart, J. (2001) *Calculus: Concepts and* In D.Tall (ed) *Advanced Mathematical Thinking.* Dordrecht: Kluwer Academic Publishers, *Contexts.* Pacific Grove CA: Brooks/Cole.

7.5

MATHEMATICAL MODELLING MODULES FOR CALCULUS TEACHING

Qiyuan Jiang[1], Jinxing Xie[1] and Qixiao Ye[2]
[1]Tsinghua University, Beijing, PR China
[2]Beijing Institute of Technology, PR China

Abstract–*This paper introduces a two-year project on incorporating ideas and methods of mathematical modelling into the teaching of main mathematical courses in Chinese universities and colleges, initiated by the National Organizing Committee (NOC) of the China Undergraduate Mathematical Contest in Modelling (CUMCM) in 2003. The importance and necessity of the project are briefly discussed. The project's emphases are put on designing mathematical modelling modules, which include the whole mathematical modelling process for solving real-world problems and should be easily understood and can be effectively used for the existing courses. The use of the modules will not disturb instructors' regular teaching, but will stimulate and raise students' interest in studying mathematics. In particular, recommendations on how and where this can be done in the existing calculus teaching are discussed. A sample module "Why a Coca Cola can takes such a shape" is presented in detail.*

1. INTRODUCTION

More and more professors and administrators from universities and the Ministry of Education in China realize that the teaching of mathematics at university level is very important. Especially, they realize that mathematical modelling (MM) techniques are very important from the very success of the China Undergraduate Mathematical Contest in Modelling (CUMCM). Mathematical modelling and associated computations and simulation are becoming critical tools in the engineering design process. Scientists and engineers rely increasingly on computational methods and must have sufficient experience in mathematical / computational methods to be able to choose the correct methods and interpret the accuracy and reliability of the results (Friedman et al., 1992). Modelling now permeates the daily lives of professionals. Mathematical modelling has become indispensable as a research tool, particularly in connection with the appearance of computers. It allows one to design new techniques to find optimal regimes for the solutions of complex scientific problems and predict new phenomena. Therefore, the mathematical education of our future engineers and scientists needs to change to reflect this new reality, in particular through its early introduction at university level, where it has become desirable, or even mandatory (Friedman et al., 1992; Hazewinkel, 1995).

In China, more than 40,000 students participate in CUMCM and similar competitions in many universities each year. They have learned a lot of MM and shown competence in their subsequent courses, projects, and later careers. We have about five million students entering into universities and colleges each year; most of them have to study calculus for at least one or two semesters. Many do not understand why they have to spend so much time in studying calculus or other mathematical courses, and why it is important for their future careers, therefore their study lacks motivation and initiative. In order to solve these kinds of problems, the NOC of CUMCM initiated a two-year project in 2003 titled "incorporating ideas and methods of mathematical modelling into the main mathematical courses in universities and colleges" (Ye, 2003). There are three main mathematical courses in universities and colleges: calculus, linear algebra, and elements of probability and statistics. The project's emphases are put on designing and writing feasible modules on MM, which include the explanation of the whole mathematical modelling process from real-world problems, and they can be embedded or effectively used for the teaching of existing courses. Most importantly, the use of the modules will not disturb instructors' regular teaching but will stimulate and raise students' interest in studying mathematics. In particular, recommendations on how and where this can be done in the existing calculus teaching will be discussed in this paper. As an example, a sample module titled "Why a Coca Cola can takes such a shape" is presented in detail.

2. AIMS OF THE PROJECT

2.1 Organizing and encouraging instructors from different institutions to write modules on MM that can be used in the teaching of calculus, linear algebra, and elements of probability and statistics.

2.2 Using these modules in the teaching of these three courses in a few institutions for testing and identifying problems. In the teaching we will emphasize not only the "know how", but also the "know why". Our expectation is that students who have learned theses modules should know why MM and mathematics are important and even critical for their careers and for strengthening their creativity and competitive ability. Especially, we hope more students will get interested in mathematics and work harder on it.

2.3 Communicating and discussing good modules at seminars for incubating and training instructors from various universities and encouraging them to use these modules in their own teaching. The collection of the best modules will be published.

2.4. We hope that the implementation of the project will go well, and more students will be interested in MM and take further courses on MM and related mathematical courses, and as a result, we will have more talented student to participate in the CUMCM.

3. PRINCIPLES OF WRITING MM MODULES

3.1 The problem should be a real-world problem and is easy to understand.

3.2 The module will embody the whole mathematical modelling process.

3.3 The module should be attractive to instructors who are teaching calculus, linear algebra, and elements of probability and statistics, so that they are willing to read and use these modules in their teaching activities.

3.4 The module should meet needs of various students.

3.5 Most important is that the use of the modules will not disturb instructors' regular teaching. In general, only less than 2 extra teaching hours is needed for teaching one module.

4. A SAMPLE MODULE - WHY A COCA COLA CAN TAKES SUCH A SHAPE?

We present a sample module for calculus teaching in detail in this section. In almost all the calculus courses, there is a section on optimization problems as applications of differentiation. A typical such problem in many calculus texts is as follows.

A cylindrical can is to be made to hold some liters of oil. Find the dimensions that will minimize the cost of the metal to manufacture the can (see, for example, Stewart, 2003). It is essentially a geometric optimal problem, that is, given a right circular cylinder and its volume, find its diameter and height that minimize its surface area. Our comment is that before using our module teachers first solve this simple problem together with their students. This is easy and the solution can be found in most calculus textbooks.

<u>Solution</u>. Denote r the radius, d the diameter $(d = 2r)$, h the height, V the volume and S the surface area of the right circular cylinder, and V is given. Then

$$V = \pi r^2 h, \ h = V / \pi r^2, \ S = 2\pi r h + 2\pi r^2. \tag{1}$$

Therefore, the optimization problem is

$$\begin{cases} \min_{r>0, h>0} S(r,h) \\ s.t. \ V = \pi r^2 h \end{cases} \tag{2}$$

<u>Method 1</u>: Reduce the constrained minimization problem to an unconstrained minimization problem. Substituting h into the expression of S, we have

$$S = 2\pi r V / \pi r^2 + 2\pi r^2 = 2(\pi r^2 + V / r). \tag{3}$$

Finding critical points as follows:

$$S' = 2(2\pi r - V / r^2) = \frac{2}{r^2}(2\pi r^3 - V) = 0, \ r = \sqrt[3]{\frac{V}{2\pi}}. \tag{4}$$

Since $S'' = 4(\pi + V / r^3) > 0$ for $r > 0$ and there is only one critical point, the surface area is minimized at $r = \sqrt[3]{V / 2\pi}$. Thus the optimal height is

$$h = \frac{V}{\pi}\sqrt[3]{\frac{(2\pi)^2}{V^2}} = \sqrt[3]{\frac{(2\pi)^2 V^3}{V^2 \pi^3}} = \sqrt[3]{\frac{2^3 V}{2\pi}} = 2r = d. \tag{5}$$

It means that the diameter equals the height of the right circular cylinder.

<u>Method 2</u>: Using the arithmetic mean and geometric mean inequality (Isoperimetric Inequality). We will use this method later, where the same conclusion $d=h$ can be obtained.

In our module, we have a sub-section on what is mathematical model and what is mathematical modelling. Its main contents are as follows.

A mathematical model is a (rough) description of a class of real-world problems or phenomena expressed using mathematical symbolism. The process of building, solving, and validation of it is called mathematical modelling. The methods of mathematical modelling are not new, and almost all the persons from ancient to modern who use mathematics to solve the real-world problems are using the ideas and methods of mathematical modelling (including some examples such as Euclid (about BC 330 ~ BC 275), Archimedes (about BC 287 ~ BC 212), Galileo, G. (1564 ~ 1642), Navier, C.M.L.H. (1785 ~ 1836), and Newton, I. (1642-1727), etc.).

Key steps of mathematical modelling (Wan, 1990):

Step 1: Observation and analysis of the key aspects of the real-world problem.

Step 2: Simplification and making reasonable assumptions (it is quite difficult).

Step 3: Determine variables and parameters.

Step 4: Using certain geometrical relations or physical laws to relate these variables and forming a concrete mathematical problem (we may call it a mathematical model at this stage) (note that the math problem might be very complicated).

Step 5: Solving this mathematical model analytically or approximately if possible (note that it might be very difficult to solve it).

Step 6: Validation: for example, by using historical or experimental data to check that if the result is right (reasonable) or not to some extent (it is also not an easy job).

Step 7: If the results are positive we can use it, otherwise we have to go back to all the previous steps to check if there is something wrong, and if so, correct them, and rerun the whole process again.

The emphasis is put on reasonable assumptions, solving math problems and validation.

Organizing the teaching. Before teaching the problem "why a Coca Cola can takes such a shape" (may be before the ending of your last lecture), ask students to review the related text and do the typical example mentioned above, and take a Coca Cola can to measure its dimensions and other data (for instance, approximately the radius of its top cover is 3cm, the radius of its middle part is 3.3cm, and its height is 13cm). The content of the drink is 355 millilitres (ml) (about 355cm^3), and there are about 10cm^3 in the can remains empty, thus the volume inside the can is 365cm^3. Then you can teach the MM process step by step together with your students in the classroom when you start your next lecture. You can show students an empty Coca Cola can and let them feel the top cover is much harder and thicker than the other parts of the can. Its central section is quite close to the shape shown in Figure 1(a).

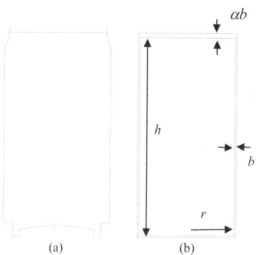

(a)   (b)
Figure 1. The shape of the Coca Cola Can.

A simplified model. In this case we have to consider the volume of the cover material. We note "the 'simple-to-elaborate' approach is so important in mathematical modelling" (see Wan, 1990).

Step 2. Simplification. Assume that the can is a right circular cylinder as shown in Figure 1(b).

Step 3. Determining variables and parameters. The radius r and height h of the can are variables, and the volume of the can V is a given known parameter. The thickness of the top cover is αb and the thickness of the other part is b (cm). Here b and α are parameters.

Step 4. Building the mathematical model. The volume of the cover material of the can (SV) and the volume inside the can (V) can be calculated as

$$SV(r,h) = h[\pi(r+b)^2 - \pi r^2] + b\pi(r+b)^2 + \alpha b\pi(r+b)^2$$
$$= 2\pi rhb + (1+\alpha)\pi r^2 b + 2(1+\alpha)\pi rb^2 + \pi hb^2 + (1+\alpha)\pi b^3, \qquad (6)$$

$$V(r,h) = \pi r^2 h. \qquad (7)$$

Because $b \ll r$, the items $2(1+\alpha)\pi rb^2 + \pi hb^2 + (1+\alpha)\pi b^3$ in (7) can be omitted. (*Is this a reasonable assumption?* We will discuss this later.) Therefore, we have

$$SV(r,h) = \pi rb[2h + (1+\alpha)r], G(r,h) = \pi r^2 h - V. \qquad (8)$$

We need to find the r and h so that the material volume $SV(r, h)$ is minimized under the constraint $G(r, h) = 0$, that is, the mathematical model is

$$\min_{r>0, h>0} SV(r,h) = \pi rb[2h + (1+\alpha)r]$$

s.t. $G(r,h) = \pi r^2 h - V = 0.$

$$(9)$$

Step 5. Solving this mathematical model. We suggest two methods here. The first one is to solve h from $G(r, h) = 0$, that is, $h = V / \pi r^2$, and substitute it into $SV(r, h)$, that is, $SV(r) = b[2V / r + (1+\alpha)\pi r^2]$. Using calculus we can find the unique critical point of $SV(r)$ as $r_0 = \sqrt[3]{V /(1+\alpha)\pi}$, and it is easy to verify that $S''(r_0) > 0$. Therefore the material volume reaches its minimum at r_0. Furthermore, if we denote the corresponding height as h_0, then $r_0 : h_0 = r_0 : (V / \pi r_0^2) = \pi r_0^3 / V = 1 : (1+\alpha)$.

Another method is to use the arithmetic mean and geometric mean inequality (elementary method, it also can be used for the high school students), that is,

$$\frac{1}{n}\sum_{i=1}^{n} a_i \geq \sqrt[n]{\prod_{i=1}^{n} a_i}, \ a_i > 0, \ i = 1,...,n \ ,$$

Equality holds if and only if $a_1 = a_2 = ... = a_n$.

Specifically, taking $n=3$ and $a_1 = a_2 = V / r, a_3 = (1+\alpha)\pi r^2$, we have

$$SV(r) = b[2V / r + (1+\alpha)\pi r^2] \geq 3b \times \sqrt[3]{(V / r)^2 (1+\alpha)\pi r^2} = 3b \times \sqrt[3]{(1+\alpha)\pi V^2}.$$ The Equality holds if and only if $V / r = (1+\alpha)\pi r^2$, that is,, $r = \sqrt[3]{V /(1+\alpha)\pi}$, so that we get the same result as in the first method.

Step 6. If $\alpha = 3$ (If one measures the thickness of the top cover and the other part of the cover, it is indeed so), then the ratio $r:h=1:4$. The measurement roughly tells us that it is right!

The students will check if the model is right by using their measurement, and discuss if they need to modify the model through their imagination, while studying why omitting $2(1+\alpha)\pi r b^2 + \pi h b^2 + (1+\alpha)\pi b^3$ is a reasonable assumption (this is not an easy problem, because if we do not omit these items, we have to find the zeros of a polynomial of degree three for finding critical points, namely, $\frac{dSV(r)}{dr} = \frac{b}{r^3}\left[-2Vr + 2(1+\alpha)\pi r^4 + 2\pi(1+\alpha)br^3 - 2bV\right] = 0$. Students can use math software for drawing pictures and obtaining numerical results to show it is indeed a reasonable assumption, especially students will get a strong impression about why they have to learn more mathematics in order to solve the problem and why math software is very helpful), and so on.

In fact, if we do not omit the higher order terms of b, we can also solve this problem analytically. Making use of $h = V / \pi r^2$, Equation (6) can be rewritten as

$$SV(r) = \pi b[(1+\alpha)(r+b)^2 + (2r+b)\frac{V}{\pi r^2}], \tag{10}$$

and therefore

$$SV'(r) = 2\pi b(r+b)((1+\alpha) - \frac{V}{\pi r^3}), \ r_0 = \sqrt[3]{\frac{V}{(1+\alpha)\pi}}, \tag{11}$$

$$SV''(r) = 2b(2\pi + \frac{(3b+2r)V}{r^4}) > 0, \ r > 0. \tag{12}$$

Finally, $r_0 = \sqrt[3]{V /(1+\alpha)\pi}$ is the unique critical point and the result is still the same.

Discussion

The students will discuss if this model is right, or if they need to modify the model, the complexity of mathematical modelling etc. It is important that we leave room for students' imagination.

5. OTHER EXERCISES

We design several exercises for students to practice the modelling process, especially, to get insight of the importance of mathematics and the necessity and convenience of using mathematical software, such as Mathematica, MATLAB, etc. In addition, we motivate students to think why we can believe the results from the use of the mathematical software to be correct. And we follow the principle: teach students in accordance with their aptitude. Two examples of the exercises are given in below.

1. If the thickness of the top cover of a beer can is four times thicker than the thickness of the other part of the can. The inside volume of the can is $V = 500 \ cm^3$. What is the ratio of r and h of the can, which makes the volume of the can material minimum? (Go to the supermarket to see if there is a beer can which has the size from your modelling result.)

2.   A Space Shuttle Water Container

Consider the space shuttle and an astronaut's water container that is stored within the shuttle's wall. The water container is formed as a sphere surmounted by a cone (like an ice cream cone), the base of which is equal to the radius of the sphere (see Figure 2). If the radius of the sphere (r) is restricted to exactly 6 ft and a surface area of 460 square feet is all that is allowed in the design, find the dimensions of the height (x_1) of the spherical cap cutting by the cone and the height (x_2) of the cone such that the volume of the container is maximized (See Giordano et al., 2003). This exercise is much harder but attracts the talented students.

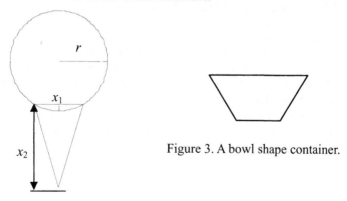

Figure 3. A bowl shape container.

Figure 2. The shape of the water container.

Assessment. (An example of the teamwork) Ask each team to write an essay on an optimization problem about a bowl shape (the frustum of a right cone) container (see Figure 3). Either one of the radiuses of the top base, lower base and the height of the frustum of the corresponding right cone is known.

Reading materials. We also designed several reading materials on what is mathematical modelling and related topics such as isoperimetric problems in order to raise their interest in studying mathematics and to strengthen their mathematical knowledge. Reading materials also include some interesting stories, for example:

The Legend of Princess Dido

According to the epic *Aeneid*, Dido (pronounced "Dee Dough") was a Phoenician princess from the city of Tyre (now part of Lebanon). Her treacherous brother, the king, murdered her husband, so she fled the city and sailed with some of her loyal subjects to Carthage, a city on the northern coast of Africa. She wished to purchase some land from the local ruler in order to begin a new life. However, he didn't like the idea of selling land to foreigners. In an attempt to be gracious and yet still spoil Princess Dido's request, the ruler said, "You may purchase as much land as you can enclose with the skin of an ox." Undaunted, Princess Dido and her subjects set about the task by slicing the ox skin into thin strips and then tying them together to form a long band of ox hide, and foiling the ruler's malicious plan (see Hildebrand & Anthony, 1985).

REFERENCES

Friedman, A., Glimm, J. and Lavery, J. (1992) The mathematical and computational sciences in emerging manufacturing technologies and management practices. *SIAM Report on Issues in the Mathematical Sciences, SIAM*, 62.

Giordano, F.R., Weir, M.D. and Fox, W.P. (2003) *A First Course in Mathematical Modelling, 3rd Edition*. Brooks/Cole, 476-478.

Hazewinkel, M. (ed) (1995) *Encyclopedia of mathematics - An updated and annotated translation of the Soviet mathematical encyclopedia*. Dordrecht: Kluwer Academic Publishers, 3, 784-785.

Hildebrand, S. and Anthony, T. (1985) *Mathematics and Optimal Form*. Scientific American Books.

Stewart, J. (2003) *Calculus, 5th Edition*. Brooks/Cole, 332-333.

Wan, F.Y.M. (1990) Mathematical models and their formulation. In C.E.Pearson, (ed) *Handbook of Applied Mathematics – Selected Results and Methods, 2nd Edition*. Van Nostrand Reinhold, 1048.

Ye, Q.X. (2003) Incorporating Ideas and Methods of Mathematical Modelling and Mathematical Experimentation into the Teaching of Calculus (in Chinese), *Engineering Mathematics*, 20, 8, 3-13.

7.6

AN EXPERIMENTAL APPROACH TO TEACHING MODELLING

Ken Houston and Mark McCartney
University of Ulster, Jordanstown, Northern Ireland, UK

Abstract–*A case study is presented in the use of a web-based resource for performing physics experiments, mostly mechanics, using graphics calculators and allied compact peripherals in the teaching of engineering mathematics.. The usefulness of such materials in developing skills in mathematical modelling, independent learning, and group work is discussed. It is concluded that such materials provide a welcome addition to the armoury of the mathematics lecturer who wishes to teach mathematics and mathematical modelling in an engaging, student centred, and, in terms of engineering mathematics, subject-specific way.*

1. INTRODUCTION

LEPLA - *Learning Environment for Physics Laboratory Activities* (LEPLA 2006) is a transnational project supported by the EU Socrates Minerva Programme within the eEurope initiative. The authors of this paper are members of the project team, which created, in six languages, web based learning materials. These enable students to carry out experiments using inexpensive and readily available equipment including graphics calculators, data loggers and PCs.

Being aware of the value to learning of a variety of approaches, we devised a teaching innovation using the LEPLA material. For the first quarter of their mathematics module, first year mechanical engineering students at the University of Ulster were required to carry out an experiment from the LEPLA resource and to report their work in writing and at a seminar. Thus, students at the start of their course would be using their linguistic, logical-mathematical, spatial, bodily-kinaesthetic and inter- and intra-personal intelligences (Gardner 1983, 1993; Houston 2001) and they would be taking an approach that involved working with algebra, numbers and pictures simultaneously (National Council of Teachers of Mathematics (NCTM), 1989).

Much of the LEPLA material is based on published, innovative teaching techniques developed by members of the project team and their colleagues (see for example Pecori & Torzo, 1998; Pecori, Torzo & Sconza, 1999; Debowska, Jakubowicz & Mazur, 1999). Students are guided through experiments, which involve the collection, analysis and interpretation of data. Each experiment is grounded in theory, and all of this is, of course, mathematical modelling. Data analysis is carried out using a graphics calculator or a spreadsheet. Students also have the opportunity to develop transferable, life-enhancing skills such as independent learning, group work and written and oral communication.

In this paper we describe in more detail the LEPLA Project (§2) and how we made use of the materials for this teaching innovation (§3). The activity was evaluated, and this is reported in §4.

2. THE LEPLA PROJECT

LEPLA - *Learning Environment for Physics Laboratory Activities* - (LEPLA, 2006) is a project which was funded from January 2003 - to January 2005 under the European Community Socrates-Minerva action. This grant-awarding program aims to promote and fund innovation in open and distance learning and the use of information and communication technologies in education.

LEPLA combines web based resources and hand held technology (that is, portable, compact data-loggers with compatible sensors, controlled by and used together with programmable graphing calculators or computers) to provide relevant, simple, but non-trivial, physics experiments with supporting educational materials. The resources are targeted at students and educators not only of physics, but also of applied mathematics and engineering studying at the upper secondary school and first year undergraduate level.

The aim is to use apparatus which is both versatile and inexpensive, to enable fundamental and creative experiments to be performed elegantly and efficiently. There is immediate access to the mathematical and data handling tools available on many graphing calculators (such as Texas Instruments-TI 83), with this latter aspect helping modelling and data analysis to become a natural part of student learning.

A wide range of sensors (for example, for measuring displacement, acceleration, force, light, temperature, sound, charge, magnetic field, current, voltage and gas pressure) are commercially available (see for example Vernier, 2006). This has enabled a correspondingly wide range of over twenty experiments to be developed in the areas of mechanics, heat, electricity and magnetism, optics and radioactivity.

The instructions for each experiment can be viewed on the LEPLA website (LEPLA, 2006) or may be downloaded in PDF format. A CD which provides a copy of all the website materials is also available. Information is given on the experimental set-up, data acquisition and data analysis for each experiment. In case a student has difficulty making measurements, specimen data are provided for the student to analyse. The underlying theory is explained and, in appropriate cases, extension work and directions to other experiments within the LEPLA website are provided. Materials are available in six languages (English, Italian, French, German, Swedish and Polish).

The authors are the UK members of the LEPLA Project team. Other team members are from University College Cork, Ireland, Universities of Bologna and Padova, Italy, University of Malmö, Sweden and the Technical University of Lodz, Poland, the latter being the coordinating institution. Recently, other partners have joined an extended team; these are the French Physics and Chemistry Teachers' Association, Lycée Maryse Bastié, Limoges and University of Münster.

The experiments, mostly in mechanics, that we considered to be most suitable for our purpose are these:

- Using a motion sensor to measure the acceleration of a toy car on an inclined plane (LEPLA experiment 1)
- Using a force plate to measure the changes when a person jumps up and down on the plate (LEPLA experiment 2)
- Using a motion sensor to observe a body oscillating vertically on the end of a spring (LEPLA experiment 3)
- Using an accelerometer to measure the acceleration of a moving car (LEPLA experiment 6)
- Using an accelerometer to measure the acceleration of a moving elevator (LEPLA experiment 7)
- Using a motion sensor to investigate the dynamics of a ball bouncing in a vertical line (LEPLA experiment 9)
- Using a temperature probe to measure the temperature of a liquid as it cools (LEPLA experiment 13)

This set of experiments presents a range of activities which were felt to be both of intrinsic interest to mechanical engineers (for example, the activities investigating the motion of an elevator and car, and the investigation of impulsive force by jumping on a force plate) and of value in the study of more 'standard' mathematical models (such as Newton's law of cooling, the coefficient of restitution and constant acceleration).

3. THE TEACHING INNOVATION

We saw this innovation as not just collecting and analysing data but also as an opportunity for students to develop their "key skills" of independent learning, working in a group, writing a report and giving a presentation. Students self-selected themselves into small groups, and choose an experiment to do from a list provided. There were six groups of four students and one group of five students. The available equipment constrained their choice, and projects were assigned on a "first come" basis. Students had access to the Internet in the university's central learning resource centre, and they had space in a laboratory to conduct their experiment. Experiments 1, 2, 6, 9 and 13 were chosen and there had to be some sharing of equipment.

Each group had one TI-83 and the following equipment was available:

3 CBR[1] devices (to be used in experiments 1, 3 and 9 to measure displacement)
1 CBL[2] device and 1 LabPro[3] device (to be used in experiments 2, 6, 7, 13)
1 Force Plate (to be used in experiment 2)
1 high "g" accelerometer (to be used in experiment 6)
1 low "g" accelerometer (to be used in experiment 7)
1 temperature probe (to be used in experiment 13)

First of all students were asked to write a paragraph on the "nature of mathematical modelling". They did this reasonably well, making use of articles on the Internet. They were given written instructions on how to access the LEPLA site and register. They were then to read the experiments and select one to do. Equipment was provided and they were left on their own, apart from a weekly meeting to review progress, to understand the theory and modelling, to carry out the experiment and collect their data, to analyse the data, and to write their report. These were submitted for marking, and at the end of the six-week period each group gave a short presentation to the whole class.

The intended learning outcomes of the exercise were given to students at the beginning. These were:

By completing this assignment successfully each student will have: -
1. demonstrated the ability to learn independently
2. demonstrated the ability to work harmoniously and effectively in a group
3. demonstrated initiative and inventiveness
4. carried out an experiment using hand held technology, which required the student to
   (i)   state the objectives of the experiment
   (ii) design the experiment using specified resources
   (iii) assemble the apparatus
   (iv) collect data
   (v)   describe, analyse, interpret and evaluate the data
5. demonstrated both written and oral communication skills through writing a (group) report and giving a (group) presentation to the rest of the class
6. carried out a self- assessment and a peer-assessment of the work of the group.

Students were advised that typical assessment criteria for written reports, oral presentations and self and peer assessment were published on the module website, along with guidelines for successful groupwork. See Haines and Dunthorne (1996) for further details. Each group was also asked to write an agreed group contract containing statements like

- Each member of the group will attend all organised group meetings unless they have a valid reason in advance not to.
- Each group member will contribute an equal amount with regards to research and the work in hand.
- All members should help each other with any problems and be willing to listen to ideas from the others.
- Everybody must meet all set deadlines.

4. EVALUATION

The evaluation methods used were a questionnaire, which included some free response questions, a study of students' written reports and presentations, and observation of some students at work and records of student queries at tutorials.

4.1 Questionnaire

Questions were posed in terms of studying "physics" as this was the principal intention of LEPLA. Students returned 18 completed questionnaires.

The responses to questions on internet use indicate that the web is increasingly being used to support teaching and learning in secondary education, with responses showing (happily) that the majority of students found the LEPLA materials user friendly, easily navigable and informative. Students stated that the experimental set up and data collection were well explained and easy to follow, even given that almost 80% of them had not used data logging technology before. Interestingly, even though the students were enrolled on an honours degree in mechanical engineering, 28% reported that they had not carried out physics experiments before, thus indicating that, disappointingly, some secondary schools still perform experiments by teacher demonstration rather than student participation.

Free response sections of the questionnaire further indicated students' broadly positive interaction with the LEPLA web site, materials and experiments. A few remarks are particularly telling. Thus, for example, in response to the question: *Which aspects of the LEPLA experiments did you find most interesting and useful for your learning of physics? Please explain:* statements such as

> "Carrying out the experiment, I learn more things if I conduct the experiment myself"

> "The fact that we get our own results and find information for ourselves rather than be told *this is what happens*"

> "Drawing the different graphs in Excel really helped me understand the physics"

indicate that students not only enjoy and learn from carrying out experiments, but also that they have perhaps felt deprived of the opportunity before coming to university. Similar comments could be made about responses to the question: *Did the LEPLA materials change your approach to scientific experiments? Please explain*: such as

> "Yes: It can also be interesting to carry out investigations"

> "Yes: It allowed me to put theory into practice"

> "Yes: It showed me that the theory we learn actually has very practical uses"

> "Yes: More interesting and more fun"

4.2 Students' Written Reports and Seminar Presentations

As noted earlier the following experiments were chosen: Motion of a toy car on an inclined plane (chosen by two groups), Newton's law of cooling, the bouncing ball (chosen by two groups), jumping on a force plate and the acceleration of a car.

The toy car experiment was not a great success as the cars used were too small to give a decent reflection of the radar. Nevertheless students recorded some data, but one group used the data provided on the web page for analysis. One group did not use linear regression to find the line of best fit, but simply selected two points and calculated the line through these. In future it will be emphasised that they use linear regression, a method that is probably new to many of the students. Indeed the use of LEPLA here provides an excellent example of an opportunity to introduce a statistical technique in context rather than, what often happens - in vacuo. One group used PowerPoint and the other used the overhead projector. They demonstrated understanding of the physics involved.

The group that chose the cooling experiment carried it out satisfactorily and were able to explain the phenomenon. However their written report was mostly plagiarised from the Internet and the LEPLA website, for which they were suitably chastised. The concept of plagiarism seems to be foreign to first year students and more instruction will be given here in future. The material for their presentation was much more "their own". They demonstrated understanding of the physics and presented their data and its interpretation in a satisfactory manner. They used an overhead projector with hand written slides.

In the bouncing ball experiments, while both groups managed to collect data, they were not happy with these, and used the data provided on the LEPLA website for their analysis. In their reports the students exhibited some ignorance of particle mechanics, even though the LEPLA material covers the fundamental ideas. The reports also betrayed their inexperience in writing. One group gave a PowerPoint show, which was well constructed and illustrated. The other seminar-presenting group used an overhead projector, and again this was well written and illustrated. The material presented was faithful to the inaccuracies of the written reports.

The group doing the accelerating car experiment had a good time driving a car around making measurements. They transferred the data to Excel and presented them in graphs, which they interpreted correctly according to the motion of the car. They did a good project and presented their work beautifully in a power point show. But their written report was hopelessly inadequate. It is hard to say this, but it seems they just got tired of the work.

The final experiment - jumping on a scale (force plate) - was carried out by a group that had degenerated into two people because others had left the course early in the semester. Since the two remaining had had difficulty setting up group meetings, the work was all done at the last minute. They conducted the experiment and collected data, which did demonstrate the characteristics of the activity, but they had difficulty, probably due to a lack of understanding, in explaining the changes in the force on the plate that they measured. The report was hurried and would have benefited from more careful writing. The seminar presentation was similarly poor.

4.3 Observation of Some Students at Work and Records of Student Queries at Tutorials

While the exercise was intended to promote independent learning, students were invited to come to see the lecturer when they were having difficulties. Apart from most groups requiring help to transfer screen shots and data from the TI-83 to the PC, only one group requested an explanation of the modelling involved in the toy car experiment and asked for some help in doing the experiment. This group had not studied particle mechanics at school, and had not realised that this would be a difficult experiment for them when they choose it.

5. CONCLUSIONS

The use of the LEPLA materials in a classroom context proved to be a positive experience both for the lecturer and the students. Student centred learning was promoted, and was clearly enjoyed. Furthermore they were given an opportunity, to perform interesting experiments, learn how to use new equipment, and last but certainly not least, to see how mathematical models provide unifying and simplifying explanations of real-world phenomena.

Notwithstanding the students' optimistic responses to the questionnaire, about finding the LEPLA material easy to follow and helpful, their performance in the experiments suggests otherwise.

The lecturers also gained valuable lessons from this experiment. First year students are still quite immature and require scaffolding to help them become independent learners. There were some instances where the LEPLA web material could be improved to help such students perform better.

We suggest that the use of experiments can enhance an engineering mathematics course if time and resources permit.

NOTES
1. A CBR (Computer Based Ranger) is a Texas Instruments sonar device, which uses radar to measure distance. It also allows the calculation of velocity and acceleration.
2. A CBL (Computer Based Laboratory) is a Texas Instruments recording device that can be used with a number of different sensors.
3. LabPro device is similar to a CBR and is manufactured by Vernier (Vernier, 2005).

REFERENCES

Debowska E., Jakubowicz S. and Mazur Z. (1999) Computer Visualisation of the Beating of a Wilberforce Pendulum. *European Journal of Physics*, 20, 89-95.

Gardner H. (1983) *Frames of Mind*. London: Heinemann

Gardner H. (1993) *Multiple Intelligences: A Theory in Practice*. New York: Basic Books

Haines, C.R. and Dunthorne S. (1996) (eds) *Mathematics Learning and Assessment: Sharing Innovative Practices*. London: Edward Arnold

Houston, S.K. (2001) The Theory of Multiple Intelligences and Mathematical Modelling. In J.F. Matos, W. Blum, S.K. Houston and S. Carreira (eds) *Modelling and Mathematics education, ICTMA 9: Applications in Science and Technology*. Chichester: Horwood Publishing, 30-38.

LEPLA (2006) http://www.lepla.edu.pl (accessed 24 March 2006)

Pecori, B. and Torzo, G. (1998) The Maxwell Wheel investigated with MBL, *The Physics Teacher*, 36, 362-366

Pecori, B. and Torzo, G. (1999) *Harmonic and Anharmonic oscillations investigated using a microcomputer based Atwood's machine*. American Journal of Physics 67, 228-235

Vernier (2006) http://www.vernier.com (accessed 24 March 2006)

7.7

MODELLING FOR PRE-SERVICE TEACHERS

Susann Mathews and Michelle Reed
Wright State University, Dayton, Ohio, USA

Abstract–*Teaching mathematical modelling has become more important as we prepare students to investigate complex phenomena that affect their lives. However, many mathematics programmes do not include modelling experiences for their future teachers even though they will be called upon to teach modelling-based mathematics to their students. At Wright State University, our nine-course programme for middle school mathematics teachers has modelling as a central theme. This paper describes four of these courses that help students develop modelling skills. Two vital practices we keep in mind when planning are reinforcing modelling skills over time throughout our courses, and aiming the mathematics and the activities at the students' ability levels.*

1. INTRODUCTION

Over the last decade at Wright State University, we have developed a nationally-recognized programme for middle school (for students aged 9-14) mathematics pre-service teachers, guided by the standards written by the (US) National Council of Teachers of Mathematics (NCTM) (1989, 1991, 2000) and the *Call for Change* (Mathematical Association of America, 1991). As a key thread of the programme, modelling helps students to develop problem-solving and mathematical skills; it creates connections between courses, and sets much of the learning in a real-world context, thus drawing upon knowledge in other disciplines. This paper describes the development of modelling abilities through the sequence of courses and models used in the programme.

As a result of NCTM standards in 1989, substantial grants have supported the development of excellent, new K-12 (primary and secondary education) mathematics curricula (textbooks, supplementary material, teacher resources, etc.) based either upon mathematical modelling or with mathematics applications as a large component. Nevertheless, school districts have been slow to adopt them. As Thomas Jefferson, our third president, is attributed with having said, "It is easier to move a graveyard than to change a curriculum." This echoes what Hamson (2003) has found in the UK -that "this [including more modelling in mathematics education] has been a slow process indeed" (p222). Before future and practising teachers will implement these new modelling-based curricula, they must not only know the mathematics itself but also have experience with mathematical modelling. Blum (2002, p161) has indicated that mathematics teacher education programmes rarely include "orientations to modelling, and the use of the modelling process in mathematics courses". Five years ago, we reorganized our programme to address this issue.

2. NEED FOR THE PROGRAMME

Through our pre-service teachers' experiences in mathematics courses, "they develop ideas about what it means to teach mathematics, beliefs about successful and unsuccessful classroom practices, and strategies and techniques for teaching particular topics. Those from whom they are learning are role models who contribute to an evolving vision of what mathematics is and how mathematics is learned" (NCTM, 1991 p127). They need to gain experience developing accessible models and discover that they can "do modelling" (Lesh & Lehrer, 2003). As with most substantive mathematical ideas, students need to develop their modelling ability over time. Therefore, our programme includes a progression of modelling and simulation efforts through a sequence of courses. With this in mind, we embed problem solving and modelling in each course and build upon this as students move from one course to the next. The capstone is an entire course on mathematical modelling. This approach is in consonance with Myer (2001) in which she indicates that learning mathematics is a constructive experience, that the learning of a concept and skill occurs over time and with increasing levels of abstraction, and that mathematical understanding is interconnected.

Most mathematics modelling textbooks have some form of the following six-step modelling process in their first chapters:
1. Construct a simplified or idealized version of the initial problem situation;
2. Construct a mathematical model of the idealized version of the problem;
3. Identify solutions within the framework of the mathematical model;
4. Interpret these solutions in terms of the idealized problem situation;
5. Verify that the solutions generated for the idealized problem are solutions to the initial, simplified problem situation;
6. Add complexity to increase fidelity and start again at step 2.

We teach this process beginning with steps 2 and 3 in Algebra and Functions, adding steps, increasing complexity, and adding reflection after each model in later courses.

The purpose of this paper is to share aspects from four of our nine courses in the programme for middle school pre-service mathematics teachers that we have developed at WSU (Figure 1) that make modelling accessible to our middle-school pre-service teachers and help them learn to do mathematical modelling over time. We highlight what all four courses share by way of teaching strategy, provide descriptions of how each course is unique, and describe what modelling skills each course addresses.

Each course meets for two-hour sessions twice each week of a ten-week quarter. This provides sufficient time in each class for investigation and discussion. All of the courses are taught in an inquiry manner with guided discovery, problem solving, group work, and whole-class discussion with lecture only when needed. With this course structure, we are able to design learning experiences that promote a depth of understanding of the content and processes (Mathews, Basista, Farrell, & Tomlin, 2003). Throughout the programme, we use appropriate technology as a tool, both to help students better understand the mathematics, and to help them model real-world problems that would be inaccessible without it. This structure naturally leads to a focus on mathematical modelling.

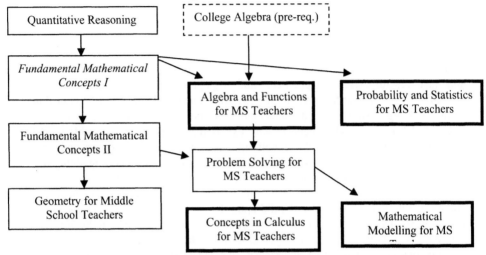

Figure 1. Middle School Pre-service Mathematics Courses at
Wright State University.

We move from small situations in which many of the parameters are given to more open-ended situations in which the students are expected to determine more parts of the model. Then we move on to more ambiguous situations in which they are expected to determine not only the model but first to determine what the problem actually is and pose the questions. In doing so, we provide our students with experience in a hierarchy of mathematical models, working from the simpler to the more complex and from the concrete to more abstract. By doing this throughout the programme, we help students experience a variety of real-world situations in which

Table 1. Alignment of Courses with their Levels of Modelling and Skills.

Course	Math Skills Needed for Modelling	Focus of Modelling: Skills and Activities	Assess-ment	Modelling Step Focus/ Level at course end
Algebra and Functions for Middle School Teachers	Functions that have appropriate shapes for various situations $\Rightarrow$ selecting appropriate functions given a situation	Relationships between variables and data; Relationships between variables; Multiple representations	Fitting functions to experiment data	Mathematisation *Level 2* Interpretation *Level 3*

Course	Topics	Skills	Assessment	Competencies
Probability and Statistics for Middle School Teachers	Regression; Stochastic situations: probabilistic simulations	Identifying relevant variables; Relationships between variables; Carrying out simulations	Final exam questions about simulations	Simplification *Level 2* Mathematisation *Level 3* Transformation *Level 2* Interpretation *Level 2 or 3*
Concepts in Calculus for Middle School Teachers	Dynamic situations; Guided modelling with calculus	Reinforcing skills from Algebra And Functions; Understanding context; Breaking a complex problem into smaller pieces; Making assumptions; Analyzing Units; Analysis of results	Structured solutions to the High Dive problem	Simplification *Level 4* Mathematisation *Level 3 or 4* Transformation *Level 3* Interpretation *Level 2 or 3* Validation *Level 4* Addition of complexity
Mathematical Modelling for Middle School Teachers	Recurrence relations in simulations and processes $\Rightarrow$ recursive models; Open ended, complex, ambiguous situations	Reinforcing previous skills; Posing questions; Less direct guidance; Assumptions; Identifying and using mathematics across areas; Pulling it all together; Model analysis	Final written report to Convoy problem	Simplification *Level 3* Mathematisation *Level 3 or 4* Transformation *Level 3* Interpretation *Level 3* Validation *Level 4* Addition of complexity

they are required to transfer their skills and understandings among different contexts and using different areas of mathematics. The modelling situations have been selected to be both interesting and accessible. Table 1 notes focus of modelling skills, steps learned, assessments, and modelling skill levels achieved in the four courses. The levels noted in the last column are the average attained level in each step. (Levels for each step are taken from a sample rubric in Hodgson, 1997).

3. BUILDING MATHEMATICAL COMPETENCY AND MODELLING SKILLS

In Algebra and Functions, students study mathematical characteristics of families of functions. After studying each class of function separately, students perform a variety of experiments and determine what type of function best models the data gathered. Graphing calculators, probes, and CBLs (Calculator- or Computer-Based Laboratories) are used to collect the data. These activities are considered inquiry-based, but have very well-defined steps in carrying out the experiments. The questions are designed to direct students' attention to the mathematics involved in the experiment and how to interpret the data, after determining what type of function best fits the data.

In Probability and Statistics, we extend the idea of fitting functions in which the data points all fall on the curve of the found function to regression, in which the function does not contain all of the data. Students begin stochastic modelling with simulations, during which they learn that these can be performed to model a situation in which relevant data needed to enable modelling the system cannot be collected. *Fathom*, software that is designed specifically for learning statistics, is used to create lines of best fit and perform simulations. The simulation problems are closely guided using the five steps created by Gnanadesikan, Scheaffer, and Swift (1987). Through being directed to define the key components of the problem, to state the underlying assumptions, to select a model to generate outcomes for a key component, to define a trial and conduct a large number of trials, and summarize the information and draw conclusions, students learn valuable modelling skills that they will use later in their modelling course. However, the problems are more clearly defined than they will be in the later courses, and students must state given assumptions instead of creating assumptions.

During the Concepts-in-Calculus course, the students use calculus to model dynamic situations. The second half of the course is devoted to solving a large modelling problem—that of determining when and where a diver in a circus act should be dropped from a turning Ferris Wheel into a tub of water carried by a moving cart (Fendel et al., 2000). Because this is the first time that they are modelling such a complex and dynamic problem, we guide them through breaking the modelling situation into do-able pieces, validating their first solution, and re-modelling the problem with fewer simplifying assumptions. Students learn to use *Mathematica* during the first half of the course and then use that along with graphing calculators and *Fathom* for representing and solving the model. Following a structured format, students submit a written report with their models and solutions.

We pick up there in Mathematical Modelling. However, even after the previous introduction to pieces of modelling in other courses, at the beginning of the class few seem to understand precisely what mathematical modelling is. Thus, in addition to creating mathematical models given less structure, students are required to read and explicitly think about modelling, its characteristics, and how they might use it in teaching mathematics. Even in this capstone course, we proceed from structured to open-ended models. Their first real-world problem is to create mathematical models for the drainage of a parking lot for a new shopping centre. The directions are detailed but they must determine how to create a model, solve it, compare it to the stated requirements of the situation, and then re-model the problem with fewer simplifying assumptions. We next model the exponential growth of different breeds of mosquitoes and use data from their simulation to develop recursive and closed-form functions to model the exponential growth. This leads to exploring logistic growth using recursive functions, followed by a study of the susceptibles-infectives-recovered (SIR) model of the spread of a disease. Students model this with less directed simulations that lead to a system of equations for describing and predicting an epidemic. As a final project, students are asked to consider the situation of Great Britain in 1917 after the Germans had re-commenced unrestricted submarine warfare. Based upon mathematical models that they create, students must determine whether to continue sending merchant ships across the Atlantic independently or in convoys. Students submit a letter to a British admiral with their recommendations, mathematical models, and explanations (Mathews, 2004).

4. SUMMARY AND CONCLUSIONS

Throughout these courses, our students quickly become engaged when they see the applicability of mathematics to many areas of their life. They are excited that they, and their future students, can apply mathematics to important topics that are at first seemingly unrelated to mathematics. Though this motivation is important, we have realized that two of our implemented practices are essential. The first is the reinforcing of skills over time throughout our courses. Conversations between the authors regarding the curriculum and students help us to keep the courses connected with regard to building skills and students' understanding. The second practice is to keep the mathematics and the activities aimed at the students' ability levels. We attempt to do this by building ideas from concrete to abstract, using technological and other tools, and continually keeping in mind our audience. We note that the students' excitement with and success in learning mathematical modelling indicate that we have made the modelling accessible to them. We have focused this paper on describing our course activities that are aimed at helping pre-service middle school teachers become mathematical modellers.

REFERENCES

Blum, W. (2002) ICMI Study 14: Applications and modelling in mathematics education—Discussion document. *Educational Studies in Mathematics*, 51(1/2), 149-171.

Fendel, D., Resek, D., Alper, L., and Fraser, S. (2000) High dive, *Interactive Mathematics Program*. Emeryville, CA: Key Curriculum Press.

Gnanadesikan, M., Scheaffer, R. L., and Swift, J. (1987) *Quantitative literacy series: The art and techniques of simulation*. Dale Seymour.

Hamson, M.J. (2003) The place of mathematical modelling in mathematics education. In S.J. Lamon, W.A. Parker, and K. Houston (eds) *Mathematical modeling: A way of life: ICTMA 11*. Chichester: Horwood Publishing, 216-226.

Hodgson, T. (1995) Secondary mathematics modeling: Issues and challenges. *School Science and Mathematics 95*(7), 351-357.

Lesh, R. and Lehrer, R. (2003) Models and modeling perspectives on the development of students and teachers. *Mathematical Thinking and Learning, 5*(2/3), 109-129.

Mathematical Association of America, (1991) *A call for change: Recommendations for the mathematical preparation of teachers*. Washington DC: Author.

Mathews, S. M. (2004) Convoying merchant ships in wartime. *Mathematics Teaching in the Middle School, 9* (7), 382-391.

Mathews, S.M., Basista, B., Farrell, A., and Tomlin, J. (2003) Meeting the challenge of middle school mathematics and science teacher preparation. *Teacher Education and Practice, 16* (4), 399-413.

Myer, M. R. (2001) Representation in realistic mathematics education. In Cuoco, A. A., and F.R. Curcio (eds) *The roles of representation in school mathematics: 2001 Yearbook of The National Council of Teachers of Mathematics* Reston, VA: National Council of Teachers of Mathematics, 238-250.

National Council of Teachers of Mathematics. (2000) *Principles and standards for school mathematics*. Reston, VA: Author.

National Council of Teachers of Mathematics. (1991) *Professional teaching standards*. Reston, VA: Author.

National Council of Teachers of Mathematics. (1989) *Curriculum and evaluation standards for school mathematics*. Reston, VA: Author.

7.8

THE FINNISH NETWORK FOR MATHEMATICAL MODELLING

Robert Piché, Seppo Pohjolainen, Kari Suomela,
Kirsi Silius, Anne-Maritta Tervakari
Tampere University of Technology, Finland

Subject–*Since 2002, ten universities and research institutes in Finland have collaborated in order to teach introductory and advanced university-level mathematical modelling courses as a pilot project of a national Virtual University programme. This pooling of resources and expertise has made it possible to offer modelling courses that would otherwise be beyond the capabilities of the individual universities. To date, five university courses, comprising 33 teaching modules prepared by 22 teachers, have been developed and they are taught annually, to 50-60 students at 7 Finnish universities. Each course features streaming video-lectures from faculties of several universities, student teamwork and peer review in a browser-based learning environment, extensive use of numerical software, and video-conferenced presentations of students' modelling projects. In this paper, some details of the courseware production, teaching methods, technology development, and student, as well as expert, evaluation reports are presented.*

1. INTRODUCTION

One of the first pilot projects funded when the Finnish Virtual University programme started in 2001, was the national network for teaching mathematical modelling (MM). The participating institutions are the Lappeenranta, Tampere, and Helsinki Universities of Technology, the Universities of Helsinki, Joensuu, Jyväskylä, Kuopio, Oulu, and Tampere, and the national Centre for Scientific Computing in Helsinki. The project is co-ordinated by the Institute of Mathematics at Tampere University of Technology (TUT). The goal of the project is to develop content, tools, and administrative structure for web-based learning and teaching of mathematical modelling for students in all the participating universities.

Mathematical modelling was a good choice as a theme for a national virtual university pilot project. The funding agencies can readily appreciate the importance of mathematical modelling and computer simulation in technology and science (Lucas, 1999), and the need for training in this field to provide competent people for industry and research institutes. However, MM is a very broad area and no single teacher commands all its methods and applications. Also, new courses in advanced, more specialised MM areas would attract too few students from a single university to justify the expense. Web teaching technology makes it possible to pool expertise, resources and students from different institutes, to offer teaching that would not otherwise be available. Finally, mathematical modelling student projects fit well with the independent study style and intensive use of new communication technology that are the hallmarks of virtual university education.

Already in its first year of operation, the network prepared and offered two virtual university courses: an introductory MM course with 9 lectures and an advanced MM course with 11 lectures. Each web-based course featured streaming video lectures, student teamwork and peer review in a browser-based learning environment, local tutoring, and video-conferenced presentations of term projects. Now the network has 32 lectures and one self-study module grouped into 5 courses that are followed by 50-60 students at 7 Finnish universities, and 3 more courses are in preparation.

In this paper we present details of the curriculum, teaching methods, and technological solutions. Student feedback as well as an external pedagogical expert evaluation reports are outlined. The paper closes with some ideas about future directions.

2. COURSES

Most of the network's teaching material production effort has gone into creating 32 2-hour video lectures and supporting material (slides, exercises, model solutions, notes, term project topics). Most of the lectures are largely independent of the others and could be used for self study or as supplementary lectures for a conventional course. They have been collected into elective courses that are offered for credit to the students of the participating universities during the school year. The purpose of these courses is to introduce mathematical modelling, to integrate knowledge from conventional mathematics courses, to show how mathematics relates to problems in the real world and to teach how to use numerical or symbolic software to solve practical problems.

Each course begins with a real time videoconference that introduces the curriculum and describes the practical arrangements. The remaining lectures are pre-recorded, with a "talking head" streaming video window synchronised to slides. Written material related to the lectures is available on-line or in the university library.

Students working alone or in groups of two or three submit solutions in PDF format to assignments related to the lectures. Each student group is then required to read and to comment online on at least two solutions from other groups. This was intended to expose students to alternative approaches in the case of open-ended questions. The teacher (usually the author of the lecture) also participates in the discussion, contributing comments on the solutions and on the student comments.

A distinguishing feature of MM education is the exposure to open-ended real-life problems. All of the courses require students to do a term project with written report and oral presentation in a videoconference at the end of the course. Students have generally reported that the project is the most demanding but also the most interesting and rewarding part of the course. Project topics for the introductory course have included

- Design a control for an outdoor fountain in a windy square
- Devise a system for overbooking of airline tickets

- Are fingerprints unique?
- Design a better Quickpass system for an amusement park
- Predict the population of Finland in 2100
- Does a sandwich tend to fall buttered side down?

The first four topics are problems from COMAP's undergraduate Mathematical Contest for Modelling (MCM). As in the MCM, the problem statements are very brief (typically one paragraph) and it is left to the students to define the scope, search literature, and find relevant data. The project instructions emphasize the importance of generating and comparing alternative solutions. We have been pleasantly surprised by the inventiveness of some project solutions. For example, in the solution presentations we have seen computer animations of water droplets, data from field trips to amusement parks, and videos of falling sandwiches.

The introductory course is given annually, and is intended for second and third year students who have already learned the basics of matrix algebra, ordinary differential equations, numerical analysis, probability theory and statistics. The course topics are:

- survey of mathematics applications,
- the modelling process,
- survey of numerical software,
- continuum models,
- differential equations,
- dimensional analysis,
- discrete models,
- fitting data,
- probabilistic models, and
- decision analysis.

The other courses are given every second year, and are intended for senior undergraduate and graduate students for whom mathematics is a major or minor subject. These courses present case studies of real modelling problems from research in university, research institutes and industry. The courses are:

Modelling with PDEs: Modelling fields, flows and waves with PDEs; solving multiphysics problems with FEM; case studies in acoustics, melting and solidification, silicon crystal growth, and impedance tomography.

Modern Data Analysis: Theory and applications of fuzzy and multivalued logic, neural networks, genetic algorithms; data mining, decision analysis.

Statistical Modelling: MCMC for process parameter estimation, statistical methods for pattern recognition, mixed models for forest harvester operation, multivariate regression and design of experiments for process optimization.

Modelling Randomness: Brownian motion applications in finance and communications, stochastic differential equations, time series analysis, Bayesian solution of inverse problems, filtering.

The following courses are planned for the 2005-06 school year:

Mathematics of visual motion: How to reconstruct two and three dimensional motion and shape from a video sequence. Feature extraction, optical flow and visual velocity field, projecting 3d rigid body motion onto image sequence, statistical calibration.

Mathematical methods for continuum modelling: Techniques for setting up and analysing ODE and PDE models of continuum processes: linearisation, asymptotics, regularisation, model fitting, model based optimization and control.

These new courses each have a single author, whereas the five existing courses are collections of material from different teachers. For example, the introductory course has lectures from nine teachers from six universities. Other than agreeing on topic headings, there was no central control, and teachers prepared material independently of the others. As a result, course material has varying notation and terminology, some overlaps and gaps, and widely different levels of prerequisite knowledge. Each course introduces a broad range of topics, and a two hour lecture can only scratch the surface of any of the theories or case studies.

3. ICT TOOLS AND SOLUTIONS

The modelling network project home page http://www.alpha.cc.tut.fi/mallinnus.html is a conventional web site that describes the pilot project and has links to all the course materials.

Students sign up for the MM courses through the A&O learning environment developed at the Tampere University of Technology's Hypermedia Laboratory. A&O provided the students with a continually updated course page from which links could be found to everything that was to be covered in the coming week. The discussion area was the most-used A&O feature. Here groups submitted, commented, and got feedback on weekly exercises and projects.

The web lectures provide video and audio recordings of the lecturer that are synchronised with the presentation of slides, and navigation controls to allow moving backwards or forwards in the lecture. They are realised using the W3C standard Synchronized Multimedia Integration Language (SMIL), by which a single XML-based file prescribes layout and chronological presentation of multimedia presentations from several media elements of different types that can be located anywhere in the net. The free Real(One)Player was chosen because of the efficient proprietary format for the streaming of text and images.

The opening lectures and the presentations of the term projects were arranged as multipoint videoconferences using the Finnish universities' fast FUNET network. Two presentation devices were used at each participation location, one to handle audio and video of speakers and the other to present the lecture slides and browser picture. Lecture slides and WWW material were presented using application distribution in Microsoft NetMeeting.

4. STUDENT FEEDBACK

At the end of every course, students were asked to answer an extensive questionnaire using the Helsinki University of Technology's Opinions-Online program (http://www.opinions.hut.fi/). In this section are summarised the results of the questionnaire for the presentation of the introductory MM course in 2004. Results from questionnaires for the other courses are similar.

The introductory MM course in 2004 was given to 29 students at 6 universities. There were 22 respondents, with all universities represented. 46% were mathematics majors. 46% were undergraduates in years 3-5, but we were surprised that 54% were doctoral students. The students of this course thus have a wide range of mathematical preparation.

Students report a wide range of "attendance" of the video lectures: 20% watched fewer than 40% of the lectures, while 30% claimed they watched at least 80% of them. The lectures are therefore quite important for the latter group. Half the students watched lectures at home, 27% watched them on university computers on their own time, and only 5% (one student) came to scheduled viewings.

Weekly exercises and the term project got reasonably positive ratings. The usefulness of student comments got a mixed rating. Most students said they devoted more time to the term project and to the course as a whole than indicated by the credits. This can be interpreted as indicating that these students found the course interesting, because this is an elective course for all of them.

In the course there were quite a lot of technical tasks for the students to perform. They had to register in the A&O learning environment, produce pdf documents, upload and download files, watch video lectures using Real Player, and participate and make presentations in a videoconference. In earlier years technical problems with the A&O learning environment and some minor course administration problems generated negative student feedback. In this realisation of the course the feedback on these issues was generally positive, reflecting the improvements constantly being made to the software and to the course planning. Fewer than 20% reported that technical issues were a significant difficulty in the course.

In the write-in portion of the questionnaire the following aspects were criticised:
- the lack of active participation in the discussion boards by some teachers,
- the course's overemphasis on physics,
- the need to write up solutions and term project reports on computer,
- some lecturers assumed too much prior knowledge.

Among the positive comments: one student appreciated how this kind of course gives access to the teachers in other universities, and another was glad for the opportunity to "attend" lectures without fixed time or place.

Just over half the respondents claimed they prefer a web-based course over a conventional mathematics course. This mixed response is also shown in the answers to similar questions (Figure 1).

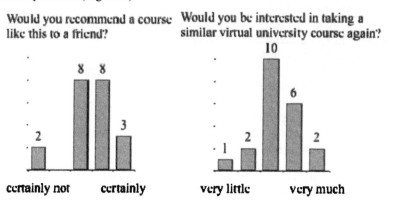

Figure 1. Some student feedback.

5. PEDAGOGICAL EVALUATION

In 2003, two of the authors (KS & AMT), who are pedagogical specialists in the TUT Hypermedia Laboratory, were asked to evaluate the MM courses. They performed this evaluation using a multidisciplinary evaluation framework of usefulness (Silius et al., 2003). The framework takes systematically into consideration the most important factors of usability, pedagogical usability, accessibility and informational quality.

The evaluation report includes sections on the reliability of information, presentation of information, visual design, support for online reading and navigation, technical realisation and use of multimedia elements and accessibility. Pedagogical usability was evaluated from the viewpoint of *meaningful learning* (Jonassen, 1995, Ruokamo & Pohjolainen, 1998). Meaningful learning emphasizes that learning should be active, constructive, collaborative, intentional, contextual and reflective, and the learning outcomes should be transferred to new contexts.

The evaluation report judged the implementation of web-based courses of mathematical modelling to be pedagogically well designed. For example, the project assignments were carried out according to the principles of meaningful learning:

- Theoretical knowledge was needed to solve the modelling problems of authentic contexts. The students solved problems from various situations in everyday life; for example, does a piece of bread usually fall butter side downward or not? This gave them the opportunity to transfer the theoretical knowledge from lectures to a new situation.
- To solve the problems students had to be active: they observed natural contexts and interacted with them.
- The students also structured their theoretical knowledge in co-operation with other students in a constructive and collaborative way.
- All the students and lecturers participated in the final session of the course, which was organised as a videoconference. From the pedagogical point of view this was an important learning experience for the students: Each student group gave an oral presentation of their project work and the others provided comments and feedback.

According to the evaluation report the implementation of a web-based learning environment also paid attention to usability:

- The structure of the web-based learning environment was simple, likewise the visual layout. The well-designed structure made it easier for students to construct a conceptual model and perceive the functions of the web-based learning environment.
- Navigation was supported by using a different colour to show which hyperlink was visited and which was not. The naming of the hyperlinks was also well thought out, which also supports navigation.
- Because the students usually use various hardware and software, attention had been paid to file types used in the learning environment. All the materials were in HTML or PDF formats, which are widely supported. The students had also been offered a freeware PDF converter to return their exercises in PDF format, which can be opened with an Acrobat Reader.

The evaluation report also pointed out some aspects of the MM courses to be improved (some of these have since been implemented):

- More information about the schedule, learning objectives, assessments of the courses etc. should be available in the course homepages outside the A&O learning environment. Students need information before the course begins in order to decide whether they will participate.
- Greater uniformity of courseware implementation is needed. At present the structures and layouts of teachers' contents and presentations differ. Harmonisation would facilitate studying.
- The accessibility of the learning material should also be improved. For example more text equivalents should be included in the video lectures. Without text equivalents a person who is deaf or hard of hearing cannot access the content of the video lectures.
- The usability of the learning material should also be improved in part: The file sizes of some multimedia elements and PDF files should be compressed. Some files were too large for downloading via a modem connection.
- From the pedagogical point of view the importance of net discussion should be emphasised. At present the students make some peer-to-peer comments on modelling projects, but analysis of discussions is lacking. The teachers should be more active in commenting the students' discussion. At present the teachers' feedback mostly concerns the grading of the modelling projects.

The evaluation report lists a number of useful ideas and best practices in the implementation of web-based courses of mathematical modelling. For example, the A&O learning platform offers tools for scheduling. The schedule mentions all weekly lectures, exercises and activities with important notices like deadlines for exercises etc. The schedule supports students in estimating the course load and their own chances of completing the course successfully. From the didactical point of view course organisation was well done: In spite of the fact that the courses were web based, all participating institutions had named face to face tutors and assistants for students. They supported learning processes in face-to-face sessions while lecturers were responsible for web based learning materials and web based teaching. A good example of well-designed web teaching was the video lectures in which the lecturers' train of thought was supported rhythmically with motion and articulation.

6. CONCLUSIONS AND FUTURE DIRECTIONS

The student feedback and expert evaluation are generally positive and indicate that the pilot project is successfully meeting its goals of combining the expertise of several institutes to create and teach new mathematical modelling courses using new methods.

After being taught for four and a half years, our mathematical modelling courses are now well integrated into the regular curriculum of the participating universities. Their integration has been further facilitated by the thorough curriculum reform carried out this year by Finnish universities in order to implement the principles of the Bologna agreement. For example, at Tampere University of Technology, the set of mathematical modelling courses now forms one of the 25 credit modules that students can choose as part of their degree requirements.

The pilot project had another objective: to build a national network of mathematical modelling practitioners and teachers. This has also succeeded, as evidenced by increased cooperation among the project partners in other areas:

- Every year, the network has sent students to participate in the Modelling Week organised by the European Consortium for Mathematics in Industry (ECMI).
- Study Groups bringing together mathematical modelling professionals and representatives from industry have been organised by pilot project members in Tampere in 2002 and in Helsinki in 2004.

The ministry's FVU funding continues for another year, during which the existing courses will be taught and new courses will be developed. What happens after this?

The courses created as part of the project will probably continue to be given. The consortium agreement gives all participating institutions user rights to all materials produced in the project, so the universities may use parts of the material on their own, or continue to offer the courses jointly as they are doing now. With the video lectures and course material already prepared, the courses can be offered at relatively low cost. As mentioned earlier, the courses are already well integrated into the curriculum.

Having successfully built a mathematical modelling teaching network spanning Finland, we are now considering how to extend the network further. International cooperation seems to be the natural direction. Although most of our teaching material is in Finnish, the content, teaching methods, and IT solutions are in place and well-tested, so adapting some of our courses for other languages seems feasible.

REFERENCES

Jonassen D.H. (1995) Supporting Communities of Learners with Technology: A Vision for Integrating Technology with Learning. *Schools Educational Technology* 35 (4), 60-63

Lucas W.F. (1999) The Impact and Benefits of Mathematical Modeling In D.R. Schier and K.T. Wallenius: *Applied Mathematical Modeling: a multidisciplinary approach.* London: Chapman & Hall.

Ruokamo H. and Pohjolainen S. (1998) Pedagogical Principles for Evaluation of Hypermedia Based Learning Environments in Mathematics, *Journal for Universal Computer Science*, 4 (3), 292-307, http://www.iicm.edu/jucs .

Silius K., Tervakari A-M and Pohjolainen S. (2003) A Multidisciplinary Tool for the Evaluation of Usability, Pedagogical Usability, Accessibility and Informational Quality of Web-based Courses. Proceedings of the Eleventh International PEG Conference: Powerful ICT for Teaching and Learning, 28 June - 1 July 2003 in St. Petersburg, Russia. http://www.virtuaaliyliopisto.tut.fi/arvo/raportit.php .

7.9

LEARNING ENVIRONMENT THROUGH MODELLING AND COMPUTING

Regina Lino Franchi
Methodist University of Piracicaba, Brazil

Abstract–*Mathematical modelling and computing are important in mathematics education and mathematical modelling because, in addition to enabling the learning of mathematics in a contextual way, it gives students the opportunity to develop their potential. Computing facilitates visualizations and offers precise and quick answers and, because it enables mathematical work through experiment form, it allows the student to exploit the concepts in an autonomous way, trying out changes and variations and drawing conclusions. The joint work of computing with modelling has brought new possibilities to modelling. Many of the difficulties in the modelling process were overcome by the ease of data collection and handling, and by the representations management with software and using the Internet. This paper presents a characterization and utilization of mathematical learning environment through mathematical modelling and computing.*

1. INTRODUCTION

The use of mathematical modelling in classes makes it possible to work with mathematics in a contextual way, so that it can contribute significantly to the process of acquisition of mathematical knowledge by the students (Franchi, 1993). In doing so it gives students the opportunity to develop their potential.

There is also a trend in education to include computing. Talking specifically about mathematics, the possibility of calculations, simulation, using visual, auditory and animating resources, the precision and speed of the answers has led to reflections about the processes of construction of mathematics concepts by the students, about the validity of working some techniques and about other possibilities of work in differentiated environments for mathematics learning.

Modelling and computing have been tried and discussed by teachers and researchers in mathematics education. There have also been attempts to evaluate the possibility to work in contexts that involve these two trends simultaneously. The present time is marked by the search of theory from the developed practice with Modelling and computing in learning situations. This paper intends to contribute to this debate by the characterization of learning environments with mathematical modelling and computing. The paper also shows examples of activities in this kind of environment.

2. CHARACTERIZATION OF LEARNING ENVIRONMENTS THROUGH MATHEMATICAL MODELLING AND COMPUTING

In spite of the great number of experiences of modelling use in learning contexts, there is no consensus amongst researchers about the characteristics of modelling in mathematical education and about the possibilities of use of this kind of strategy.

From the point of view of applied mathematics it is said that the characterization of the modelling is clear. The main objective for applied mathematics is the model. Thus, the activity is successful when it gets solutions that are validated and that can give answers to questions with accuracy.

For mathematical education, the process of construction of the model acquires special importance. Modelling can be thought as a strategy of mathematical learning. In this case, mathematical concepts are worked by its relation with the phenomenon in study. The concepts are introduced through the modelling process, inserted in the context of the studied problem and then the study of techniques and deepening of theories may be carried out.

It is also possible to evaluate the modelling process from its relation with the development of the student's potentialities, such as: capacity to identify, to formulate and to solve problems, capacity to search for information, to use resources varied in the search for solutions, creativity, cooperation, team work, initiative, capacity to analyze and to compare possible solutions, capacity do decide and to evaluate the consequences of the carried out actions and so forth. These abilities are stimulated, demanded and developed during the different stages of the model construction process.

The use of modelling can still contribute for the social-critical students' formation if the possibility to reflect about the context of the phenomenon that has been studied is considered. Modelling can contribute for the understanding of the reality and at the same time make it possible to act on it.

Some researchers have tried to identify what, in fact, characterizes the modelling presence in learning situations. They search for indications of modelling presence by the development of thematic projects external to mathematics (that do not always result in the construction of a model) or by the use of hypotheses that may simplify the representation and the use of approaches culminating with the effective model creation (see Bean, 2001 p56).

Barbosa (2001) gives an important contribution for this debate characterizing what he called learning environments with modelling. Based on different types of activities described in the research literature on modelling and education and based on Skovsmose's definition (2000) of learning environments, he defines: "Modelling is a learning environment in which the students are invited to inquire and/or to investigate situations with reference in the reality, by means of the mathematics" (Barbosa, 2001 p31). This characterization includes many different possibilities of modelling use that have already been included in such literature.

Relating to computing, there is a certain agreement that the development of activities using the computer in classroom characterizes a computing environment. However, the use of this resource does not necessarily imply in changes of pedagogical position. We may essentially have a lecture, in spite of the use of the computer. Borba and Penteado (2001), discoursing about educational environments with the presence of computing, point to the importance of educative practices that emphasize the construction acquiring of meaning by those involved.

Based on the above statements, we characterize, in a form analogous to Barbosa: an environment of mathematical learning through computing is the one in which the students are invited to inquire and/or to investigate mathematics objects, by means of computing. The situations that stimulate the investigation, experimentation and formation of hypotheses can come from an internal or external context of mathematics (Franchi, 2002).

It is also important to consider a combined work of computing with modelling. Computing has brought new possibilities for modelling. Many of the difficulties of the modelling process were overcome by the facility of data collection and treatment and by the ease of representations possible through particular software. The model can be constructed with more freedom without being concerned about complex mathematics, or that that mathematics might be difficult for certain school levels.

So, we can characterize a learning environment through mathematical modelling and computing as the one in which the students are invited to inquire and/or to investigate situations with reference on the reality, by means of mathematics, using computing in the stages of modelling process or yet the one in which the students are invited to inquire and/or investigate mathematics objects, by means of computing, motivated by situations with reference on the reality (Franchi, 2002).

On the first case we can consider, for instance, a study of a real situation, by means of mathematics, using computing to organize and to represent collected data or to solve formulated equations. On the second case we can consider, for instance, a study of a mathematical subject that can characterize a phenomenon of the reality, using software to work with characteristics and relative content proprieties.

3. WORK POSSIBILITIES IN LEARNING ENVIRONMENT THROUGH MATHEMATICAL MODELLING AND COMPUTING

The following description is about a study carried out about the work possibilities with the theme "Dengue" in a learning environment through mathematical modelling and computing. More than the knowledge of the subject, the main objective of the description is to give an example of how the study of a reality theme (of interest to some community) makes possible the creation of a mathematical learning environment through mathematical modelling and computing.

Dengue is a disease transmitted by the bite of the mosquito *Aedes Aegypti*. The mosquito reproduces in any container used to store water in shady or sunny areas. Cans, tyres, pots of aquatic plants or water boxes are common places for these mosquitoes to reproduce themselves.

Its life cycle is divided into two phases: the aquatics phase (egg, eclosion, larva and pupa) and the aerial phase (birth and adult).

The main factors that can influence the mosquito propagation are the temperature, humidity, regional topography, and the human activities related to the places where the mosquito may reproduce.

The usual ways used to control the mosquito propagation are: mechanical control (elimination of the places where they may reproduce), chemical control (like spray) and biological control (cohabit of animal species that feeds on the mosquito larva).

There are many work possibilities related to the theme. For instance, the growth of the mosquito population can be studied (from the aquatic phase to the adult phase), the biological larva control or the control of the places where they may reproduce.

Related to the control of the places where they may reproduce it is possible to study the conservation of water pools, pots or tanks. In the following example, a set of pools with water movement (by cascades) was considered, and the periodic chlorine addition in the tanks that make up the set. The formulated question was: how to control the amount of chlorine in each tank of the set?

Time (hours)	chlorine (grams)
0	1000.0
1	900.0
2	810.0
3	729.0
4	656.1
5	590.5
6	531.4
7	478.3
8	430.5
9	387.4
10	348.67

Table 1. Amount of Chlorine.

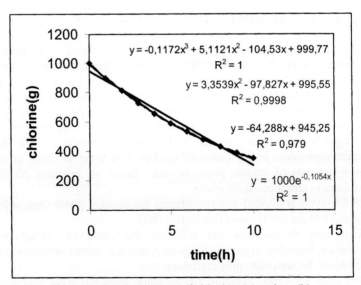

Figure 1. Graph Amount of chlorine (g) x time (h).

Hypotheses considered were: the volume of water in the tank remains the same (the rate of water entering is equal to the exit rate); the chlorine is all added to the tank just once at the beginning of the process and then mixed to the water; the chlorine is eliminated only by the dilution due to the water renewal in the tank.

From experimental data, found by research, the following problem was formulated: a tank is completely full and contains one hundred cubic metres of water with 1 Kg of chlorine. Pure water is placed in the tank in a constant outflow while an escape valve eliminates the water excess. How is the amount of chlorine in the tank at time t determined?

At first, the experimental data (Table 1) were considered using mathematical software. The graphs were plotted and fitting produced some possibilities of mathematical expressions (Figure 1).

Any of the expressions was considered applicable between the initial and final data. However, it is not possible to extrapolate about the phenomena beyond the data. So, a theoretical study was necessary.

We considered $A(t)$ the amount of chlorine in the tank in instant t and K the rate of pure water entering and chlorinated water exiting by the escape valve. $(A/100)$. K kg chlorine goes out of the tank in each instant, having therefore a dilution along the process. After one hour, the amount of chlorine in the tank is:

$$A(1) = A(0) - \frac{A(0)}{100}.K \qquad (1)$$

$$\frac{\Delta A}{\Delta t} = -\frac{K}{100}.A(t) \qquad (2)$$

Transforming this discrete analysis into a continuous analysis ($\Delta t \rightarrow 0$):

$$\frac{dA}{dt} = -\frac{K}{100}.A \qquad (3)$$

The expression for A is then:

$$A = 1000.e^{-0.1053.t} \qquad (4)$$

This expression is equivalent of the exponential function found by the software. Testing this solution with the data of the table we get very similar values.

The retrieve function is an approach of what happens in the reality. Mathematically the solution indicates the t-axis as a horizontal asymptote. This means that $A \rightarrow 0$ when $t \rightarrow \infty$. Actually, within a determined tolerance range, it is possible to find a value of t for which the amount of chlorine is practically zero. It is possible to make predictions with this model for beyond the data set shown in the table. For example, an analysis of the conditions of conservation of the tank: after how many hours will the amount of chlorine in the water be less than $2g/m^3$? Or yet: with what regularity must chlorine be added in order to keep the water of the tank in a level considered acceptable for the treatment in question?

Another approach for the same problem (discrete analysis) can be considered and the expression found (also equivalent to the previous ones) is:

$$A(n) = A(0).(1-K/100)^n \qquad (5)$$

A simpler analysis about the adequacy of the curves found graphically can be made by taking in account the characteristics of the problem. The second degree polynomial curve is not adjusted because, according to this function, after some time the amount of chlorine will increase and it is not in accordance with the problem. The third degree curve also is not adjusted because this curve tends to assume negative values and it is not in accordance with the problem either.

The activity can be characterized as a learning environment through modelling and computing. The students were invited to inquire and to investigate that the situation studied accorded with reality, in doing so both mathematics and computing were used as tools. Initially the computer was used to look for information about the theme. After the problem was formulated, the computer was used to organize and to represent collected data. Computing resources enabled us to fit data and obtain different expressions for possible solutions. In order to interpret the answers found by software, a theoretical study was carried out and the results were compared with reality. In this specific problem it was not necessary to use mathematical software to solve the equation but, if the problem had not been so simple, this recourse would have been used. In these kinds of activities, the different approaches, analyses and comparisons necessitate a frequent movement from modelling to computing and from computing to modelling. It can enrich the processes of mathematical knowledge construction, contributing to the learning development of the students.

It is possible to perceive in the description, different levels of complexity in treating the same problem and the different mathematical concepts that can be systemized by the process of model construction. The choice of this, or that, approach depends on the context of the activity development (type of course or school level). For an average level course, for example, the polynomial and exponential functions could be studied using the software recourses. For the mathematical model, a discrete approach could be made which results in an exponential function whose expression can be compared with the one found using software. In case of an undergraduate course, a continuous treatment makes it possible the study of derivative concepts, differential equation and integral. The derivative concept is used to equate the variation rate of the amount of chlorine, the types of differential equations and the processes of resolution explored. The concept of integration as inverse of the differentiation is also applied.

The work with the theme can be completed by the development of activities related to other questions already mentioned. All the studied problems may contribute for the knowledge and understanding of the reality related to the chosen theme allowing to critical analysis and action on this reality.

4. FINAL CONSIDERATIONS

The previous description is an example of the countless possibilities of work in mathematical learning environment through mathematical modelling and computing. In the described situation the students cover the modelling process in all of its stages, since the problem formulation about the theme, the construction of the model and its validation.

This kind of activity contributes much to the development of the student's potential, stimulating and facilitating the activity within a reality. So such activities are interesting and useful within mathematical education.

However, the characterization of learning environments through modelling and computing admits the possibility to work with activities that do not involve the whole modelling process or that do not result in complete model construction. Certain activities in these environments can be equally interesting in some situations, contributing for the student's intellectual development. It is admits the possibility that students could be invited to investigate a real situation in which the questions have already been formulated. In this case they have to be encouraged to search for different resources (like mathematics and computing) for the resolution. It is also possible to inquire of the proper mathematical content, with reference in something external to mathematics, using computing resources. In this case the model is not a result of the activity, but it is previously given and used as motivation for the study.

In learning environments through modelling and computing, teachers and students are co-participants of the process. The performance of each of them in the organization of the activities can be more or less intense depending on the type of the developed activity.

The choice of the way to implement these environments in curricular activities depends on the context, varying in accordance with different possibilities and limitations of it.

REFERENCES

Barbosa, J. C.(2001) *Modelagem matemática: concepções e experiências de futuros professores*. 2001. 253 f.Tese (Doctoral Thesis in Mathematics Education), Instituto de Geociências e Ciências Exatas, Universidade Estadual Paulista, Rio Claro.

Bassanezi, R. C.(1994) Modelagem Matemática. *Dynamis, Blumenau*, 2, 7, 55-83.

Bean, D.(2001) O que é modelagem matemática? *Educação matemática em revista. São Paulo: SBEM*, 8, 9/10.

Borba, M. C., Penteado, M. G.(2001) *Informática e educação matemática*. Belo Horizonte: Autêntica.

Franchi, R. H. O. L. (1993) *Modelagem Matemática como estratégia de aprendizagem do Cálculo Diferencial e Integral nos cursos de Engenharia*. Masters disseration in Mathematics Education, Universidade Estadual Paulista, Rio Claro.

Frachi, R. H. O. L. (2002) *Uma proposta curricular de matemática para cursos de engenharia utilizando modelagem matemática e informática*. 189 f. Tese (Doctoral Thesis in Mathematics Education), Universidade Estadual Paulista, Rio Claro.

Skovsmose, O. (2000) Cenários de investigação. *Boletim de educação matemática, Rio Claro*, 14, 66-91.

7.10

MODELLING IS FOR REASONING

Luís Soares Barbosa[1] and Maria Helena Martinho[2]
Minho University, Braga, Portugal

Abstract–*In a broad sense, computing is an area of knowledge from which a popular and effective technology emerged long before a solid, specific, scientific methodology, let alone formal foundations, had been put forward. This might explain some of the weaknesses in the software industry, on the one hand, as well as an excessively technology-oriented view which dominates computer science training at pre-university and even undergraduate teaching, on the other. Modelling, understood as the ability to choose the right abstractions for a problem domain, is consensually recognised as essential for the development of true engineering skills in this area, as it is in all other engineering disciplines. But, how can the basic problem-solving strategy, one gets used to from school physics: understand the problem, build a mathematical model, reason within the model, calculate a solution, be taken (and taught) as the standard way of dealing with software design problems? This paper addresses this question, illustrating and discussing the interplay between modelling and reasoning.*

1. INTRODUCTION

The use of computer-based systems to support mathematical modelling is a recurring theme in the practice and research of most of our readers. On the one hand, computers have radically expanded the range of problem-solving and decision-making situations that can be effectively tackled. On the other, they play a fundamental role in training modelling skills and promoting associated competences. In this paper, however, we take the dual viewpoint: we will not be concerned with computers as modelling aids, but instead with the use of mathematics to model and reason about computer-based systems. Maybe such a shift of concern deserves some explanation.

The exponential increase of both the availability of processor power and the complexity of the problems computers are requested to solve, is unprecedented in any other engineering domain. Even so, software remains hard to develop, it is often unreliable ('faulty goods delivered over budget and behind schedule'), difficult to re-use and excessively costly to modify and maintain. Traditional design methods emphasising diagrammatic or textual descriptions, with an informal semantics, have created the illusion that software development was little more than a balanced compromise of intuition and craft.

As a result, *conceptual* questions are often relegated to a secondary level of attention, and the mastering of particular, often ephemeral, technologies appears as a decisive requirement, for example, on recruitment policies. Often, in industry, the whole software production is totally biased to a specific technology or programming system, encircling, as a long term effect, the company's culture in quite strict limits. In a broad sense computing is an area of knowledge from which a popular and effective technology emerged long before a solid, specific, scientific methodology, let alone formal foundations, have been put forward.

This situation has to be contrasted, however, with the increasing demand for *quality certified* software, namely in safety-critical systems, which requires development approaches in which a system would be unacceptable unless accompanied by a *guarantee* that it respects a rigorously *specified* behaviour. This is the point where, from our perspective, mathematics comes into the picture. Or, more precisely, where software development is framed as a *mathematical modelling activity*.

In fact we are beginning to collect the fruits of more than four decades of intensive, even if sometimes neglected, basic research on the foundations of computation and programming semantics, upon which a true engineering discipline for software design can be based. Such research helped, in particular, to shed light on the underlying mathematical structures and reasoning principles, and to establish the connection between mathematics and computing at a foundation, rather than application (as, *for example,* in numerical analysis), level.

In such a context, the starting point of this paper is that to become a mature engineering discipline, software design has to adopt the basic problem-solving strategy one gets used to from school physics:

- understand the problem,
- build a mathematical model of it,
- reason in such a model,
- upgrade the model whenever necessary,
- calculate a solution, which, in this domain, means a *program*.

Moreover we would like to underline two concepts in this strategy:

- *Modelling*, understood as the ability to choose the right abstractions for a problem domain;
- *Calculation*, in the sense that such abstractions should be expressed in a mathematically rich framework to enable rigorous reasoning both to establish models' properties and to transform models towards effective implementations.

The context for this research was the assessment of two concrete educational experiments specifically designed to introduce modelling as a central issue in the computer science curriculum. They were conducted at both undergraduate and professional, post-graduation, training levels.

Some lessons learnt from these experiments are reported in §4, which concludes and raises some topics for further research. Before that, §2 and §3 discuss, respectively, the role of modelling and calculation in software design.

2. SOFTWARE DESIGN AS A MODELLING ACTIVITY

Software design is concerned with the engineering of computer-based solutions to real-world problems in which information, its acquisition, flow and transformation, plays a key role. The starting point is often a collection of more or less structured requirements, usually stated in plain English. Consider for example the following fragment of a statement of requirements placed by a mobile phone manufacturer:

> *For each row of calls stored in the mobile phone (for example, numbers dialled, SMS messages, lost calls, etc.) the store operation should work in a way such that (a) the more recently a call is made the more accessible it is; (b) no number appears twice in a list; (c) only the last 10 entries in each list are stored.*

In the analysis of an information system a fundamental distinction is drawn between *entities*, which represent information sources, and *transformations* upon them. A similar distinction appears in the definition of *algebras* (as collections of sets and functions) which makes them interesting models for this sort of systems. An elementary grammatical analysis of the requirements above provides the basic ingredients:

- *Nouns* (such as "call" or "raw of calls") lead to the identification of the fundamental information structures, or data domains, which, at a latter stage, will originate what is known as the program *data types*.
- *Verbs* (such as store) identify the services to be made available from the system. They denote processes which transform information, and therefore can be modelled by *functions* or, more generally, by *relations*.
- *Integrated sentences* (such as requirements (a) to (c)) identify *properties* or *constraints*, corresponding to predicates which tune the model to its specific purpose.

Software designs can be naturally expressed in the centenary notation of set theory. Notions such as set, sequence, Cartesian product, function or relation have the potential to provide expressive, but rigorous, descriptions of a design. For several years in their past education students (and professionals) have become familiar with this sort of notation as a *tool to think with*. Our courses in software design intend to build on such a background.

In the example considered here, a "row of calls" can be modelled as a sequence of whatever a "call" is. Registering a new call in the row amounts then to place the former in front of the latter, a well-known operation in the algebra of sequences that we denote by the: (read *append*) combinator. This leads to elementary model shown in Figure 1.

$$CallRow = Call^*$$
$$store : Call \times CallRow \longrightarrow CallRow$$
$$store\,(c, l) = c : l$$

Figure 1. An elementary model.

Note that data domain *Call* is left unspecified: the initial requirements do not place any restriction on this structure, so, at the level of abstraction of this model, it is considered a primitive notion, *that is,* an unstructured element in the universe of discourse. Are we done? Is this model acceptable? Let us address these questions going through the problem constraints already identified:

- Constraint (a) is guaranteed by construction, *that is,* it is a direct consequence of the definition of service *store*.
- Constraint (b), on the other hand, can be stated as equation

$$\#{\bullet}elems = length \tag{1}$$

that is, the number of calls in a row (as measured by function *length*) is equal to the cardinal (#) of the set of its elements (as computed by *elems*). Note that regarding a sequence as a set eliminates all duplicated occurrences of its elements.

- Finally, constraint (c) imposes a limit on the length of the sequence used to model a row of calls. It can be documented by the inequality

$$length {\leq} 10 \tag{2}$$

Therefore, to meet constraints (b) and (c), operation *store* has to be modified: when storing element x in sequence l, l must be first depurated of any occurrence of x, and, after appending, the whole sequence reduced to its first 10 elements. Formally,

$$store(x,l) = take_{10} \, (x : filter_{\neq x} l) \tag{3}$$

where $take_n$ and $filter_\phi$ are combinators in the algebra of sequences. The former returns the first n elements of a sequence, the latter filters out elements which violate predicate ϕ.

This small example illustrates the *iterative* character of the modelling process: one starts with a very bare, but precise, model which evolves as the understanding of the problem increases. It also shows the fundamental role of the simple notion of a function in modelling software engineering problems. To be precise, not only the notion of a function, but that of the whole *algebra of functions*. As any other algebra, this defines the ways in which functions can be combined. And there are a number of different ones. Function composition provides a *pipeline* connection between functions with the right types:

$$A \xrightarrow{\;f\;} B \xrightarrow{\;g\;} C$$

Often in design classes one explores an analogy between composition of functions and multiplication of numbers. In this way, students soon realise that non commutativity of composition leads to *two*, rather then *one*, division problems for functions[3]. What is interesting, from our point of view, is that each of them has a particular modelling potential. Consider, for illustration purposes, diagrams in Figure 2, in the context of the mobile phones example. In each case, a function x, acting as an unknown, has to be found to close the triangle. In the first case x computes the amount to be paid for a call in a particular price plan. This means that through x function *type*, which assigns a price plan to each call, determines the cost of a call (given by *cost*). In the other example, x associates a call to a particular network. Closing the diagram means that for each call a network is to be found such that the origin of the call and the location of the used network coincide.

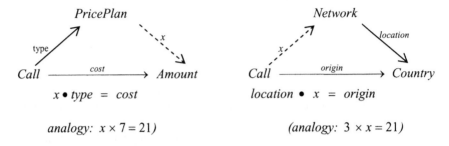

$$x \bullet type = cost \qquad\qquad location \bullet x = origin$$

analogy: $x \times 7 = 21$) *(analogy:* $3 \times x = 21$)

Figure 2. Composition vs Multiplication.

Combinator $\bullet$ provides a gluing scheme for functions, but it is not the only one. In particular, functions with a *common domain* can be glued by a *pairing* construction

$$C \xrightarrow{\ \langle f,g \rangle\ } A \times B$$

where $\langle f,g \rangle x = (fx,gx)$ Similarly, functions sharing a *common codomain* can be put together by an *alternative* construction

$$A + B \xrightarrow{\ [f,g]\ } C$$

which applies f or g depending on the argument coming from A or B ($A + B$ stands for the disjoint union of sets A and B).

The relevance of these composition patterns is that they correspond to different ways of composing services in a (model of a) computational system. Moreover, each of them entails a different way of modelling information aggregation. For example, *pipelining* leads to the notion of *function space* $A \rightarrow B$, which models functional dependencies. On its turn, *pairing* leads to *Cartesian product* $A \times B$ modelling *spatial aggregation* of information. Finally, alternative leads to *disjoint union* $A + B$, which models *choice* or *aggregation in the temporal dimension*.

In this section, we have highlighted the use of *functions* as a modelling tool. Of course, other problems may require different conceptual tools. For example, one may resort to *partial functions*, to model problems which remain undefined for a number of situations, or *relations*, whenever the outcome of a service is non deterministic (and therefore a functional dependency does not exist between its input and output). Or one may resort to some form of *automata* to express the dynamics of a computation. In contrast to other, more classical engineering disciplines, *continuous* mathematics is not the primary problem-solving tool here. Actually, classical models are manipulated either analytically or numerically, often resorting to some sort of testing on a physical model of the problem. Software design, by contrast, resort to *discrete* mathematics, which is easy to understand and animate in a computer, but usually no physical models are available: computer science deals essentially with non tangible mathematical models (what Henderson (2003) calls *mental models*).

Similarly, however, to what happens in other engineering disciplines, the purpose of a model in software design is double: to provide insight into the problem/system structure, and to form a basis upon which one can *reason* about such structure. The latter is a fundamental step: it is the ability of calculating within design models that paves the way to the possibility of transforming them into effective programs and computational systems. This leads directly to the second topic of this paper.

3. MODELLING IS FOR REASONING

There are two ways in which the title of this section can be understood in the context of computer science education. In one sense it means that a model should be amenable to *experimentation*. In the other that it must provide a basis for *effective calculation*, for example to verify the equivalence between two designs or to transform one into the other by controlled introduction of detail. Let us discuss each of them separately.

Although mathematical notation is a very good way of expressing requirements and of communicating among the design team, it requires more and more precision from people. Furthermore, writing mathematics does not mean to write everything perfect at the first time. So, there is a need for tools for validating mathematical descriptions. Moreover, educational practice has shown that to be effective the whole modelling process must be supported by some sort of animation tool. That is, a computer-based tool which understands an elementary language of sets and functions and executes designs.

There is a variety of available animation tools used either in educational or professional contexts (see for example, Fitzgerald & Larsen, 1998; Abrial, 1996; Almeida et al., 1997). Such tools build a *prototype* out of a (formal) model, which can be executed, tested and modified on-the-fly. This is also an old idea in Engineering. Think, for example, in a wind tunnel test of an aircraft, where performance in checked against theory, or a mock-up for a building, in which design features are checked for usability. From our experience the use of prototypes provides:

- Early feedback on the model.
- Increased confidence in the models developed achieved by a check on its self-consistency and general sensibleness.
- More effective communication among the design team.

Furthermore it emphasises the incremental and iterative character of the software design process. Prototypes develop side by side with formal models, from the very beginning until a stable and detailed design is found. Each iteration is formally documented, and, what is more, such a document is executable.

But when is a software design equivalent to another? Or to a sub-model thereof? How can a real program be extracted from a design model? How can a particular property be shown to hold of a given model? How is a model built to satisfy a set of properties? To answer these sort of questions is the purpose of a design calculus. But what calculus?

The definition of a *calculation style* to reason about software models is an essential ingredient to the success of the modelling approach. Actually, there is a well-established reasoning style in mathematics, the *theorem-followed-by-proof*, which is quite inadequate for the construction of computer systems. The reason is that it reflects a *guess-and-verify* approach which supposes the system is first built (out of the blue?) and then formally verified. If one has a model, however, the reasonable attitude is to use it to calculate the system, making progress through a whole chain of progressively more concrete models. A number of authors have discussed the dichotomy *verification-oriented* and *calculational-driven* styles of reasoning (see for example, Gries & Schneider, 1993; Zeitz, 1999; Backhouse, 2001) and concluding on the ineffectiveness of the former to Computer Science. There is also an extensive body of research on calculi for transforming software design models (see for example, Bird & Moor, 1997; Backhouse, 2003), which we actually use in the modelling classes.

Although this is not the proper place to introduce such calculi, we would like to briefly comment on a related issue which is often neglected: *notation*. Actually expressiveness in modelling and suitability for calculation may seem potentially conflicting aims. Mathematical modelling requires *descriptive* notations, often domain-specific, and hopefully intuitive. Calculation, on the other hand, requires notations that are *generic, concise and precise* (Backhouse, 2003) or, to put it in another way, *elegant*, in the sense the word has in the writings of Dijsktra: *simple and remarkably effective* (Dijsktra & Scholten, 1990), *that is,* easy to manipulate.

The extensive use of nested quantifiers in a logic formula, for example, may provide what one may think of as an intuitive description of a problem, but makes manipulation of such descriptions an uneasy, even overwhelming task.

Such a trend for *notational economy* is well-known throughout the history of Mathematics, as a sort of "natural language implosion". The driven force has always been the same: facilitate formulae manipulations, therefore enriching its suitability for calculation. Contrast, for example, formula:

$$.60.\tilde{p}.2.ce \quad son \quad yguales \quad a \quad .30.co$$

used by Pedro Nunes, a Portuguese mathematician of the 16th century, in his *Libro de Algebra*, published in Coimbra, in 1567, with nowadays $60 + 2x^2 = 30x$.

Again the history of mathematics is full of examples in which not only different notations, but also different, although interrelated, conceptual domains are used for *modelling* and *calculation*. The former emphasises expressiveness and closeness to intuition, the latter manipulation simplicity. A classical example is the Laplace transform, which allows an expressive but complex model to be converted into a less intuitive but simpler (*that is,* linear) one.

Is there a similar *transform* to reason about software designs? The answer turns out to be very simple: just *avoid the variables*. In particular, in the algebra of functions briefly discussed in the previous section, replace function application by function composition and look for definitions in terms of generic properties rather than *ad hoc* representations. The reader may recognise here the whole discipline of category theory (Mac Lane, 1971), but we will not elaborate further on that.

4. CONCLUDING REMARKS

As reported in the Introduction, the context for this paper was a reflection on two concrete experiments at Minho University. Both experiments have been conducted for five years now at two quite different levels: *first year undergraduate students in a computer science degree* and *professional training* at post-graduation courses for software engineers.

Although students' age, backgrounds and motivations are quite different between these two groups, we have found extremely relevant the explicit incorporation of modelling in the computer science curriculum. In particular we have been able to assess how this contributes

- To emphasise the *conceptual* rather than the *instrumental* aspects of an engineering carrier[4].
- To develop *design literacy*: reasoning flexibility and, as Lesh and Doerr (2003) put it, *a handful of models in your hip pocket*.
- To enhance both *communication* and *teamwork* skills.

From a technical point of view modelling and reasoning are intertwined. Moreover emphasis should be placed on the *construction* rather than the *verification* level, a point that has often been neglected in research. Another lesson learnt was that, in computer science as in mathematics, notations are not neutral. Well designed notations do make the difference when one has to reason upon a model. Also, as already commented, the crucial need for tool support, in particular for prototyping systems.

Formal concepts of the kind required by computer science, and both modelling and problem-solving skills develop slowly along long periods of time. Our experience with professional engineers that return to the University to participate in this sort of seminars, suggests such training adds up and is probably effective even when initiated later in life.

Modelling in software design, as in any other domain of application, enhances what is known as *mathematical fluency* (see Lesh, 1996; Kaput & Shaffer, 2002), which is at the heart of what it means to understand. In more general terms, however, assessing to what extent mathematics education, at both university and pre-university levels, is centred on the on going construction and revision of models rather than on the acquisition of self-contained (?) bodies of knowledge remains an open question. We believe there is still a long way to go in that direction. Actually, acquisition of facts, results and procedures are merely surface manifestations of what goes on when people learn. As Devlin (2000) points out, *we know they are surface phenomena since we generally forget them soon after the last exam is over*.

Finally, a word on the role of the 'teacher'. Our experience, however limited it is, suggests she/he is more likely to be expected to act as *coacher*, than as repository of pre-framed knowledge. The insistence on new educational practices would not be effective without an assessment of how typical university lecturers feel about that and how this interacts with their own images of their profession. Also at this level, further research is certainly needed.

NOTES

1. DI-CCTC, Minho University.
2. IEP, Minho University.
3. See Lawvere & Schanuel (1997) for a detailed discussion.
4. The following opening statement of Paul Halmos autobiography (Halmos, 1985) is particularly elucidative, written as it was by a mathematician, which in the 1950's, was director of doctoral studies in what was then one of the top Mathematics Departments of the world, in the University of Chicago: *I like words more than numbers, and I always did (...) This implies, for instance that in Mathematics I like the conceptual more than the computational. To me the definition of a group is far clearer and more important and more beautiful than the Cauchy integral formula.*

REFERENCES

Abrial, J. R. (1996) *The B Book: Assigning Programs to Meanings.* Cambridge: CUP.

Almeida, J. J., Barbosa, L. S., Neves, F. L., and Oliveira, J. N. (1997) CAMILA: Prototyping and Refinement of Constructive Specifications. In M. Johnson (ed) *6th International Conference on Algebraic Methods and Software Technology.* Sydney: Springer Lecture Notes in Computer Science (1349), pp554–559.

Backhouse, R. (2001) Mathematics and Programming. a Revolution in the Art of Effective Reasoning. Inaugural Lecture, School of Computer Science and IT, University of Nottingham.

Backhouse, R. (2003) *Program Construction.* Chichester: John Wiley and Sons, Inc.

Bird, R. and Moor, O. (1997) *The Algebra of Programming.* Series in Computer Science. Hemel Hempstead: Prentice-Hall International.

Devlin, K. (2000) *The Math Gene: How Mathematical Thinking Evolved and Why Numbers Are Like Gossip.* Basic Books.

Dijkstra, E. W. and Scholten, C. S. (1990) *Predicate Calculus and Program Semantics.* New York: Springer Verlag.

Fitzgerald, J. and Larsen, P. G. (1998) *Modelling Systems: Practical Tools and Techniques in Software Development.* Cambridge: CUP.

Gries, D. and Schneider, F. (1993) *A Logical Approach to Discrete Mathematics.* New York: Springer Verlag.

Halmos, P. R. (1985) *I Want to Be a Mathematician.* New York: Springer Verlag.

Henderson, P. (2003) The Role of Modelling in Software Engineering Education. *33rd ASEE/IEEE Frontiers in Education Conference.* Boulder.

Kaput, J. and Shaffer, D. (2002) On the Development of Human Representational Competence from an Evolutionary Point of View. In K. Gravemeijer, R. Lehrer, B. v. Oers and L. Verschaffel (eds) *Symbolizing, Modeling and Tool Use in Mathematics Education.* Amsterdam: Kluwer Academic Publishers.

Lawvere, F. W. and Schanuel, S. H. (1997) *Conceptual Mathematics.* Cambridge: CUP.

Lesh, R. (1996) Mathematizing: The real need for representational fluency. In C. Janvier (ed) *20th Conference of the International Group for the Psychology of Mathematics Education.* Valencia: Universitat do Valencia, 3–13.

Lesh, R. and Doerr, H. M. (2003) Foundations of a Models and Modeling Perspective on Mathematics Teaching, Learning, and Problem Solving. In Lesh, R. and Doerr, H. (eds) *Beyond Constructivism*. London: Lawrence Erlbaum Associates Publishers.

Mac Lane, S. (1971) *Categories for the Working Mathematician*. New York: Springer Verlag.

Zeitz, P. (1999) *The Art and Craft of Problem Solving*. Chichester: John Wiley and Sons, Inc.

Authors' contact email addresses
(at date of publication)

Sergei	ABRAMOVICH	abramovs@potsdam.edu
Burkhard	ALPERS	balper@fh-aalen.de
Marta	ANAYA	manaya@fi.uba.ar
Jussara	ARAÚJO	jussara@mat.ufmg.br
Shuki	AROSHAS	Igor Verner
Jonei	BARBOSA	joneicb@uol.com.br
Luis	BARBOSA	lsb@di.uminho.pt
Kate	BARKER	kate.barker@bankofengland.co.uk
Jacques	BÉLAIR	jacques.belair@umontreal.ca
Abraham	BERMAN	Igor Verner
Maria Sallet	BIEMBENGUT	salett@furb.br
Werner	BLUM	blum@mathematik.uni-kassel.de
Cinzia	BONOTTO	bonotto@math.unipd.it
Rita	BORROMEO FERRI	borromeo@erzwiss.uni-hamburg.de
Astrid	BRINKMANN	astrid.brinkmann@math-edu.de
Klaus	BRINKMANN	Astrid Brinkman
Jill	BROWN	j.brown@patrick.acu.edu.au
Andreas	BÜCHTER	andreas.buechter@uni-dortmund.de
Hugh	BURKHARDT	hugh.burkhardt@nottingham.ac.uk
France	CARON	france.caron@umontreal.ca
Enrique	CASTRO	ecastro@ugr.es
María	CAVALLARO	micavall@fi.uba.ar
Rafael	COLÁS	Leticia Corral
Leticia	CORRAL	evelynmarabel@yahoo.com.mx
Rosalind	CROUCH	r.m.crouch@herts.ac.uk
María	DOMÍNGUEZ	mcdomin@fi.uba.ar
Ian	EDWARDS	ie@luther.vic.edu.au
Ludwik	FINKELSTEIN	l.finkelstein@city.ac.uk
Regina	FRANCHI	regina.franchi@uol.com.br
Peter	GALBRAITH	p.galbraith@mailbox.uq.edu.au
Martin	GOEDHART	Gerrit Roorda
Kenneth	GRATTAN	k.t.v.grattan@city.ac.uk
Norbert	GRUENWALD	n.gruenwald@mb.hs-wismar.de
Christopher	HAINES	c.r.haines@city.ac.uk
Pasi	HÄKKINEN	Robert Piché
Graham	HARDY	graham.hardy@manchester.ac.uk
Nelson	HEIN	hein@furb.br
Hans-Wolfgang	HENN	wolfgang.henn@mathematik.uni-dortmund.de

Antonino	HERNÁNDEZ	Leticia Corral
Yvonne	HILLIER	y.g.j.hillier@brighton.ac.uk
Mikael	HOLMQUIST	mikael.holmquist@ped.gu.se
Kenneth	HOUSTON	sk.houston@ulster.ac.uk
Celia	HOYLES	c.hoyles@ioe.ac.uk
Julian	HUNT	jcrh@cpom.ucl.ac.uk
Toshikazu	IKEDA	ikeda@ed.ynu.ac.jp
John	IZARD	john.izard@rmit.edu.au
Tomas	JENSEN	thje@dpu.dk
Qiyuan	JIANG	qjiang@math.tsinghua.edu.cn
Michael	JONES	jonesm@mail.montclair.edu
Cyril	JULIE	cjulie@uwc.ac.za
Gabriele	KAISER	gabriele.kaiser@uni-hamburg.de
Sanowar	KHAN	s.h.khan@city.ac.uk
Sergiy	KLYMCHUK	sergiy.klymchuk@aut.ac.nz
Andrei	KOLYSHKIN	akoliskins@rbs.lv
Jerry	LEGÉ	glege@exchange.fullerton.edu
Dominik	LEIß	dleiss@mathematik.uni-kassel.de
Francesco	LEONETTI	leonetti@matematica.univaq.it
Thomas	LINGEFJÄRD	thomas.lingefjard@ped.gu.se
Katja	MAAß	katjamaass@ph-freiburg.de
Sarah	MAOR	Igor Verner
Helena	MARTINHO	mhm@iep.uminho.pt
Susann	MATHEWS	susann.mathews@wright.edu
Akio	MATSUZAKI	makio@human.tsukuba.ac.jp
Mark	McCARTNEY	m.mccartney@ulster.ac.uk
Arup	MUKHERJEE	mukherjeea@mail.montclair.edu
Richard	NOSS	r.noss@ioe.ac.uk
José	ORTIZ	ortizjo@cantv.net
Robert	PICHÉ	robert.piche@tut.fi
Seppo	POHJOLAINEN	seppo.pohjolainen@tut.fi
Michelle	REED	michelle.reed@wright.edu
Luis	RICO	lrico@ugr.es
Adolf	RIEDE	riede@mathi.uni-heidelberg.de
Gerrit	ROORDA	g.roorda@rug.nl
Gabriele	SAUERBIER	g.sauerbier@mb.hs-wismar.de
Kirsi	SILIUS	kirsi.silius@tut.fi
ManMohan	SODHI	m.sodhi@city.ac.uk
Byung-Gak	SON	b.g.son@city.ac.uk
Max	STEPHENS	m.stephens@unimelb.edu.au
Gloria	STILLMAN	g.stillman@unimelb.edu.au
Kari	SUOMELA	kari.suomela@tut.fi
Anne-Maritta	TERVAKARI	anne.tervakari@tut.fi

Igor	VERNER	ttrigor@techunix.technion.ac.il
Pauline	VOS	f.p.vos@rug.nl
Michael	VOSKOGLOU	voskoglou@teipat.gr
Geoff	WAKE	geoff.wake@manchester.ac.uk
Jinxing	XIE	jxie@math.tsinghua.edu.cn
Qixiao	YE	yeqx@bit.edu.cn
Tatyana	ZVERKOVA	tsklim@rambler.ru

Printed and bound by CPI Group (UK) Ltd, Croydon, CR0 4YY

08/05/2025

01864842-0001